T0296734

Cognitive Intelligence with Neutrosophic Statistics in Bioinformatics

Cognitive Data Science in Sustainable Computing

Cognitive Intelligence with Neutrosophic Statistics in Bioinformatics

Edited by

Florentin Smarandache

Mathematics, Physical and Natural Sciences Division, University of New Mexico, Gallup, NM, United States

Muhammad Aslam

Department of Statistics, Faculty of Science, King Abdulaziz University, Jeddah, Saudi Arabia

Series Editor

Arun Kumar Sangaiah

Academic Press is an imprint of Elsevier
125 London Wall, London EC2Y 5AS, United Kingdom
525 B Street, Suite 1650, San Diego, CA 92101, United States
50 Hampshire Street, 5th Floor, Cambridge, MA 02139, United States
The Boulevard, Langford Lane, Kidlington, Oxford OX5 1GB, United Kingdom

ISBN: 978-0-323-99456-9

For Information on all Academic Press publications
visit our website at https://www.elsevier.com/books-and-journals

Publisher: Mara E. Conner
Editorial Project Manager: Zsereena Rose Mampusti
Production Project Manager: Sajana Devasi P. K.
Cover Designer: Christian J. Bilbow

Typeset by MPS Limited, Chennai, India

Contents

List of contributors

Tahir Abbas Department of Statistics, Government College University, Lahore, Pakistan

Mohammad Abobala Department of Mathematics, Tishreen University, Latakia, Syria

Usama Afzal School of Microelectronics, Tianjin University, Tianjin, P.R. China

Kadri Ulaş Akay Department of Mathematics, Faculty of Science, İstanbul University, İstanbul, Turkey

A. Aleeswari Department of Management, PSNA College of Engineering and Technology, Dindigul, Tamil Nadu, India

Zahid Ali Services Hospital, Lahore, Pakistan

M. Arockia Dasan Department of Mathematics, St. Jude's College, Manonmaniam Sundaranar University, Kanyakumari, Tamil Nadu, India

Muhammad Aslam Department of Statistics, Faculty of Science, King Abdulaziz University, Jeddah, Saudi Arabia

Bhimraj Basumatary Department of Mathematical Sciences, Bodoland University, Kokrajhar, Assam, India

E. Bementa PG and Research Department of Physics, Arulanandar College, Madurai, Tamil Nadu, India

Said Broumi Laboratory of Information Processing, Faculty of Science, Ben M'Sik, University Hassan II, Casablanca, Morocco; Regional Center for the Professions of Education and Training (C.R.M.E.F), Casablanca-Settat, Morocco

Alok Dhital Mathematics, Physical and Natural Science Division, University of New Mexico, Gallup, NM, United States

Volkan Duran Faculty of Science and Arts, Iğdır University, Iğdır, Turkey

S.A. Edalatpanah Department of Applied Mathematics, Ayandegan Institute of Higher Education, Tonekabon, Iran

Muhammad Gulistan Department of Mathematics and Statistics, Hazara University, Mansehra, Pakistan

S. Jafari Department of Mathematics, College of Vestsjaelland South, Slagelse, Denmark

Azhar Ali Janjua Government Graduate College, Hafizabad, Pakistan

Darjan Karabasevic University Business Academy, Novi Sad, Serbia

S. Jegan Karuppiah Department of Rural Development Science, Arul Anandar College (Autonomous), Karumathur, Madurai, Tamil Nadu, India

Muhammad Kashif Department of Mathematics and Statistics, University of Agriculture Faisalabad, Faislabad, Pakistan

Jeevan Krishna Khaklary Department of Mathematics, CIT, Kokrajhar, Assam, India

Muhammad Imran Khan Department of Mathematics and Statistics, University of Agriculture Faisalabad, Faislabad, Pakistan

Nasrullah Khan Department of Statistics and Auctorial Science, University of the Punjab, Lahore, Pakistan

Zahid Khan Department of Mathematics and Statistics, Hazara University, Mansehra, Pakistan

Katrina Khadijah Lane Krebs Higher Education Division, Central Queensland (CQ) University Rockhampton, Australia

U. Pavithra Krishnan Department of Food Science and Technology, Arul Anandar College (Autonomous), Karumathur, Madurai, Tamil Nadu, India

Ushna Liaquat Dental Section, Faisalabad Medical University, Faisalabad, Pakistan

V.F. Little Flower Department of Mathematics, St. Jude's College, Manonmaniam Sundaranar University, Kanyakumari, Tamil Nadu, India

Nivetha Martin Department of Mathematics, Arul Anandar College (Autonomous), Karumathur, Madurai, Tamil Nadu, India

Rohan Mishra Department of Statistics, Institute of Science, Banaras Hindu University, Varanasi, Uttar Pradesh, India

Giorgio Nordo MIFT-Department of Mathematical and Computer Science, Physical Sciences and Earth Sciences, Messina University, Italy

Ion Patrascu Mathematics Department, Fratii Buzesti College, Craiova, Romania

R. Radha Department of Mathematics, Nirmala College for Women, Coimbatore, Tamil Nadu, India

Gadde Srinivasa Rao Department of Mathematics and Statistics, University of Dodoma, Dodoma, Tanzania

Amna Riaz Department of Statistics, University of Gujrat, Gujrat, Pakistan

Rafael Rojas Universidad Industrial de Santander, Bucaramanga, Santander, Colombia

Sultan Salem Department of Economics, Birmingham Business School, College of Social Sciences, University of Birmingham, Edgbaston, England, United Kingdom

Rehan Ahmad Khan Sherwani College of Statistical and Actuarial Sciences, University of the Punjab, Lahore, Pakistan

Abhishek Singh Department of Mathematics and Computing, Indian Institute of Technology (ISM) Dhanbad, Dhanbad, Jharkhand, India

Rajesh Singh Department of Statistics, Institute of Science, Banaras Hindu University, Varanasi, Uttar Pradesh, India

Florentin Smarandache Mathematics, Physical and Natural Sciences Division, University of New Mexico, Gallup, NM, United States

A. Stanis Arul Mary Department of Mathematics, Nirmala College for Women, Coimbatore, Tamil Nadu, India

Dragisa Stanujkic Technical Faculty in Bor, University of Belgrade, Bor, Serbia

Ferhat Taş Department of Mathematics, Faculty of Science, İstanbul University, İstanbul, Turkey

Selçuk Topal Department of Mathematics, Bitlis Eren University, Bitlis, Turkey

X. Tubax School of Information Technology, Cyryx College, Male, Maldives

Sami Ullah College of Agriculture, University of Sargodha, Sargodha, Pakistan

Maikel Yelandi Leyva Vazquez Universidad Regional Autónoma de los Andes (UNIANDES), Babahoyo, Los Ríos, Ecuador

Gajendra K. Vishwakarma Department of Mathematics and Computing, Indian Institute of Technology (ISM) Dhanbad, Dhanbad, Jharkhand, India

Nijwm Wary Department of Mathematical Sciences, Bodoland University, Kokrajhar, Assam, India

Mohamed Bisher Zeina Department of Mathematical Statistics, University of Aleppo, Aleppo, Syria

About the editors

Florentin Smarandache, PhD, PostDoc, is an emeritus professor of mathematics at the University of New Mexico, United States. He is the founder of neutrosophy (generalization of dialectics), neutrosophic set, neutrosophic logic, neutrosophic probability, and neutrosophic statistics since 1995 and has published hundreds of papers and books on multispace and multistructure, hypersoft set, IndetermSoft Set, TreeSoft Set, degree of dependence and independence between neutrosophic components, refined neutrosophic set, neutrosophic over–under off-set, plithogenic set, neutrosophic triplet and duplet structures, quadruple neutrosophic structures, extension of algebraic structures to NeutroAlgebras and AntiAlgebras, NeutroGeometry and AntiGeometry, SuperHyperAlgebra, SuperHyperGraph, SuperHyperTopology, Dezert–Smarandache Theory, and many peer-reviewed international journals and books. He also presented papers and plenary lectures to many international conferences and reputed institutions around the world (such as NASA, the University of California, Berkeley, CalTech, and *Doctor Honoris Causa* from the Beijing Jiaotong University).

Muhammad Aslam introduced the area of neutrosophic statistical quality control (NSQC) for the first time. He is the founder of neutrosophic inferential statistics and NSQC. He contributes to the development of neutrosophic statistics theory for inspection, inference, and process control. He originally developed the theory in these areas under the neutrosophic statistics. He extended the classical statistics theory to neutrosophic statistics originally in 2018. His areas of interest include industrial statistics, neutrosophic inferential statistics, neutrosophic statistics, neutrosophic quality control, neutrosophic applied statistics, and classical applied statistics.

Chapter 1

Introduction and advances to neutrosophic probability and statistics and plithogenic probability and statistics and their applications in bioinformatics and other fields (review chapter)

Florentin Smarandache
Mathematics, Physical and Natural Sciences Division, University of New Mexico, Gallup, NM, United States

1.1 Introduction

The classical statistics operates only with determinate data and determinate inference methods. But, in the real world, many indeterminate (unclear, vague, partially unknown, contradictory, incomplete elements, etc.) elements occur. A way to deal with indeterminacy is the neutrosophic statistics, that is, data that has degrees of indeterminacy, and inference methods with degrees of indeterminacy also. One may have inexact or ambiguous arguments and values instead of crisp arguments and values for the probability distributions, charts, diagrams, algorithms, or functions.

The neutrosophic sciences were founded by Florentin Smarandache, in 1998, who developed the field by introducing the Neutrosophic Descriptive Statistics in 2014. Supplementally, Muhammad Aslam introduced in 2018 the Neutrosophic Inferential Statistics, Neutrosophic Applied Statistics, and Neutrosophic Statistical Quality Control. See Ref. [1−9] for theoretic works in neutrosophic sciences and neutrosophic statistics.

The neutrosophic statistics is as well a generalization of the interval statistics: while interval statistics is based on interval analysis, neutrosophic

Cognitive Intelligence with Neutrosophic Statistics in Bioinformatics.
DOI: https://doi.org/10.1016/B978-0-323-99456-9.00013-1

statistics is based on set analysis (all kinds of sets, not only intervals). Neutrosophic Statistics is an extension of the Interval Statistics, since it deals with all types of indeterminacies (with respect to the data, inferential procedures, probability distributions, graphical representations, etc.), it allows the reduction of indeterminacy by using the neutrosophic numbers of the form $N = a + bI$, where a is the determinate part of N, while bI is the indeterminate part of N, and it uses the neutrosophic probability that is more general than imprecise and classical probabilities, and has more detailed corresponding probability density functions. While Interval Statistics *only* deals with indeterminacy that can be represented by intervals. We show that not all indeterminacies (uncertainties) may be represented by intervals. Also, in some applications, we should better use hesitant sets (that have less indeterminacy) instead of intervals.

Neutrosophic statistics is more elastic than classical and interval statistics. If all data and inference methods are determinate, then neutrosophic statistics coincides with classical statistics.

Since in our world we have much more indeterminate data than determinate data, therefore more neutrosophic statistical procedures are needed than the procedures in classical statistics. The neutrosophic datasets, when the data have various degrees of indeterminacy, are utilized in neutrosophic statistics.

The neutrosophic numbers are of the form $N = a + bI$, where a and b are real or complex numbers, which is interpreted as: "a" the determinate part of the number N and "bI" the indeterminate part of the number N, defined by W. B. Vasantha Kandasamy and F. Smarandache in 2003 [3].

The neutrosophic quadruple numbers are of the form $NQ = a + bT + cI + dF$, where a, b, c, d are real or complex numbers, and "a" is the known part of NQ, while "$bT + cI + dF$" is the unknown part of NQ; further on, the unknown part is split into three subparts: degree of confidence (T), degree of indeterminacy of the confidence (I), and degree of no confidence (F). The four-dimensional vector NQ can also be written as $NQ = (a, b, c, d)$, introduced by Smarandache in 2015 [10,11].

Neutrosophic probability is a generalization of the classical probability and imprecise probability in which the chance that an event A occurs is $t\%$ true—where t varies in the subset T, $i\%$ indeterminate, where i varies in the subset I—and $f\%$ false—where f varies in the subset F.

In classical probability, the sum of all space probabilities is equal to 1, while in neutrosophic probability it is equal up to 3.

The probability of an event, in imprecise probability, is a subset T in [0, 1], not a number p in [0, 1], while what is left is supposedly the opposite, subset F (also from the unit interval [0, 1]); there is no indeterminate subset I in imprecise probability [9].

The neutrosophic and plithogenic statistics can be and already were successfully applied in many fields, for example, in the medical field.

Inconsistency and contradictions in medical data are largely accepted as unavoidable, and indeterminacy must be integrated as appropriately as possible into the behavior of biological systems and their characterization. The danger of fallacies due to misplaced precision is impending when there exists unfitness to capture fuzzy or incomplete knowledge.

The practitioner is eager to detect the disease in the early stages, but sometimes the diagnosis test does not clearly indicate the presence of the disease or the practitioner is not sure about the test results or the true diagnosis. As an application of neutrosophic statistics in the medical field, Muhammad Aslam et al. introduced in Ref. [12] the neutrosophic diagnosis tests (NDTs), a diagnosis table under the neutrosophic statistical interval method considering a test for diabetes to assess the status of the sugar level in the patients.

Employing the neutrosophic statistical analysis, Carralero et al. presented in Ref. [13] a structure in the application of physical exercises in the rehabilitation of the adult with gonarthrosis, a chronic, degenerative, progressive joint condition of the knee that results from mechanical and biological events that destabilize the normal coupling of the joint.

To study and analyze the risk of heart disease or stroke issue, Muhammad Aslam suggested in Ref. [14] to check the patient's heart conditions—such as pulse rate (PR), blood systolic pressure, and diastolic pressure (DP)—the application of Grubbs's test under neutrosophic statistics for testing the null hypothesis that the sample has the outlier versus the alternative hypothesis that sample has no outlier, also generally recommending applying the proposed test in medical sciences and biostatistics for the analysis of the data obtained in an uncertainty environment.

Most studies on prostate cancer use classical statistics when analyzing the social or biological data, and as a consequence, those researches are not helpful when the data are recorded in an indeterminate environment. In Ref. [15], Muhammad Aslam and Mohammed Albassam looked for the relationship between prostate cancer and dietary fat level using the neutrosophic interval method, aiming to present the analysis for variables using data collected from 30 countries. This paper shows that there is a strong relationship between the prostate cancer rate and dietary fat and presents a neutrosophic regression model that provides the estimation of the parameters in the indeterminacy interval range for the estimation and forecasting of prostate cancer death.

1.2 Definitions

Neutrosophy [1995] is a branch of philosophy, introduced by Florentin Smarandache in 1998, which studies the origin, nature, and scope of neutralities, as well as their interactions with different ideational spectra, considering a proposition, theory, concept, event, or entity, A in relation to its opposite anti$A$, and that which is neither A nor anti$A$, denoted by neut$A$.

Neutrosophy is an extension of the international movement called paradoxism [16,17], based on contradictions in science, literature, and arts [1980]. Neutrosophic principles constitute the basis for neutrosophic logic, neutrosophic probability, neutrosophic set, and neutrosophic statistics.

A proposition is t true, i indeterminate, and f false, where t, i, and f are real values from the ranges T, I, F, with no restriction on T, I, F, or the sum $n = t + i + f$. Neutrosophic logic thus generalizes the intuitionistic logic, which supports incomplete theories (for $0 < n < 1$, $0 \leq t,i,f \leq 1$); but also the fuzzy logic (for $n = 1$ and $i = 0$, and $0 \leq t,i,f \leq 1$); the Boolean logic as well (for $n = 1$ and $i = 0$, with t,f either 0 or 1); multivalued logic (for $0 \leq t,i, f \leq 1$); the paraconsistent logic (for $n > 1$, with both $t,f < 1$); dialetheism, which says that some contradictions are true (for $t = f = 1$ and $i = 0$; some paradoxes can be denoted this way).

Compared with all other logics, neutrosophic logic introduces a percentage of "indeterminacy" (or "neutrality") due to unexpected hidden parameters in some propositions. It also allows each component t,i,f to "boil over" 1 or "freeze" under 0. For example, in some tautologies $t > 1$, called "overtrue".

Neutrosophic set is a generalization based on neutrosophic principles of the intuitionistic set, classical set, fuzzy set, paraconsistent set, dialetheist set, paradoxist set, and tautological set. An element $x(T, I, F)$ belongs to the set in the following way: it is t true in the set, i indeterminate in the set, and f false, where t, i, and f are real numbers taken from the sets T, I, and F with no restriction on T, I, F or on their sum $n = t + i + f$.

Smarandache enlarged the neutrosophic set in 2013 to the refined neutrosophic set [18], where each neutrosophic component was refined/split into subcomponents; that is, T was refined/split into $T_1, T_2, \ldots, T_p$; I was refined/split into $I_1, I_2, \ldots, I_p$; and F was refined/split into $F_1, F_2, \ldots, F_s$, where p, r, $s \geq 0$ are integers and at least one of p, r, s is ≥ 2, with all T_j, I_k, F_l as subsets of [0, 1] and no other restrictions.

Neutrosophic set is a generalization of crisp set, fuzzy set, intuitionistic fuzzy set, inconsistent intuitionistic fuzzy set (picture fuzzy set, ternary fuzzy set), Pythagorean fuzzy set, Fermatean set, q-rung orthopair fuzzy set, spherical fuzzy set, and n-hyperspherical fuzzy set. Neutrosophic set has been further extended to refined neutrosophic set [16].

Neutrosophy, as a new branch of philosophy, has been also extended to refined neutrosophy, and consequently, neutrosophication was extended to refined neutrosophication. Whence, regret theory, grey system theory, and three-ways decisions are particular cases of neutrosophication and neutrosophic probability. We have further extended the three-ways decision to n-ways decision, and the last one is a particular case of refined neutrosophy.

1.3 Neutrosophic probability

Neutrosophic probability is a generalization of the classical probability and imprecise probability, based on neutrosophy, in which the chance that an event A occurs is $t\%$ true—where t varies in the subset T, $i\%$ indeterminate, where i varies in the subset I—and $f\%$ false—where f varies in the subset F. In classical probability, the sum of all space probabilities is equal to 1, while in neutrosophic probability it is up to 3.

The function that models the neutrosophic probability of a random variable x is called neutrosophic distribution: $NP(x) = (T(x), I(x), F(x))$, where $T(x)$ represents the probability that the value x occurs, $F(x)$ represents the probability that value x does not occur, and $I(x)$ represents the indeterminate/unknown probability of value x to occur or not [2].

1.4 Neutrosophic statistics

Neutrosophic statistics is the analysis of events described by neutrosophic probability.

Neutrosophic statistics refers to a set of data, such that the data or a part of it are indeterminate to some degree, and to methods used to analyze the data.

Neutrosophic statistics is also a generalization of interval statistics, because, among others, while interval statistics is based on interval analysis, neutrosophic statistics is based on set analysis (meaning all kinds of sets, not only intervals).

Neutrosophic statistics is more elastic than classical statistics. If all data and inference methods are determinate, then neutrosophic statistics coincides with classical statistics.

1.4.1 Plithogenic probability and statistics

Plithogeny is the genesis or origination, creation, formation, development, and evolution of new entities from dynamics and organic fusions of contradictory and/or neutrals and/or noncontradictory multiple old entities. Plithogeny pleads for the connections and unification of theories and ideas in any field.

By “entities,” it is meant “knowledges” in various fields, such as soft sciences, hard sciences, arts, and letters theories. Plithogeny is the dynamics of many types of opposites, and/or their neutrals, and/or non-opposites and their organic fusion, while plithogenic means what is pertaining to plithogeny. Etymologically, plithogeny comes from: (Gr.) πλήθος (plithos) = crowd, many while -geny < (Gr.) -γενιά (-geniá) = generation, the production of something.

Plithogeny is a generalization of dialectics (dynamics of one type of opposites: <A> and <antiA>), neutrosophy (dynamics of one type of opposites and their neutrals: <A> and <antiA> and <neutA>), since plithogeny studies the dynamics of many types of opposites and their neutrals and non-opposites (<A> and <antiA> and <neutA>, <B> and <antiB> and <neutB>, etc.), and many non-opposites (<C>, <D>, etc.), all together.

In the plithogenic probability, each event E from a probability space U is characterized by many chances of the event to occur [not only one chance of the event to occur: as in classical probability, imprecise probability, and neutrosophic probability], chances of occurrence calculated with respect to the corresponding attributes' values that characterize the event E.

Plithogenic probability is a generalization of the classical probability [since a single event may have more crisp probabilities of occurrence], imprecise probability [since a single event may have more subunitary subset probabilities of occurrence], and neutrosophic probability [since a single event may have more triplets of subunitary subset probabilities of occurrence, subunitary subset probabilities of indeterminacy (not clear if occurring or not occurring), and subunitary subset probabilities of nonoccurring].

As a generalization of classical statistics and neutrosophic statistics, plithogenic statistics is the analysis of events described by the plithogenic probability [19].

The classical multivariate analysis (MVA) studies a system, which is characterized by many variables, or one may call it a system-of-systems. The variables, that is, the subsystems, and the system as a whole are also classical (i.e., they do not deal with indeterminacy). Many classical measurements are needed, and the classical relations between variables are to be determined. This system-of-systems is generally represented by a surrogate approximate model. The plithogenic variate analysis (PVA) is an extension of the classical multivariate analysis, where indeterminate data or procedures, that are called neutrosophic data and, respectively, neutrosophic procedures, are allowed. Therefore PVA deals with neutrosophic/indeterminate variables, neutrosophic/indeterminate subsystems, and neutrosophic/indeterminate system-of-systems as a whole.

The most general definition is the following: plithogenic probability of an event to occur consists of the chances that the event occurs with respect to all random variables (parameters) that determine it. Each variable is described by a probability distribution (density) function, which may be a classical, (T,I,F)-neutrosophic, I-neutrosophic, (T,F)-intuitionistic fuzzy, (T, N,F)-picture fuzzy, (T,N,F)-spherical fuzzy, or (other fuzzy extensions) distribution function.

The plithogenic probability is a generalization of the classical multivariate probability. The analysis of the events described by the plithogenic probability is the plithogenic statistics.

Plithogenic statistics (PS) encompasses the analysis and observations of the events studied by the plithogenic probability. Plithogenic statistics is a generalization of classical multivariate statistics, and it is a simultaneous analysis of many outcome neutrosophic/indeterminate variables, and it as well is a multi-indeterminate statistics.

1.5 Applications of neutrosophic statistics in the medical field and other areas

Over 100 papers have been published on neutrosophic probability and statistics and plithogenic probability and statistics, including many journals by Elsevier and Springer of high impact factor, most of them offering applications in the bioinformatics and related fields, and several books [20,21]. It follows a review of some of them.

Classical probability, Bayes theory, Dempster–Shafer theory, certainty factor, and fuzzy set approaches are not enough to express uncertain problems in expert systems for managing uncertainty data. In Ref. [22], Nouran M. Radwan, M. Badr Senousy, and Alaa El Din M. Riad suggested that the best models for handling uncertainty in order to derive decisions are the multivalued logic models, intuitionistic fuzzy set, vague set, and neutrosophic set. Radwan et al. analyzed the relationships between them and provided insights into their application in expert systems for evaluating learning management systems.

The sampling plans have been widely used for the inspection of a product lot. In practice, the measurement data may be imprecise, uncertain, unclear, or fuzzy. When there is uncertainty in the observations, the sampling plans designed using classical statistics cannot be applied for the inspection of a lot of the product. Neutrosophic statistics can be used when data are not precise, uncertain, unclear, or fuzzy.

In analyzing lot sentencing, classical statistics assumes that data are determinate. Consequently, the imprecise or intermediate data lead to indecision about the quality of a specific lot of the product. Muhammad Aslam proposed a new attribute sampling plan using the neutrosophic interval method in Ref. [23], computing the lot acceptance, rejection, and indeterminate probabilities using the neutrosophic binomial distribution and adding an example to explain the proposed sampling plan.

In another paper [24], Muhammad Aslam extracted the neutrosophic operating characteristic (NOC) from the neutrosophic normal distribution and propounded an optimization solution for the proposed plan under the neutrosophic interval method.

In Ref. [25], Muhammad Aslam, Nasrullah Khan, and Ali Hussein Al-Marshadi designed the variable sampling plan under the Pareto distribution using neutrosophic statistics. They used the symmetry property of the normal distribution, assuming uncertainty in measurement data and sample size required for the inspection of a lot of the product. The authors determined the neutrosophic plan parameters using the neutrosophic optimization problem, offering some tables for practical use.

Group acceptance sampling plans occupy an important place in the literature. The aim of reducing the inspection cost using acceptance sampling can be achieved by employing the features of allocating more than one sample item to a single tester. In Ref. [26], designing a group acceptance sampling plan is considered to provide assurance on the product's mean life. The authors submitted the proposed plan based on neutrosophic statistics under the assumption that the product's lifetime follows a Weibull distribution, determining the optimal parameters using two specified points on the operating characteristic curve.

In industry, in order to advance the quality of a product, the process capability index (PCI) is widely used. Aslam and Albassam recommended a variable sampling plan for the PCI using neutrosophic statistics [27], determining the neutrosophic plan parameters by the use of the neutrosophic optimization solution.

When security is concerned, one must deal with uncertainty. The intrusion detection systems are challenged due to imprecise knowledge in classifying normal or abnormal behavior patterns. Kavitha et al. introduced in Ref. [28] an emerging approach for intrusion detection system using Neutrosophic Logic Classifier, which is capable of handling fuzzy, vague, incomplete, and inconsistent information under one framework, offering an increase in detection rate and a significant decrease in false alarm rate. This method classifies the dataset into normal, abnormal, and indeterministic based on the degrees of membership of truth, indeterminacy, and falsity. The Neutrosophic Logic Classifier generates the neutrosophic rules to determine the intrusion in progress.

The uncertainties in image segmentation problems derive from gray-level and spatial ambiguities. Accurate segmentation of text regions in mixed documents is also a difficult issue to overcome. Soumyadip Dhar and Malay K. Kundu [29] proposed a method based on digital shearlet transform (DST). The proposed method uses the DST coefficients as input features to a segmentation process block to capture the anisotropic features of the text regions, blocks designed by employing neutrosophic set. This method is experimentally verified both quantitatively and qualitatively using a benchmark dataset.

Pierpaolo D'Urso illustrated the connection between information and uncertainty from the perspective of the so-called informational paradigm [30], according to which information is constituted by "informational ingredients," specifically the "empirical information," represented by statistical data, and "theoretical information" consisting of background knowledge and basic modeling assumptions. Focusing on the uncertainty associated with a particular statistical methodology, that is, cluster analysis, and adopting as theoretical platform the informational paradigm, Pierpaolo D'Urso systematically reviewed the literature of different uncertainty-based clustering approaches, such as fuzzy clustering, possibilistic clustering, shadowed clustering, rough sets-based clustering, intuitionistic fuzzy clustering, evidential clustering, credibilistic clustering, type-2 fuzzy clustering, neutrosophic clustering, hesitant fuzzy clustering, interval-based fuzzy clustering, and picture fuzzy clustering, showing how all these clustering approaches are able of managing in different ways the uncertainty associated with the two components of the informational paradigm.

The control charts based on failure-censored (type-II) reliability tests are designed using classical statistics, which is applied for monitoring the process when observations in the sample or the population are determined, while neutrosophic statistics can be applied when there is uncertainty in the sample or population.

In Ref. [31], Muhammad Aslam, Nasrullah Khan, and Mohammed Albassam proposed a control chart for failure-censored (type-II) reliability tests, by the use of neutrosophic statistics, with the design of a control chart for the Weibull distribution. The proposed control chart was used to monitor the neutrosophic mean and neutrosophic variance, which are related to the neutrosophic scale parameter.

A control chart methodology for monitoring the mean time between two events using the belief estimator under the neutrosophic gamma distribution is put forward for consideration by Aslam, Bantan, and Khan in Ref. [32]. The proposed control chart coefficients and the neutrosophic average run length (NARL) are determined using different process settings, while the chart's performance is compared with the control chart under classical statistics in terms of NARL using the simulation data and real example. The comparisons drive the conclusion that the chart is more efficient for use in an uncertain environment than the classical charts.

Another relevant paper is Ref. [33]. The estimation of the ratio of two means is studied within the neutrosophic theory framework. The variable of interest Y is measured in a sample of units, and the auxiliary variable X is obtainable for all units using records or predictions. They are correlated, and the sample is selected using simple random sampling. The indeterminacy of the auxiliary variable is considered and is modeled as a neutrosophic variable. The bias and variance of the proposed estimator are derived.

The mechanical properties of geological bodies are complex. Their various parameters are vague, incomplete, imprecise, and indeterminate, impossible to be expressed by crisp values from classical probability and statistics. There is a need in geotechnical engineering to approximate exact values in indeterminate environments. Determining effectively the joint roughness coefficient (JRC) is a key parameter in the shear strength between rock joint surfaces. Chen et al. came up with a solution in Ref. [34], introducing the neutrosophic interval probability (NIP) and defining the confidence degree based on the cosine measure between NIP and the ideal NIP. By combining the neutrosophic number with the confidence degree to express indeterminate statistical information, the researchers proposed a new neutrosophic interval statistical number (NISN). The results demonstrated that NISNs are suitable and advantageous for JRC expressions.

In the same area of geological bodies, Jun Ye et al. investigated the anisotropy and scale effect of indeterminate JRC values by neutrosophic number (NN) functions [35]. The NN is composed of a determinate part and an indeterminate part, being suitable for the expression of JRC data with determinate and/or indeterminate information. The researchers considered the lower limit of JRC data as the determinate information, and the difference between the lower and upper limits as the indeterminate information. To reflect the anisotropy and scale effect of JRC values, the NN functions of the anisotropic ellipse and logarithmic equation of JRC were developed. Additionally, the NN parameter ψ was defined in order to quantify the anisotropy of JRC values, and a two-variable NN function was suggested based on the factors of both the sample size and measurement orientation. Further on, the changing rates in various sample sizes and/or measurement orientations were investigated by their derivative and partial derivative NN functions.

Since the nature of the rock mass is indeterminate and incomplete to some extent, one cannot always express rock JRC by a certain/exact number. Therefore, Wenzhong Jiang et al. introduced the NISNs based on the concepts of neutrosophic numbers and neutrosophic interval probability to express JRC data of the rock mass in the indeterminate setting [36]. The authors presented the calculational method of the neutrosophic average value and standard deviation of NISNs based on neutrosophic statistics. In an actual case, the neutrosophic average value and standard deviation of the rock JRC NISNs are employed to analyze the scale effect and anisotropy of the rock body corresponding to different sample lengths and measuring directions.

Other researchers suggested the neutrosophic diagnosis test (NDT), with an application to diabetic patients [12]. NDT can be put at work effectively and adequately in medical science, biostatistics, decision, and classification analysis, anytime when some or all observations are not determined.

A new control chart for monitoring reliability using sudden death testing under neutrosophic statistics is carried forth in Ref. [37]. For evaluating the quick detection ability for small and moderate shifts, Arif and Aslam determined the average run lengths of the in-control and the out-of-control processes.

Sudden death testing to reduce testing time is also implemented by parts manufacturers. As proved in Ref. [38], neutrosophic fuzzy statistics can be applied for the testing of manufacturing parts when observations are fuzzy, imprecise, and incomplete. The authors projected a neutrosophic fuzzy sudden death testing plan for the inspection/testing of the electronic parts manufacturing. The neutrosophic fuzzy operating function is enforced to determine the neutrosophic fuzzy plan parameters through a neutrosophic fuzzy optimization problem.

Khan et al., in Ref. [39], proposed a new neutrosophic exponentially weighted moving average (NEWMA) control chart based on the moving average statistic. A NEWMA design is applied with moving average statistic to get a sensitive control chart for the quick detection of any unwanted process variations. The efficacy of the chart is evaluated with the average run lengths computed with Monte Carlo simulations.

In Ref. [13], a neutrosophic statistical analysis of the application of physical exercise for the rehabilitation of patients with knee osteoarthritis is proposed. Physical exercises for the rehabilitation of the elderly with gonarthrosis were applied to a selected sample in order to verify the effectiveness of the treatment.

The participation of young people in the electoral processes of a country has a great impact on political life. Analyzing the incidence of optional voting and the responsibility that the use of the right implies in electoral democracy constitute an important task to guarantee citizen rights. These elements can be analyzed in the scientific literature through statistical studies. In Ref. [40], Molina Manzo et al. aim to carry out an analysis of the incidence of the facultative vote of young people between 16 and 18 years old in the electoral process of Ecuador. Neutrosophic numbers are used to model the uncertainty in the proposed analysis.

Another paper [41] treats the fuzzy random multi-objective unbalanced transportation data problems. The fuzzy programming technique when the sources and destination parameters are fuzzy random variables in inequality type of constraints is employed. Introducing the concept of linear membership function of fuzzy programming, Salama et al. focused on the solution procedure of the specified transportation data problems where the objective functions are minimization type and supplies; also, demands are replaced by the fuzzy random variables.

The analysis of means test under the neutrosophic statistics is approached in Ref. [42]. Using wind power data, Aslam explained here the proposed neutrosophic analysis of means and compared it with the existing analysis of means under classical statistics. The comparison conducted made the author to come to the conclusion that the proposed neutrosophic analysis of means is more effective, informative, and flexible than the existing analysis of means applied in the area of renewable energy.

In Ref. [43], an improved group-sampling plan based on time-truncated life tests for Weibull distribution under neutrosophic statistics has been developed. Aslam et al. developed improved single and double group-sampling plans and obtained design neutrosophic plan parameters by satisfying simultaneously both producer's and consumer's risks under neutrosophic optimization solution.

For quality and reliability, the Weibull distribution is widely used. When all the observations in quality and reliability work are exactly determined, Anderson–Darling test under classical statistics is applied, while when data are indeterminate or fuzzy, the Anderson–Darling test under neutrosophic statistics could be administered, as in Ref. [44].

Neutrosophic statistics is used in Ref. [45] to analyze the income of some TV channels (Ducky Bahi TV, Hasi TV, and Haqeeqat TV) operating on the YouTube website. Not only the value of earnings—which is directly proportional to the number of views, likes, and subscribers—is under examination, but also the constancy of the income.

An application of the time-truncated plan for the Weibull distribution under the indeterminacy is given using wind speed data in Ref. [46]. From the wind speed example, it is concluded that the proposed plan is helpful to test the average wind speed at smaller values of sample size as compared to the existing sampling plan.

To detect the indeterminacy effect in the manufacturing process, attribute control chart using neutrosophic Weibull distribution is proposed in Ref. [47]. To make the attribute control chart more efficient for persistent shifts in the industrial process, an attribute control chart using the Weibull distribution has been proposed. A neutrosophic Weibull distribution-based attribute control chart has been developed for efficient monitoring of the process. The indeterminacy effect was studied with the control chart's performance using characteristics of run length, and the proposed chart effectively detected shifts in uncertainty.

The Durbin–Watson test under neutrosophic statistics is suggested in Ref. [48] and applied to test autocorrelation existing in error terms when neutrosophy is presented in the data taken from metrology. It is concluded that the test is effective in metrology data in the presence of indeterminacy.

Cochran's tests work to detect outliers when observations in the data are determinate. But, when some or all observations are not precise, the existing Cochran's tests cannot be applied to detecting outliers in the data. In Ref. [49], the authors introduced Cochran's tests and modification of Cochran's tests based on Levene's test under neutrosophic statistics. The performances of tests are evaluated using the power curves for various non-normal distributions. The comparison of both Cochran's tests is given and explained with the aid of a real example.

For testing the means of two non-normal populations, the Mann–Whitney test is applied. In Ref. [50], the Mann–Whitney test is adhibitted in neutrosophic environment in order to design and implement a test under neutrosophic statistics rules. The application of the proposed test is given with the aid of melting points of alloy data.

In real life, the data detected from complex situations or uncertain conditions do not contain all determined observations. Given the case, the data are recorded in an indeterminacy interval, which can be properly judged using neutrosophic statistics. That is why Muhammad Aslam and Mohammed Albassam [51] modified the existing least significant difference test, Bonferroni test, and Scheffe test under neutrosophic statistics rules, obtaining post hoc flexible and informative modified tests under the neutrosophic statistical interval method (NSIM).

In another paper [52], the indefatigable researcher in neutrosophic statistics, Professor Muhammad Aslam, introduced the neutrosophic analysis of variance (NANONA), which is an extension of the classical ANOVA and performs the NANONA to test teaching methods using data collected from the university students.

The paper [53] develops a two-stage process loss using neutrosophic statistics. Neutrosophic plan parameters are obtained using nonlinear optimization using the neutrosophic statistical interval method for the constraints.

A repetitive group-sampling control has been introduced for the neutrosophic statistics under the Conway–Maxwell–Poisson (COM–Poisson) distribution in Ref. [54]. Using simulated data generated from neutrosophic COM–Poisson distribution, the chart has been compared with the existing plan, and the practical implementation has been illustrated by data from electric circuit boards.

The designing of the V-test for testing the randomness of angles under neutrosophic statistics is presented in Ref. [55], which can be employed when the decision-maker is uncertain about the sample size or the measurement of angles. The operational procedure for testing the randomness of the radar angles is given under an indeterminate environment.

In Ref. [56], a repetitive sampling control chart for the gamma distribution under the indeterminate environment is explained. Using the neutrosophic interval method, the average run lengths, the control chart coefficients, and the probability of in-control and out-of-control are determined under the assumption of the symmetrical property of the normal distribution. The performance is evaluated by the average run length measurements under different process settings. A real-life example from health care is included for practical application.

Muhammad Aslam suggests in Ref. [57] two measures of skewness and kurtosis and evaluates the nature of the wind speed in neutrosophic numbers, discussing the importance of the proposed measures of skewness and kurtosis under neutrosophic statistics. The author recommends the application of a heavy-tailed distribution, like the Cauchy distribution, for forecasting and estimation of energy produced by wind speed.

The same author generalizes the Anderson–Darling test under neutrosophic statistics in Ref. [58]. The designing and operating procedure of neutrosophic Anderson–Darling is presented following the neutrosophic normal distribution with data from the renewable energy field.

Professor Aslam also introduces the designing of Grubbs's test under neutrosophic statistics in Ref. [14], also presenting the operational procedure of the proposed test under the neutrosophic statistical interval method, with the application of real data from the medical field.

To evaluate medical correlation, Aslam et al. presented an epidemiological study [15] on the dietary fat that may cause prostate cancer. To analyze this relationship under the indeterminate environment, data from 30 countries are selected for the prostate cancer death rate and dietary fat level in the food, fitting with data of the neutrosophic correlation and regression line, and proving that the prostate cancer death rate increases as the dietary fat level in the people increases. This paper concludes that the neutrosophic regression is a more effective model under uncertainty than the regression model under classical statistics.

A flexible method for the forecasting of wind speed is projected in Ref. [59], and the semi-average method under neutrosophic statistics is introduced.

Another study [60] investigates the climate impact on rice yield. Temperature and rain are the indicators for climate variation on stages of rice growth as independent variables and the yield of rice as dependent variable. The results show that the crop is more vulnerable to variation in temperature than rain and that the climate variability is negatively impacting the rice yield.

An extension of the concept of neutrosophic normal distribution is attempted in Ref. [61], by defining it in various forms and providing relevant examples. Properties such as mean and mean deviation, variance and cumulants, moment-generating function and characteristics function are derived under neutrosophic normal probability distribution. Moreover, the neutrosophic quantile function and neutrosophic Q–Q plot of the distribution are computed in different success probabilities, and maximum likelihood estimates, entropy, and Fisher information matrix are provided as well for the neutrosophic normal probability distribution.

In Ref. [62], the beta distribution is considered in the indeterminate environment, obtaining the neutrosophic beta distribution. Distributional properties are derived, such as variance, mean, moment-generating function, r-th moment order statistics including smallest order statistics, largest order statistics, joint order statistics, and median order statistics, while the parameters are estimated via maximum likelihood method, with an application on two real datasets assessed through AIC and BIC criteria.

To efficiently monitor real-life situations, such as weather forecasting and stock prices, Aslam et al. proposed a new cumulative sum (CUSUM) X-chart under the assumption of uncertainty employing neutrosophic statistics [63], investigating the performance of the chart in terms of neutrosophic run length properties, with the Monte Carlo simulations approach by the implementation of the NCUSUM X-chart for simulated petroleum and meteorological data.

In Ref. [64], a control chart for neutrosophic exponentially weighted moving average (NEWMA) is designed by repetitive sampling. An application to monitor road traffic crashes (RTCs) is performed.

Shapiro–Wilk test is largely performed to verify the normality of the data, and in Ref. [65], it is improved by considering the neutrosophic statistics. The necessary equations are given under the indeterminacy environment. The testing procedure is explained with the help of chemical data.

A new attribute control chart for monitoring the blood components under neutrosophic statistics is examined in Ref. [66]. The design of the proposed control chart is given under the neutrosophic statistical interval method. The applications of these control charts demonstrate that the proposed control charts are quite effective, adequate, flexible, and informative for monitoring the blood components in an uncertain environment.

A study on the neutrosophic measurement system analysis model (NMSAM) is accomplished by Aslam et al. in Ref. [67]. The NMSAM gives assistance to industrial engineers to understand the capability of gauge in measuring the product in an uncertain environment. Various gauge capability evaluation criteria when indeterminacy is presented are also discussed.

A neutrosophic exponentially weighted moving average (NEWMA) statistic for the attribute data is developed in Ref. [68]. The authors use the NEWMA to design an attribute control chart and employ the neutrosophic Monte Carlo simulation (NMCS) to find the neutrosophic average run length (NARL). Two examples of having neutrosophic parameters are given to explain the proposed control chart, which clearly works best in an uncertain environment.

In Ref. [69], a new np control chart for the multiple dependent state (MDS) sampling employing neutrosophic statistics is introduced. The aim is to monitor efficiently the number of defective items in any production process or the customer services agencies. The coefficients of the control limits of the control chart are determined using the neutrosophic algorithms, and the efficiency is determined under different process settings at different process shift levels by computing the neutrosophic average run lengths with respect to different false alarm rates.

Another control chart scheme is proposed under the neutrosophic statistics in Ref. [70] for the mean monitoring using gamma distribution for belief statistics using MDS sampling. The coefficients of the control chart and the neutrosophic average run lengths are estimated under various process conditions for specific false alarm probabilities, by considering simulation and real data.

A new sampling plan using the neutrosophic approach for the process loss function is proposed in Ref. [71], with the application of the neutrosophic nonlinear to determine the neutrosophic plan parameters of the proposed sampling plan.

In Ref. [72], Aslam et al. design an S^2 control chart under the neutrosophic interval methods, offering the complete structure and the necessary measures of neutrosophic S^2 control chart, determined through the neutrosophic algorithm, with tables for practical use.

Assuming that the proportion of defective items is fuzzy and following the Birnbaum–Saunders distribution, an acceptance sampling plan based on binomial distribution is developed for the fuzzy operating characteristic curve in Ref. [73]. The real-life example considered shows the behavior of curves with different combinations of parameters of Birnbaum–Saunders distribution.

An exponentially weighted moving average (EWMA) scheme is normally performed to detect small shifts, but when fuzzy data are present, Khan et al. suggested in Ref. [74] a fuzzy EWMA (F-EWMA) control chart to detect small shifts in the process mean, illustrated by real-life data.

In Ref. [75], the neutrosophic coefficient of variation (NCV) is introduced, and a sampling plan based on it is designed. The NOC function is then given and employed to determine the neutrosophic plan parameters under some constraints, determined through the neutrosophic optimization solution.

A new neutrosophic exponentially weighted moving average (NEWMA) combined with a neutrosophic logarithmic transformation chart for monitoring the variance with neutrosophic numbers is presented in Ref. [76], while the computation of the neutrosophic control chart parameters is performed through the NMCS.

The symmetry property of normal distribution using the neutrosophic exponentially weighted moving average statistics is employed for the designing of the X-bar control chart in Ref. [77] for monitoring the data under an uncertain environment, and again, the NMCS is performed to determine the neutrosophic average run length.

In Ref. [78], the diagnosis of the manufacturing process under the indeterminate environment is presented. The similarity measure index is employed to find the probability of the in-control and the out-of-control of the process. For various values of specified parameters, the average run length (ARL) is also computed, with an example from a juice company considered under the indeterminate environment with the aim of eliminating the nonconforming items and alternatively increasing the profit.

Another control chart methodology for monitoring the mean time between two events using the belief estimator under the neutrosophic gamma distribution is proposed in Ref. [79], while the chart coefficients and the neutrosophic average run length (NARL) are determined using different process settings.

Another sampling plan for the measurement error using neutrosophic statistics is designed in Ref. [80], with sample size and acceptance number as neutrosophic parameters. The neutrosophic plan parameters are determined by the neutrosophic optimization problem, and the neutrosophic operating function is given.

In Ref. [81], Aslam et al. designed a Shewhart attribute control chart under the neutrosophic statistical interval method. To study the performance of the proposed chart, some neutrosophic measures are given, and the neutrosophic control chart coefficients are determined by a neutrosophic algorithm.

A new test of independence under neutrosophic statistics in order to test the association between two criteria of classification is presented in Ref. [82], associated with the necessary contingency tables for the neutrosophic population and the neutrosophic sample. A real example from education is selected to explain the proposed test.

The W/S test under neutrosophic statistics is held forth by Albassam et al. in Ref. [83]. The neutrosophic statistical interval method is applied to the Monte Carlo simulation toward the sensitivity of various neutrosophic statistical distributions, while the power of test curves for neutrosophic distributions is presented.

In Ref. [84], the neutrosophic COM−Poisson (NCOM−Poisson) distribution is introduced, and the design of the attribute control chart using the NCOM−Poisson distribution is given, together with the algorithm to determine the ARL under the neutrosophic statistical interval system, with the purpose of minimizing the defective production.

In Ref. [85], the Hotelling T-squared statistic under neutrosophic statistics is applied, and an application of the neutrosophic Hotelling T-squared statistic is suggested with the aid of data.

In Ref. [86], the Kolmogorov−Smirnov (K−S) tests are generalized under the neutrosophic statistics to neutrosophic Kolmogorov−Smirnov tests, and their necessary measures and procedures are presented.

Bartlett's and Hartley's tests under the neutrosophic statistics are offered in Ref. [87].

In Ref. [88], the central tendency and dispersion measures are stretched under neutrosophic statistics. The purpose is to analyze the data which has been measured under uncertain environments, focusing on neutrosophic arithmetic mean, neutrosophic median, neutrosophic geometric mean, neutrosophic harmonic mean, neutrosophic mode, and the relationship between these measures.

To maintain the quality of the product at the specified standard level is critically important in current extremely competitive environment in almost all businesses. Strategic priorities for reducing the variability of the product and meeting customer expectations are efficiently and quickly adopted for maximizing the profits. The quality of the manufacturing product is, thus, focused on using all available physical, aesthetic, intellectual, and strategic resources.

In Ref. [89], the variability in the product is monitored by developing a control chart scheme using MDS sampling which allows to combine the information of the current and the preceding sample when the scenario of the interested quality characteristic is uncertain, unclear, or vague in nature. Employing the neutrosophic statistical interval method, the control chart coefficients are estimated.

1.6 Conclusions

The extensions from classical probability and statistics to neutrosophic probability and statistics and further on to plithogenic probability and statistics brought the mathematical modeling to a better representation of our real world. It is a progress from the classical crisp probability distribution to the multi-probability distribution that deals with indeterminate (vague, unclear, conflicting, etc.) data and inference procedures in neutrosophic and plithogenic probability and statistics, which is of much help for representing medical knowledge, for example, in diagnosis, disease probability distributions, health-improving charts. Many applications have been proposed in various other fields, such as image segmentation, cognitive science, clustering computing.

References

[1] F. Smarandache, A unifying field in logics: neutrosophic logic, Neutrosophy, Neutrosophic Set, Neutrosophic Probability and Statistics, sixth ed., InfoLearnQuest, 2007. Available from: http://fs.unm.edu/eBook-Neutrosophics6.pdf.

[2] F. Smarandache, Introduction to Neutrosophic Statistics, Sitech & Education Publishing, 2014. Available from: http://fs.unm.edu/NeutrosophicStatistics.pdf.

[3] W.B. Vasantha Kandasamy, F. Smarandache, Fuzzy cognitive maps and neutrosophic cognitive maps, Xiquan (2003). Available from: http://fs.unm.edu/NCMs.pdf.

[4] F. Smarandache, N. Overset, N. Underset, N. Offset, Similarly for neutrosophic over-/under-/off- logic, probability, and statistics, Pons Ed. (2016). Available from: http://fs.unm.edu/NeutrosophicOversetUndersetOffset.pdf.

[5] M.L. Vázquez, F. Smarandache, Neutrosofía: Nuevos avances en el tratamiento de la incertidumbre, Pons Ed. (2018). Available from: http://fs.unm.edu/NeutrosofiaNuevosAvances.pdf.

[6] T.V. Gutierrez Quinonez, F.A. Espinoza, I.K. Giraldo, A.S. Asanza, M.D. Montenegro, Estadistica y Probabilidades: Una Vision Neutrosofica desde el Aprendizaje Basado en Problemas en la Construccion del Conocimiento, Pons Ed. (2020). Available from: http://fs.unm.edu/EstadisticaYProbabilidadNeutrosofica.pdf.

[7] F. Smarandache, Neutrosophic statistics vs. classical statistics, section in Nidus Idearum/Superluminal Physics, 7, 2019, p. 117. http://fs.unm.edu/NidusIdearum7-ed3.pdf.

[8] F. Smarandache, Nidus Idearum/de Neutrosophia, Pons Ed. (2016–2019) 1–7. Available from: http://fs.unm.edu/ScienceLibrary.htm.

[9] F. Smarandache, Introduction to neutrosophic measure, neutrosophic integral, and neutrosophic probability, Sitech (2013). Available from: http://fs.unm.edu/NeutrosophicMeasureIntegralProbability.pdf.

[10] F. Smarandache, Neutrosophic quadruple numbers, refined neutrosophic quadruple numbers, absorbance law, and the multiplication of neutrosophic quadruple numbers, Neutrosophic Sets Syst. 10 (2015) 96–98. Available from: https://doi.org/10.5281/zenodo.571562. Available from: http://fs.unm.edu/NSS/NeutrosophicQuadrupleNumbers.pdf.

[11] F. Smarandache, A. Rezaei, A.A.A. Agboola, Y.B. Jun, R.A. Borzooei, B. Davvaz, et al., On NeutroQuadrupleGroups, in: 51st Annual Mathematics Conference, Kashan, 16–19 February 2021. http://fs.unm.edu/OnNeutroQuadrupleGroups-slides.pdf.

[12] M. Aslam, O.H. Arif, R.A. Khan Sherwani, New diagnosis test under the neutrosophic statistics: an application to diabetic patients, BioMed. Res. Int. (2020). Available from: https://doi.org/10.1155/2020/2086185. ID 2086185.

[13] A.C. Yumar Carralero, D.M. Ramírez Guerra, G.P. Iribar, Análisis estadístico neutrosófico en la aplicación de ejercicios físicos en la rehabilitación del adulto mayor con gonartrosis, Neutrosophic Comput. Mach. Learn. 13 (2020) 9.

[14] M. Aslam, Introducing Grubbs's test for detecting outliers under neutrosophic statistics - an application to medical data, J. King Saud. Univ.-Sci. 32 (2020) 2696–2700.

[15] M. Aslam, M. Albassam, Application of neutrosophic logic to evaluate correlation between prostate cancer mortality and dietary fat assumption, Symmetry 11 (2019) 330. Available from: https://doi.org/10.3390/sym11030330.

[16] (a) F. Smarandache, Neutrosophic set is a generalization of intuitionistic fuzzy set, inconsistent intuitionistic fuzzy set (picture fuzzy set, ternary fuzzy set), pythagorean fuzzy set (atanassov's intuitionistic fuzzy set of second type), q-rung orthopair fuzzy set, spherical fuzzy set, and n-hyperspherical fuzzy set, while neutrosophication is a generalization of regret theory, grey system theory, and three-ways decision (revisited), Journal of New Theory 29 (2019) 1–35; and arXiv, Cornell University, New York City, NY, USA, pp. 1–50, 17–29 November 2019, https://arxiv.org/ftp/arxiv/papers/1911/1911.07333.pdf; and The University of New Mexico, Albuquerque, USA, Digital Repository, pp. 1–50, https://digitalrepository.unm.edu/math_fsp/21.
(b) C. Le, Preamble to Neutrosophy and Neutrosophic Logic, Multiple Valued Logic/An Int. J. 8 (3) (2002) 285–295. http://fs.unm.edu/ParadoxismMVL.pdf.

[17] C. Le, pARadOXisM in English – The Last Vanguard of the Second Millennium. http://fs.unm.edu/a/paradoxism-en.htm.

[18] F. Smarandache, n-Valued refined neutrosophic logic and its applications in physics, Prog. Phys. 4 (2013) 143–146. Available from: https://arxiv.org/ftp/arxiv/papers/1407/1407.1041.pdf.

[19] F. Smarandache, Plithogenic probability & statistics are generalizations of multivariate probability & statistics, Neutrosophic Sets Syst. 43 (2021) 280–289. Available from: https://doi.org/10.5281/zenodo.491489.

[20] F. Smarandache, M. Abdel-Basset (Eds.), Neutrosophic Operational Research, Methods and Applications, Springer, 2021. Available from: https://link.springer.com/book/10.1007/978-3-030-57197-9.

[21] F. Smarandache, M. Abdel-Basset (Eds.), Optimization Theory Based on Neutrosophic and Plithogenic Sets, Elsevier, 2020. Available from: https://www.elsevier.com/books/optimization-theory-based-on-neutrosophic-and-plithogenic-sets/smarandache/978-0-12-819670-0.

[22] N.M. Radwan, M. Badr Senousy, A. El Din, M. Riad, Approaches for managing uncertainty in learning management systems, Egypt. Comput. Sci. J. 40 (2) (2016) 10.

[23] M. Aslam, A new attribute sampling plan using neutrosophic statistical interval method, Complex. Intell. Syst. 7 (2018). Available from: https://doi.org/10.1007/s40747-018-0088-6.

[24] M. Aslam, A variable acceptance sampling plan under neutrosophic statistical interval method, Symmetry 11 (2019) 114. Available from: https://doi.org/10.3390/sym11010114.

[25] M. Aslam, N. Khan, A. Hussein Al-Marshadi, Design of variable sampling plan for pareto distribution using neutrosophic statistical interval method, Symmetry 11 (2019) 80. Available from: https://doi.org/10.3390/sym11010080.

[26] M. Aslam, P. Jeyadurga, S. Balamurali, A. Hussein Al-Marshadi, Time-truncated group plan under a weibull distribution based on neutrosophic statistics, Mathematics 7 (2019) 905. Available from: https://doi.org/10.3390/math7100905.

[27] M. Aslam, M. Albassam, Inspection plan based on the process capability index using the neutrosophic statistical method, Mathematics 7 (2019) 631. Available from: https://doi.org/10.3390/math7070631.

[28] B. Kavitha, S. Karthikeyan, P. Sheeba Maybell, An ensemble design of intrusion detection system for handling uncertainty using neutrosophic logic classifier, Knowl. Syst. 28 (2012) 88–96. Available from: https://doi.org/10.1016/j.knosys.2011.12.004.

[29] S. Dhar, M.K. Kundu, Accurate segmentation of complex document image using digital shearlet transform with neutrosophic set as uncertainty handling tool, Appl. Soft Comput. 61 (2017) 412–426. Available from: https://doi.org/10.1016/j.asoc.2017.08.005.

[30] P. D'Urso, Informational Paradigm, management of uncertainty and theoretical formalisms in the clustering framework: a review, Inf. Sci. 400–401 (2017) 30–62. Available from: https://doi.org/10.1016/j.ins.2017.03.001.

[31] M. Aslam, N. Khan, M. Albassam, Control chart for failure-censored reliability tests under uncertainty environment, Symmetry 10 (2018) 690. Available from: https://doi.org/10.3390/sym10120690.

[32] M. Aslam, R.A.R. Bantan, N. Khan, Monitoring the process based on belief statistic for neutrosophic gamma distributed product, Processes 7 (2019) 209. https://doi.org/10.3390/pr7040209.

[33] C.N. Bouza-Herrera, M. Subzar, Estimating the ratio of a crisp variable and a neutrosophic variable, Int. J. Neutrosophic Sci. 11 (1) (2020) 9–21.

[34] J. Chen, J. Ye, S. Du, R. Yong, Expressions of rock joint roughness coefficient using neutrosophic interval statistical numbers, Symmetry 9 (2017) 123. Available from: https://doi.org/10.3390/sym9070123.

[35] J. Ye, J. Chen, R. Yong, S. Du, Expression and analysis of joint roughness coefficient using neutrosophic number functions, Information 8 (2017) 69. Available from: https://doi.org/10.3390/info8020069.

[36] W. Jiang, J. Ye, W. Cui, Scale effect and anisotropic analysis of rock joint roughness coefficient neutrosophic interval statistical numbers based on neutrosophic statistics, J. Soft Comput. Civ. Eng. 2–4 (2018) 62–71. Available from: https://doi.org/10.22115/SCCE.2018.143079.1086.

[37] O.H. Arif, M. Aslam, A new sudden death chart for the Weibull distribution under complexity, Complex. Intell. Syst. 9 (2021). Available from: https://doi.org/10.1007/s40747-021-00316-x.

[38] M. Aslam, O.H. Arif, Testing of grouped product for the weibull distribution using neutrosophic statistics, Symmetry 10 (2018) 403. Available from: https://doi.org/10.3390/sym10090403.

[39] N. Khan, M. Aslam, A. Arshad, A. Shafqat, Tracking temperature under uncertainty using EWMA-MA control chart, MAPAN 36 (2021) 497–508.

[40] A.D. Molina Manzo, R.L. Maldonado Manzano, B.E. Brito Herrera, J.I. Escobar Jara, Análisis estadístico neutrosófico de la incidencia del voto facultativo de los jóvenes entre 16 y 18 años en el proceso electoral del Ecuador, Neutrosophic Comput. Mach. Learn. 11 (2020) 6.

[41] A.A. Salama, M. Elsayed Wahed, E. Yousif, A multi-objective transportation data problems and their based on fuzzy random variables, Neutrosophic Knowl. 1 (2020) 41–49.

[42] M. Aslam, Analyzing wind power data using analysis of means under neutrosophic statistics, Soft Comput. 25 (2021) 7087–7093.

[43] M. Aslam, G. Srinivasa Rao, N. Khan, Single-stage and two-stage total failure-based group-sampling plans for the Weibull distribution under neutrosophic statistics, Complex. Intell. Syst. 7 (2021) 891–900. Available from: https://doi.org/10.1007/s40747-020-00253-1.

[44] M. Aslam, A new goodness of fit test in the presence of uncertain parameters, Complex. Intell. Syst. 7 (2021) 359–365. Available from: https://doi.org/10.1007/s40747-020-00214-8.

[45] I. Shahzadi, M. Aslam, H. Aslam, Neutrosophic statistical analysis of income of YouTube channels, Neutrosophic Sets Syst. 39 (2021).

[46] M. Aslam, Testing average wind speed using sampling plan for Weibull distribution under indeterminacy, Nat. Res.-Sci. Rep. 11 (2021) 7532. Available from: https://doi.org/10.1038/s41598-021-87136-8.

[47] A.H. Al-Marshadi, A. Shafqat, M. Aslam, A. Alharbey, Performance of a new time-truncated control chart for weibull distribution under uncertainty, Int. J. Comput. Intell. Syst. 14 (1) (2021) 1256–1262. Available from: https://doi.org/10.2991/ijcis.d.210331.001.

[48] M. Aslam, On testing autocorrelation in metrology data under indeterminacy, MAPAN 36 (2021) 515–519.

[49] M. Aslam, N. Khan, Normality test of temperature in jeddah city using Cochran's test under indeterminacy, MAPAN 36 (2021) 589–598.

[50] M. Aslam, M. Sattam Aldosari, Analyzing alloy melting points data using a new Mann-Whitney test under indeterminacy, J. King Saud. Univ. Sci. 32 (2020) 2831–2834. Available from: https://doi.org/10.1016/j.jksus.2020.07.005.

[51] M. Aslam, M. Albassam, Presenting post hoc multiple comparison tests under neutrosophic statistics, J. King Saud. Univ. Sci. 32 (6) (2020) 2728–2732.

[52] M. Aslam, Neutrosophic analysis of variance: application to university students, Complex. Intell. Syst. 5 (2019) 403–407. Available from: https://doi.org/10.1007/s40747-019-0107-2.

[53] M. Aslam, G. Srinivasa Rao, N. Khan, L. Ahmad, Two-stage sampling plan using process loss index under neutrosophic statistics, J. Taibah Univ. Sci. (2019). Available from: https://doi.org/10.1080/03610918.2019.1702212.

[54] M. Aslam, G. Srinivasa Rao, A. Shafqat, L. Ahmad, R.A. Khan Sherwani, Monitoring circuit boards products in the presence of indeterminacy, Measurement 168 (2021) 108404.

[55] M. Aslam, Radar data analysis in the presence of uncertainty, Eur. J. Remote. Sens. 54 (1) (2021) 140–144. Available from: https://doi.org/10.1080/22797254.2021.1886597.

[56] A.M. Almarashi, M. Aslam, Process monitoring for gamma distributed product under neutrosophic statistics using resampling scheme, J. Math. 12 (2021). Available from: https://doi.org/10.1155/2021/6635846. Article ID 6635846.

[57] M. Aslam, A study on skewness and kurtosis estimators of wind speed distribution under indeterminacy, Theor. Appl. Climatol. 143 (2021) 1227–1234.

[58] M. Aslam, A. Algarni, Analyzing the solar energy data using a new anderson-darling test under indeterminacy, Int. J. Photoenergy 6 (2020). Available from: https://doi.org/10.1155/2020/6662389. Article ID 6662389.

[59] M. Aslam, Forecasting of the wind speed under uncertainty, Nat. Res. Sci. Rep. 10 (2020) 20300. Available from: https://doi.org/10.1038/s41598-020-77280-y.

[60] A.A. Janjua, M. Aslam, N. Sultana, Evaluating the relationship between climate variability and agricultural crops under indeterminacy, Theor. Appl. Climatol. 142 (2020) 1641–1648.

[61] R.A. Khan Sherwani, M. Aslam, M.A. Raza, M. Farooq, M. Abid, M. Tahir, Neutrosophic normal probability distribution—a spine of parametric neutrosophic statistical tests: properties and applications, in: F. Smarandache, M. Abdel-Basset (Eds.), Neutrosophic Operational Research, Springer, Cham, 2021. Available from: https://doi.org/10.1007/978-3-030-57197-9_8.

[62] R.A. Khan Sherwani, M. Naeem, M. Aslam, M.A. Raza, M. Abid, S. Abbas, Neutrosophic beta distribution with properties and applications, Neutrosophic Sets Syst. 41 (2021) 209–214.

[63] M. Aslam, A. Shafqat, M. Albassam, J.C. Malela-Majika, S.C. Shongwe, A new CUSUM control chart under uncertainty with applications in petroleum and meteorology, PLoS One 16 (2) e0246185. https://doi.org/10.1371/journal.pone.0246185.

[64] M. Aslam, Monitoring the road traffic crashes using NEWMA chart and repetitive sampling, Int. J. Injury Control. Saf. Promot. 28 (1) (2021). Available from: https://doi.org/10.1080/17457300.2020.1835990.

[65] M. Aslam, Analysing gray cast iron data using a new shapiro-wilks test for normality under indeterminacy, Int. J. Cast. Met. Res. 34 (1) (2021). Available from: https://doi.org/10.1080/13640461.2020.1846959.

[66] N. Khan, M. Aslam, P. Jeyadurga, S. Balamurali, Monitoring of production of blood components by attribute control chart under indeterminacy, Nat. Res. – Sci. Rep. 11 (2021) 922. Available from: https://doi.org/10.1038/s41598-020-79851-5.

[67] M. Aslam, R.A.R. Bantan, A study on measurement system analysis in the presence of indeterminacy, Measurement 166 (2020) 108201. Available from: https://doi.org/10.1016/j.measurement.2020.108201.

[68] M. Aslam, R.A.R. Bantan, N. Khan, Design of NEWMA np control chart for monitoring neutrosophic nonconforming items, Soft Comput. 24 (2020) 16617–16626.

[69] M. Albassam, M. Aslam, Monitoring non-conforming products using multiple dependent state sampling under indeterminacy-an application to juice industry, IEEE Access. 8 (2020). Available from: https://doi.org/10.1109/ACCESS.2020.3024569.

[70] A.I. Shawky, M. Aslam, K. Khan, Multiple dependent state sampling-based chart using belief statistic under neutrosophic statistics, J. Math. (2020). Available from: https://doi.org/10.1155/2020/7680286. Article ID 7680286.

[71] M. Aslam, A new sampling plan using neutrosophic process loss consideration, Symmetry 10 (2018) 132. Available from: https://doi.org/10.3390/sym10050132.

[72] M. Aslam, N. Khan, M. Zahir Khan, Monitoring the variability in the process using neutrosophic statistical interval method, Symmetry 10 (2018) 562. Available from: https://doi.org/10.3390/sym10110562.

[73] M. Zahir Khan, M. Farid Khan, M. Aslam, A.R. Mughal, Design of fuzzy sampling plan using the Birnbaum-Saunders distribution, Mathematics 7 (2019) 9. Available from: https://doi.org/10.3390/math7010009.

[74] M. Zahir Khan, M. Farid Khan, M. Aslam, S.T. Akhavan Niaki, A.R. Mughal, A fuzzy EWMA attribute control chart to monitor process mean, Information 9 (2018) 312. Available from: https://doi.org/10.3390/info9120312.

[75] M. Aslam, M.S. Aldosari, Inspection strategy under indeterminacy based on neutrosophic coefficient of variation, Symmetry 11 (2019) 193. Available from: https://doi.org/10.3390/sym11020193.

[76] M. Aslam, R.A.R. Bantan, N. Khan, Design of S2N-NEWMA control chart for monitoring process having indeterminate production data, Processes 7 (2019) 742. Available from: https://doi.org/10.3390/pr7100742.

[77] M. Aslam, A.H. Al-Marshadi, N. Khan, A new X-bar control chart for using neutrosophic exponentially weighted moving average, Mathematics 7 (2019) 957. Available from: https://doi.org/10.3390/math7100957.

[78] M. Aslam, O.H. Arif, Classification of the state of manufacturing process under indeterminacy, Mathematics 7 (2019) 870. Available from: https://doi.org/10.3390/math7090870.

[79] M. Aslam, R.A.R. Bantan, N. Khan, Monitoring the process based on belief statistic for neutrosophic gamma distributed product, Processes 7 (2019) 209. Available from: https://doi.org/10.3390/pr7040209.

[80] M. Aslam, Product acceptance determination with measurement error using the neutrosophic statistics, Adv. Fuzzy Syst. 8 (2019). Available from: https://doi.org/10.1155/2019/8953051. Article ID 8953051.

[81] M. Aslam, R.A.R. Bantan, N. Khan, Design of a new attribute control chart under neutrosophic statistics, Int. J. Fuzzy Syst. 21 (2019) 433–440.

[82] M. Aslam, O.H. Arif, Test of association in the presence of complex environment, Complexity 6 (2020). Available from: https://doi.org/10.1155/2020/2935435. Article ID 2935435.

[83] M. Albassam, N. Khan, M. Aslam, The W/S test for data having neutrosophic numbers: an application to USA village population, Complexity 8 (2020). Available from: https://doi.org/10.1155/2020/3690879. Article ID 3690879.

[84] M. Aslam, A.H. Al-Marshadi, Design of a control chart based on COM-poisson distribution for the uncertainty environment, Complexity 9 (2019). Available from: https://doi.org/10.1155/2019/8178067. Article ID 8178067.

[85] M. Aslam, O.H. Arif, Multivariate analysis under indeterminacy: an application to chemical content data, J. Anal. Methods Chem. 6 (2020). Available from: https://doi.org/10.1155/2020/1406028. Article ID 1406028.

[86] M. Aslam, Introducing Kolmogorov-Smirnov tests under uncertainty: an application to radioactive data, ACS Omega 5 (2020) 914–917.

[87] M. Aslam, Design of the Bartlett and Hartley tests for homogeneity of variances under indeterminacy environment, J. Taibah Univ. Sci. 14 (1) (2020) 6–10. Available from: https://doi.org/10.1080/16583655.2019.1700675.

[88] R.A. Khan Sherwani, M. Aslam, H. Shakeel, K. Abbas, F. Jamal, Neutrosophic statistics for grouped data: theory and applications, in: F. Smarandache, M. Abdel-Basset (Eds.), Neutrosophic Operational Research, Springer, 2021. Available from: https://doi.org/10.1007/978-3-030-57197-9_14.

[89] N. Khan, L. Ahmad, M. Azam, M. Aslam, F. Smarandache, Control chart for monitoring variation using multiple dependent state sampling under neutrosophic, in: F. Smarandache, M. Abdel-Basset (Eds.), Neutrosophic Operational Research, Springer, 2021. Available from: 10.1007/978-3-030-57197-9_4.

Chapter 2

Neutrosophic Weibull model with applications to survival studies

Zahid Khan[1], Muhammad Gulistan[1], Katrina Khadijah Lane Krebs[2] and Sultan Salem[3]

[1]*Department of Mathematics and Statistics, Hazara University, Mansehra, Pakistan,*
[2]*Higher Education Division, Central Queensland (CQ) University Rockhampton, Australia,*
[3]*Department of Economics, Birmingham Business School, College of Social Sciences, University of Birmingham, Edgbaston, England, United Kingdom*

2.1 Introduction

The term "survival time" refers to the time it takes for a certain event to occur. Depending on the situation, this event might be the onset of a disease, a reaction to a medication, a recovery, or the loss of life. Survival data may contain information on patient characteristics associated with response, survival, and the genesis of a disease, as well as information on the length of time a patient survives [1]. In analyzing survival statistics, researchers have concentrated on estimating the likelihood of response, survival, or mean lifespan, comparing the survival distributions, and determining the risk factors [2]. To describe survival time in survival studies, many theoretical distributions have been commonly used [3]. Among these, the Weibull model has broad applications in survival data modeling [4]. The Weibull distribution is prominent in life testing and survival studies because it correctly models real-world survival events and is flexible enough when considering two parameters [5]. From a statistical viewpoint, the wide adaptability of the Weibull model in lifetime analysis is utilization to simulate a variety of distributions, including normal, exponential, and Rayleigh [6]. The Weibull distribution works well for estimating monotonous hazard rates [7]. Nevertheless, when the failure rate is bathtub-shaped or unimodal, the Weibull distribution is not applicable. The Weibull model is frequently used to represent data with extreme values [8]. Some of the studies in which Weibull distribution is frequently employed to fit the failure data under the extreme values are in

Cognitive Intelligence with Neutrosophic Statistics in Bioinformatics.
DOI: https://doi.org/10.1016/B978-0-323-99456-9.00007-6

[9–11]. The extreme values in these studies are the expiration date of a product or patient survival times. Frequently, studies concerning patient survival and product failure are censored for a specific sample of participants. The use of censored data in lifetime studies is justified because it also provides valuable information on the subjects' survival times and the reliability of the products [12]. It would be inappropriate just to discard these data points. The Weibull model is used to model failure times in a complete sample and is applicable in censored data modeling [13]. Wind speed analysis is another domain where Weibull is also effective [14]. Wind energy, compared to other renewable energy sources such as tidal energy or solar, has diffuse energy flow and more variability, according to [15]. Because of this fluctuation, the Weibull model is often utilized to represent wind speed variation in order to optimize the benefits gained from wind energy. It is therefore well tested and highly regarded.

It is vital to note that the Weibull model and its generalized forms are not novel, discussed briefly in the literature [16]. The classical approach of the Weibull distribution in statistical distribution theory is based on the assumptions that the fitted data are completely specified or at least the parameters of the distribution do not contain any type of uncertainty [17]. However, the enormous quantity of data we encounter in survival analysis is inconclusive or ambiguous in some way. For example, in medical research, it is essential to know how long people with potentially deadly illnesses might expect to live after being diagnosed or at any other point in time. Survival data gained on disease may contain ambiguity due to the partial diagnosis process or other potential factors such as human or machine errors. Likewise, laboratory animals are administered by the pharmaceutical agent and subsequently monitored for tumor development in a typical carcinogenic drug study. The period from dosing to tumor appearance is an important variable that may contain incomplete and imprecise information because the timescale is not necessarily literal or chronological, particularly when it comes to machinery or equipment. In short, a lifetime analysis specifies the event (such as failure or death) that defines a subject's lifespan, as well as a timescale and origin. In most cases, it is difficult to pinpoint when an event happens. Neutosophy, a new school of thought specializing in dealing with uncertain circumstances in real-world problems, is one method used to address this challenge [18]. If the data or at least some part of the data are uncertain, particularly neutrosophic statistics is the approach utilized to analyze them [19]. The primary contrast between classical and neutrosophic statistics is that all data are precisely defined in classical statistics [20–22]. When there is no ambiguity in data, the neutrosophic statistical method and classical approach provide equivalent results [23]. In this way, the neutrosophic statistical approach allows us to analyze the data that may include certain indeterminacies to uncover the underlying data generation process.

Despite extensive coverage of the Weibull model in the literature, it is infrequently employed in regular medical data analysis and reporting. This work aims to investigate the usage and implementation of the neutrosophic Weibull distribution (NWD) in the analysis of medical data from survival studies and to demonstrate the practical advantages of the suggested model.

The rest of this work is organized as follows: Section 2.2 includes a description of the proposed and other key features. In Section 2.3, a brief discussion on the theoretical properties of the NWD is given. The estimation process under the neutrosophic logic is provided in Section 2.4. Simulation studies, including the quantile function of the NWD, are described in Section 2.5. In Section 2.6, a concise explanation of significant theoretical findings is followed by a real-world example. The results of the study are concluded in Section 2.7.

2.2 Proposed model with some associated functions

If the random variable $\mathscr{T}$ with two parameters ρ_n and η_n follows the Weibull model, then the density function (DF) of the proposed distribution is defined as:

$$\hbar_n(\mathscr{T}) = \frac{\rho_n}{\eta_n^{\rho_n}} \mathscr{T}^{\rho_n - 1} e^{-\left(\frac{\mathscr{T}}{\eta_n}\right)^{\rho_n}}, \text{ for } \rho_n > 0 \text{ and } \eta_n > 0, \tag{2.1}$$

where $\rho_n = [\rho_l, \rho_u]$ and $\eta_n = [\eta_l, \eta_u]$ are the neutrosophic scale and shape parameters, respectively, of the NWD.

Note that the proposed model differs from the existing structure of the classical Weibull model, where shape and scale parameters are precisely determined. When the indeterminate part is considered zero in the proposed model, that is, $\rho_l = \rho_u = \rho$ and $\eta_l = \eta_u = \eta$, it becomes equivalent to the classical model. It is also worth noting that at $\rho_n = [1, 1]$, the proposed distribution tends to the neutrosophic exponential model [24].

Various values of ρ_n and η_n result in different density curves as given in Figs. 2.1 and 2.2 with neutrosophic scale and shape parameter values, respectively.

Fig. 2.1 shows that different indeterminate values of scale parameter resulted in different sturdy curves of the NWD. In contrast, Fig. 2.2 depicts the density curves of the NWD with various interval values of the shape parameter. The cumulative function (CF) of any density is another interesting feature in applications of survival studies. The CF is a jointly coupled form of the DF given by:

$$H_n(\mathscr{T}) = 1 - e^{-\left(\frac{\mathscr{T}}{\eta_n}\right)^{\rho_n}} \tag{2.2}$$

The function $H_n(\mathscr{T})$ indicates the likelihood that a random variable would have a value less than a specified value. The CF curves at various interval values of shape parameter are given in Fig. 2.3.

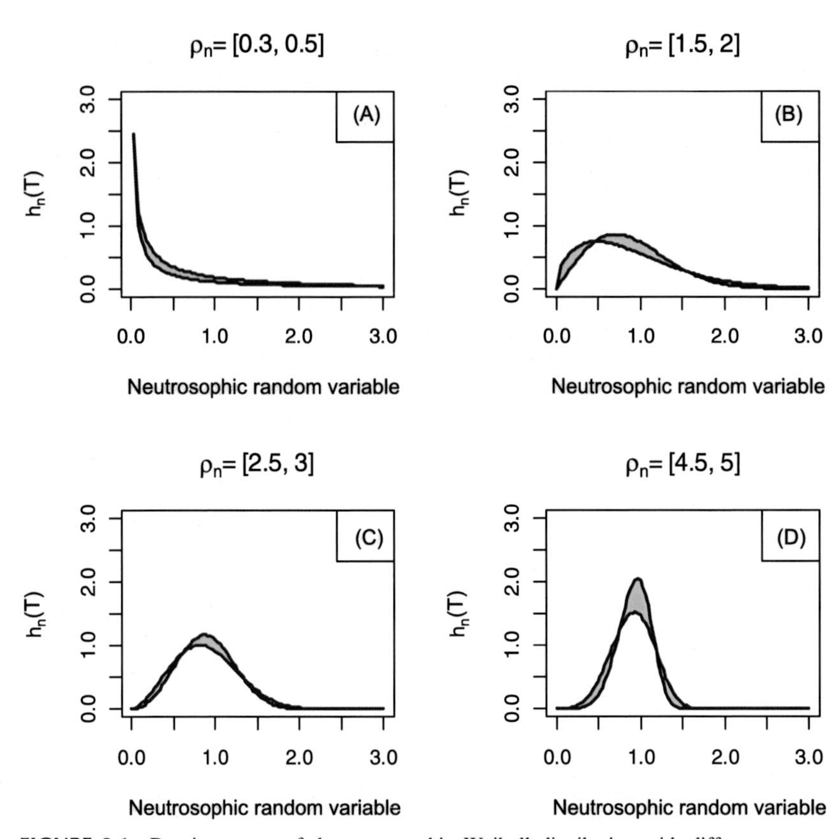

FIGURE 2.1 Density curves of the neutrosophic Weibull distribution with different neutrosophic scale parameter values.

Fig. 2.3 shows the cumulative densities of the NWD at different interval values of shape and fixed value of scale parameter. The CF curve in each panel of Fig. 2.3 is nondecreasing and ranges from 0 to 1. The fact that the CF is nondecreasing indicates that the DF cannot be negative and true for every distribution. The likelihood that the individual life surpasses a specific duration is another valuable function in lifetime analysis. This function is known as survival function (SF) or simple as the survival rate. The SF for the proposed model can be expressed as:

$$\mathbb{S}_n(\mathcal{T}) = e^{-\left(\frac{\mathcal{T}}{\eta_n}\right)^{\rho_n}} \tag{2.3}$$

The graph of SF is known as a survival curve. For the proposed NWD, the survival curve is shown in Fig. 2.4. The steep curve can demonstrate a short survival period, or a low survival rate can be shown by the steep curve, as shown in Fig. 2.4C and D. A flat or progressive survival curve indicates a longer survival rate, as shown in Fig. 2.4A and B.

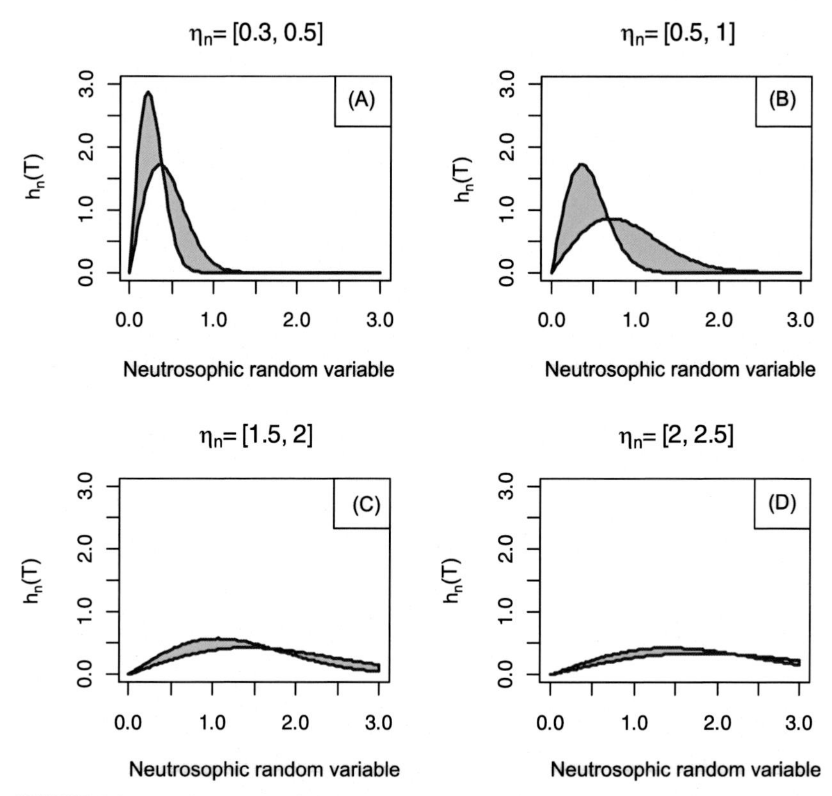

FIGURE 2.2 Density curves of the neutrosophic Weibull distribution with different neutrosophic shape parameter values.

The hazard function (HF), also known as the immediate failure rate, is another important function in lifetime analysis. It is a ratio of the survival function and density function and can be derived for the proposed model as:

$$\varrho_n(\mathscr{T}) = \frac{\hbar_n(\mathscr{T})}{\mathbb{S}_n(\mathscr{T})} = \frac{\mathscr{T}^{\rho_n - 1}\rho_n}{{\eta_n}^{\rho_n}} \tag{2.4}$$

The function $\varrho_n(\mathscr{T})$ provides the failure probability of an individual or item for a minimal time. The HF may increase, decrease, stay constant, or reflect a more complex process. The graphical behavior of the hazard curve is shown in Fig. 2.5.

Fig. 2.5 provides the hazard curves of NWD at the fixed value of the shape parameter and interval values of the scale parameter. For $\rho_n = [1, 1]$ and $\eta_n = [1, 1]$, hazard curve is represented by a straight line whereas $\rho_n > 1$ and $\rho_n < 1$ provide increasing and decreasing trends of the hazard curves.

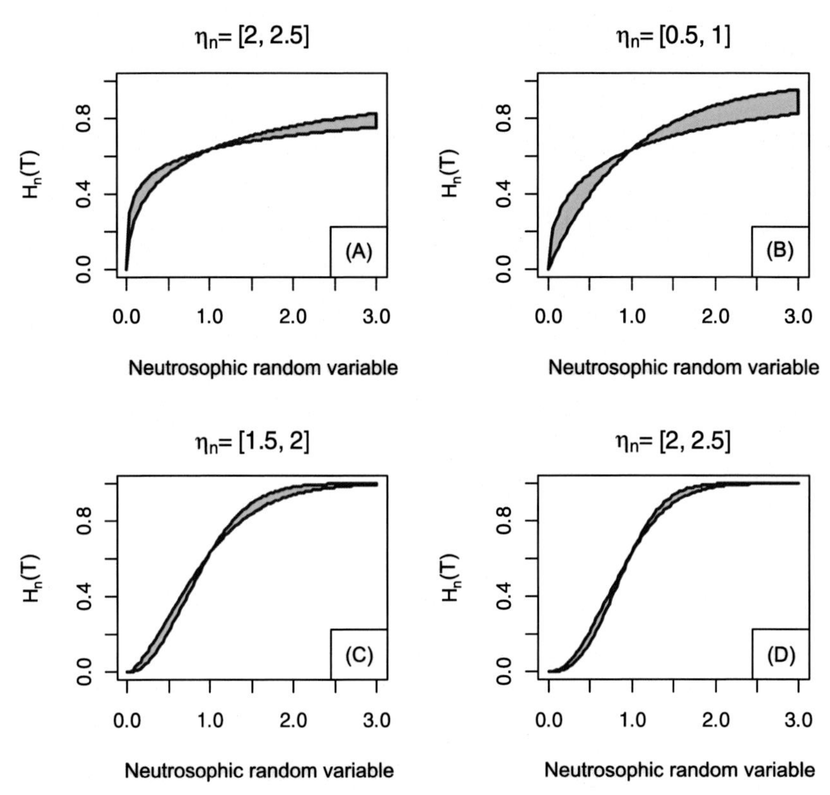

FIGURE 2.3 Cumulative distribution of the neutrosophic Weibull distribution with different neutrosophic shape parameter values.

2.3 Theoretical background

In this section, we investigate the theoretical background and present some key distributional properties of the proposed NWD in the context of neutrosophic logic. The distributional properties subject to parameterization as given in (2.1) are given below:

Theorem 2.1: shows that the mean of the neutrosophic Weibull distribution is

$$\eta_n \Gamma\left(1 + \frac{1}{\rho_n}\right)$$

Proof: By definition, mean is given by:

$$\mho_n(\mathscr{T}) = \int_0^{\infty} \mathscr{T} \frac{\rho_n}{\eta_n^{\rho_n}} \mathscr{T}^{\rho_n - 1} e^{-\left(\frac{\mathscr{T}}{\eta_n}\right)^{\rho_n}} d\mathscr{T}$$

$$= \left[\int_0^{\infty} \mathscr{T} \frac{\rho_l}{\eta_l^{\rho_l}} \mathscr{T}^{\rho_l - 1} e^{-\left(\frac{\mathscr{T}}{\eta_l}\right)^{\rho_l}} d\mathscr{T}, \int_0^{\infty} \mathscr{T} \frac{\rho_u}{\eta_u^{\rho_u}} \mathscr{T}^{\rho_u - 1} e^{-\left(\frac{\mathscr{T}}{\eta_u}\right)^{\rho_u}} d\mathscr{T}\right] \quad (2.5)$$

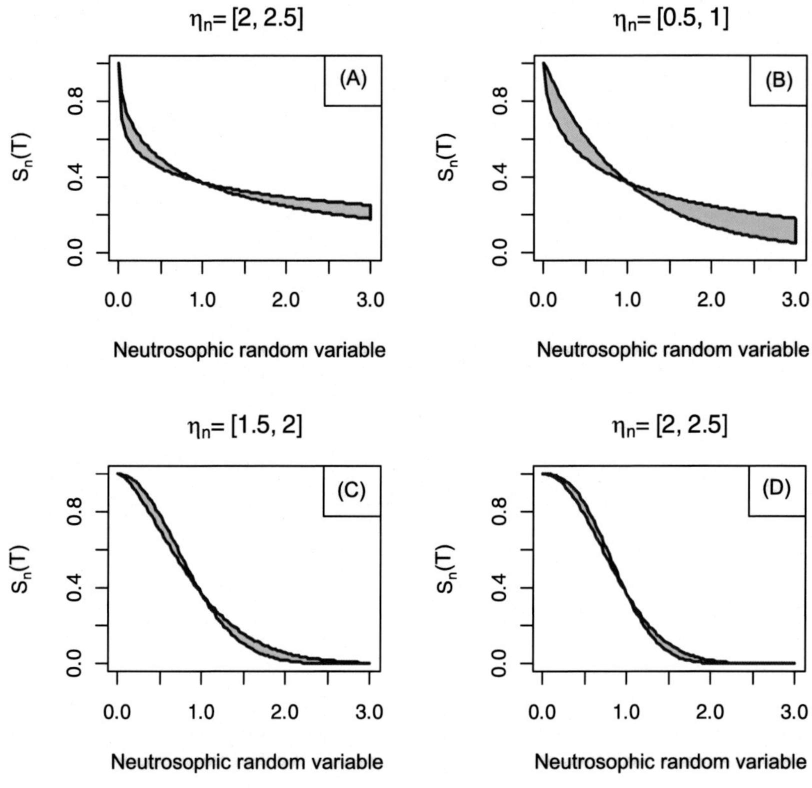

FIGURE 2.4 Survival curves of the neutrosophic Weibull distribution.

By substituting

$$\mathcal{W} = \left(\frac{\mathcal{T}}{\eta_n}\right)^{\rho_n}$$

in (2.5) yielded:

$$\frac{\eta_l^{\rho_l+1}}{\eta_l^{\rho_l}} \int_0^\infty \mathcal{W}^{\frac{1}{\rho_l}} e^{-\mathcal{W}} d\mathcal{W} = \eta_l \Gamma\left(1 + \frac{1}{\rho_l}\right)$$

and

$$\frac{\eta_u^{\rho_u+1}}{\eta_u^{\rho_u}} \int_0^\infty \mathcal{W}^{\frac{1}{\rho_u}} e^{-\mathcal{W}} d\mathcal{W} = \eta_u \Gamma\left(1 + \frac{1}{\rho_u}\right)$$

so,

$$\mho_n(\mathcal{T}) = \left[\eta_l \Gamma\left(1 + \frac{1}{\rho_l}\right), \eta_u \Gamma\left(1 + \frac{1}{\rho_u}\right)\right] = \eta_n \Gamma\left(1 + \frac{1}{\rho_n}\right)$$

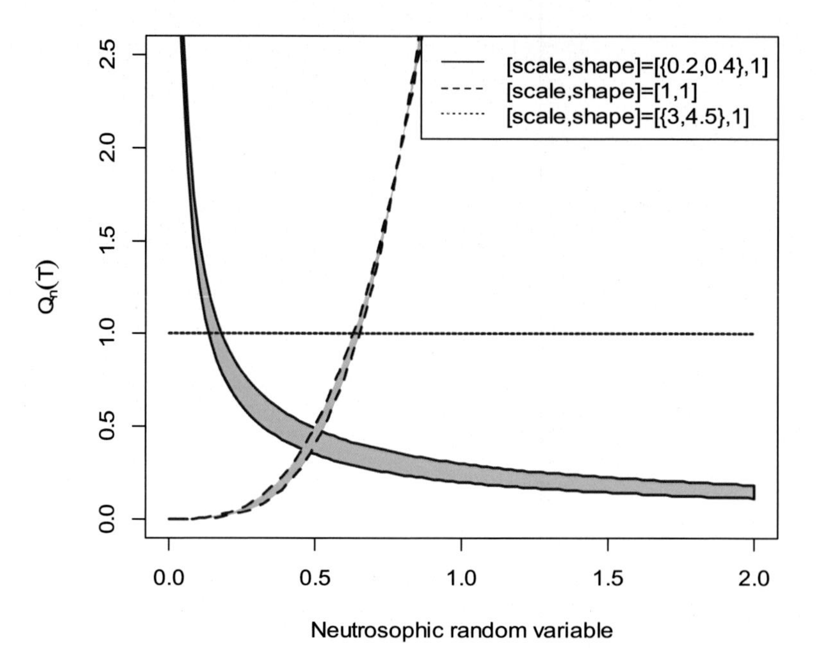

FIGURE 2.5 Hazard curves for the proposed neutrosophic Weibull distribution.

Theorem 2.2: shows that the variance of the NWD is

$$\eta_n^2\left[\Gamma\left(1+\frac{2}{\rho_n}\right)-\Gamma^2\left(1+\frac{1}{\rho_n}\right)\right]$$

Proof: Noting that variance is given by:

$$\tilde{V}_n(\mathscr{T})=\left[\mho_n(\mathscr{T}^2)\right]-\left[\mho_n(\mathscr{T})\right]^2 \tag{2.6}$$

Now

$$\mho_n(\mathscr{T}^2)=\int_0^\infty \mathscr{T}^2\frac{\rho_n}{\eta_n^{\rho_n}}\mathscr{T}^{\rho_n-1}e^{-\left(\frac{\mathscr{T}}{\eta_n}\right)^{\rho_n}}d\mathscr{T}$$

$$=\left[\int_0^\infty \mathscr{T}^2\frac{\rho_l}{\eta_l^{\rho_l}}\mathscr{T}^{\rho_l-1}e^{-\left(\frac{\mathscr{T}}{\eta_l}\right)^{\rho_l}}d\mathscr{T},\int_0^\infty \mathscr{T}^2\frac{\rho_u}{\eta_u^{\rho_u}}\mathscr{T}^{\rho_u-1}e^{-\left(\frac{\mathscr{T}}{\eta_u}\right)^{\rho_u}}d\mathscr{T}\right]$$

Further simplification provides:

$$=\left[\eta_l^2\Gamma\left(\frac{2}{\rho_l}+1\right),\eta_l^2\Gamma\left(\frac{2}{\rho_l}+1\right)\right]=\eta_n^2\Gamma\left(\frac{2}{\rho_n}+1\right)$$

Eq. (2.6) becomes:

$$=\left[\eta_l^2\Gamma\left(1+\frac{2}{\rho_l}\right),\eta_u^2\Gamma\left(1+\frac{2}{\rho_u}\right)\right]-\left[\eta_l^2\Gamma^2\left(1+\frac{1}{\rho_l}\right),\eta_u^2\Gamma^2\left(1+\frac{1}{\rho_u}\right)\right]$$

Hence

$$= \left[\eta_l^2\left\{\Gamma\left(1+\frac{2}{\rho_l}\right),\Gamma^2\left(1+\frac{1}{\rho_l}\right)\right\},\eta_u^2\left\{\Gamma\left(1+\frac{2}{\rho_u}\right)-\Gamma^2\left(1+\frac{1}{\rho_u}\right)\right\}\right]$$

so,

$$\tilde{V}_n(\mathcal{T}) = \eta_n^2\left[\Gamma\left(1+\frac{2}{\rho_n}\right)-\Gamma^2\left(1+\frac{1}{\rho_n}\right)\right]$$

Theorem 2.3: shows that the *k*th raw moment of the NWD is

$$\eta_n^k\Gamma\left(1+\frac{k}{\rho_n}\right)$$

Proof: By definition:

$$\mu''_{kn} = \int_0^\infty \mathcal{T}^k h(\mathcal{T})d\mathcal{T}$$

$$= \int_0^\infty \mathcal{T}^k \frac{\rho_n}{\eta_n^{\rho_n}}\mathcal{T}^{\rho_n-1}e^{-(\frac{\mathcal{T}}{\eta_n})^{\rho_n}}d\mathcal{T}$$

$$= \left[\int_0^\infty \mathcal{T}^k \frac{\rho_l}{\eta_l^{\rho_l}}\mathcal{T}^{\rho_l-1}e^{-(\frac{\mathcal{T}}{\eta_l})^{\rho_l}}d\mathcal{T}, \int_0^\infty \mathcal{T}^k \frac{\rho_u}{\eta_u^{\rho_u}}\mathcal{T}^{\rho_u-1}e^{-(\frac{\mathcal{T}}{\eta_u})^{\rho_u}}d\mathcal{T}\right] \tag{2.7}$$

Simplification of (2.7) provided:

$$\int_0^\infty \mathcal{T}^k \frac{\rho_l}{\eta_l^{\rho_l}}\mathcal{T}^{\rho_l-1}e^{-(\frac{\mathcal{T}}{\eta_l})^{\rho_l}}d\mathcal{T} = \eta_l^2\Gamma 1+\frac{k}{\rho_l}$$

$$\int_0^\infty \mathcal{T}^k \frac{\rho_u}{\eta_u^{\rho_u}}\mathcal{T}^{\rho_u-1}e^{-(\frac{\mathcal{T}}{\eta_u})^{\rho_u}}d\mathcal{T} = \eta_u^2\Gamma 1+\frac{k}{\rho_u}$$

Thus (2.7) becomes

$$\mu''_{kn} = \left[\eta_l^k\Gamma 1+\frac{k}{\rho_l},\eta_u^k\Gamma 1+\frac{k}{\rho_u}\right]$$

Hence,

$$\mu''_{kn} = \eta_n^k\Gamma\left(1+\frac{k}{\rho_n}\right)$$

where k = 1,2,3,... is a general expression for the *k*th moment about the origin of the NWD.

By using the following relations, moments about the mean for the NWD can be obtained as:

$$\mu'_{1n} = \mu''_{1n} = \eta_n \Gamma\left(1 + \frac{1}{\rho_n}\right)$$

$$\mu'_{2n} = \mu''_{2n} - \left(\mu''_{1n}\right)^2 = \eta_n^2 \Gamma 1 + \frac{2}{\rho_n}$$

$$\mu'_{3n} = \mu''_{3n} - 3\mu''_{2n}\mu''_{1n} + 2\left(\mu''_{1n}\right)^3 = \eta_n^3 \Gamma 1 + \frac{3}{\rho_n} - 3\mu''_{1n}\sigma_n^2 - (\mu''_{1n})^3$$

$$\mu'_{4n} = \mu''_{4n} - 4\mu''_{3n}\mu''_{1n} + 6\mu''_{2n}\left(\mu''_{1n}\right)^2 - 3\left(\mu''_{1n}\right)^4$$

$$= \eta_n^4 \Gamma 1 + \frac{4}{\rho_n}$$

$$-4\eta_n^4\left(\Gamma 1 + \frac{1}{\rho_n}\right)\left(\Gamma 1 + \frac{3}{\rho_n}\right) + 6\eta_n^4\left(\Gamma 1 + \frac{2}{\rho_n}\right)\left(\Gamma 1 + \frac{1}{\rho_n}\right)^2 - 3\eta_n^4\left(\Gamma 1 + \frac{1}{\rho_n}\right)^4$$

Theorem 2.4: shows that the coefficient of skewness for the NWD is

$$\frac{\eta_n^3 \Gamma 1 + \frac{3}{\rho_n} - 3\mu''_{1n}\sigma_n^2 - (\mu''_{1n})^3}{\sigma_n^3}$$

Proof: The coefficient of skewness of the NWD is given by:

$$\breve{\xi}_{1n} = \frac{\mu'_{3n}}{\left(\mu'_{2n}\right)^{3/2}} \tag{2.8}$$

where

$$\mu'_{3n} = \eta_n^3 \Gamma 1 + \frac{3}{\rho_n} - 3\mu''_{1n}\sigma_n^2 - (\mu''_{1n})^3 \text{ and } \mu'_{2n} = \eta_n^2 \Gamma 1 + \frac{2}{\rho_n}$$

Substituting in (2.8) provides:

$$\breve{\xi}_{1n} = \frac{\eta_n^3 \Gamma 1 + \frac{3}{\rho_n} - 3\mu''_{1n}\sigma_n^2 - (\mu''_{1n})^3}{\sigma_n^3}$$

where

$$\breve{\xi}_{1n} = [\breve{\xi}_{1l}, \breve{\xi}_{1u}].$$

Theorem 2.5: shows that the coefficient of kurtosis for NWD is

$$\frac{-6\Gamma^4 1 + 12\Gamma^2 1 \Gamma 2 - 3\Gamma^2 2 - 4\Gamma 1 \Gamma 3 + \Gamma 4}{\left[\Gamma 2 - \Gamma^2 1\right]^2}$$

Proof: By definition, the coefficient of kurtosis of NWD is given by:

$$\breve{\xi}_{2n} = \frac{\mu'_{4n}}{\mu'^{2}_{2n}} \tag{2.9}$$

where

$$\mu'_{4n} = \eta_n^4\left[\Gamma 4 - 4\Gamma 1\Gamma 3 + 6\Gamma 2\Gamma^2 1 - 3\Gamma^4 1\right] \text{ and } \mu'_{2n} = \eta_n^2\Gamma 1 + \frac{2}{\rho_n}$$

Substituting in (2.9) and further simplification implies:

$$= \frac{\eta_n^4\left[\Gamma 4 - 4\Gamma 1\Gamma 3 + 6\Gamma 2\Gamma^2 1 - 3\Gamma^4 1\right]}{\eta_n^4[\Gamma 2 - \Gamma^2 1]^2} - 3$$

$$\breve{\xi}_{2n} = \frac{-6\Gamma^4 1 + 12\Gamma^2 1\Gamma 2 - 3\Gamma^2 2 - 4\Gamma 1\Gamma 3 + \Gamma 4}{\left[\Gamma 2 - \Gamma^2 1\right]^2}$$

where

$$\breve{\xi}_{2\mathrm{n}} = [\breve{\xi}_{2l}, \breve{\xi}_{2u}].$$

Theorem 2.6: shows that CF of the NWD is $\left(1 - e^{-\left(\frac{\mathscr{T}}{\eta_n}\right)^{\rho_n}}\right)$

Proof: By definition, the CF of NWD is given by:

$$H_n(\mathscr{T}) = \int_0^{\mathscr{T}} \frac{\rho_n}{\eta_n^{\rho_n}} \mathscr{T}^{\rho_n - 1} e^{-\left(\frac{\mathscr{T}}{\eta_n}\right)^{\rho_n}} d\mathscr{T} \tag{2.10}$$

$$= \left[\int_0^{\mathscr{T}} \frac{\rho_l}{\eta_l^{\rho_l}} \mathscr{T}^{\rho_l - 1} e^{-\left(\frac{\mathscr{T}}{\eta_l}\right)^{\rho_l}} d\mathscr{T}, \int_0^{\mathscr{T}} \frac{\rho_u}{\eta_u^{\rho_u}} \mathscr{T}^{\rho_u - 1} e^{-\left(\frac{\mathscr{T}}{\eta_u}\right)^{\rho_u}} d\mathscr{T}\right]$$

By substituting

$$y = \left(\frac{\mathscr{T}}{\eta_n}\right)^{\rho_n}$$

Thus (2.10) becomes:

$$= \left[1 - e^{-\left(\frac{\mathscr{T}}{\eta_l}\right)^{\rho_l}}, 1 - e^{-\left(\frac{\mathscr{T}}{\eta_u}\right)^{\rho_u}}\right]$$

where

$$\left[1 - e^{-\left(\frac{\mathscr{T}}{\eta_l}\right)^{\rho_l}}, 1 - e^{-\left(\frac{\mathscr{T}}{\eta_u}\right)^{\rho_u}}\right] = \left(1 - e^{-\left(\frac{\mathscr{T}}{\eta_n}\right)^{\rho_n}}\right)$$

Hence proved.

Theorem 2.7: shows that the survival function of the NWD is $\left(e^{-\left(\frac{\mathscr{T}}{\eta_n}\right)^{\rho_n}}\right)$

Proof: By definition, the survival function of NWD is given by:

$$S_n(t) = 1 - \mathscr{G}_n(t) \tag{2.11}$$

$$= 1 - \left[1 - e^{-\left(\frac{\mathscr{T}}{\eta_n}\right)^{\rho_n}}\right]$$

Thus becomes

$$= \left[e^{-\left(\frac{\mathscr{T}}{\eta_l}\right)^{\rho_l}}, e^{-\left(\frac{\mathscr{T}}{\eta_u}\right)^{\rho_u}}\right]$$

where

$$\left[e^{-\left(\frac{\mathscr{T}}{\eta_l}\right)^{\rho_l}}, e^{-\left(\frac{\mathscr{T}}{\eta_u}\right)^{\rho_u}}\right] = e^{-\left(\frac{\mathscr{T}}{\eta_n}\right)^{\rho_n}}$$

Theorem 2.8: shows that the hazard rate of NWD is $\frac{\rho_n}{\eta_n}\left(\frac{\mathscr{T}}{\eta_n}\right)^{\rho_n - 1}$

Proof: By definition, the DF of NWD is given by:

$$Q_n(t) = \frac{\wp'_n(\mathscr{T})}{1 - \wp_n(\mathscr{T})} \tag{2.12}$$

where

$$\wp'_n(\mathscr{T}) = \frac{\rho_n}{\eta_n}\left(\frac{\mathscr{T}}{\eta_n}\right)^{\rho_n - 1} e^{-\left(\frac{\mathscr{T}}{\eta_n}\right)^{\rho_n}} \text{ and } \wp_n(\mathscr{T}) = 1 - e^{-\left(\frac{\mathscr{T}}{\eta_n}\right)^{\rho_n}}$$

Substituting in (2.12) provides:

$$= \frac{\rho_n}{\eta_n}\left(\frac{\mathscr{T}}{\eta_n}\right)^{\rho_n - 1}$$

where

$$\frac{\rho_n}{\eta_n}\left(\frac{\mathscr{T}}{\eta_n}\right)^{\rho_n - 1} = \frac{\rho_l}{\eta_l}\left(\frac{\mathscr{T}}{\eta_l}\right)^{\rho_l - 1}, \frac{\rho_u}{\eta_u}\left(\frac{\mathscr{T}}{\eta_u}\right)^{\rho_u - 1}.$$

Theorem 2.9: shows that the median of NWD is $\eta_n[\ln(2)]^{\frac{1}{\rho_n}}$

Proof: By definition:

$$H_n(\mathscr{M}) = \int_0^{\mathscr{M}} \hbar(\mathscr{T}) d\mathscr{T} = \frac{1}{2}$$

$$= \int_0^{\mathscr{M}} \frac{\rho_n}{\eta_n^{\rho_n}} \mathscr{T}^{\rho_n - 1} e^{-\left(\frac{\mathscr{T}}{\eta_n}\right)^{\rho_n}} d\mathscr{T} = \frac{1}{2}$$

By substituting

$$\begin{aligned} \mathscr{W} &= \left(\frac{\mathscr{T}}{\eta_n}\right)^{\rho_n} \\ &= \int_0^{\left(\frac{\mathscr{M}}{\eta_n}\right)^{\rho_n}} e^{-\mathscr{M}} d\mathscr{M} = \frac{1}{2} \\ &= e^{-\left(\frac{\mathscr{M}}{\eta_n}\right)^{\rho_n}} = \frac{1}{2} \end{aligned} \tag{2.13}$$

Further simplification of (2.13) yields:

$$\mathscr{M} = \eta_n [\ln(2)]^{\frac{1}{\rho_n}}$$

2.4 Estimation method

In this section, a well-known approach of maximum likelihood (ML) has been adopted to find the neutrosophic parameters of the proposed NWD. The ML approach is defined by letting our parameters unknown and computing the joint density of all observations in a dataset that is considered identical and independently distributed. Once the likelihood of the NWD is established, the maxima of the function is determined. These ML estimators are essential in the statistical viewpoint because of minimal variance and asymptotic unbiasedness properties. Let $\tau_1, \tau_2, \ldots \tau_m$ are identical and independently survival times from the m subjects which follow the parametric model given in (2.1); then, the joint density of observing the survival times is given by:

$$\begin{aligned} \mathscr{L}\left(\eta_n, \rho_n | \mathscr{T}\right) &= \prod_{i=1}^{m} \mathscr{h}\left(\mathscr{T}_i | \eta_n, \rho_n\right) \\ &= \prod_{i=1}^{m} \left[\frac{\rho_n}{\eta_n^{\rho_n}} \mathscr{T}^{\rho_n - 1} e^{-\left(\frac{\mathscr{T}}{\eta_n}\right)^{\rho_n}}\right] \\ &= \frac{\rho_n}{\eta_n^{\rho_n}} \prod_{i=1}^{m} \mathscr{T}_i^{\rho_n - 1} e^{-\left(\frac{\mathscr{T}_i}{\eta_n}\right)^{\rho_n}} \end{aligned} \tag{2.14}$$

Taking the logarithm of (2.14) and symbolizing it by $\omega_n\left(\mathscr{T}_i | \eta_n, \rho_n\right)$,

$$\omega_n\left(\mathscr{T}_i | \eta_n, \rho_n\right) = \log\left[\frac{\rho_n}{\eta_n^{\rho_n}} \prod_{i=1}^{m} \mathscr{T}_i^{\rho_n - 1} e^{-\left(\frac{\mathscr{T}_i}{\eta_n}\right)^{\rho_n}}\right] \tag{2.15}$$

Simplification of (2.15) yielded:

$$\omega_n\left(\mathscr{T}_i | \eta_n, \rho_n\right) = \log(\rho_n) - \log\left(\eta_n^{\rho_n}\right) + (\rho_n - 1) \sum_{i=1}^{m} \log(\mathscr{T}_i) - \sum_{i=1}^{m} \left(\frac{\mathscr{T}_i}{\eta_n}\right)^{\rho_n} \tag{2.16}$$

Partially differentiating (2.16) by unknown values and equating to zero implies:

$$\left[\frac{\delta\omega_n(\mathcal{T}_i,\eta_n)}{\delta\eta_n},\frac{\delta\omega_n(\mathcal{T}_i,\rho_n)}{\delta\rho_n}\right]=[0,0] \tag{2.17}$$

so,

$$\left.\begin{array}{l}\left[\dfrac{\delta\omega_n(\mathcal{T}_i,\eta_n)}{\delta\eta_l},\dfrac{\delta\omega_n(\mathcal{T}_i,\rho_n)}{\delta\eta_u}\right]=[0,0]\\[2ex]\left[\dfrac{\delta\omega_n(\mathcal{T}_i,\eta_n)}{\delta\rho_l},\dfrac{\delta\omega_n(\mathcal{T}_i,\rho_n)}{\delta\rho_u}\right]=[0,0]\end{array}\right\} \tag{2.18}$$

Simultaneous Eq. (2.18) for $\hat{\rho}_n$ and $\hat{\eta}_n$ has no closed form; therefore numerical iterative methods such as Newton–Raphson can be utilized to find the solution for estimated values. This can be done easily by using the estimation package in R software.

2.5 Simulation study

In this part, a Monte Carlo approach for generating the random numbers that are assumed to follow NWD has been used. In general, the Monte Carlo approach refers to any strategy for finding solution to the problem utilizing random outcomes. The aim of the work here is to use the Monte Carlo technique to simulate random samples from the NWD with known parameter values in order to validate the theoretical results described in Section 2.3. The inverse DF approach has been employed as the most straightforward technique to simulate random numbers from the proposed model. This approach enables us to make use of a computer-built-in pseudo-random number generator for generating random numbers. The inverse DF of the proposed model is given by:

$$\tau_p=\rho_n[-ln(1-u_i)]^{\frac{1}{\eta_n}} \tag{2.19}$$

where u_i is randomly generated numbers from the uniform distribution and τ_p is desired percentile value of the proposed NWD. Let 10^5 random samples are drawn according to inverse CF method from the proposed model with $\rho_n=[1,1.5]$ and $\eta_n=[0.5,0.8]$. Analytical outcomes based on the analytical results given in Section 2.3 are calculated with baseline parameter values. Estimated values of different distribution properties along with exact results are provided in Table 2.1.

Table 2.1 shows the descriptive metrics of the proposed model for the known values of distributional parameters. The descriptive measures of the

TABLE 2.1 A summary of distribution properties of the proposed model based on simulated data.

Distribution properties	Simulated outcomes	Analytical outcomes
Mean	[1.107, 3.021]	[1.133, 3.000]
Median	[0.630, 0.717]	[0.632, 0.720]
Standard deviation	[1.444, 6.710]	[1.428, 6.708]
Skewness	[2.764, 6.826]	[2.720, 6.845]
Kurtosis	[15.545, 98.364]	[15.246, 98.672]

simulated data using the proposed model are in intervals due to assumed indeterminacies in defined parameters. The close agreement between simulated and analytical outcomes validates the theoretical framework of the proposed model.

2.6 Illustrative example

In this part, real medical data are utilized to demonstrate how the suggested NWD can be implemented. A dataset on remission periods (in weeks) of leukemia patients is used for this analysis [25]. The distribution fitting package in R software has been employed to determine first the suitability of Weibull distribution on remission periods data. Basic probability plots along with empirical density are shown in Fig. 2.6.

When evaluating the systematic departure of the points from the straight line in each graph, it is determined that the Weibull distribution is an acceptable model for this dataset. Although remission times are originally crisp numbers, for illustration, we consider some patients' remission times as uncertain sample values, as given in Table 2.2.

Analyzing such type of data by using the classical approach of the Weibull model is not appropriate due to the existence of imprecise values. However, the proposed NWD can easily be implemented to describe data with indeterminacies. The descriptive summary of remission time data by using NWD is given in Table 2.3.

Table 2.3 provides the neutrosophic metrics estimated based on the proposed model. All the estimated values are interval because of indeterminacies considered in the analyzed sample. Thus the proposed model is more flexible and can analyze the data efficiently when data or estimation method involves incomplete or imprecise information.

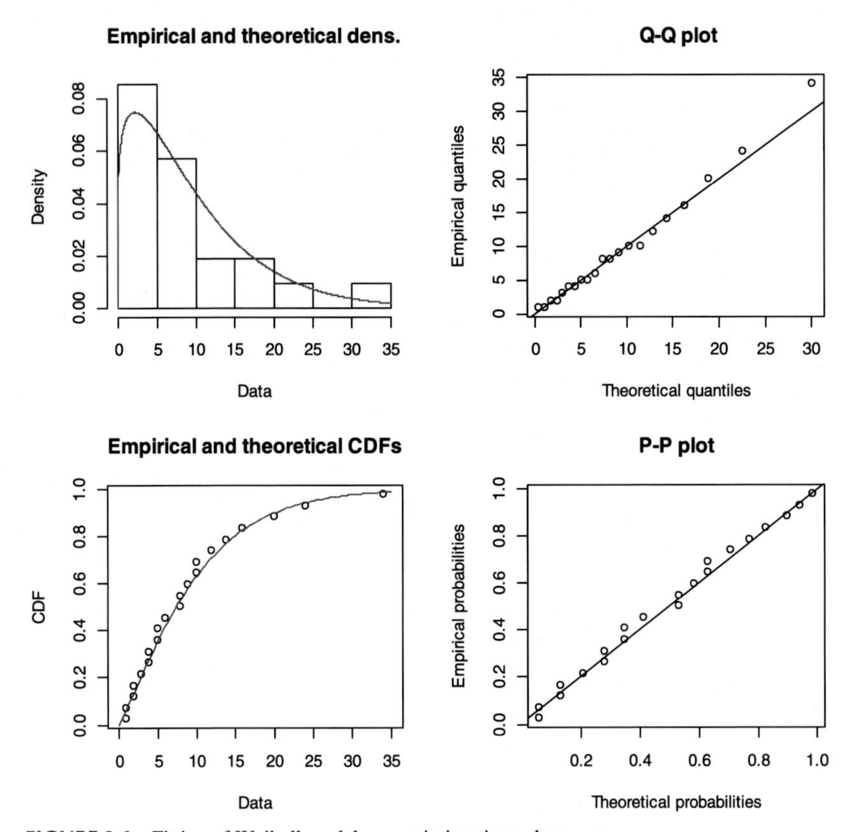

FIGURE 2.6 Fitting of Weibull model on remission times data.

TABLE 2.2 Remission times for leukemia patients with uncertainties.

Remission times (in weeks) for 21 patients
[5, 5.5], 5, [6, 6.4], [8, 8.5], 8, 1, [0.5, 1], 2, [2, 2.5], [14, 14.5], [2.5, 3]
[16, 16.2], [20, 20.6], 4, [3.5, 4], 10, 10, 24, 34, [8.5, 9], [12, 12.5]

TABLE 2.3 Analysis of remission periods dataset using the proposed model.

Descriptive summary	Estimated values
Scale parameter	[9.65, 10.91]
Shape parameter	[1.14, 1.29]
Mean	[9.20, 10.09]
Standard deviation	[7.88, 8.09]

2.7 Conclusions

The neutrosophic framework of the Weibull model and its applications in the field of survival analysis have been presented in this work. Statistical characteristics of the newly proposed model, so-called NWD under the neutrosophic logic, are widely investigated. The estimation method of ML in the neutrosophic environment has been discussed. All the derived properties of the proposed model have been validated using the strategy of the Monte Carlo simulation. The usefulness of the proposed NWD has been demonstrated by using a real dataset on the remission times of leukemia patients.

References

[1] H.P. Zhu, X. Xia, C.H. Yu, A. Adnan, S.F. Liu, Y.K. Du, Application of Weibull model for survival of patients with gastric cancer, BMC Gastroenterol. 11 (2011).

[2] K. C.C. clinical trials and undefined, On the Use and Utility of the Weibull Model in the Analysis of Survival Data, *Elsevier*, 2003.

[3] S. Tuljapurkar, W. Zuo, T. Coulson, C. Horvitz, J.M. Gaillard, Skewed distributions of lifetime reproductive success: beyond mean and variance, Ecol. Lett. 23 (4) (2020) 748–756.

[4] J.K. Starling, C. Mastrangelo, Y. Choe, Improving Weibull distribution estimation for generalized Type I censored data using modified SMOTE, Reliab. Eng. Syst. Saf. 211 (2021) 107505.

[5] A.E.B.A. Ahmad, M.G.M. Ghazal, Exponentiated additive Weibull distribution, Reliab. Eng. Syst. Saf. 193 (2020) 106663.

[6] X. Jia, Reliability analysis for Weibull distribution with homogeneous heavily censored data based on Bayesian and least-squares methods, Appl. Math. Model. 83 (2020) 169–188.

[7] M. Azizmohammad Looha, E. Zarean, F. Masaebi, M.A. Pourhoseingholi, M.R. Zali, Assessment of prognostic factors in long-term survival of male and female patients with colorectal cancer using non-mixture cure model based on the Weibull distribution, Surg. Oncol. 38 (2021) 101562.

[8] L.D. Jones, L.J. Vandeperre, T.A. Haynes, M.R. Wenman, Theory and application of Weibull distributions to 1D peridynamics for brittle solids, Comput. Methods Appl. Mech. Eng. 363 (2020) 112903.

[9] Z. Zhu, H. Tian, R. Wang, G. Jiang, B. Dou, G. Mei, Statistical thermal damage constitutive model of rocks based on Weibull distribution, Arab. J. Geosci. 14 (6) (2021) 1–14.

[10] G. Liu, et al., Extreme values of storm surge elevation in Hangzhou Bay, Ships Offshore Struct. (2019) 431–442. Available from: https://doi.org/10.1080/17445302.2019.1661618.

[11] B. Silahli, K.D. Dingec, A. Cifter, N. Aydin, Portfolio value-at-risk with two-sided Weibull distribution: evidence from cryptocurrency markets, Financ. Res. Lett. 38 (2021) 101425.

[12] L. Wang, C. Zhang, Y.M. Tripathi, S. Dey, S.J. Wu, Reliability analysis of Weibull multicomponent system with stress-dependent parameters from accelerated life data, Qual. Reliab. Eng. Int. 37 (6) (2021) 2603–2621.

[13] L. Xie, J. Ren, J. Song, N. Wu, A Knowledge Synthesis Method for Weibull Distribution Estimation with Four Right-Censored Life Data, in: 2020 Asia-Pacific International Symposium on Advanced Reliability and Maintenance Modeling, APARM 2020, August 2020.

[14] V.E.L.A. Duca, T.C.O. Fonseca, F.L. Cyrino, Oliveira, “A generalized dynamical model for wind speed forecasting, Renew. Sustain. Energy Rev. 136 (2021) 110421.

[15] A. Serban, L.S. Paraschiv, S. Paraschiv, Assessment of wind energy potential based on Weibull and Rayleigh distribution models, Energy Rep. 6 (2020) 250–267.
[16] M. Nassar, A.Z. Afify, M.K. Shakhatreh, S. Dey, On a new extension of Weibull distribution: properties, estimation, and applications to one and two causes of failures, Qual. Reliab. Eng. Int. 36 (6) (2020) 2019–2043.
[17] R.A.K. Sherwani, T. Arshad, M. Albassam, M. Aslam, S. Abbas, Neutrosophic entropy measures for the Weibull distribution: theory and applications, Complex. Intell. Syst. 7 (6) (2021) 3067–3076.
[18] F. Smarandache, A Unifying Field in Logics: Neutrosophic Logic. Neutrosophy, Neutrosophic Set, Neutrosophic Probability, American Research Press, Rehoboth, 1999.
[19] Smarandache, Neutrosophical Statistics, Sitech & Education Publishing, 2014.
[20] Z. Khan, M. Gulistan, W. Chammam, S. Kadry, Y. Nam, A new dispersion control chart for handling the neutrosophic data, IEEE Access. 8 (2020) 96006–96015.
[21] Z. Khan, M. Gulistan, R. Hashim, N. Yaqoob, W. Chammam, Design of S-control chart for neutrosophic data: an application to manufacturing industry, J. Intell. Fuzzy Syst. 38 (4) (2020) 4743–4751.
[22] Z. Khan, M. Gulistan, N. Kausar, C. Park, Neutrosophic rayleigh model with some basic characteristics and engineering applications, IEEE Access. 9 (2021) 71277–71283.
[23] M. Aslam, N. Khan, M. Khan, Monitoring the variability in the process using neutrosophic statistical interval method, Symmetry 10 (11) (2018) 562.
[24] W.-Q. Duan, Z. Khan, M. Gulistan, A. Khurshid, Neutrosophic exponential distribution: modeling and applications for complex data analysis, Complexity 2021 (2021) 1–8.
[25] E.T. Lee, J.W. Wang, Statistical Methods for Survival Analysis, third ed., John Wiley & Sons, 2003.

Chapter 3

Using the four-valued Rasch model in the preparation of neutrosophic form of risk maps for the spread of COVID-19 in Turkey

Volkan Duran[1], Selçuk Topal[2], Florentin Smarandache[3] and Muhammad Aslam[4]

[1]Faculty of Science and Arts, Iğdır University, Iğdır, Turkey, [2]Department of Mathematics, Bitlis Eren University, Bitlis, Turkey, [3]Mathematics, Physical and Natural Sciences Division, University of New Mexico, Gallup, NM, United States, [4]Department of Statistics, Faculty of Science, King Abdulaziz University, Jeddah, Saudi Arabia

3.1 Introduction and motivation

When it comes to neutrosophic philosophy, it is all about taking a fresh look at the world and then modifying that perspective to account for the indeterminacy that exists in daily life. Neutrosophy proposes a third logical alternative to the binary paradigm of true or false, which is referred to as neutrals. As a whole, neutrosophy replaces the binary approach in logic by adding indeterminacy so as to expand it into a richer logical space. When Smarandache introduced neutrosophy in 1998, it was considered revolutionary at the time. It has developed greatly since then due to the addition of logical extensions like measure, sets, and graphs, as well as practical applications in a range of areas [1].

The quantum theory is very significant to the neutrosophic philosophy since they both deal with phenomena that are found on comparable principles. The uncertainty principle established by Heisenberg, for example, is quite close to the indeterminacy of neutrosophic logic. In the same way that classical probability theory provides a framework for building cognitive models that are based on the mathematics of quantum probability theory, quantum cognition can provide one such framework. Quantum probability

Cognitive Intelligence with Neutrosophic Statistics in Bioinformatics.
DOI: https://doi.org/10.1016/B978-0-323-99456-9.00008-8

theory is, like classical probability theory, a mathematical framework for assigning probabilities to occurrences. Decision-making provides a new foundation for grasping a wide variety of phenomena via the use of an agreed-upon set of axiomatic notions [2].

In this respect, it should be important to review the basic quantum principles as well. These can be summarized as the three basic principles of quantum mechanics [3]:

- The degree of freedom may be discrete or continuous, and it can range from one to an unlimited number of degrees of freedom. The degree of freedom is symbolized by the symbol φ; the collection of all of its values is denoted by the symbol $\mathcal{F}$.
- The state space V is referred to as a Hilbert space in quantum physics. When dealing with systems that do not conserve probability, the state space might be far bigger than the Hilbert space. The state space is indicated by the letter V, and an element is denoted by the letter, where V: $\mathcal{F} \rightarrow V$. The dual space V_D consists of all mappings, denoted by $\langle\chi|$ of V to the complex numbers $\mathbb{C}$. The expression $\langle\chi|\psi\rangle = \langle\psi|\chi\rangle^* \in \mathbb{C}$ is the scalar or inner product.
- Operators $\hat{O}$ that act on V and map it to itself $\hat{O}: V \rightarrow V$. The space of operators is denoted by $Q \equiv V \otimes V_D$. The tensor or outer product of two state vectors is given by $|\psi\rangle \otimes \langle\chi| \equiv |\psi\rangle\langle\chi| \in V \otimes V_D$.

In summary, quantum mechanics consists of the mathematical triple $\{F, V, Q\}$. The Hermitian conjugation of an operator, denoted by i, is defined by $\hat{O}_i^\dagger$ as

$$\langle\chi|\hat{O}_i|\psi\rangle * = \langle\psi|\hat{O}_i|\chi\rangle *.$$

All eigenvalues of a Hermitian operator are real and can represent physically observed quantities. Hence, physical observations are represented by Hermitian operators.

$$\hat{O}_i^\dagger = \hat{O}_i, i = 1, 2, \ldots ..., N\text{:Hermitian}$$

In general, $[\hat{O}_i, \hat{O}_j] \neq 0$. The physically observed value of a physical quantity Q, such as position, energy, is given by $\langle\psi|Q|\psi\rangle$, where $|\psi\rangle$ represents the quantum state of the physical entity.

Let us consider the eigenfunctions and eigenvalues of a Hermitian operator given by

$$\hat{O}|\psi_n\rangle = \lambda_n|\psi_n\rangle; \langle\psi_m|\psi_n\rangle = \delta_{m-n}$$

All Hermitian operators have the following spectral decomposition in terms of their eigenvalues and eigenfunctions

$$\hat{O} = \sum_n \lambda_n|\psi_n\rangle\langle\psi_n|$$

The collection of all the eigenfunctions of a Hermitian operator yields a complete set of basis states and yields the completeness equation for V given by

$$\mathbb{1} = \sum_n |\psi_n\rangle\langle\psi_n| = \sum_n \Pi_n;\ \mathbb{1}^2 = \mathbb{1}$$

where $\mathbb{1}$ is the unit operator on V.

If π_1 and π_2 are projection operators that commute, that is, $\pi_1\pi_2 = \pi_2\pi_1$, then their product is also a projection operator. Suppose that a given vector space has n dimensions and a basis given by $|1\rangle$, $|2\rangle$, …, $|n\rangle$ [4]. Define projection operators Π_n

$$\Pi_n = |\psi_n\rangle\langle\psi_n| \rightarrow \Pi_n{}^2 = \Pi_n$$

and

$$\hat{\text{O}} = \sum_n \lambda_n \Pi_n.$$

Every state vector has the following decomposition that can be given as follows:

$$|\chi\rangle = \mathbb{1}\,|\chi\rangle = \sum_n c_n|\psi_n\rangle \rightarrow c_n = \langle\psi_n|\chi\rangle.$$

It then follows:

$$\langle\chi|\chi\rangle = 1 \rightarrow 1 = \sum_n |c_n|^2 \rightarrow |c_n|^2 \in [0, 1].$$

Note that the expectation value Π_n for the state vector $|\chi\rangle$ is given by

$$E_\chi[\Pi_n] = \langle\chi|\Pi_n|\psi\rangle = P_n.$$

Note that

$$P_n = E_\chi[\Pi_n] = \langle\chi|\Pi_n|\psi\rangle = |\langle\chi|\psi_n\rangle|^2 \geq 0.$$

Furthermore

$$0 \geq E_\chi\left[(\Pi_n - P_n)^2\right] = E_\chi\left[\Pi_n{}^2\right] - P^2{}_n = E_\chi[\Pi_n] - P^2{}_n = P_n - P^2{}_n$$

$$\rightarrow P_n{}^2 \leq P_n \rightarrow P_n \leq 1$$

The completeness of the basis $|\psi_n\rangle$ from $|\chi\rangle = \mathbb{1}\,|\chi\rangle = \sum_n c_n|\psi_n\rangle \rightarrow c_n = \langle\psi_n|\chi\rangle$ yields

$$\sum_n \Pi_n = \mathbb{1} \rightarrow E_x\left[\sum_n \Pi_n\right] = E_\chi[\,\mathbb{1}\,] \rightarrow \sum_n P_n = 1$$

Hence, the coefficients have the important property that

$$0 \leq P_n \leq 1; \sum_n P_n = 1$$

It shows that P_n has the interpretation of the probability of an event labeled by n. It is important to note that the quantum theory of measurement dictates that only one of the detectors, denoted by the letter Π_n, will detect the quantum state. This is also called the *collapse of the wave function* [3].

These mathematical tools are important in terms of conveying such a mathematical structure into the different disciplines where it is especially important to use this tool in decision-making or risk analysis.

In this study, we have created a vectorial form of risk map based on COVID-19 data. The data show that it has a moderate risk level in Turkey in terms of the risk map. Apart from its output, this study contributes to the literature, especially in terms of its methodology and assumptions. The first contribution of this study is to take the risks based on the cognitive and perceptive bases rather than defining them as a danger-level threshold in the context of scalar data. In traditional risk studies, most of the research focuses on the data according to predefined variables and predicts the risk levels according to data concerning those hierarchies. For instance, the capacity of the governments to deal with a virus such as the number of hospitals, doctors, and present vaccines or medicines can be taken as a basis for the risk hierarchy, the spread of the virus and the number of deaths can be taken as main data, and the risk threshold can be defined based on these variables. This is a machine-like view of the risk concept in this situation. However, in the actual world, we are not living in such rigid data in which the membership of truth, falsity, and indeterminacy values of the data comes to the fore. To avoid exaggeration or devaluation of that risks, we should consider the importance of the membership values of the data in the analysis. Therefore, we can use four-valued logic in such cases, especially where we make the backings of the arguments based on the square of opposition. In this respect, the study is novel by taking the vectorial form of four-valued logic to analyze the data to determine the concept of the risk.

The second contribution of this study is that it deals with incomplete information and even contradictory information (probability) rather than complete information (probability) of the states. As mentioned above, most of the studies assume that their data have complete information or at least they make their arguments based on these assumptions. Therefore, they sometimes dismiss contradictory or incomplete information, and they might restrict themselves to a rigid space. In this respect, this study gives another perspective to project the possible boundaries of the risks and provide more objective scenarios.

Third, this study is a projection study rather than based on mere estimation or prediction. Estimation is mostly based on observed perceptive and phenomenal data, whereas prediction is mostly based on assumptions and inferences based on the data. A projection study encompasses both the estimation and prediction in this respect. In other words, a projection study includes both factual and counterfactual probabilities. The factual and counterfactual distinction can be more clearly examined by Huszár's descriptions given as follows [4]:

> *This is an example David brought up during the Causality Panel, and I referred back to this in my talk. I am including it here for the benefit of those who attended my MLSS talk:*
>
> *Given that Hillary Clinton did not win the 2016 presidential election, given that she did not visit Michigan 3 days before the election, and given everything else we know about the circumstances of the election, what can we say about the probability of Hillary Clinton winning the election, had she visited Michigan 3 days before the election?*
>
> *Let us try to unpack this. We are interested in the probability that:*
>
> - *She hypothetically wins the election conditioned on four sets of things:*
>   - *She lost the election*
>   - *She did not visit Michigan*
>   - *Any other relevant observable fact*
>   - *She hypothetically visits Michigan*
>
> *It is a weird beast: you are simultaneously conditioning on her visiting Michigan and not visiting Michigan. And you are interested in the probability of her winning the election given that she did not. WHAT?*

We want to know why Hillary lost the election and how much of the failure can be ascribed to her inability to visit Michigan 3 days before the election. It is helpful to quantify this since it may aid political advisers in future decision-making. Because counterfactuals cannot be scientifically tested, they are typically seen as unscientific. Even though counterfactual assertions are untested and difficult to comprehend, people employ them often and instinctively believe that they are advantageous for intelligent behavior. Therefore, projection studies are more advantageous by taking both factual and counterfactual information at the same time [4,5].

Fourth, this study contributes to the literature by applying neutrosophic logic, syllogisms, and sign–signifier–signified concepts of the semiotics in the argumentation model of Toulmin. Semiotics has an important role in the definition of risk because we mostly create the meaning of the risk based on sign–signifier–signified relationships. Therefore, the mathematical structure of neutrosophic is compatible with semiotics, and we show its compatibility to some extent in this study. Additionally, argumentation is also important for any kind of risk analysis, and we show its mathematical application in the context of Toulmin's argumentation model and the square of the opposition.

Finally, the application of the quantum decision-making model in the context of risk analysis is a significant contribution to the literature. When dealing with scalar data, the bulk of descriptive statistics and analyses are carried out. Additionally, scalar data are often used in the production of visuals. The descriptive analysis of scenarios that are more strongly related to cognitive and semantic characteristics, on the other hand, may provide more in-depth conclusions than the descriptive analysis of scenarios that are more closely associated with scalar characteristics. This leads to the demonstration that the analysis of COVID-19 spreads may be carried out using the four-valued vectorial versions of probability functions.

3.2 Background, definitions, and notations

3.2.1 Quantum decision-making model in the context of neutrosophic philosophy

The method by which neutrosophic logic and quantum decision-making were developed resulted in the development of a new theory of probabilistic and dynamic systems that is broader than the prior classic theory. In the neutrosophic logic, there are three probabilities given $P(S)$ = the chance particular trial results in a success, $P(F)$ = the chance particular trial results in a failure, and finally $P(I)$ = the chance particular trial results in an indeterminacy [5]. When $P(S) + P(I) + P(F) = 1$, this case is called complete probability. But for incomplete probability (where there is missing information):

$$0 \leq P(S) + P(I) + P(F) < 1$$

While in the paraconsistent probability (which has contradictory information):

$$1 < P(S) + P(I) + P(F) \leq 3$$

Because risk perception is mostly based on conflicting information, paraconsistent probability is particularly valuable for the study of risky situations. Furthermore, only the scalar form of this kind of probability, however, may not be sufficient since, in certain situations, the noncommutative aspects of human perception come to force. Therefore, we define the quantum version of this as a probability bra–ket like this

$$\alpha|1\rangle + \beta|2\rangle + \delta|3\rangle + \Omega|4\rangle = |P\rangle$$

Based on three ways, we can get the neutrosophic interpretation of quantum states

- Where $[\langle P|P\rangle = \|\alpha^2\| + \|\beta\|^2 + \|\delta\|^2 + \|\Omega\|^2] = 1$ in which there is complete information (probability) of the states
- Where $0 \leq [\langle P|P\rangle = \|\alpha\|^2 + \|\beta\|^2 + \|\delta\|^2 + \|\Omega\|^2] < 1$ in which there is incomplete information (probability) part of the states
- Where $0 \leq [\langle P|P\rangle = \|\alpha\|^2 + \|\beta\|^2 + \|\delta\|^2 + \|\Omega\|^2] \leq 3$ in which there is a contradictory information (probability) part of the states

Pitfall Prevention: *It should be noted that we modify* $c_n \in [0, 1]$ *interval so that we get a neutrosophic form of the quantum decision-making model* [6].

One could wonder how this interpretation differs from the classical interpretation of quantum mechanics. The fact that there exist superposed states even in quantum theory is not surprising; yet, they are never seen due to the collapse of the wave function. Fundamental to quantum physics is the paradoxical fact that the fundamental quantum entity, that is, the degree of freedom, can never be seen by any experiment, no matter how sophisticated the technology is used. Furthermore, two orthogonal projection operators Π_n and Π_m can never observe the same state function at the same time since they are orthogonal to one another. Measurement results in the state function collapsing to either the state $|\psi_n\rangle = \Pi_n|\psi_n\rangle$ or $|\psi_m\rangle = \Pi_m|\psi_m\rangle$, the state vector $|\psi\rangle$ is never simultaneously seen by both the projection operators, and the state function collapses to one of these states. A state function is seen simultaneously by two orthogonal projection operators in any experiment. If this occurs in any experiment, the present (Copenhagen) interpretation of quantum mechanics comes to an end [3]. When it comes to the contradictory information (probability) part of the states in the neutrosophic case, however, such a thing can happen that is compatible with humans' cognition because the human mind is capable of dealing with contradictory states that are either occurring simultaneously or at least occurring at similar time intervals. The assumptions behind earlier cognitive and decision science models were derived from classical probabilistic dynamical systems, which are now considered to be out of date.

Understanding the operations of the brain is regarded to be one of humanity's major problems today. According to the scientists, the brain develops a mental model of the world. According to the definition of the term "model," knowledge is not just kept as a collection of facts but is instead structured to represent the world's overall structure, as well as everything that exists inside it. For example, we do not memorize a list of information about bicycles to understand what a bicycle is. The model of a bicycle that our brain produces includes the distinct pieces, how they are positioned about one another, and how the individual parts move and operate together as a whole. It is necessary to first understand what something looks and feels like before we can identify it, and it is equally necessary to first learn how objects in the world normally behave when we engage with them before we can accomplish our objectives. Some researchers recently have concluded that the neocortex retains all we know, all of our information, in the form of reference frames. It is possible to think about maps as different types of models. For example, the map of a town is a model of the town, and the grid lines, such as the lines of latitude and longitude, are a form of reference frame. The grid lines on a map, which serve as the map's reference frame, define the map's overall structure. Reference frames may be used to determine where items are positioned concerning one another, and they can also guide how to accomplish objectives such as traveling from one area to another. It has been discovered that the brain's representation of the world is constructed using reference frames that resemble maps, not just one, but hundreds of thousands of different reference frames. According to current knowledge, the majority of the cells in the neocortex are devoted to the creation and manipulation of reference frames, which the brain utilizes to plan and think. The brain generates a model that may be used to anticipate the future. The brain can either use estimation and prediction of events in the context of its past, present, or future. After an event has occurred, an estimate is made, which is known as a posterior probability. A prediction is a form of assessment before the occurrence of the event, that is, a priori probability. This simply implies that the brain makes estimations and predictions about what its inputs will be constantly. Estimations and predictions are not something that the brain accomplishes now and then; rather, it is an innate quality that never ceases to exist, and it plays an important part in the process of learning and development. When the estimations and predictions of the brain are confirmed, it indicates that the brain's model of the world is correct. When you make a mistake in estimation or prediction, you are forced to correct the mistake and update the model [7]. Therefore, we can conclude that the brain learns via a change over time. This change is based on a map or reference frame but it does not just stay on this frame of reference; it can also estimate, predict, and even create counterfactuals to understand what is going on in the cases. According to [8], all learners benefit from focused attention, active engagement, error feedback, and a cycle of daily rehearsal and nightly consolidation which are defined as the "four pillars" of learning (Fig. 3.1) [8].

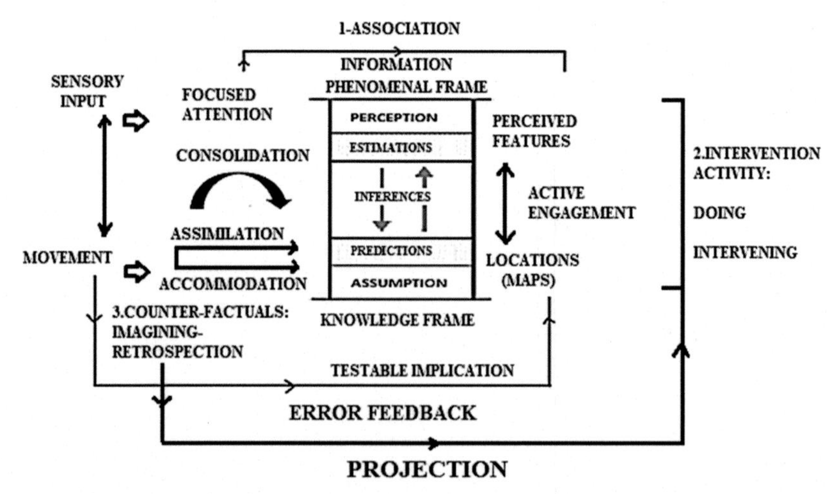

FIGURE 3.1 The basic stages of learning. *Modified from J. Hawkins, A Thousand Brains: A New Theory of Intelligence, Hachette UK, 2021; S. Dehaene, How We Learn? Penguin Random House LLC, 2020; J. Pearl, Causality: Models, Reasoning, and Inference, Cambridge University Press, UK, 2000.*

This illustration depicts two layers of neurons (represented by the shaded boxes) in a single cortical column. The top layer is responsible for receiving sensory information into the column. When an input is received, many hundred neurons become activated as a result of the stimulation. The deliberate learning process which requires feedback, conscious thought, and interaction with the environment begins with focused attention. In this phase, the information is processed via association by seeing and observing or more generally sensing the sensory input. The top layer reflects what you see when you are in a particular place, such as a fountain. The current position in a reference frame is represented by the bottommost layer. A novel item, such as a teacup, is largely learned by learning the connections that exist between the two levels, shown by the vertical arrows in the diagram below. The following is the fundamental flow of information which can be defined as the intervention phase: when sensory input is received, it is represented by the neurons in the top layer of the pyramidal tract. Intervention ranks higher than association because it involves not just seeing but changing what is. Interventions are something we conduct regularly in our everyday lives, albeit we do not normally refer to them by such a formal name. For example, when we take aspirin to relieve a headache, we are interfering with one variable (the number of aspirin in our body) to influence another variable (the severity of the headache) (our headache status). If we are accurate in our causal belief about aspirin, the "outcome" variable will change from "headache" to "no headache" as a result of our causal belief being true. This activates the place in the lower layer that is related to the input that was specified before. As soon as a movement happens, such as the movement of a finger, the bottom layer shifts to the predicted new position, which causes a prediction of the next input to be generated in the top layer. For ambiguous inputs, such as the coffee shop, the network activates various locations in the bottom layer—for example, all of the sites where a coffee shop is located. This is what occurs when you use one finger to touch the rim of a coffee cup. Because many items have a rim, you are not always sure what you are touching when you first contact them. When you move, the bottom layer alters all of the potential locations, which then causes the higher layer to generate different predictions based on the changes in the lower layer. This process is related to counterfactuals as well. Because data are, by definition, a reality, the link between counterfactuals and data is especially hard to navigate. There is no way for them to predict what would happen in a counterfactual or imagined world in which certain observable facts are categorically denied. Nonetheless, the human mind is capable of making such explanation-seeking conclusions in a consistent and repeatable manner. Finally, in the learning process, the next input will eliminate any locations that do not match the previous one in the reference frame. This is achieved by the error feedback of the information from the movement. The reference frame is modified and adjusted by assimilation or accommodation of the information which is consolidated in practice [7–9].

In such a learning process, humans use formal or informal syllogistic arguments based on reasoning and logic in addition to our intuitions and emotions. Syllogisms are arguments about the properties of entities. They consist of two premises and a conclusion, which can each be in one of four "moods": All *S* are *P*, Some *S* are *P*, No *S* is *P*, and Some *P* are not *P*. Their logical analysis started with Aristotle, and their psychological investigation began more than a century ago with Freud. According to the classic Scholastic description of syllogisms in the Middle Ages, a syllogism consists of two premises and a conclusion, and each statement is in one of four moods, two of which are positive and two of which are negative, with each assertion being in one of the four moods. These can be represented as follows [10]:

All *S* are *P*. Affirmative universal (abbreviated as "A")
Some *S* are *P*. Affirmative existential (abbreviated as "I")
No *S* is *P*. Negative universal (abbreviated as "E")
Some *S* are not *P*. Negative existential (abbreviated as "O")

The square of opposition is a collection of theses that is represented by a diagram. The graphic is not required for the theses; it is only a convenient technique to keep them organized and separate. The theses are concerned with logical relationships between four different logical forms. In the square, these are represented as follows [11]:

- *"Every S is P" and "Some S is not P" are contradictories.*
- *"No S is P" and "Some S is P" are contradictories.*
- *"Every S is P" and "No S is P" are contraries.*
- *"Some S is P" and "Some S is not P" are subcontraries.*
- *"Some S is P" is a subaltern of "Every S is P."*
- *"Some S is not P" is a subaltern of "No S is P."*

These theses were supplemented with the following explanations:

- *Two propositions are contradictory if they cannot both be true and they cannot both be false.*
- *Two propositions are contraries if they cannot both be true but can both be false.*
- *Two propositions are subcontraries if they cannot both be false but can both be true.*
- *A proposition is a subaltern of another if it must be true if its superaltern is true, and the superaltern must be false if the subaltern is false.*

We can combine the four moods of the square of opposition with syllogisms that can be used to analyze learning frames. In this regard, we might construct cognitive frames based on the probabilistic categorization based on the categories of sign, signified, and signifier developed by the Swiss linguist Ferdinand de Saussure, where signified corresponds to the "plane of content" mostly in mental forms and signifier corresponds to the "plane of expression" mostly in material forms, respectively. The membership relation between signifier and signified in the context of sign can result in three possibilities such as truth, falsity, and indeterminacy as emphasized by neutrosophic philosophy. This provides more flexibility when assigning probabilities to occurrences since it does not necessitate the creation of all possible joint probabilities, which we believe is a trait essential to appreciate the full complexity of human cognition and decision.

The relationship among neutrosophic logic, syllogisms, and sign–signifier–signified concepts of the semiotics can be understood in the argumentation model of Toulmin (Fig. 3.2). In there, the claim is defined as the position or claim being argued for; the conclusion of the argument is the projection of the estimation and prediction in this respect. Grounds are regarded as reasons or supporting evidence that bolsters the claim which is defined as the phenomenal frame of perception side of the reference frame of the "sign" in which association and connotation have the predominant roles. Warrants are the principles, provisions, or chains of reasoning that connect the grounds/reason to the claim that stands for the knowledge frame where the assumptions, deduction, and inferences have the dominant roles. The backing is defined as the support, justification, and reasons to back up the warrant that is represented by the square of opposition there. Rebuttals/reservations are the exceptions to the claim; description and rebuttal of counter-examples

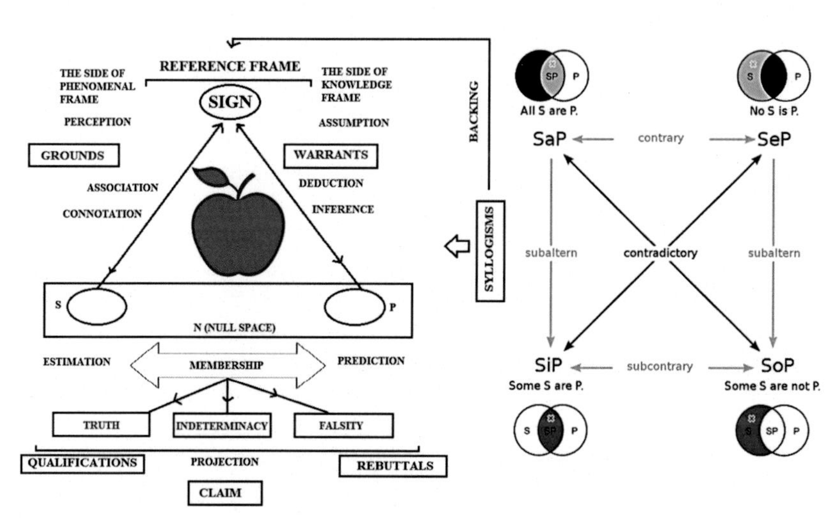

FIGURE 3.2 The concepts of sign, signified, and signifier were developed by the Swiss linguist Ferdinand de Saussure in the context of neutrosophic logic (left) and the square of opposition (right [12]).

and counter-arguments are mostly related to the membership of falsity. Qualifications are specifications of limits to claim, warrant, and backing which are mostly related to the membership of truth [13].

In most mathematical modeling of cognitive schemes, researchers focus on scalar data. However, the information in the cognitive process does not always have a scalar form, but it has a direction, that is, it might have vectorial forms. In this respect, the mathematical structure of quantum theory gains significance in this area because it also benefited from the vector spaces. Additionally, quantum theory is useful because human judgments are based on indefinite states which are also used in quantum theory in the superposition states. Second, judgments disturb each other, introducing uncertainty. Order effects are also responsible for introducing uncertainty into a person's judgments. This can be explained by quantum modeling because a question is asked and the response to that question changes the state of a person from one that is indefinite to one that is more definite about the question that was asked, according to quantum theory. Judgments do not always obey classic logic which is also true in quantum theory [14,15].

3.3 Literature review and state of the art

The Rasch model is a model based on the definition of P/(1-P), known as the "log odds ratio" or "logit." Rasch model was originally specified in terms of odds [probability/(1 − probability)] or log odds (the natural log of the odds, also called logits), but it is now often specified in terms of probability. We can use the Rasch model for the analysis of the COVID-19 spread [16]. The degree of discrepancy between observed item performance and expected item performance can be quantified using goodness-of-fit statistics. Unweighted (outfit) and weighted (infit) mean square statistics are calculated by comparing the observed data to the model probability matrix. We can consider odds [probability/(1 − probability)] as the ability of the virus, and we can make diagrams in terms of the probability of success rates. It should be noted that [probability/(1 − probability)] should be changed as [probability/(3 − probability)] because there is a piece of contradictory information rather than complete information [17].

Based on the cumulative probabilities of COVID-19, a model was proposed based on the quantum decision-making model by the qudit states. In our analysis, first we define the basis states as given which are defined as qudit states in quantum information theory as follows:

$$|\boldsymbol{T}\rangle = \alpha\begin{bmatrix}1\\0\\0\\0\end{bmatrix} = \alpha|\mathbf{1}\rangle, |\boldsymbol{T}'\rangle = \beta\begin{bmatrix}0\\1\\0\\0\end{bmatrix} = \beta|\mathbf{2}\rangle, |\boldsymbol{F}'\rangle = \delta\begin{bmatrix}0\\0\\1\\0\end{bmatrix} = \delta|\mathbf{3}\rangle, |\boldsymbol{F}\rangle = \boldsymbol{\Omega}\begin{bmatrix}0\\0\\0\\1\end{bmatrix} = \boldsymbol{\Omega}|\mathbf{4}\rangle$$

In purest complete information cases, the sum of the coefficients should be 1 as follows:

$$\alpha + \Omega = 1$$

$$\alpha + \beta = 1$$

$$\alpha + \delta = 1$$

$$\delta + \Omega = 1$$

This can be represented in the traditional square of oppositions as follows. Such a plane can be called a knowledge plane because all the information is rigidly known by the observer and there is no indeterminacy or uncertainty in such a plane (Fig. 3.3).

However, life is not a great endeavor to enable us to know the complete information stated. Therefore, we should create a phenomenal plane where the sum of the probabilities is greater than 1 because it has contradictory information. It is called as phenomenal plane because it is based on perceptible information by the senses or through immediate experience. Our bases vectors in such a plane can be represented by different symbols to prevent confusion between the planes.

$$|T\rangle = \lambda \begin{bmatrix} 1 \\ 0 \\ 0 \\ 0 \end{bmatrix} = \lambda|\mathbf{1}\rangle, |T'\rangle = \sigma \begin{bmatrix} 0 \\ 1 \\ 0 \\ 0 \end{bmatrix} = \sigma|\mathbf{2}\rangle, |F'\rangle = \mu \begin{bmatrix} 0 \\ 0 \\ 1 \\ 0 \end{bmatrix} = \mu|\mathbf{3}\rangle, |F\rangle = \nu \begin{bmatrix} 0 \\ 0 \\ 0 \\ 1 \end{bmatrix} == \nu|\mathbf{4}\rangle$$

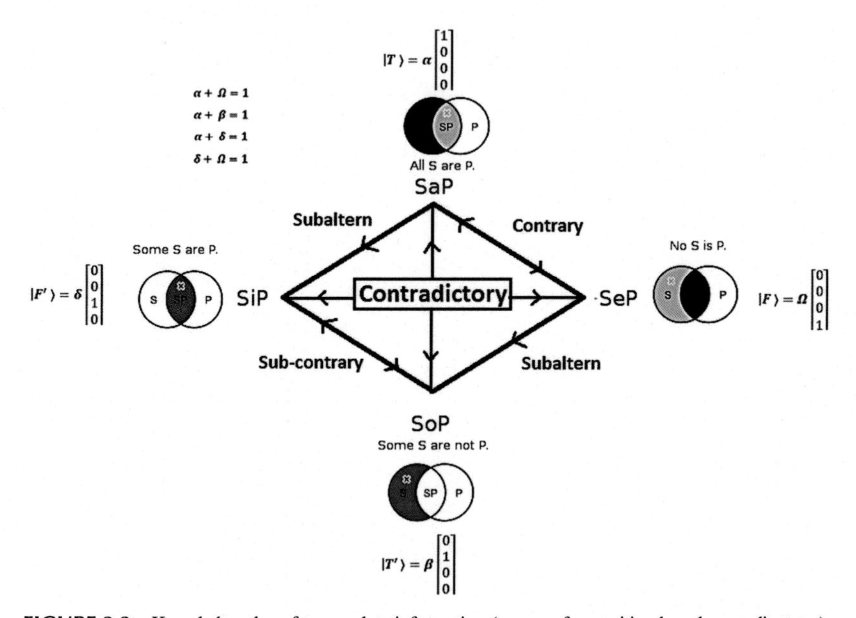

FIGURE 3.3 Knowledge plane for complete information (square of opposition based on qudit states).

In such plane, probabilities can be any positive value between 0 and n where n is any positive number.

$$0 \le \lambda + \nu \le n$$

$$0 \le \lambda + \sigma \le n$$

$$0 \le \lambda + \mu \le n$$

$$0 \le \mu + \nu \le n$$

To assess the phenomenal plane, we should have a knowledge plane as given in Fig. 3.4 as a basis for the assessment of the probabilities. It should be noted that probabilities are expanded from the knowledge plane to the phenomenal plane because of the contradictory information that are depicted in Fig. 3.5. However, just as encountered in many daily situations we do not have such a knowledge plane, and we must create hypothetical phenomenal planes to assess or at least guess the cases.

So, we assume that we have two hypothetical vectors for the upper and lower phenomenal planes as $|P_{\text{upper}}\rangle$ and $|P_{\text{lower}}\rangle$:

$$|P_{\text{upper}}\rangle = \alpha|1\rangle + \beta|2\rangle + \delta|3\rangle + \Omega|4\rangle$$

$$|P_{\text{lower}}\rangle = \lambda|1\rangle + \sigma|2\rangle + \mu|3\rangle + \nu|4\rangle$$

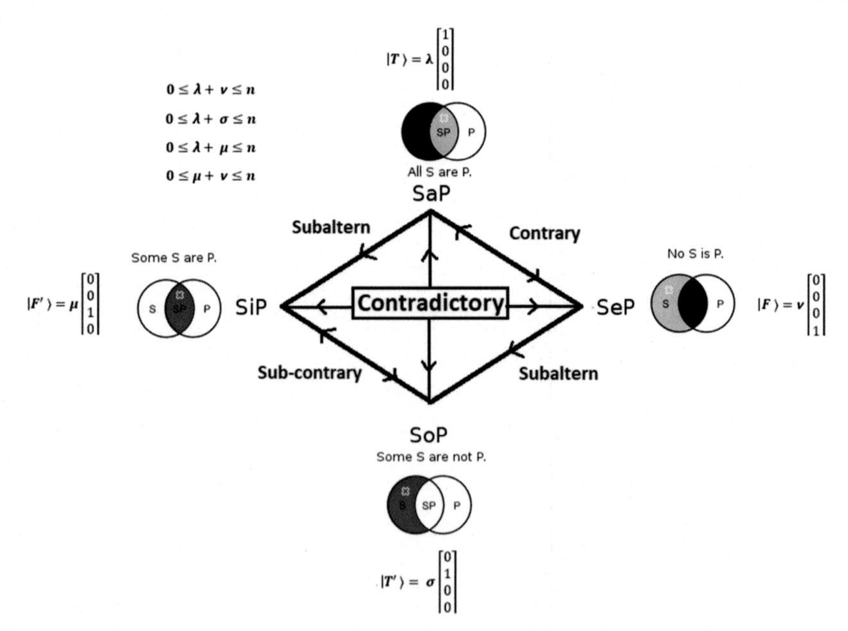

FIGURE 3.4 Phenomenal plane for the contradictory information.

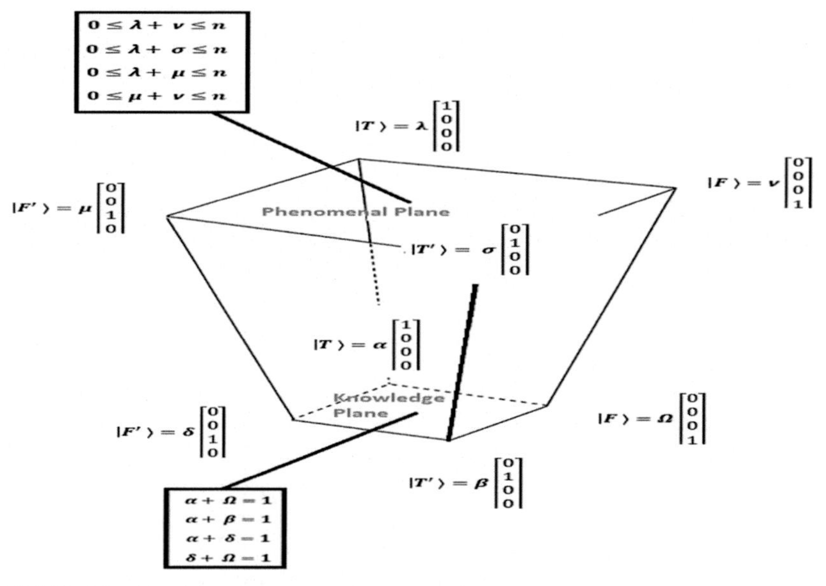

FIGURE 3.5 Knowledge plane versus phenomenal plane.

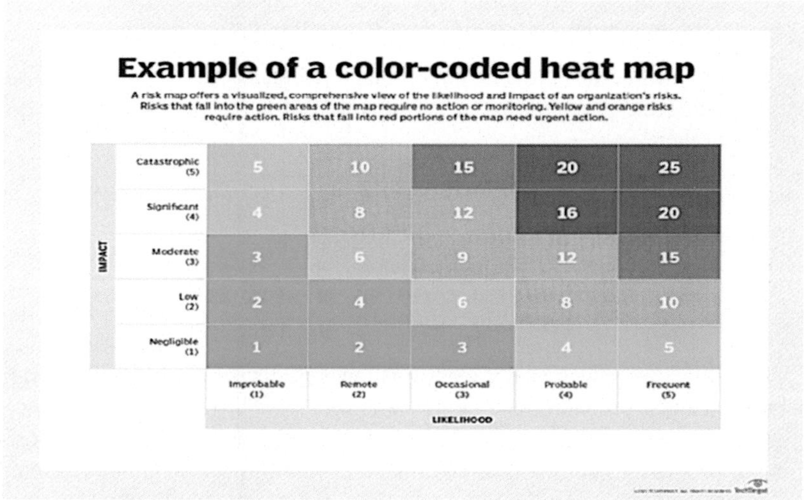

FIGURE 3.6 An example of a risk map [18].

We can use such vectors for the creation of risk maps (Fig. 3.6). Risk maps are most successful when researchers properly evaluate the numerous categories of risk they face, as well as the multiple risks inside each of those categories and the potential probabilities of occurrence and the potential impacts. Risk maps are normally square; however, they may also be rectangular or

TABLE 3.1 The possible risk map in the range of 0 and 1.

		0,25	0,50	0,75	1
	·	$\|T\rangle$	$\|F'\rangle$	$\|T'\rangle$	$\|F\rangle$
0,25	$\|T\rangle$	0,06	0,125	0,1875	0,25
0,50	$\|F'\rangle$	0,125	0,25	0,375	0,50
0,75	$\|T'\rangle$	0,1875	0,375	0,5625	0,75
1	$\|F\rangle$	0,25	0,50	0,75	1
		Negligible	Moderate	Significant	Catastrophic

circular. Although they are most often graphs with an x−y axis, they may also be split into quadrants with the upper-right block denoting the most serious dangers. Several maps utilize color coding indicating whether dangers are large, moderate, or low-level concerns. However, other maps use changing shades of the same hue to signify different degrees of risk. There are also more presentation options, such as the option to portray the risk map as a bar graph, that may be used [15].

To create a risk map based on state vectors for the planes defined as

$$\left|P_{\text{upper}}\right\rangle = \alpha|1\rangle + \beta|2\rangle + \delta|3\rangle + \Omega|4\rangle$$

$$|P_{\text{lower}}\rangle = \lambda|1\rangle + \sigma|2\rangle + \mu|3\rangle + \nu|4\rangle$$

We can use the dot product for the higher dimensional vectors since we only focus on the probabilities in the direction of bases vectors. So that dot product in there

$$\left|P_{\text{upper}}\right\rangle \cdot |P_{\text{lower}}\rangle = \alpha\lambda|1\rangle|1\rangle + \beta\sigma|2\rangle|2\rangle + \delta\mu|3\rangle|3\rangle + \Omega\nu|4\rangle|4\rangle$$

where any dot product of basis vectors is defined as

$$\begin{cases} a|i\rangle \cdot b\left|j\right\rangle \rightarrow a.b = 1 & if \quad i = j \\ a|i\rangle \cdot b\left|j\right\rangle \rightarrow a.b = 0 & if \quad i \neq j \end{cases}$$

Therefore, we can make a classification say, for example, in the range between 0 and 1 probabilities like in Table 3.1.

Therefore, we use the vector forms of the basis of the square of opposition for providing neutrosophic risk maps in this respect. It is neutrosophic because it includes indeterminacy factors based on the combination of four moods as well.

3.4 Problem/system/application definition

The main problem of this study is to make a projection of risk zones of COVID-19 based on the current data from the sources by estimating and predicting the range of boundary values of COVID cases. Therefore, first, in the

TABLE 3.2 The formulas for the interpretation of the analysis at firsthand.

Data type	Variable	Formula
Real	***P*(observed)**	$\frac{\text{Observed Infected Individuals}}{\text{Observed Cumulative Sample}}$
Upper boundaries	$P(R_{min})$	$\frac{\text{Expected Infected Individuals based on Rminvalue}}{\text{Observed Cumulative Sample}}$
	$P(R_{max})$	$\frac{\text{Expected Infected Individuals based on Rmaxvalue}}{\text{Observed Cumulative Sample}}$
Lower boundaries	$P(R_{min})$	$\frac{\text{Observed Infected Individuals}}{\text{Expected Cumulative Sample based on Rminvalue}}$
	$P(R_{max})_I$	$\frac{\text{Observed Infected Individuals}}{\text{Expected Cumulative Sample based on Rmaxvalue}}$

analysis, the data regarding COVID-19 cases in Turkey were taken from https://turCovid19.com/acikveri/ which is between 27 March 2020 and 09 February 2021. In the first analysis, we take two samples, one is real data for infected individuals and the hypothetical data which is called imaginary in there for the possible infected individuals based on the average reproduction number of novel coronavirus. We take the reproduction number as 1755 which is found to be as the confidence intervals of the reproduction number (R) values calculated for Turkey range from 1.30 (lowest) to 2.21 (highest) [19].

Second, we find the cumulative probabilities based on real data; the lowest is 1.30 for Turkey, and the highest is 2.21 for Turkey based on the following formulas representing the possible scenarios given in Table 3.2 to determine the range and upper and lower boundaries in our cognitive scheme.

Finally, we can produce a risk map based on the upper and lower boundaries of the cumulative probabilities at the end of the march of events.

3.5 Proposed solution

Many researchers believe that the notion of potentiality is critical to understanding the concept of risk. "Risk" is a phrase used to describe circumstances in which negative outcomes are potential but do not manifest themselves as such. A concept that has been coined to describe this concept is the word "possibility." In terms of their semantic relationship, the words "risk" and "possibility" might be seen to be a form of hyponym for one another in some situations. Some of the possible occurrences that may occur, that is, the possibilities, are desired, while others are despised by the participants. Finally, the word risk refers to the likelihood of an undesired occurrence occurring, which may be either true or false, as in "the risk of a nuclear meltdown is one in a million." The measure of potentiality and uncertainty is probabilities [20]. In this regard, it is necessary to clarify what kind of uncertainty is associated with the idea of risk. Therefore, we need to develop projections allowing for additional hypothetical inputs that may be utilized for what-if situations. To do so, we must first categorize the uncertain situations in terms of vectorial forms based on probabilities, and then we may proceed. The probabilistic forms that we employ as base states in this form are the four-valued logic based on the square of the opposition. However, the square of the opposition is not a rigid vector rather it has indeterminate forms as strongly emphasized by the neutrosophic philosophy which defines the neutrosophic definition of truth, indeterminacy, and falsity membership probability functions. Human cognition is situational; that is, it shows a vectorial kind of activity, which necessitates the development of such a categorization. It should be noted that human cognition is based on prediction. In other words, our brains learn by using predictive models of the world. To do that, our brains use reference frames which are the phenomenal logical planes consisting of four-valued logic. As emphasized in [10], a reference frame aids in the learning of something's structure by the brain. A coffee cup is a thing because it is made up of a collection of characteristics and surfaces that are organized in space concerning one another. A face, in the same way, is made up of the nose, eyes, and mouth, all of which are positioned in relative locations. To express the relative locations and structure of objects, you must use a reference frame. First and foremost, the brain can alter the whole thing by defining it in terms of a reference frame. Therefore, we try to make a model based on vectorial hypothetical reference frames to predict the probabilistic nature of COVID-19 in terms of contradictory information.

3.6 Analysis

The graph of the boundaries of cumulative probabilities based on different reproduction numbers is given in Fig. 3.7. It can be seen that the ceil of the highest probability ended up with 2.21 while the lowest one ended up with 0.45.

Logit model for the real cumulative probability data is given in Fig. 3.8. It should be noted that this model is based on complete probability.

However, we focus on paraconsistent probability (which has contradictory information); therefore the formula in logit model $(\ln(p/(1-p))$ is replaced by $(\ln(p/(3-p))$ (Fig. 3.9).

Logit models for different data sets based on upper and lower boundaries based on paraconsistent probabilities show that the worst scenario is for the data set having an upper cumulative probability of 2.21 in the upper boundaries of our data. The least risky scenario is the lower cumulative probability of 1.30 for lower boundaries of the data. We can interpret these values in terms of four-valued vectorial logic forms as well (Fig. 3.10).

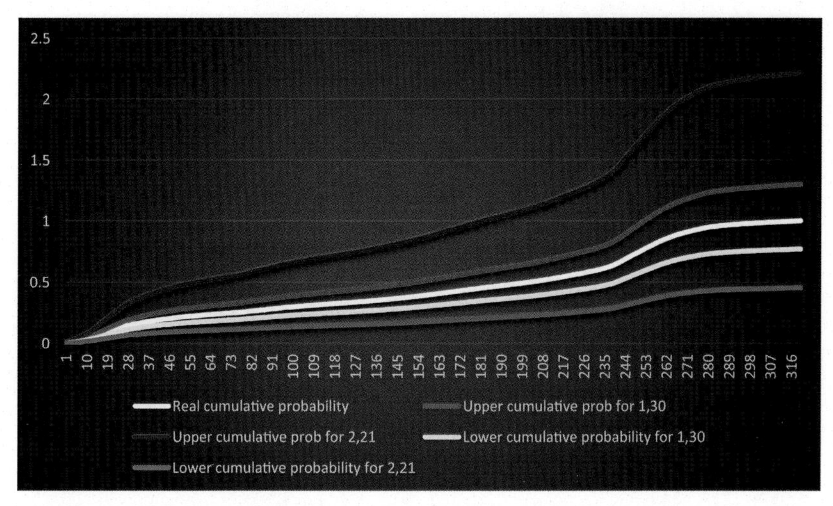

FIGURE 3.7 The graph of upper and lower cumulative probability values based on different R values and a real cumulative value.

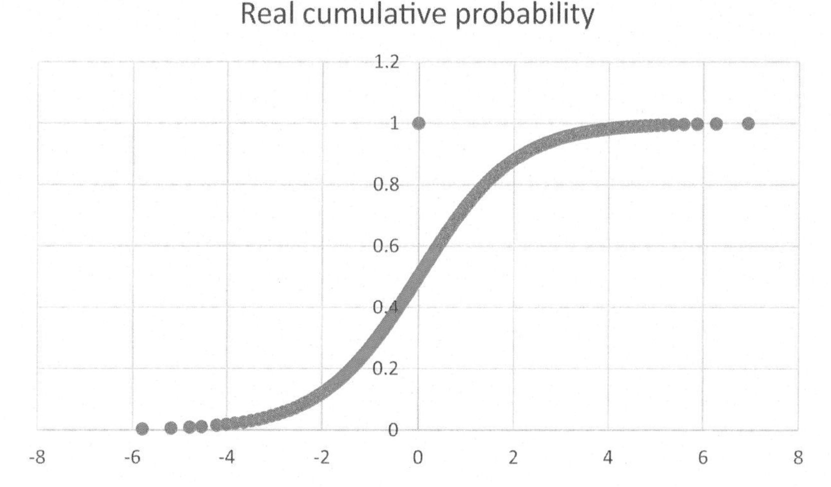

FIGURE 3.8 Logit model for real data based on complete probability.

3.7 Use cases

First, we create COVID-19 data based on real values and estimated R_{max} and R_{min} (reproduction number) for Turkey. Second, we arrange the cumulative end probabilities and locate them in two dimensions as upper and lower bounds. We have four cumulative probabilities in this data set in terms of maximum and minimum probabilities of the highest and lowest R_0 values. We arrange $P(R_{max}$ for the highest R_0 value) and $P(R_{max}$ for the lowest R_0 value) and them into the upper bound. Similarly, we arrange $P(R_{min}$ for the highest

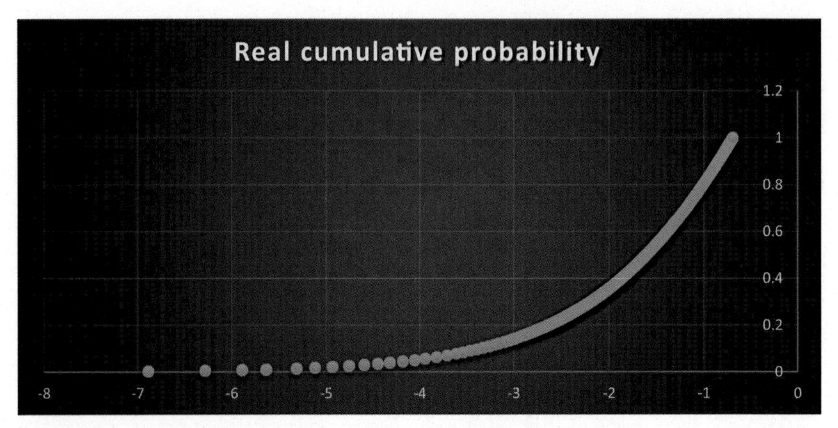

FIGURE 3.9 Logit model for real data based on paraconsistent probability.

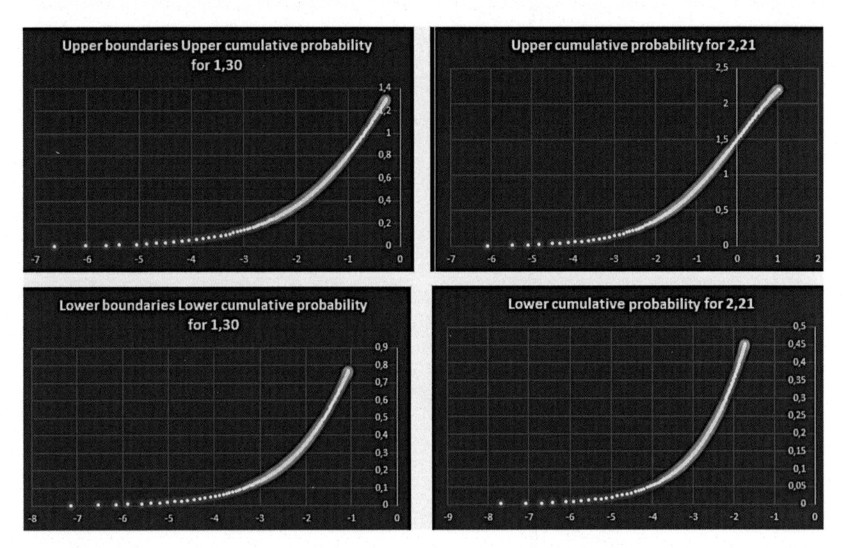

FIGURE 3.10 Logit models for different data sets based on upper and lower boundaries based on paraconsistent probabilities.

R_0 value) and $P(R_{min}$ for the lowest R_0 value) and them into the lower bound. We categorize upper bound and lower bound and label these values as supremum and infimum for each boundary. In the context of the square of opposition, we label S set beginning from inf (2) to sup (1) while our targeting P set remains in sup (2). Similarly, while the S set remains in inf (1) we label the P targeting set from sup (1) to inf (2). Based on this algorithm (Fig. 3.11), we get probability vectors for the hypothetical upper and lower planes to calibrate or at least guess the risk posed by the data. Finally, using the dot product defined in this article, we put the states on the risk map.

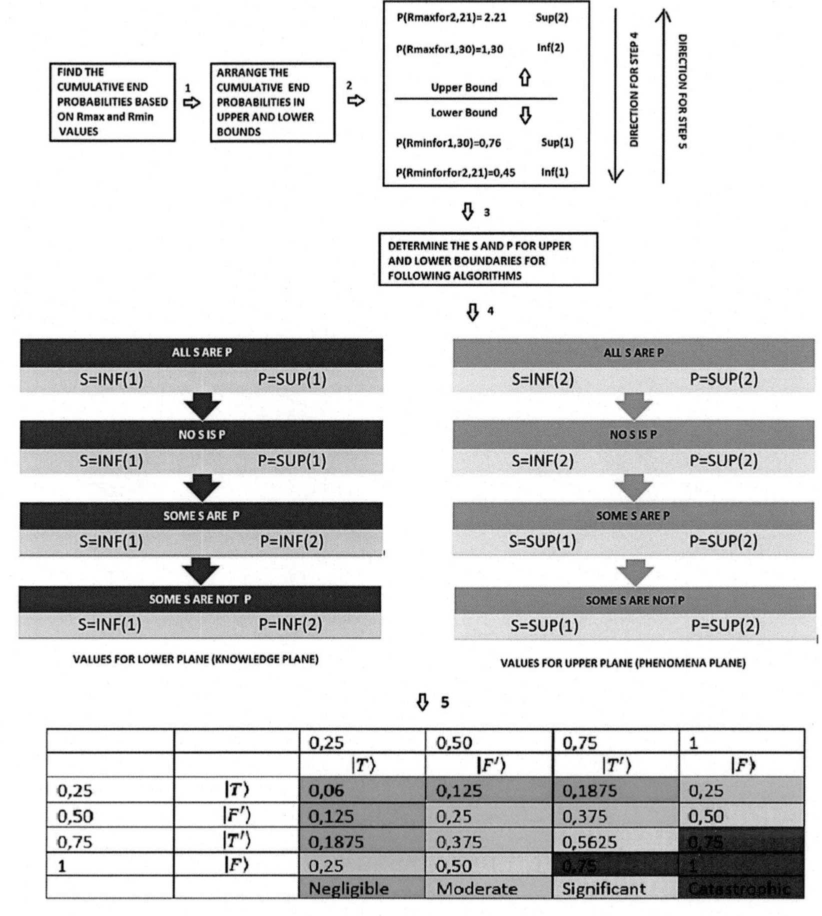

FIGURE 3.11 Strategy for the solution.

3.8 Applications

Based on the values, we can calculate the cumulative end probabilities for upper and lower planes as in Table 3.1. In Table 3.1, we found the values of λ as 0,58, $\nu = 0{,}37$, $\mu = 0{,}21$, and $\sigma = 0{,}15$ given in Table 3.3.

The vector form of the upper plane can be written as follows:

$$P_{\text{upper}} = 0,58|T\rangle + 0,37|F\rangle + 0,21|F'\rangle + 0,15|T'\rangle$$

In Table 3.4, we found the values of α as 0,59, Ω as 0,37, δ as 0,45, and β as 0,37.

$$P_{\text{upper}} = 0,58|T\rangle + 0,37|F\rangle + 0,21|F'\rangle + 0,15|T'\rangle$$

$$P_{\text{lower}} = 0,59|T\rangle + 0,37|F\rangle + 0,45|F'\rangle + 0,37|T'\rangle$$

TABLE 3.3 The cumulative end probabilities for the upper plane.

Attribution		S	P	$S \cup P$		
$\|T\rangle$	All *S* are *P*	1,30	2,21	2,21	 SP P All S are P.	$\lambda = 1{,}30/2{,}21 = 0{,}58$
$\|F\rangle$	No *S* is *P*	1,30	2,21	3,51	 S P No S is P.	$\nu = 1{,}30/3{,}51 = 0{,}37$
$\|F'\rangle$	Some *S* are *P*	0,76	2,21	3,51	 Some S are P. S SP P	$\mu = 0{,}76/3{,}51 = 0{,}21$
$\|T'\rangle$	Some *S* are not *P*	0,54	2,21	3,51	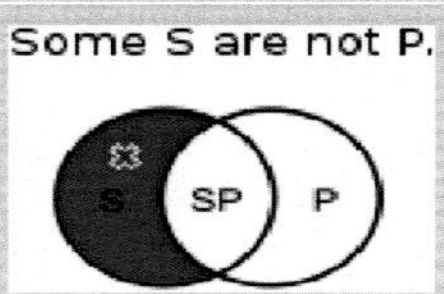 Some S are not P. S SP P	$\sigma = 0{,}54/3{,}51 = 0{,}15$

TABLE 3.4 The cumulative end probabilities for the lower plane.

Attribution		S	P	$S \cup P$		
$\|T\rangle$	All S are P	0,45	0,76	0,76	 SP P All S are P.	$\alpha = 0{,}45/0{,}76 = 0{,}59$
$\|F\rangle$	No S is P	0,45	0,76	1,21	 S P No S is P.	$\Omega = 0{,}45/1{,}21 = 0{,}37$
$\|F'\rangle$	Some S are P	0,45	0,54	0,99	 Some S are P. S SP P	$\delta = 0{,}45/0{,}99 = 0{,}45$
$\|T'\rangle$	Some S are not P	0,45	0,76	1	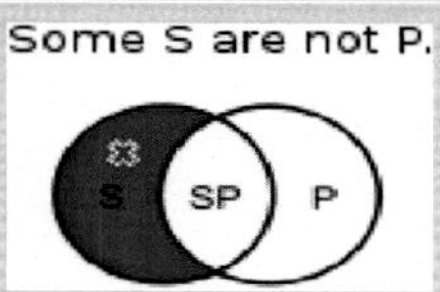 Some S are not P. S SP P	$\beta = 0{,}45/1{,}21 = 0{,}37$

TABLE 3.5 Risk map created based on the upper and lower vectorial planes.

		0,25	0,50	0,75	1
		$\|T\rangle$	$\|F'\rangle$	$\|T'\rangle$	$\|F\rangle$
0,25	$\|T\rangle$	$\|T'\rangle$	$\|F'\rangle$	$\|F\rangle$	$\|T\rangle$
0,50	$\|F'\rangle$	$\|F'\rangle$	$\|T\rangle$	$\|T\rangle$	
0,75	$\|T'\rangle$	$\|F\rangle$	$\|T\rangle$		
1	$\|F\rangle$	$\|T\rangle$			
		Negligible	Moderate	Significant	Catastrophic

We sum two vectors

$$P_{\text{upper}} + P_{\text{lower}} = 0,58|T\rangle + 0,37|F\rangle + 0,21|F'\rangle + 0,15|T'\rangle$$

$$+ 0,59|T\rangle - 0,37|F\rangle - 0,45|F'\rangle - 0,37|T'\rangle$$

$$= 1,17|T\rangle + 0,74|F\rangle + 0,66|F'\rangle + 0,52|T'\rangle$$

We can look at their difference as well =

$$P_{\text{upper}} - P_{\text{lower}} = 0,58|T\rangle + 0,37|F\rangle + 0,21|F'\rangle + 0,15|T'\rangle$$

$$+ 0,59|T\rangle - 0,37|F\rangle - 0,45|F'\rangle - 0,37|T'\rangle$$

$$= - 0,01|T\rangle - 0,24|F'\rangle - 0,22|T'\rangle$$

We can find the maximum and minimum values of our vectors as follows:

$$P_{\max} = 1,17|T\rangle + 0,74|F\rangle - 0,66|F'\rangle + 0,52|T'\rangle$$

$$P_{\min} = - 0,01|T\rangle - 0,24|F'\rangle - 0,22|T'\rangle$$

Finally, we can create a risk map based on the dot product of the vectors as follows (Table 3.5):

$$P_{\text{upper}} \cdot P_{\text{lower}} = 0,34|T\rangle + 0,13|F'\rangle + 0,05|T'\rangle$$

The data show that it have a moderate risk level in Turkey in terms of the risk map because most of the $|T\rangle$ vectors are cumulated in moderate cells.

3.9 Discussion

In this paper, we try to propose four-valued logic based on the square of opposition to assess the COVID-19 data in Turkey. First, we analyze the data in terms of the Rasch model. Many psychological tests, including questionnaires and assessment exams, are organized according to a ranking scale framework. Standard analytical techniques, such as summing the scores for the responses provided to the questions, carry out the testing procedure on the assumption that the intervals between the choices are equal. In practice, however, the intervals between choices are not necessarily the same distances apart. It has been shown via some research that the disparities between the choices are not equal. While questionnaires and tests are used to evaluate concepts such as attitude, skill, amount of knowledge, and degree of disability, the items in these surveys and tests are not all placed on the same point on the scale. To put it another way, not all things are similar in difficulty since some items require a greater depth of knowledge and/or ability than others. In contrast, the fact that all of the elements are at the same position on the scale does not result in any extra information being gathered about the variable under consideration. Because the questions are not all similar in difficulty, making a generalization based on the raw scores received from these items may be deceptive. When the difficulty levels of transition between categories are computed using Rasch analysis in this context, it allows for the incorporation of uneven differences across categories. Classical test theory provides test and item statistics that are depending on the group under investigation. Item and test statistics derived from the item response theory and similar models (such as the Rasch model, two-parameter and three-parameter logistic models) are not reliant on the group to which they are applied, but rather are based on a specific distribution assumption derived from the theory. The Rasch model, like the classical model, presupposes, as does the classical model, that the teacher or student, or patient or doctor, has full knowledge of himself or the topic with which he is working. If the uncertainty is not taken into consideration, the conclusion may be unrealistic. This is because, in the traditional method, when the probability of a feature is known, the probability of not considering it is also known. This is where our four-valued logic based on the square of opposition comes into play since it is a more realistic approach. In this approach, we create two frames of reference consisting of four-vector bases. Just as in the neutrosophic logic, we focus on the indeterminacy of the states and try to make use of them to understand the data at hand.

3.10 Conclusions

In this study, we have created a vectorial form of risk map based on COVID-19 data. The data show that it have a moderate risk level in Turkey in terms of the risk map. Apart from its output, this study contributes to the literature, especially in terms of its methodology and assumptions. An important initial addition of this research is that it considers the risks on a cognitive and perceptual basis, rather than just using a scalar threshold to define them. Data based on predefined factors are the main focus of conventional risk studies, which then utilize the data in order to estimate risk based on that data. Data membership values should be taken into account to prevent overstating or undervaluing the risks. Therefore, four-valued logic may be used in such instances, particularly when the arguments are based on the opposition square.

The second contribution of this work is that it deals with incomplete information and even contradictory information (probability) rather than complete information (probability) of the states. As indicated before, most of the researchers presume that their data provide complete information or at least they develop their arguments based on these assumptions. Therefore, they may disregard conflicting or insufficient information, and they could constrain themselves in a rigid zone. In this regard, this research presents another viewpoint to project the probable bounds of the risks and provide more objective possibilities.

Third, this work contributes to the literature by incorporating neutrosophic logic, syllogisms, and the sign–signifier–signified ideas of semiotics into Toulmin's argumentation model. When it comes to the concept of risk, semiotics plays a crucial role since we construct the meaning of risk primarily on the basis of sign–signifier–signified interactions. The mathematical structure of neutrosophic is thus consistent with semiotics, and we demonstrate this compatibility to some degree in our paper. Additional to this, argumentation is critical for any kind of risk analysis, and we demonstrate how it may be used mathematically in the context of Toulmin's argumentation model and the square of oppositions.

Finally, this research contributes to the literature by the implementation of the quantum decision-making model in risk analysis. The majority of descriptive statistics and analyses are performed on scalar data. Scalar data are also often used in the creation of visualizations. On the other hand, descriptive analysis for scenarios that are more closely associated with cognitive and semantic features may provide more in-depth findings than descriptive analysis for situations that are more closely tied to scalar data. As a result, we demonstrate that the analysis of COVID-19 spreads may be conducted using the four-valued vectorial version of probability functions. When researching societal understandings of risk and safety, it is standard practice to do so without first gaining a clear understanding of the conceptual foundations of the "risk" phenomena under inquiry [18]. Quantification of the scalar data and interpreting it in vectorial forms may give us more clues about the analysis of the risky situations because the risk is a cognitive perception based on the reference frame of the human brain in terms of its prediction of a likelihood of an event and labeling and classifying it in the context of the cognitive map.

Considering the above-mentioned contributions, our article may be useful for future AI research because most artificial intelligence research is based on pattern recognition, and they define learning in this context. However, as mentioned above, counterfactuals are an important part of human cognition, as well as the direction or vectorial structure of the data in making argumentation and projection. The realistic artificial intelligence models should mimic the human mind in this respect, and this study is thought to contribute to the literature especially by giving an intuitive methodological perspective.

3.11 Outlook and future work

This research is based on the application of four-valued vectorial logic. However, higher-valued logic can also be useful in the analysis of such data. Also, a new type of optimization method of an n-valued vectorial logic to decide the best n value for the analysis of the data can be proposed.

References

[1] F. Smarandache, B. Said, Neutrosophic Theories in Communication, Management, and Information Technology, NOVA Science Publisher, USA, 2020.

[2] J.M. Yearsley, J.R. Busemeyer, Quantum cognition and decision theories: a tutorial, J. Math. Psychol. (2015).

[3] B.E. Baaquıe, Quantum Field Theory for Economics and Finance, Cambridge University Press, UK, 2018.

[4] <https://www.inference.vc/causal-inference-3-counterfactuals/> Retrieving from 29.11.2021.

[5] J. Pearl, D. Mackenzie, The Book of Why: The New Science of Cause and Effect, Basic Books, 2018.

[6] F. Smarandache, Introduction to Neutrosophic Statistics, Sitech & Education Publishing, USA, 2014.

[7] J. Hawkins, A Thousand Brains: A New Theory of Intelligence, Hachette UK, 2021.

[8] S. Dehaene, How We Learn? Penguin Random House LLC, USA, 2020.

[9] J. Pearl, Causality: Models, Reasoning, and Inference, Cambridge University Press, UK, 2000.

[10] S. Khemlani, P. Johnson-Laird, Theories of the syllogism: a meta-analysis, Psychol. Bull. 138 (2012) 427–457. Available from: https://doi.org/10.1037/a0026841.

[11] <https://plato.stanford.edu/entries/square/> Retrieved from 01.12.21.

[12] <https://en.wikipedia.org/wiki/Square_of_opposition> Retrieved from 01.12.21.

[13] <https://web.ics.purdue.edu/~pbawa/421/THE%20TOLUMIN%20MODEL.htm> Retrieved from 01.12.21.

[14] J.R. Busemeyer, P.D. Bruza, Quantum Models of Cognition and Decision, Cambridge University Press, Cambridge, 2012.

[15] A. Łukasik, Quantum models of cognition and decision, Int. J. Parallel. Emergent. Distrib. Syst. 33 (3) (2018) 336–345. Available from: https://doi.org/10.1080/17445760.2017.1410547.

[16] C. DeMars, Item Response Theory, Understanding Statistics, Measurement, Oxford University Press, Inc, New York, 2010.

[17] <https://www.publichealth.columbia.edu/research/population-health-methods/rasch-modeling> Retrieving from 29.11.21.

[18] <https://searchcompliance.techtarget.com/definition/risk-map> Retrieving from 29.11.21.

[19] S.K. Köse, E. Demir, G. Aydoğdu, Estimation of the time dependent reproduction number of novel coronavirus (COVID 19) for Turkey in the late stage of the outbreak, Türkiye Klinikleri Biyoistatistik Derg. 13 (2021) 1.

[20] M. Boholm, The semantic field of risk, Saf. Sci. 92 (2017) 205–216.

Chapter 4

Neutrosophic linear models for bioinformatics

Muhammad Kashif[1], Muhammad Imran Khan[1], Sami Ullah[2] and Muhammad Aslam[3]

[1]Department of Mathematics and Statistics, University of Agriculture Faisalabad, Faislabad, Pakistan, [2]College of Agriculture, University of Sargodha, Sargodha, Pakistan, [3]Department of Statistics, Faculty of Science, King Abdulaziz University, Jeddah, Saudi Arabia

4.1 Introduction and motivation

The discipline of statistics mainly helps with learning from data. It assists in collecting data, exploring the salient features of data, and drawing conclusions for the parameter(s) of interests using relevant statistic(s) [1]. The subject of statistics usually has two main branches: descriptive and inferential. Descriptive statistics deals with the presentation of data in tabular and graphical forms, while summarization of data has been done through measures of locations and measures of dispersions. Inferential statistics focuses on making statistical decision about the population's characteristic(s) of interest, commonly called parameter, using the statistic(s) computed from the data. All sorts of inferential statistical procedures, estimation and testing of hypothesis, are based on some probability of significance [2] along with an estimator. In the case of estimation, point estimation is less likely to provide rational estimation than interval estimation. The testing of the hypothesis approach helps to decide on a specific postulation about the parameter(s) of interest. Often testing of hypothesis for more than two population means is required, which has been done using analysis of variance (ANOVA) [3]. The ANOVA is commonly considered by the researchers [4–8] of different fields such as social, agricultural, natural, physical, engineering, medical, and management sciences. The ANOVA will only be applied when population means will be normally and independently distributed with homogeneity of variances and no multicollinearity. These classical assumptions cannot be feasible when the data values of each of the variables are not measured exactly. In such situation, fuzzy logic is more suitable. ANOVA with fuzzy approach has been considered by [9] for consumer data, and [10] applied fuzzy ANOVA for functional data.

Cognitive Intelligence with Neutrosophic Statistics in Bioinformatics.
DOI: https://doi.org/10.1016/B978-0-323-99456-9.00005-2

Smarandache [11] introduced the idea of neutrosophic statistics (NS) and showed that neutrosophic logic is better than fuzzy logic, and the former is a generalization of the latter, as the former studied all of truth, false, and indeterminacy. The NS can only be applied when the parent population from whom sample data are drawn has uncertain observations. The researchers have used NS for different situations parallel to what already had been done for classical logic of statistics. For example, Chen et al. [12,13] utilized the neutrosophic numbers to study rock engineering issues. Aslam [14] has considered first the neutrosophic statistical quality control (NSQC) and has also provided a summarized table to explain the differences among fuzzy, neutrosophic, and classical approaches of statistics. The same is presented in Table 4.1 here with the permission of [14].

TABLE 4.1 Differences among fuzzy, neutrosophic, and classical approaches of statistics.

Classical statistics	Fuzzy statistics	Intuitionistic fuzzy statistics	Neutrosophic statistics
Classical statistics is applied to analyze the data when all observations/ parameters in the sample or the population are precise and determined	The fuzzy statistics is applied to analyze the data having imprecise, uncertain, and fuzzy observations/ parameters. The statistics is based on fuzzy statistics and does not consider the measure of indeterminacy	Intuitionistic fuzzy (IF) is the extension of the classical fuzzy logic. It considers membership and nonmembership, belonging to the real unit interval. Therefore, statistics based on IF are the extension of fuzzy statistics	The neutrosophic statistics (NS) is based on the idea of neutrosophic logic. The neutrosophic logic is the extension of the fuzzy logic and considers the measure of indeterminacy. Therefore, the NS is the extension of classical statistics which deals to analyze under uncertainty
Limitations: Classical statistics can be only applied when all data or parameters are determined and precise	The fuzzy statistics will be applied only when some observations/ parameters are fuzzy	The IF will be applied only when membership and nonmembership belong to the real unit interval	The NS is applied under the uncertainty environment. It reduces to classical statistics when all observations/ parameters are determined

A thorough investigation of the literature revealed that only [15] has recently introduced the concept of ANOVA under the NS and no work has been published so far regarding the neutrosophic ANOVA (NANOVA) using medical or bioinformatics data. The objective of this chapter is to the application of linear model for comparing population means for any uncertain environment. The application of the neutrosophic linear model is illustrated by analyzing gene expression data. It is hoped that this piece of research will provide new research avenues in the domain of neutrosophic statistics.

4.2 Background, definitions, and notations

The classical linear model for the gene expression data [16] can be written as

$$Y_{n\times 1} = X_{n\times k} \quad \beta_{k\times 1} + \in_{n\times 1} \tag{4.1}$$

where $Y = n\times 1$ column vector of observations on the dependent variable $Y, X = n\times k$ matrix given n observation on $k-1$ variables and its first column of 1's representing the intercept term. The matrix X is known as the data matrix, $\beta = k\times 1$ column vector of the unknown parameters $\beta_1, \beta_2, \cdots\cdots, \beta_k, \in = n\times 1$ column vector of n disturbances $\in_i$ or errors

In Eq. (4.1), the term $X\beta$ represents mean of the gene expression Y_i. The gene expression model is known as "linear" because all unknown parameters in vector β have single degree. The important assumption of the above model is that the error variables $\in_1, \in_2, \cdots\cdots, \in_n$ are independent and normally distributed with zero mean, that is, $\in \sim N(0, \sigma^2)$.

Let $X_N \asymp \langle X_L, X_U\rangle$ and $Y_N \asymp \langle Y_L, Y_U\rangle$ represent two neutrosophic random variables; then, the neutrosophic linear model for the gene expression data can be extended as

$$Y^N_{n\times 1} = X^N_{n\times k}\beta^N_{k\times 1} + \in^N_{n\times 1} \tag{4.2}$$

With the assumption that

$$E\left(\in^N\right) = 0_{n\times 1} \text{ and } \mathrm{Cov}\left(\in^N\right) = \sigma^{2N}_{n\times n} I^N$$

where $Y^N = n\times 1$ column vector of neutrosophic observations on the dependent variable $Y, X^N = n\times k$ matrix given n neutrosophic observations on $k-1$ neutrosophic variables and its first column of 1's representing the neutrosophic intercept terms. The matrix X^N is known as neutrosophic data matrix, $\beta^N = k\times 1$ column vector of the unknown neutrosophic parameters $\beta_{N1}, \beta_{N2}, \cdots\cdots, \beta_{Nk}, \in^N = n\times 1$ column vector of n neutrosophic disturbances $\in_i$ or neutrosophic errors.

Before the responses $Y^{N/} = \left[Y^N_1, Y^N_2, \cdots\cdots, Y^N_n\right]$ are observed, the error $\in^{N/} = \left[\in^N_1, \in^N_2, \cdots\cdots, \in^N_n\right]$ is random, and we can write the neutrosophic linear model defined in Eq. (4.2) as

$$Y^N = X^N\beta^N + \in^N \tag{4.3}$$

where

$$Y^N \in \left(\begin{bmatrix} Y_1{}^L \\ \vdots \\ Y_n{}^L \end{bmatrix}, \begin{bmatrix} Y_1{}^U \\ \vdots \\ Y_n{}^U \end{bmatrix} \right), X^N = \begin{bmatrix} 1 & x_{11} \\ 1 & x_{21} \\ \vdots & \vdots \\ 1 & x_{n1} \end{bmatrix}, \beta^N = \begin{bmatrix} \beta_{N1} \\ \beta_{N2} \\ \vdots \\ \beta_{Nk} \end{bmatrix}, \in^N = \begin{bmatrix} \in_{N1} \\ \in_{N2} \\ \vdots \\ \in_{Nk} \end{bmatrix}$$

The least-square estimate of β^N is defined as the generalization of the classical least-square estimator [17,18] of β.

$$\hat{\beta^N} = \left(X^{N/} \times X^N\right)^{-1} X^{N/} \times Y^N \tag{4.4}$$

The two-variable neutrosophic linear model can be written as

$$Y_{Ni} = \beta_{N0} + \beta_{N1} X_{Ni} + \varepsilon_{Ni} \text{ for i} = 1, 2, \ldots, n \tag{4.5}$$

The above equation is called a linear neutrosophic regression model, where X_N and Y_N represent the independent and dependent variables, respectively, and $\beta_{N0} \asymp \langle \beta_{L0}, \beta_{U0} \rangle$ and $\beta_{N1} \asymp \langle \beta_{L0}, B_{U0} \rangle$ represent the neutrosophic intercept and rate of change per unit, respectively. The above neutrosophic models defined in Eqs. (4.3) and (4.4) are useful for testing neutrosophic hypotheses about group means, but the case of overall test for the equality of means of gene expression data under uncertainty is of great interest, especially in bioinformatics. For this purpose, analysis of variance (ANOVA) can be used using neutrosophic data sets. The detail of neutrosophic ANOVA calculation is given as follows.

4.2.1 Neutrosophic analysis of variance

As $X_N \asymp \langle X_L, X_U \rangle$ represents the neutrosophic random variable and suppose it follows neutrosophic normal distribution, that is, $X_N \approx \langle \mu_N, \sigma_N \rangle$, where $\mu_N \in [\mu_L, \mu_U]$ and $\sigma_N \in [\sigma_L, \sigma_U]$ represent the neutrosophic mean (NM) and standard deviation (NSD), respectively.

In order to compare three or more population means simultaneously, the neutrosophic analysis of variance technique is used which will be considered as an extension of the classical analysis of variance. The detailed discussion can be found in [15]. The one-way NANOVA procedure is explained as follows:

Let k group of patients be available with neutrosophic measurements in the form of gene expression values. The null and alternate hypotheses about equality of neutrosophic population means can be defined as

$$H_{NO}{:}\mu_{N1} = \mu_{N2} = \mu_{31} = \ldots\ldots = \mu_{Nk}$$

H_{N1} At least one of the k_N-neutrosophic population means is not equal

TABLE 4.2 NANOVA table.

Source	NSS	ndf	NMS	F_N
Treatment	SS Treatment$_N$	K_N-1	S^2_{NT} = SS Treatment$_N$/K_N-1	$\frac{S^2_{NT}}{S^2_{NE}}$
Error	SS Error$_N$	n_{Nk}-t_k	S^2_{NE} = SS Error$_N$/n_{Nk}-t_k	
Total	SS Total$_N$	n_{Nk}-1		

To reject the null hypothesis about equality of neutrosophic population means, the neutrosophic sum of squares can be calculated as

$$\text{SS Total}_N = 1' \times \left(Y^N.Y^N\right) - \frac{(1' \times Y^N).(1' \times Y^N)}{1'1} \tag{4.6}$$

$$\text{SS Treatment}_N = 1'_Z \times \hat{\beta^N}(X'_N \times Y) - \frac{(1' \times Y^N).(1' \times Y^N)}{1'1} \tag{4.7}$$

$$\text{SS Error}_N = 1' \times \left(Y^N.Y^N\right) - 1'_Z \times \hat{\beta^N}\left(X'_N \times Y^N\right) \tag{4.8}$$

where $\hat{\beta^N}$ is defined in Eq. (4.4), "X^N" denotes the design matrix, Y^N denotes the data matrix, $\boldsymbol{1}$ is the matrix of ones with order $(n \times 1)$, and 1_x denotes the matrix of ones of order $[(t+1) \times 1]$. Finally, neutrosophic ANOVA (NANOVA) is defined in Table 4.2.

4.3 Application of proposed neutrosophic linear models

Example 1

The neutrosophic linear model concept is illustrated using the following examples. Suppose we have the following artificial neutrosophic gene expressing values 2, [3, 5],1,2, of Group 1,8,7,9, [8, 10] of Group 2, and 11,12,13, [12, 14] of Group 3. These neutrosophic values are presented in Table 4.3.

From the above values, it can be observed that the number of expression values per group was four and the total number of data values was in which three values showed neutrosophic response. The boxplot of gene expressions for different groups is shown in Fig. 4.1.

The summary statistics of the neutrosophic gene expression data reported in Table 4.3 is presented in Table 4.4.

The single mean testing of each group can be performed using t-test under neutrosophic environment. For this purpose, "gl" function in R

TABLE 4.3 Artificial neutrosophic gene expression data value.

Group	Neutrosophic gene expression values
1	2
1	[3,5]
1	1
1	2
2	8
2	7
2	9
2	[8,10]
3	11
3	12
3	13
3	[12,14]

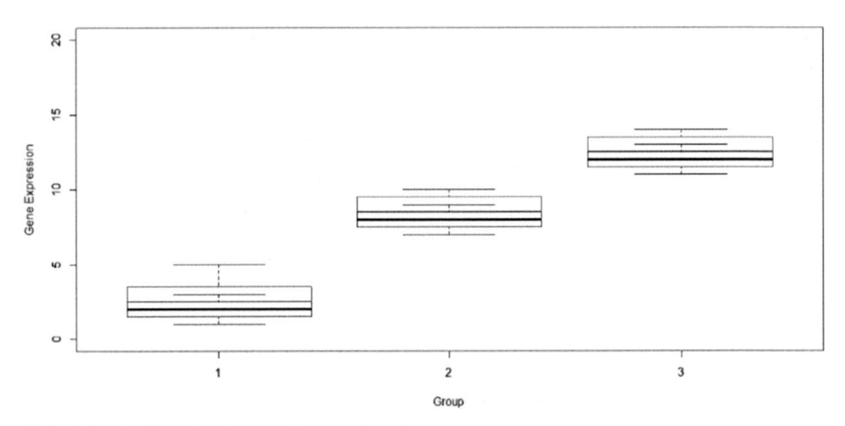

FIGURE 4.1 Boxplots for gene expression data.

software can be used. In order to estimate the results, we used the following R function

```
b<-gl(3,4)#Generating three treatments with 4 replications
model.matrix(y~b-1)#Design Matrix
summary(lm(y~b-1))#Summary of the fitted model
```

TABLE 4.4 Summary statistics of artificial neutrosophic gene expression data.

Disease stage	Mean	S.D.
$Group_1$	$[2 \quad 2.5]$	$[0.8165 \quad 1.7321]$
$Group_2$	$[8 \quad 8.5]$	$[0.8165 \quad 1.2910]$
$Group_3$	$[12 \quad 12.5]$	$[0.8165 \quad 1.2910]$

TABLE 4.5 Estimation of artificial neutrosophic gene expression data.

Coefficients	Estimates	Std. error	*T.* Value	Pr(> |*t*|)
b1	[2.00,2.25]	[0.4082,0.7265]	[4.89,3.441]	[0.0089,0.0073]
b2	[8.00,8.50]	[0.4082,0.7285]	[19.59,11.70]	[1.09e^08,9.55e-07]
b3	[12.00,12.50]	[0.4082,0.7285]	[29.39,17.21]	[2.98e-10, 3.41e-09]

The function $y \sim b - 1$ is used as model equation to be estimated, and -1 indicates that the model is estimated without intercept. By using the above R functions, the results from neutrosophic gene expression data are reported in Table 4.5.

The "estimates" column in Table 4.5 shows the estimated neutrosophic mean of each group whereas their neutrosophic standard errors are reported in the third column of Table 4.5. The estimated neutrosophic t-statistic and corresponding p-values are reported in the last two columns of Table 4.5. From the p-values, it can be concluded that we reject the null hypothesis of testing the neutrosophic mean of each group. The above representation of the linear model is useful for testing hypotheses about single group means. In the case of testing the overall significance of equality of several population means which is commonly required in bioinformatics, we proceed as follows.

The first step is the construction of the hypothesis, and for this illustration, let us continue with the previous neutrosophic data reported in Table 4.3. The objective is to compare the neutrosophic means of three groups with the following hypothesis.

$$H_{NO}{:}\mu_{N1} = \mu_{N2} = \mu_{31}$$

H_{N1}: At least one of the k_N-neutrosophic populations means is not equal

In order to calculate the neutrosophic F-statistic, we require the following calculation steps. For data given in Table 4.3, design matrix, data matrix, and matrices of ones are

$$X^N = \begin{bmatrix} 1 & 1 & 0 & 0 \\ 1 & 1 & 0 & 0 \\ 1 & 1 & 0 & 0 \\ 1 & 1 & 0 & 0 \\ 1 & 0 & 1 & 0 \\ 1 & 0 & 1 & 0 \\ 1 & 0 & 1 & 0 \\ 1 & 0 & 1 & 0 \\ 1 & 0 & 0 & 1 \\ 1 & 0 & 0 & 1 \\ 1 & 0 & 0 & 1 \\ 1 & 0 & 0 & 1 \end{bmatrix} \quad Y^N = \begin{bmatrix} 2 & 2 \\ 3 & 5 \\ 1 & 1 \\ 2 & 2 \\ 8 & 8 \\ 7 & 7 \\ 9 & 9 \\ 8 & 9 \\ 11 & 11 \\ 12 & 12 \\ 13 & 13 \\ 12 & 14 \end{bmatrix} \quad 1 = \begin{bmatrix} 1 \\ 1 \\ 1 \\ 1 \\ 1 \\ 1 \\ 1 \\ 1 \\ 1 \\ 1 \\ 1 \\ 1 \end{bmatrix} \quad 1_x = \begin{bmatrix} 1 \\ 1 \\ 1 \\ 1 \end{bmatrix}$$

$$1' \times 1 = 12$$

$$1' \times Y = \begin{bmatrix} 88 & 94 \end{bmatrix}$$

$$1' \times \left(Y^N.Y^N\right) = \begin{bmatrix} 1 \\ 1 \\ 1 \\ 1 \\ 1 \\ 1 \\ 1 \\ 1 \\ 1 \\ 1 \\ 1 \\ 1 \end{bmatrix}' \times \begin{bmatrix} 2 & 2 \\ 3 & 5 \\ 1 & 1 \\ 2 & 2 \\ 8 & 8 \\ 7 & 7 \\ 9 & 9 \\ 8 & 9 \\ 11 & 11 \\ 12 & 12 \\ 13 & 13 \\ 12 & 14 \end{bmatrix} \cdot \begin{bmatrix} 2 & 2 \\ 3 & 5 \\ 1 & 1 \\ 2 & 2 \\ 8 & 8 \\ 7 & 7 \\ 9 & 9 \\ 8 & 9 \\ 11 & 11 \\ 12 & 12 \\ 13 & 13 \\ 12 & 14 \end{bmatrix} = \begin{bmatrix} 854 & 900 \end{bmatrix}$$

$$(1' \times Y^N).(1' \times Y^N) = \begin{bmatrix} 7744 & 8836 \end{bmatrix}$$

$$\frac{(1' \times Y^N).(1' \times Y^N)}{1'1} = \frac{1}{12}\begin{bmatrix} 7744 & 8836 \end{bmatrix} = \begin{bmatrix} 645.33 & 736.33 \end{bmatrix}$$

$$\text{Total SS} = 1' \times \left(Y^N.Y^N\right) - \frac{(1' \times Y^N).(1' \times Y^N)}{1'1}$$

$$\text{Total SS} = \begin{bmatrix} 854 & 958 \end{bmatrix} - \begin{bmatrix} 645.33 & 736.33 \end{bmatrix}$$

$$\text{Total SS} = \begin{bmatrix} 208.67 & 221.67 \end{bmatrix}$$

$$\hat{\beta^N} = \left(X^{N/} \times X^N\right)^{-1} X^{N/} \times Y^N$$

where $\left(X^{N/} \times X^{N}\right)^{-1}$ is the generalized inverse of $X^{N/} \times X^{N}$.

$$X^{N/} \times X^{N} = \begin{bmatrix} 12 & 4 & 4 & 4 \\ 4 & 4 & 0 & 0 \\ 4 & 0 & 4 & 0 \\ 4 & 0 & 0 & 4 \end{bmatrix}$$

Generalized inverse of $\left(X^{N/} \times X^{N}\right)$ is computed by using R statistical language command *MASS::ginv(t(X)%*%X)*

$$\left(X^{N/} \times X^{N}\right)^{-1} = \begin{bmatrix} 0.0469 & 0.156 & 0.0156 & 0.0156 \\ 0.156 & 01719 & -.0781 & -.0781 \\ 0.156 & -.0781 & 0.1719 & -.0781 \\ 0.156 & -0.0781 & -0.0781 & 0.1719 \end{bmatrix}$$

$$X^{N/} \times Y^{N} = \begin{bmatrix} 88 & 94 \\ 8 & 10 \\ 32 & 34 \\ 48 & 50 \end{bmatrix}$$

$$\hat{\beta^{N}} = \left(X^{N/} \times X^{N}\right)^{-1} X^{N/} \times Y^{N} = \begin{bmatrix} 5.5 & 5.875 \\ -3.5 & -3.375 \\ 2.5 & 2.625 \\ 6.5 & 6.625 \end{bmatrix}$$

$$\text{SS Treatment} = 1'_x \times \hat{\beta}.\left(X^{N'} \times Y^{N}\right) - \frac{(1' \times Y^{N}).(1' \times Y^{N})}{1'1}$$

$$\text{SS Treatment} = \begin{bmatrix} 1 & 1 & 1 & 1 \end{bmatrix} \begin{bmatrix} 5.5 & 5.875 \\ -3.5 & -3.375 \\ 2.5 & 2.625 \\ 6.5 & 6.625 \end{bmatrix} . \begin{bmatrix} 88 & 94 \\ 8 & 10 \\ 32 & 34 \\ 48 & 50 \end{bmatrix} - \begin{bmatrix} 645.33 & 736.33 \end{bmatrix}$$

$$\text{SS Treatment} = \begin{bmatrix} 848 & 939 \end{bmatrix} - \begin{bmatrix} 645.33 & 736.33 \end{bmatrix} = \begin{bmatrix} 202.67 & 202.67 \end{bmatrix}$$

$$\text{SS Error} = \text{Total SS} - \text{Treatment SS}$$

$$\text{SS Error} = \begin{bmatrix} 208.67 & 221.67 \end{bmatrix} - \begin{bmatrix} 202.67 & 202.67 \end{bmatrix} = \begin{bmatrix} 6 & 19 \end{bmatrix}$$

After all the calculations, the resultant NAVOVA table is reported in Table 4.6.

According to [11], the null hypothesis will be rejected if the maximum neutrosophic *p-value* is less than the given level of significance, that is, $\max p_N < \alpha$ where α is the given level of significance which is normally 5%. By following the above rule, it is concluded that we reject the hypothesis that the means of the three groups are equal at 5% level of significance.

TABLE 4.6 NANOVA table for artificial gene expression.

S.O.V	DF	SS	MSS	*F*-value	*pN*-value
Treatment	2	[202.67 202.67]	[101.33 101.33]	[151.24 48.24]	[0.000 0.000]
Error	9	[6 19]	[0.67 2.11]		
Total	11				

Example 2

To illustrate the analysis of variance using real data, a simulated data is generated by following the "B-cell ALL: 1866_g_at" data reported in [16] under some neutrosophic observations. The simulated neutrosophic data are reported in Table 4.7.

The boxplot of the data is reported in Fig. 4.2. From this figure, it can be observed that the expression levels differ between the disease stages.

To illustrate the neutrosophic analysis of variance for this data, the null hypothesis is to test the equality of three neutrosophic expression means in each stage. To test this hypothesis, we proceed as follows and require the following calculations.

The summary statistics of the above data is reported in Table 4.8.

The hypothesis for this data is

$$H_{NO}:\mu_{BN1} = \mu_{BN2} = \mu_{\mathrm{BN3}}$$

H_{N1} At least one of the k_N-neutrosophic populations means is not equal

Following the above calculation procedure performed in Example 1, we have

$$1' \times Y = \begin{bmatrix} 325.0805 & 328.2305 \end{bmatrix}$$

$$(1' \times Y).(1' \times Y) = \begin{bmatrix} 105677.3 & 107735.2 \end{bmatrix}$$

$$\frac{(1' \times Y).(1' \times Y)}{1'1} = \frac{1}{78}\begin{bmatrix} 105677.3 & 107735.2 \end{bmatrix} = \begin{bmatrix} 1354.837 & 1381.221 \end{bmatrix}$$

$$\text{Total SS} = 1' \times (Y.Y) - \frac{(1' \times Y).(1' \times Y)}{1'1}$$

$$\text{Total SS} = \begin{bmatrix} 1370.603 & 1400.491 \end{bmatrix} - \begin{bmatrix} 1354.837 & 1381.221 \end{bmatrix}$$

$$\text{Total SS} = \begin{bmatrix} 15.76533 & 19.27004 \end{bmatrix}$$

$$\hat{\beta} = (X' \times Z)^{-} \times (X' \times Y)$$

TABLE 4.7 Neutrosophic B-cell patients expression data.

Stage	Expression level	Stage	Expression level	Stage	Expression level
B1	4.748315599	B2	4.160715817	B2	4.098489114
B1	[5.32920979, 4.32920979]	B2	4.465892877	B2	4.662688299
B1	[4.136771795, 4.936771795]	B2	4.026947096	B2	4.290779066
B1	4.651865601	B2	[3.873398586, 4.873398586]	B3	3.958671445
B1	5.532406694	B2	3.801749979	B3	[5.15663079, 5.00663079]
B1	4.382222827	B2	4.003345402	B3	[3.525424557, 4.525424557]
B1	4.339951722	B2	4.396185768	B3	3.740191692
B1	5.01872793	B2	4.278970501	B3	3.900447549
B1	5.633289097	B2	3.99138404	B3	4.017776598
B1	4.48206522	B2	4.584728984	B3	3.599750487
B1	[4.057957571, 5.057957571]	B2	4.061969352	B3	3.511419815
B1	4.260202093	B2	3.718707949	B3	3.819948454
B1	4.003934937	B2	3.909847573	B3	3.660851554
B1	4.049365436	B2	4.827191257	B3	3.603216915
B1	4.132384231	B2	4.025477142	B3	3.666252072
B1	4.189772401	B2	4.29826332	B3	3.97194264
B1	4.60053732	B2	4.032224016	B3	[3.425854721, 5.425854721]
B1	4.620995253	B2	4.347705444	B3	[4.124991022, 3.124991022]
B1	4.892131432	B2	3.889400535	B3	[3.973701895, 4.973701895]
B2	4.025540188	B2	3.554626825	B3	3.773181212
B2	4.315583185	B2	3.732435511	B3	4.146652422
B2	[4.503295673, 4.003295673]	B2	4.553980462	B3	3.546062214

(Continued)

TABLE 4.7 (Continued)

Stage	Expression level	Stage	Expression level	Stage	Expression level
B2	[4.124303557, 3.124303557]	B2	4.07684283	B3	3.878578594
B2	4.144300838	B2	4.07366059	B3	4.074514996
B2	3.850644015	B2	4.008399944	B3	3.985025478
B2	4.612701101	B2	3.909420171	B3	3.725480072

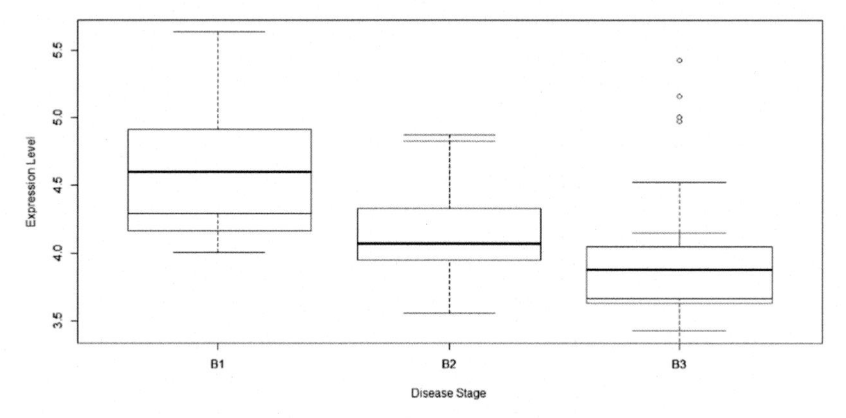

FIGURE 4.2 Boxplot for expression level data for the selected disease stages.

TABLE 4.8 Mean and standard deviation of neutrosophic expression levels data at different disease stages.

Disease stage	Mean	S.D
B_1	$[4.5822 \quad 4.6243]$	$[0.5026 \quad 0.4654]$
B_2	$[4.1453 \quad 4.1314]$	$[0.2943 \quad 0.3552]$
B_3	$[3.8603 \quad 3.9842]$	$[0.3523 \quad 0.5352]$

where $(X' \times X)^{-}$ is the generalized inverse of $X' \times X$.

$$X' \times X = \begin{bmatrix} 78 & 19 & 36 & 23 \\ 19 & 19 & 0 & 0 \\ 36 & 0 & 36 & 0 \\ 23 & 0 & 0 & 23 \end{bmatrix}$$

$$\left(X' \times X\right)^{-} = \begin{bmatrix} 0.0077 & 0.0054 & -0.0008 & 0.0031 \\ 0.0054 & 0.0341 & -0.0124 & -0.0163 \\ -0.0008 & -0.0124 & 0.0216 & -0.0101 \\ 0.0031 & -0.0163 & -0.0101 & 0.0295 \end{bmatrix}$$

$$X' \times Y = \begin{bmatrix} 325.08047 & 328.23047 \\ 87.06211 & 87.86211 \\ 149.23180 & 148.73180 \\ 88.78657 & 91.63657 \end{bmatrix}$$

$$\hat{\beta} = \left(X' \times X\right)^{-} \quad \left(X' \times Y\right) = \begin{bmatrix} 3.1470 & 3.1850 \\ 1.4353 & 1.4393 \\ 0.9984 & 0.9464 \\ 0.7133 & 0.7992 \end{bmatrix}$$

$$\text{SS Treatment} = 1_Z' \times \hat{\beta}.\left(X' \times Y\right) - \frac{(1' \times Y)(1' \times Y)}{1'1}$$

$$\text{SS Treatment} = \begin{bmatrix} 1 & 1 & 1 & 1 \end{bmatrix} \begin{bmatrix} 3.1470 & 3.1850 \\ 1.4353 & 1.4393 \\ 0.9984 & 0.9464 \\ 0.7133 & 0.7992 \end{bmatrix} \begin{bmatrix} 325.08047 & 328.23047 \\ 87.06211 & 87.86211 \\ 149.23180 & 148.73180 \\ 88.78657 & 91.63657 \end{bmatrix} - \begin{bmatrix} 1354.837 & 1381.221 \end{bmatrix}$$

$$\text{SS Treatment} = \begin{bmatrix} 1360.294 & 1385.877 \end{bmatrix} - \begin{bmatrix} 1354.837 & 1381.221 \end{bmatrix} = \begin{bmatrix} 5.4563 & 4.6562 \end{bmatrix}$$

$$\text{SS Error} = \text{Total SS} - \text{Treatment SS}$$

$$\text{SS Error} = \begin{bmatrix} 15.76533 & 19.27004 \end{bmatrix} - \begin{bmatrix} 5.4563 & 4.6562 \end{bmatrix} = \begin{bmatrix} 10.3091 & 14.6139 \end{bmatrix}$$

All the above calculations are summarized in Table 4.9 which represents the NANOVA table for the real data set of Example 2.

TABLE 4.9 NANOVA table for neutrosophic expression levels data at different disease stages.

S.O.V	DF	SS	MSS	*F*-value	*pN*-value
Between	2	[5.4563, 4.6562]	[2.7281 2.3281]	[19.85 11.95]	[0.0000 0.0000]
Error	75	[10.3091, 14.6139]	[0.1375 0.1949]		
Total	77				

From the above table, it can be concluded that we reject the hypothesis that the means of three disease stages are equal at 5% level of significance.

4.4 Discussion

As mentioned above, the objective of this chapter is to introduce the parametric hypothesis testing procedure for comparing three or more population means related to bioinformatics data having some neutrosophic observations. The proposed test was the modification of classical linear models and helpful to test the hypothesis of equality of more than two means when the data have some complex or uncertain observations. For illustration purpose, artificial and real gene expression data sets were used. The results of the proposed methods indicate that the measure of indeterminacy cannot be ignored and should be handled using neutrosophic statistical approach. Moreover, it can be concluded that working under neutrosophic environment is flexible as compared to fuzzy logic because fuzzy logic only considers measure of truth and the measure of falseness. The other advantage of working with neutrosophic data is that it reduces to classical statistics when all observations/parameters are determined.

4.5 Conclusions

In this chapter, linear model along with one-way ANOVA using indeterminant data for testing three or more than three population means is presented. The proposed NANOVA can be viewed as the generalization of the existing analysis of variance (ANOVA) approach under classical statistics. For illustration purpose, two data sets were analyzed using R software. The matrix approach is used for calculation purposes. The results indicate that the proposed methods have the ability to deal effectively with complex and uncertain observation in the field of bioinformatics.

References

[1] S.M. Ross, Introductory Statistics, Academic Press, 2017.

[2] B. Illowsky, S. Dean, Introductory Statistics, 2018.

[3] R. Fisher, Accuracy of observation, a mathematical examination of the methods of determining, by the mean error and by the mean square error, Mon. Not. R. Astron. Soc. 80 (1920) 758–770.

[4] R.A. Armstrong, F. Eperjesi, B. Gilmartin, The application of analysis of variance (ANOVA) to different experimental designs in optometry, Ophthalmic Physiol. Opt. 22 (3) (2002) 248–256.

[5] T. Niedoba, P. Pięta, Applications of ANOVA in mineral processing, Min. Sci. 23 (2016).

[6] J. Tarrío-Saavedra, S. Naya, M. Francisco-Fernández, R. Artiaga, J. López-Beceiro, Application of functional ANOVA to the study of thermal stability of micro–nano silica epoxy composites, Chemometr. Intell. Lab. Syst. 105 (1) (2011) 114–124.

[7] U. Ulusoy, Application of ANOVA to image analysis results of talc particles produced by different milling, Powder Technol. 188 (2) (2008) 133–138.

[8] E. Borgonovo, M.D. Morris, E. Plischke, Functional ANOVA with multiple distributions: implications for the sensitivity analysis of computer experiments, SIAM/ASA J. Uncertain. Quantif. 6 (1) (2018) 397–427.

[9] P.C. Lin, N. Arbaiy, I.R. A. Hamid, One-way ANOVA model with fuzzy data for consumer demand, in: International Conference on Soft Computing and Data Mining, Springer, 2016.

[10] G. González-Rodríguez, A. Colubi, M.Á. Gil, Fuzzy data treated as functional data: a one-way ANOVA test approach, Comput. Stat. Data Anal. 56 (4) (2012) 943–955.

[11] F. Smarandache, "Introduction to neutrosophic statistics: Infinite Study, Romania-Educational Publisher, Columbus, OH, 2014.

[12] J. Chen, J. Ye, S. Du, Scale effect and anisotropy analyzed for neutrosophic numbers of rock joint roughness coefficient based on neutrosophic statistics, Symmetry 9 (10) (2017) 208.

[13] J. Chen, J. Ye, S. Du, R. Yong, Expressions of rock joint roughness coefficient using neutrosophic interval statistical numbers, Symmetry 9 (7) (2017) 123.

[14] M. Aslam, A new sampling plan using neutrosophic process loss consideration, Symmetry 10 (5) (2018) 132.

[15] M. Aslam, Neutrosophic analysis of variance: application to university students, Complex. Intell. Syst. 5 (4) (2019) 403–407.

[16] W.P. Krijnen, Applied Statistics for Bioinformatics Using R, Institute for Life Science and Technology, Hanze University, 2009.

[17] R.A. Johnson, D. Wichern, Applied multivariate statistical analysis, Statistics 6215 (10) (2015) 10.

[18] D.N. Gujarati, D.C. Porter, S. Gunasekar, Basic Econometrics, Tata McGraw-Hill Education, 2012.

Chapter 5

Modeling epidemics based on quantum decision-making model by the qutrit states and employing neutrosophic form of percolation analysis

Volkan Duran[1], Selçuk Topal[2], Kadri Ulaş Akay[3], Ferhat Taş[3] and Florentin Smarandache[4]

[1]Faculty of Science and Arts, Iğdır University, Iğdır, Turkey, [2]Department of Mathematics, Bitlis Eren University, Bitlis, Turkey, [3]Department of Mathematics, Faculty of Science, İstanbul University, İstanbul, Turkey, [4]Mathematics, Physical and Natural Sciences Division, University of New Mexico, Gallup, NM, United States

5.1 Introduction and motivation

Neutrosophy is all about taking a new look at the world and then adapting that viewpoint to accommodate for the indeterminacy that exists. There is a third logical alternative to the binary paradigm of true or false, which goes by the name of neutrals, that is offered by neutrosophy. Overall, neutrosophy substitutes the binary technique in logics with indeterminacy, which may be construed in several ways, such as ambiguousness. Neutrosophy was first proposed by Smarandache in 1998, and its development has progressed significantly since then, thanks to the incorporation of logical extensions such as measures, sets, and graphs, as well as practical applications in a variety of fields [1].

Quantum theory is highly relevant to the neutrosophic philosophy because both of them deal with phenomena based on similar principles. For instance, Heisenberg's uncertainty principle is very similar to the indeterminacy of neutrosophic logic. As with classical probability theory, quantum cognition is a framework for developing cognitive models that are based on the mathematics of quantum probability theory. Quantum probability theory is itself a mathematical framework for assigning probabilities to events, in a similar way to classical probability theory. It establishes a new basis for

Cognitive Intelligence with Neutrosophic Statistics in Bioinformatics.
DOI: https://doi.org/10.1016/B978-0-323-99456-9.00016-7

comprehending a broad range of phenomena in decision-making by using a shared set of axiomatic concepts [2].

In this respect, it should be important to review the basic quantum principles as well. These can be summarized as the three basic principles of quantum mechanics [3]:

- The degree of freedom may be discrete or continuous, and it can range from one to an unlimited number of degrees of freedom. The degree of freedom is symbolized by the symbol φ; the collection of all of its values is denoted by the symbol $\mathcal{F}$.
- The state space is referred to as a Hilbert space in quantum physics. When dealing with systems that do not conserve probability, as is the situation in economics and finance, the state space might be far bigger than the Hilbert space. The state space is indicated by the letter V, and an element is denoted by the letter, where V: $\mathcal{F} \rightarrow V$. The dual space V_D consists of all mappings, denoted by $\langle \chi |$ of V to the complex numbers $\mathbb{C}$. The expression $\langle \chi | \psi \rangle = \langle \psi | \chi \rangle^* \in \mathbb{C}$ is the scalar or inner product.
- Operators $\hat{O}$ act on V and map it to itself $\hat{O}$: $V \rightarrow V$. The space of operators is denoted by $Q \equiv V \otimes V_D$. The tensor or outer product of two state vectors is given by

$$|\psi\rangle \otimes \langle\chi| \equiv |\psi\rangle\langle\chi| \in V \otimes V_D$$

In summary, quantum mechanics consists of the mathematical triple $\{F, V, Q\}$. The Hermitian conjugation of an operator, denoted by i, is defined by $\hat{O}_i^\dagger$ as

$$\langle\chi|\hat{O}_i|\psi\rangle^* = \langle\psi|\hat{O}_i|\chi\rangle^*$$

All eigenvalues of a Hermitian operator are real and can represent physically observed quantities. Hence, physical observations are represented by Hermitian operators

$$\hat{O}_i^\dagger = \hat{O}_i,\ i = 1, 2, \ldots, N\text{:Hermitian}$$

In general, $[\hat{O}_i, \hat{O}_j] \neq 0$. The physically observed value of a physical quantity Q, such as position, energy, is given by $\langle\psi|Q|\psi\rangle$, where $|\psi\rangle$ represents the quantum state of the physical entity.

Let us consider the eigenfunctions and eigenvalues of a Hermitian operator given by

$$\hat{O}|\psi_n\rangle = \lambda_n|\psi_n\rangle; \langle\psi_m|\psi_n\rangle = \delta_{m-n}$$

All Hermitian operators have the following spectral decomposition in terms of their eigenvalues and eigenfunctions

$$\hat{O} = \sum_n \lambda_n |\psi_n\rangle\langle\psi_n|.$$

The collection of all the eigenfunctions of a Hermitian operator yields a complete set of basis states and yields the completeness equation for V given by

$$\mathbb{I} = \sum_n |\psi_n\rangle\langle\psi_n| = \sum_n \Pi_n; \mathbb{I}^2 = \mathbb{I}$$

where $\mathbb{I}$ is the unit operator on V.

If π_1 and π_2 are projection operators that commute, that is, $\pi_1\pi_2 = \pi_2\pi_1$, then their product is also a projection operator. Suppose that a given vector space has n dimensions and a basis given by $|1\rangle$, $|2\rangle$, ..., $|n\rangle$ [4]. Define projection operators Π_n

$$\Pi_n = |\psi_n\rangle\langle\psi_n| \to \Pi_n{}^2 = \Pi_n$$

and

$$\hat{O} = \sum_n \lambda_n \Pi_n.$$

Every state vector has the following decomposition that can be given as follows:

$$|\chi\rangle = \mathbb{I}|\chi\rangle = \sum_n c_n|\psi_n\rangle \to c_n = \langle\psi_n|\chi\rangle$$

It then follows:

$$\langle\chi|\chi\rangle = 1 \to 1 = \sum_n |c_n|^2 \to |c_n|^2 \in [0, 1]$$

Note that the expectation value Π_n for state vector $|\chi\rangle$ is given by

$$E_\chi[\Pi_n] = \langle\chi|\Pi_n|\chi\rangle = P_n$$

Note that

$$P_n = E_\chi[\Pi_n] = \langle\chi|\Pi_n|\psi\rangle = |\langle\chi|\psi_n\rangle|^2 \geq 0$$

Furthermore

$$0 \geq E_\chi\left[(\Pi_n - P_n)^2\right] = E_\chi\left[\Pi_n{}^2\right] - P^2{}_n = E_\chi[\Pi_n] - P^2{}_n = P_n - P^2{}_n$$

$$\to P_n{}^2 \leq P_n \to P_n \leq 1$$

The completeness of the basis $|\psi_n\rangle$ from $|\chi\rangle = \mathbb{I}|\chi\rangle = \sum_n c_n|\psi_n\rangle \to c_n = \langle\psi_n|\chi\rangle$yields

$$\sum_n \Pi_n = \mathbb{I} \to E_x\left[\sum_n \Pi_n\right] = E_\chi[\mathbb{I}] \to \sum_n P_n = 1$$

Hence, the coefficients have the important property that

$$0 \leq P_n \leq 1; \sum_n P_n = 1$$

It shows that Pn has the interpretation of the probability of an event labeled by n. It is important to note that the quantum theory of measurement dictates that only one of the detectors, denoted by the letter Π_n, will detect the quantum state. This is also called the *collapse of the wave function* [3].

The first contribution of this paper is the incorporation of mathematics into the labeling of different situations in terms of their mental states. A language is an important tool for doing science and making inferences. By using language and its tools, we can differentiate and classify the phenomena around us. Labeling different things and classes, for instance, can be regarded as the first step in analytical thinking. This was emphasized by the Sapir–Whorf hypothesis which simply states that language determines thought or that language influences thought. As Whorf put it, "We cut nature up, organize it into concepts, ascribe significance as we do, largely because we are parties to an agreement to organize it in this way—an agreement that holds throughout our speech community and is codified in the patterns of our language." As the most famous illustration of the Whorf hypothesis, Whorf said that Eskimos had multiple terms for snow, meaning that since they live in a snowy environment, the Eskimos needed to distinguish between various sorts of snow. However, the example is not as exceptional as it may seem at first since skill leads to greater vocabulary for some domains, as shown by the fact that American skiers have distinct phrases for snow [5]. Furthermore, reindeer has up to 1000 terms in the Sami language. These include elements like the reindeer's physical condition, demeanor, and the form of its antlers, among other things. No one should be surprised by this type of linguistic enthusiasm since languages change to fit the thoughts and requirements of their speakers. Whether or not ice is safe to walk on, these individuals surrounded by ice need to know its classification. Every language manages to convey the information it has to provide in some manner. In the end, it is not how much terminology there are that matters, but rather the competence they convey [6]. Similar arguments can be made for the term risk. When we are using the term "risk," do we imply "danger" or "threat" or else? Moreover, how the term risk can be encapsulated with its results. Is risk a surprising event, an unexpected outcome, or an ambiguous situation? Therefore, we must clarify the terminology of the risk in this respect. In this paper, we try to expand the meaning of the risk by assuming the probable situations of the cases in the context of neutrosophic logic as T, F, and I values. We label those three states as vectorial forms and view them in the context of quantum decision-making theory. The noncommutative nature of those states gives different outputs so that we can enrich the meaning of the risk in a higher space.

The second contribution of this paper is to encapsulate the likelihood of an event with the corresponding definition of the cognitive states although the main limitation is that we just label the cognitive states by their definitions rather than based on an experimental study. There is a close relationship between the likelihood of an event and its riskiness. Casti [7] categorized possibilities into "definitely possible" and "unlikely." In this sense, he defined an X-event as an event that will have grave ramifications for our way of life if we are not prepared. The possibility of an X-event may be split into five categories according to his hierarchy (Fig. 5.1) [7]:

Virtually certain: It is almost likely that events like an asteroid strike, a major earthquake, or a financial collapse will occur. We have enough evidence to infer those similar events have occurred in the past and that they will occur again in our geographic and historical records.

Possible: Situations have occurred in the past or that have evidence that they may presently be taking place. Ice age or pandemic or global nuclear holocaust are all included in this category, as is the depletion of the Earth's ozone layer.

Unlikely: Non-feasible events that we do not know about and are not projected to happen; when possible, this subset is characterized by either a nano-cancer or a notable cultural deterioration.

Very remote: Situations in which human civilization would have little or no impact at all. In this chain of events, a time traveler walks on an ancient animal and later becomes the first human forefather, demonstrating the prospect of the universe being "reconfigured."

Impossible to say: Here are numerous compelling theories for a violent extraterrestrial invasion or the conquering of human civilization by an intelligent robotic force.

We can add the term obvious also as the clearest and observed state in this hierarchy as well. In addition to the likelihood of an event, cognitive states described by various words such as erroneous, dubious, ambiguous,

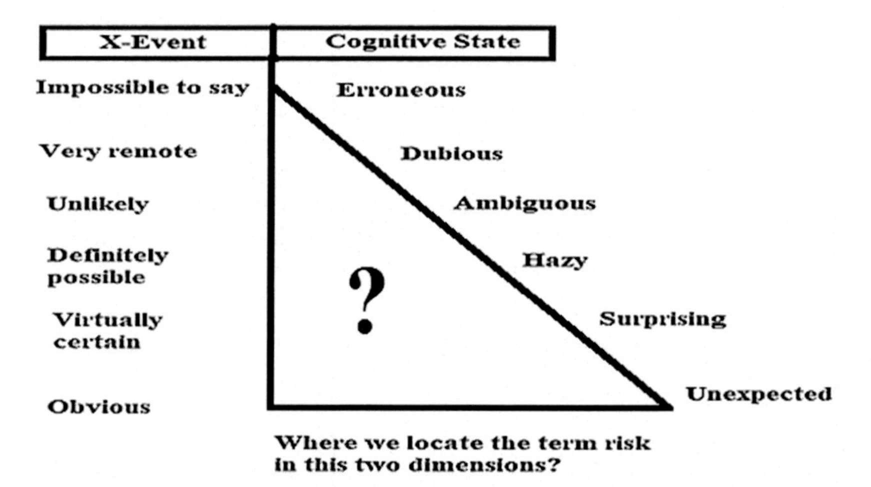

FIGURE 5.1 The term "risk" in the context of two dimensions.

hazy, surprising, or unexpected are also aligned with the concept of risk. Although we do not find the correlation between those two dimensions in the context of semantics and probability, we try to give an application of a mathematical model to reveal how those states can be incorporated into each other in the context of semiotics and neutrosophic logic.

The third contribution of this paper is the encapsulation of percolation theory in the analysis of COVID cases. Percolation theory is vital for understanding disease transmission patterns on temporal mobility networks [8]. Even though we do not use the percolation theory as the main tool, we created real and hypothetical data to show the possible risk zones in the context of incomplete information so that we get possible projection zones for the risky situations.

5.2 Background, definitions, and notations

5.2.1 Quantum decision-making model in the context of neutrosophic philosophy

The method by which neutrosophic logic and quantum decision-making were developed resulted in the development of a new theory of probabilistic and dynamic systems that is broader than the prior classic theory. In the neutrosophic logic, there are three probabilities given as $P(S)$ = the chance particular trial results in a success, $P(F)$ = the chance particular trial results in a failure, and finally $P(I)$ = the chance particular trial results in an indeterminacy [9]. When $(S) + (I) + P(F) = 1$, this case is called complete probability. But for incomplete probability (where there is missing information):

$$0 \leq (S) + (I) + P(F) < 1.$$

While in the paraconsistent probability (which has contradictory information):

$$1 \leq (S) + (I) + P(F) \leq 3.$$

Paraconsistent probability is especially useful for the analysis of risky situations because risk perception is mainly related to contradictory information. However, just a scalar version of this kind of probability might not be enough because in some cases noncommutative characteristics of human perception come to the force which is explained in the section entitled Judgments Create Rather than Record. Therefore, we define the quantum version of this as a probability ket vector as

$$\alpha|P(S)\rangle + \beta|P(I)\rangle + \delta|P(F)\rangle = |P\rangle$$

Based on three ways, we can get the neutrosophic interpretation of quantum states

- Where $[\langle P|P\rangle = \|\alpha\|^2 + \|\beta\|^2 + \|\delta\|^2] = 1$ in which there is complete information (probability) of the states

- Where $0 \leq [\langle P|P\rangle = \|\alpha\|^2 + \|\beta\|^2 + \|\delta\|^2] < 1$ in which there is incomplete information (probability) part of the states
- Where $0 \leq [\langle P|P\rangle = \|\alpha\|^2 + \|\beta\|^2 + \|\delta\|^2] \leq 3$ in which there is a contradictory information (probability) part of the states

Pitfall prevention: *We modify* $c_n \in [0, 1]$ *interval so that we get a neutrosophic form of quantum decision-making model.*

One might ask how this interpretation differs from classic quantum theory. First of all, even in quantum theory, there are superposed states, and they are never observed because of the collapse of the wave function. The basic paradox of quantum mechanics is that the basis of the quantum entity, namely, the degree of freedom, can never be seen by any experiment, regardless of how advanced the technology is. Furthermore, two orthogonal projection operators Π_n and Π_m can never observe the same state function at the same time since they are orthogonal to one another. Measurement results in the state function collapsing to either the state $|\psi_n\rangle = \Pi_n|\psi_n\rangle$ or $|\psi_m\rangle = \Pi_m|\psi_m\rangle$, the state vector $|\psi\rangle$ is never simultaneously seen by both the projection operators, and the state function collapses to one of these states. A state function is seen simultaneously by two orthogonal projection operators in any experiment. If this occurs in any experiment, the present (Copenhagen) interpretation of quantum mechanics comes to an end [3]. However, in the neutrosophic case of contradictory information (probability) part of the states, such a thing happens which is compatible with human cognition because the human mind can deal with contradictory states at the same time or at least similar time intervals.

Earlier cognitive and decision science models relied on assumptions from classical probabilistic dynamical systems, now deemed obsolete. Let us take a look at these concepts and see whether they can be applied to the disciplines of cognition and decision-making in general and examine the six reasons for a neutrosophic and quantum decision-making approach to cognition and decision (Fig. 5.2) [11].

5.2.1.1 Judgments are based on indefinite states

According to many formal models commonly used in cognitive and decision science, the cognitive system is in a specified state at any given time concerning a choice. Formal cognitive models presume that you are in a certain state concerning guilt at any one time—for example, a state in which you pick a value p such that $p > 0.50$ or a state in which you generate a value p such that $p \leq 0.50$ (in other words, p is a function of the current state of the system). Because the model does not know your current condition, it can only give a probability of you replying with $p > 0.50$. The model is called stochastic since it does not know your exact trajectory (the specified state at each time point). Stochastic models have a sample space of trajectories and a measure that assigns a probability to trajectories. In a stochastic model,

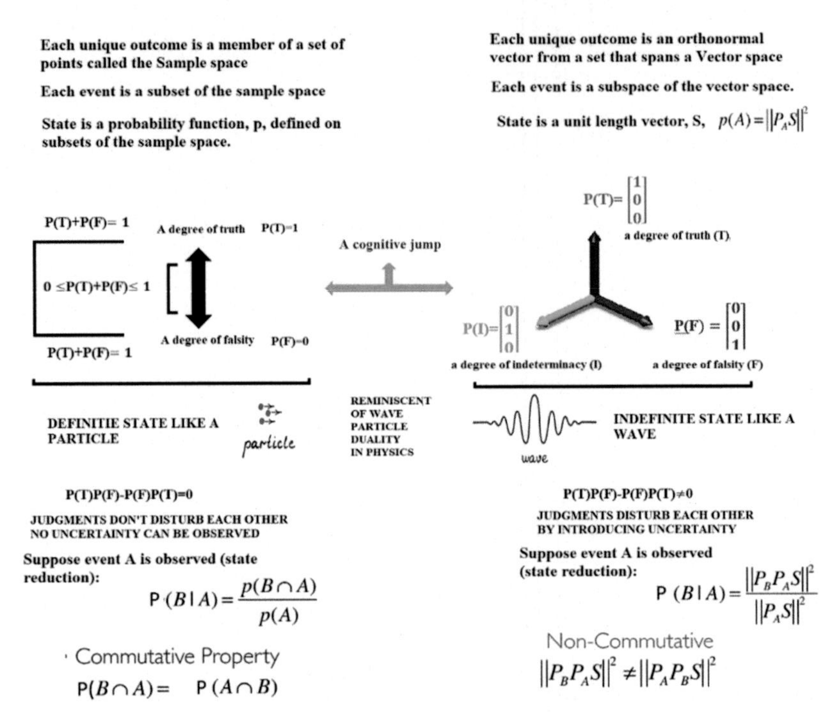

FIGURE 5.2 Difference between classical logic and neutrosophic logic based on quantum principles [10]. *Modified from J.R. Busemeyer, Quantum models of cognition and decision, http://www.physics.smu.edu/web/seminars/seminar_archive/Spring2016/busemeyer.pdf Retrieved from 24.10.21.*

however, once a trajectory is sampled (for example, once a seed is chosen to start a computer simulation), the system either jumps from one definite state (for example, respond with $p > 0.50$) to another (for example, respond with $p < 0.05$) or stays in one place throughout the simulation. Therefore, cognitive and decision sciences based on this convention increasingly represent the cognitive system as a particle with a determined sample path through state space. Unlike conventional theory, quantum theory permits you to remain in an indefinite state (superposition state) before choosing a decision. Being in an indefinite or superposition state means the model cannot assume you are either (1) guilty ($p > 0.50$) or (2) not guilty ($p < 0.50$). You may be in an indefinite condition that allows both of these definite states to exist at the same time. In an indefinite state, you do not certainly think the person is guilty, but you also do not necessarily feel the person is not guilty. Instead, you are in a state of superposition, unsure of true guilt. Both answers are possible at any moment, but the possibility for guilt is greater than the potential for innocence. These potentials (amplitudes) may change from moment to moment, but both responses are possible. In quantum theory, instead of a single trajectory or sample route through time, there is a blurring of potentials between states that flows across time. Quantum theory allows one to describe the cognitive system as a wave going across time and state space until a decision is made. Nevertheless, after a choice has been made, and all ambiguity has been addressed, the state becomes definitive, as if the wave had collapsed into a point, similar to that of a particle. As a result, both wave (indefinite) and particle (definite) perspectives of a cognitive system are required for quantum systems.

5.2.1.2 Judgments create rather than record

The cognitive system, according to many formal models, may be in a state of flux from instant to moment, but what we record at a given time represents the state of the system as it was shortly before we enquired about it. So, for example, formal cognitive models assume that if a person watches a disturbing scene and we ask the person a question such as "Are you afraid?" then the answer reflects the state of the person regarding that question just before we asked it. If instead, we asked the person "Are you excited?" then the answer again reflects the state regarding this other question just before we asked it. One of the most fascinating quantum theory insights is that measuring a system develops a property rather than recording it. This is the basis of modern psychological theories of emotion. Experts in decision science think that thoughts and preferences are created online rather than read from memory. Someone may be undecided about which picture to select from a group of paintings on display, but when forced to choose, they may alter their mind. A quantum system may be in an indefinite state just before a query is asked. After seeing a frightening situation, for example, the individual may be unsure about his or her sentiments. It is the interplay between an indeterminate state and the question that we ask that creates the response we get from a quantum system. From an indefinite state, a definite state is created via this interaction. We contend that the quantum principle of constructing reality from an interaction between a person's indefinite state and the question being asked more closely matches psychological intuition for complex judgments than the assumption that the answer simply reflects a preexisting state for complex judgments.

5.2.1.3 Judgments disturb each other, introducing uncertainty

According to quantum theory if one starts in an indefinite state and then is asked a question, the response to the question will transform the state from an indefinite one to one that is more definite about the question that was asked and so on until the state is no longer indeterminate. However, this shift in state after the first question leads one to reply differently to subsequent questions, such that the sequence in which questions are asked becomes critical to understanding the situation. Take, for example, the following well-known social psychology case study. If a teenage male is questioned directly, "How happy are you?" he would most likely respond, "Everything is fantastic." If, on the other hand, this youngster is initially asked, "When was the last time you went on a date?" he or she would most likely respond, "Seems like a long time ago." Following this sobering response, a subsequent query regarding happiness is more likely to elicit a second response that is less than bright and pink. As a result, the first question establishes a context that influences the response to the next question. So, we are unable to determine the joint probability of responses to questions A and B together and can only assign a probability to the sequence of answers to questions A and B after they have been determined. In quantum physics, if A and B are two measurements and the probability of the outcomes depends on the order of the measurements, then the two measurements are noncommutative, and the probabilities of the outcomes depend on the order of the measurements. If we take physics as an example, measurements of position and momentum in the same direction are noncommutative, but measurements of position and momentum along the horizontal and vertical coordinates are commutative. It was Heisenberg's renowned uncertainty principle that led to the development of a probabilistic model for noncommutative measurements, and here is where many of the mathematical features of quantum theory originated. As well as infusing uncertainty into a person's decisions, order effects are also accountable for this. It is possible that a response to the first question A will result in the creation of a definite state for that question, but that the state formed by A will result in the creation of an indefinite state to another question B. As a result, it may be difficult to be in a definite state about two separate questions at the same time since a definite state (technically speaking, an eigenstate) for one question is an indefinite state (superposition) for the other. In this scenario, the questions are said to be incompatible, and the mathematical implementation of the incompatibility of questions is provided by the noncommutativity of quantum measurements in this case. In other words, the classical logic is commutative ($P(T)P(F)\text{-}P(F)P(T)=0$), but the quantum one is not ($P(T)P(F)\text{-}P(F)P(T)\neq 0$). Therefore, judgments disturb each other, introducing uncertainty which is defined as a risk in this paper, and the more complicated forms are defined as chaotic cases are catastrophes as shown in Fig. 5.2.

5.2.1.4 Judgments do not always obey classic logic

The Kolmogorov axioms serve as the foundation for the classical probability theory that is employed in modern cognition and decision models. These axioms attribute probability to occurrences that may be represented as sets of events. As a result, the family of sets in the Kolmogorov theory obeys the axioms of logic known as Boolean algebra. The distributive axiom of Boolean logic is one of its most essential axioms: if $\{G, T, F\}$ are events, then $G \cap (T \cup F) = (G \cap T) \cup (G \cap F)$. Consider the thought of a boy being good (G), as well as the pair of conceptions that the boy spoke the truth (T) and that the boy did not tell the truth (falsehood, F). Boolean logic states that the event G can only happen in one of the two ways: either $(G \cap T)$ occurs or $(G \cap F)$ exclusively. For you to consider the boy to be excellent, there are only two mutually exclusive and exhaustive options: either he is good and truthful, or he is good but not truthful. The von Neumann axioms serve as the foundation for quantum probability theory. Following these axioms, occurrences specified as subspaces of a vector space are assigned probability. Definite states serve as the foundation for the vector space, while an indefinite or superposition state may be found at any point in the space represented by the vector. The use of subspaces has many major consequences, one of which is that the logic of subspaces does not satisfy the distributive axiom of Boolean logic. When attempting to determine if a boy is good without first knowing whether or not he is truthful, quantum logic says that you are not constrained to have just two ideas: he is good and he is truthful or he is good and he is not truthful. Instead, you may have any number of thoughts. Other ambiguous ideas may be represented by a superposition over the true or untruthful qualities, and they are called ambiguous thoughts. The fact that quantum logic does not always adhere to the distributive axiom indicates that the quantum model does not always adhere to the rule of total probability, as stated before. Thus, the quantum model may explain the outcomes of psychological experiments such as the disjunction experiment and physics experiments such as the two-slit experiment.

5.2.1.5 *Judgments do not obey the principle of unicity*

In the traditional (Kolmogorov) probability theory, which is employed in many modern cognitive and decision models, the principle of unicity is a key concept to understand. Using classical theory's Boolean algebra, this follows: if A is an event and B is another event from the same experiment, then A and B must also be an event, and repeated application of this concept results in intersections that cannot be broken down any further (the atoms or elements or points of the sample space). Unions of atoms or elements or locations in the sample space may be used to describe any occurrence in the sample space. Quantum probability does not operate on the assumption of the principle of unicity. This premise is violated as soon as we introduce incompatible questions into the theory, which results in measurements that are not commutative in nature. Incompatible questions cannot be assessed on the same basis as each other, and as a result, they must be evaluated on distinct sample spaces. A partial Boolean algebra can be used in quantum theory to answer the first set of questions in a Boolean way, and another sample space can be used to answer a different set of questions in a Boolean way; however, the results of the two Boolean subalgebras are pasted together in a

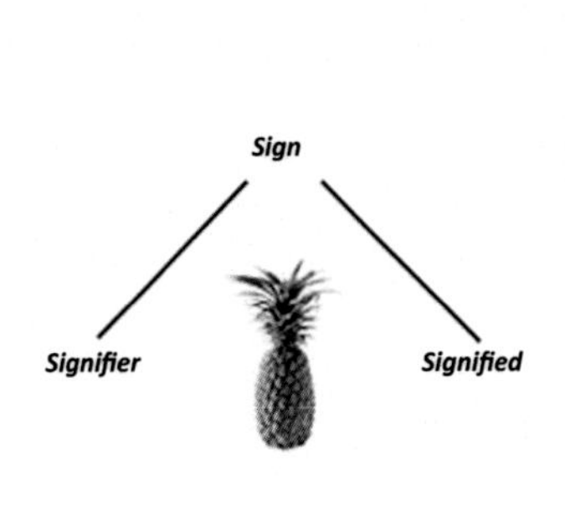

Signifier	*Signified*	*State*
$\|T\rangle = \begin{bmatrix}1\\0\\0\end{bmatrix}$	$\|F\rangle = \begin{bmatrix}0\\0\\1\end{bmatrix}$	|Surprising >
$\|F\rangle = \begin{bmatrix}0\\0\\1\end{bmatrix}$,	$\|T\rangle = \begin{bmatrix}1\\0\\0\end{bmatrix}$	|Unexpected >
$\|T\rangle = \begin{bmatrix}1\\0\\0\end{bmatrix}$	$\|I\rangle = \begin{bmatrix}0\\1\\1\end{bmatrix}$,	|Dubious >
$\|I\rangle = \begin{bmatrix}0\\0\\1\end{bmatrix}$	$\|T\rangle = \begin{bmatrix}1\\0\\0\end{bmatrix}$	|Risky(Hazy) >
$\|F\rangle = \begin{bmatrix}0\\0\\1\end{bmatrix}$	$\|I\rangle = \begin{bmatrix}0\\0\\1\end{bmatrix}$	|Ambiguous >
$\|I\rangle = \begin{bmatrix}0\\0\\1\end{bmatrix}$	$\|F\rangle = \begin{bmatrix}0\\0\\1\end{bmatrix}$	|Erroneous >

(A) Logic of semantic classification based on sign-signifier and signified

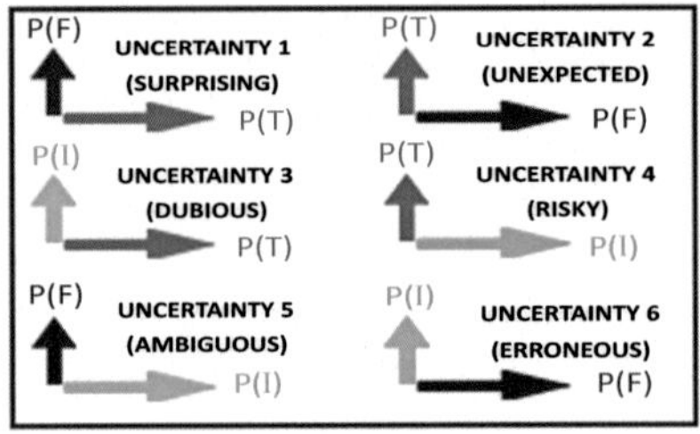

(B) Uncertianty cases

FIGURE 5.3 (A) The logic of semantic classification is based on sign−signifier and signified. (B) Uncertainty (risks).

coherent but non-Boolean way, as shown in Fig. 5.3. When assigning probabilities to events, this allows more flexibility since it does not require creating all conceivable joint probabilities, which we feel is a quality required to grasp the entire complexity of human cognition and choice. In this respect, we can create a semantic classification based on the Swiss linguist Ferdinand de Saussure's notions of sign, signified, signifier, where signified pertains to the "plane of content," while signifier is the "plane of expression." Such a classification based on the terms sign−signifier−signified is important because we define the degree of risk or safety based on our mental constructions. For instance, if the sign is true but the signifier is false, we might be surprised based on the criteria of the sign. In the real world, there is no risk or safe, but we define and create those meanings for our mental reality. For example, for the coronavirus, increase of infected individuals is not a risky situation (Fig. 5.3).[1]

1. https://www.exploratorium.edu/complexity/CompLexicon/catastrophe.html Retrieved from 24.10.21.

5.2.1.6 Cognitive phenomena may not be decomposable

In cognitive research, as well as in other fields of science, reductionism has long been recognized as a significant philosophical underpinning of model building. This refers to the concept that phenomena may be studied by evaluating each of its constituent parts separately and then synthesizing the information gathered from each component. Decomposable phenomena or systems are those that can be broken down into their constituent parts. Non-decomposable systems, on the other hand, cannot be so easily comprehended. The vast majority of models in cognitive science are decomposable, meaning that they can be broken down into their component elements and the relationships between these parts. The appearance of quantum correlations in cognitive events suggests that this assumption may at the very least be called into question in particular instances. There is a school of thought that when a word is investigated in a memory experiment, a word's associative network emerges in synchronization with the word that is being researched, as an example. When it comes to quantum entanglement, the thinking underlying this approach is extremely similar to that of quantum entanglement—the study word and its associations are acting as if they are one entity. Given the presence of quantum correlations, it is reasonable to conclude that the cognitive model in issue is not decomposable in the manner we first anticipated and specified using a specific set of random variables. It pushes us to reconsider the nature of the phenomena under consideration in a fundamentally new way.

Risks are not just related to factual data in a statistical sense but are also related to cognition in terms of psychology. A comparable scenario seems to be occurring in current cognitive psychology to that which happened in physics at the beginning of the twentieth century—several paradoxes have been reported in the literature on the topic of judgments and choices made in the face of risk and uncertainty. At least if we embrace the traditional paradigm of rationality, which requires classical logic and classical probability calculus, human judgments and choices often seem to be irrational. However, this "irrationality" is likely merely visible and that it represents the fact that human thinking comprises two layers or two levels: the classical level, which is defined by classical logic and classical probability calculus, and the quantum level, in which judgments and decision processes involve the existence of generally quantum phenomena, such as superposition of beliefs, contextuality of beliefs, and interference effects recognized in quantum probability theorems. If this is the case, then components of the formalism in quantum mechanics may be used outside of the realm of physics—for example, in the context of cognition and decision-making [12].

5.3 Literature review and state of the art

The trickiest issue in a percolation problem is to determine whether or not there is a continuous channel from one end of the system to the other for a certain system. Percolation is, at its most basic level, a method of distributing items over a vast number of empty areas. For technical reasons, percolation theory is only valid for arrays that are indefinitely big, which means that the systems of interest must be huge for percolation theory to be applicable. The array does not have to be rectangular, nor does it have to be two-dimensional: certain phenomena are better represented with a one-dimensional array, while others are best characterized with a three-dimensional array, and still others are best modeled with higher-dimensional arrays. To fix ideas, however, it is simplest to see a huge two-dimensional array of N cells, similar to an expanded checkerboard, to visualize the solution. Assume that each cell in an array has a chance of being filled with a value p. There is no relationship between any of the cells—just because a specific cell happens to be filled does not imply that the nearby cells are more or less likely to be inhabited. $P \times N$ of the cells will be filled, and $(1\text{-}p) \times N$ of the cells will be left empty. If the probability p is big, the array will include a large number of filled cells; if the probability p is small, the array will contain a small number of filled cells. Fig. 5.4 shows four 8×8 arrays that were created by a computer. For each cell in (A), the likelihood of occupancy is 30%; for each cell in (B), the probability is 40%; for each cell in (C), the probability is 50%; and for each cell in (D), the probability is 60% (Fig. 5.4). Two occupied cells that are next to each other are called *neighbors*, and groups of neighbors are called *clusters*. In the two-dimensional array seen in the illustration, each cell, except those on the edges, may have four neighbors: the cells directly above and below it, as well as the cells to the left and right of it. Percolation theory is concerned primarily with how these neighbors and clusters interact with one another, as well as how the density of these neighbors and clusters influences the particular phenomenon under consideration. When it comes to percolation theory, a cluster that spans the length or width (or both) of an array is especially crucial to consider. The spanning cluster, also known as the percolation cluster, is a kind of cluster. A spanning cluster exists only when the probability p is greater than a critical number p_c for an infinite lattice, and this is not always the case [13].

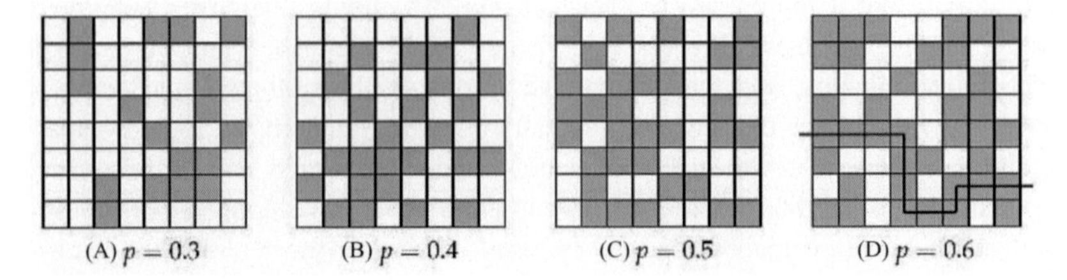

FIGURE 5.4 The cells in each of these four arrays have been darkened (i.e., they have been occupied) in a random manner. Each cell in (A) has a 30% probability of being occupied at any one time. Each cell in (D) has a 60% probability of being occupied at any one time. Even in (A), there exist "clusters," which are instances in which two or more nearest-neighbor cells are occupied at the same time. (The nearest neighbor of a cell is the cell that is immediately above, below, to the left or right of the cell in which the cell is located.) "Spanning cluster": a route via closest neighbors from one end of the array to the other is shown in (D) [13].

Such a p_c value can be used as ***"risk criteria"*** of evaluating the spread of viruses or other phenomena in this respect. Pc is dependent on the dimensionality of the space (in this example, two dimensions in Fig. 5.4) and the connectivity. In the case of Pc, colonization of the viruses will always come to an end after a certain number of colonies. Growth will occur in "clusters," with virus areas, especially those that are not in quarantine or blocked with different precautions on the outside of each cluster consisting of humans who are not infected. Alternatively, in the case of $P > \text{Pc}$, clusters will develop endlessly until they completely occupy the available space. However, small voids will exist, bounded on the inside by non-colonizing civilizations. If every given point is an isolated unoccupied point, then the chance of it being such is equal to PN; the likelihood of bigger clusters of isolated points occurring is proportional to higher powers of P. Finally, for critical percolation, with P near the value of Pc, clumps evolve into fractal structures with irregular shapes as they become larger. There are arbitrarily huge filled areas and arbitrarily big vacant parts in the world [14].

Therefore, rather than relying on complex mathematical models to explain the chance of infection by COVID-19 or any other illness, we use basic ideas from percolation theory, which predicts the likelihood that percolation will occur in terms of non-blocked p and blocked $(1-p)$ connections. Additionally, different models via different materials can be used by modifying those models into the COVID-19 spread. When the experimental findings were rationalized, two theoretical methods were developed, each of which discussed the state of a system during the percolation process. A static model proposes that, during the process of percolation, the charge carriers join with their neighbors, resulting in the formation of a network of charged particles that exchange charges. In addition, a dynamic model in which the particles carrying the charges are constantly colliding combines randomly at the percolation threshold and swaps the charges considered to be feasible. For instance, for electrical percolation in reverse micelles systems, induced by temperature variation, the equations can be modified like [15]:

$$I = \beta(t_p - t)^{-s}, \; t \ll t_p$$

$$I = \gamma(t - t_p)^{-t_s}, \; t_p \ll t$$

where β and γ are free parameters. S and t_s are called the critical exponents, and I refers to the evolution of the cumulative number of infected people. The distinction between static percolation and dynamic percolation may be seen in the critical exponent values of the equations. If we look at it from an epidemiological perspective, the static percolation model correlates to the establishment of viral transmission networks that are geographically distributed. The dynamic model, on the other hand, represents transmission networks that are arbitrarily scattered in space [15].

5.4 Problem/system/application definition

First, in the analysis, the data regarding COVID-19 cases in Turkey were taken from https://turCovid19.com/acikveri/ which is between 27 March 2020 and 09 February 2021. We take two samples, one is real data for infected individuals and the other is the hypothetical data which is called imaginary in there for the possible infected individuals based on the average reproduction number of novel coronavirus (Table 5.1). We take the reproduction number as 1755 which is found to be as the confidence intervals of the reproduction number (R) values calculated for Turkey range from 1.30 (lowest) to 2.21 (highest) [16]. We have six variables in this context such as $P(F)_{\text{real}}$, $P(T)_{\text{real}}$, and $P(I)_{\text{real}}$ and $P(F)_{\text{imaginary}}$, $P(T)_{\text{imaginary}}$, and $P(I)_{\text{imaginary}}$. It should be noted that the observed value is labeled as $P(F)$ or false because, in reality, the real number of infected individuals may be higher than the actual cases because of not reaching all the samples by tests or determining each infected individual in a similar interval. Therefore, the ideal cases based on R values are labeled as $P(R)$ in this respect although they might not reflect the actual cases. However, this labeling is just a symbolic interpretation and does not affect the calculations.

Based on these definitions, we use curve fitting by using MATLAB for the values of $P(F)$ real, $P(T)_{\text{real}}$, and $P(I)_{\text{real}}$ and $P(F)_{\text{imaginary}}$, $P(T)_{\text{imaginary}}$, and $P(I)$*imaginary*.

Many models have been created, with some of them taking into consideration characteristics such as birth and death processes, age structure, vaccination, and so on. It has been discovered by mathematical models of epidemic processes that there is a shift from the healthy to the epidemic regime, as seen in the figure below. An order parameter transition, such as the proportion of infected individuals in a particular system, happens at the critical point of a control parameter, such as a critical basic reproduction number in a system (Fig. 5.5). This trend exemplifies the inherent relationship that exists between phase transitions in statistical physics and epidemic transitions in infectious diseases. Further research revealed that there are first-order transitions in other spreading models as well, which are especially detrimental because even little fluctuations in the control parameter may produce catastrophic transitions in the order parameter [17].

Let us now consider probabilities $p > p_c$, where we will always find a spanning cluster (for sufficiently large systems). With increasing p, the order parameter approaches a finite value and grows in proportion to p. When analyzing the behavior of the order parameter, we find that $P(p)$ behaves like a power law in the region close to the percolation threshold

$$P(p) \propto (p - p_c)^{\beta},$$

with a dimension-dependent exponent β. This exponent neither depends on the type of lattice nor the detailed connectivity rule (bond or site percolation) and is therefore considered "universal" [18].

TABLE 5.1 The formulas for the interpretation of the analysis at firsthand.

Data type	Variable	Formula
Real	$P(F)_{real}$	$\frac{\text{Observed Infected Individuals}}{\text{Observed Cumulative Sample}}$
	$P(T)_{real}$	$\frac{\text{Expected Infected Individuals based on R value}}{\text{Observed Cumulative Sample}}$
	$P(I)_{real}$	$\frac{\lvert\text{Observed Infected Individuals} - \text{Expected Infected Individuals based on R value}\rvert}{\text{Observed Cumulative Sample}}$
Hypothetical	$P(F)_{imaginary}$	$\frac{\text{Observed Infected Individuals}}{\text{Expected Cumulative Sample}}$
	$P(T)_{imaginary}$	$\frac{\text{Expected Infected Individuals based on R value}}{\text{Expected Cumulative Sample}}$
	$P(I)_{imaginary}$	$\frac{\lvert\text{Observed Infected Individuals} - \text{Expected Infected Individuals based on R value}\rvert}{\text{Expected Cumulative Sample}}$
Death rate		$\frac{\text{Observed death rate}}{\text{Observed cumulative sample of deaths}}$

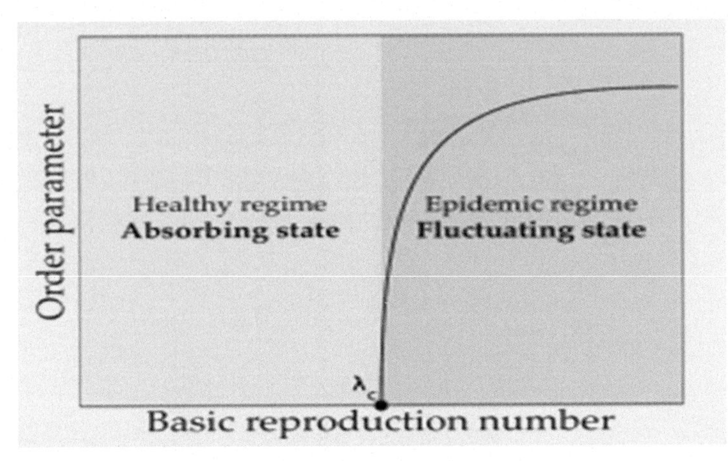

FIGURE 5.5 Epidemic phase transition. The control parameter (e.g., fraction of infected or individuals with a certain opinion) undergoes a transition from a "noninfected" state to an epidemic regime for $\lambda > \lambda c$ where λ is a control parameter such as the basic reproduction number [17].

In the fourth phase, a model was proposed for the epidemics employing a neutrosophic form of quantum decision-making model by the qutrit states. In our analysis, first we define the basis states as given which are defined as qutrit states in quantum information theory as follows:

$$|P(S)\rangle = \begin{bmatrix} 1 \\ 0 \\ 0 \end{bmatrix}, \ |P(I)\rangle = \begin{bmatrix} 0 \\ 1 \\ 0 \end{bmatrix}, \ |P(F)\rangle = \begin{bmatrix} 0 \\ 0 \\ 1 \end{bmatrix},$$

where $P(S)$ stands for the basis of the chance a particular trial results in getting a positive result of COVID-19 and other words being infected. $P(F)$ stands for the chance a particular trial results in being not infected. Finally, $P(I)$ = the chance a particular trial results in an indeterminacy regarding being infected or not. We can consider those states in a three-dimensional grid because our analysis will be in these three-dimensional grids (Fig. 5.6).

Therefore, we define the quantum version of this as a probability ket vector as

$$\alpha|P(S)\rangle + \beta|P(I)\rangle + \delta|P(F)\rangle = |P\rangle$$

In the first case, we will make our analysis in the case of

$$[\langle P|P\rangle = \|\alpha\|^2 + \|\beta\|^2 + \|\delta\|^2] = 1$$

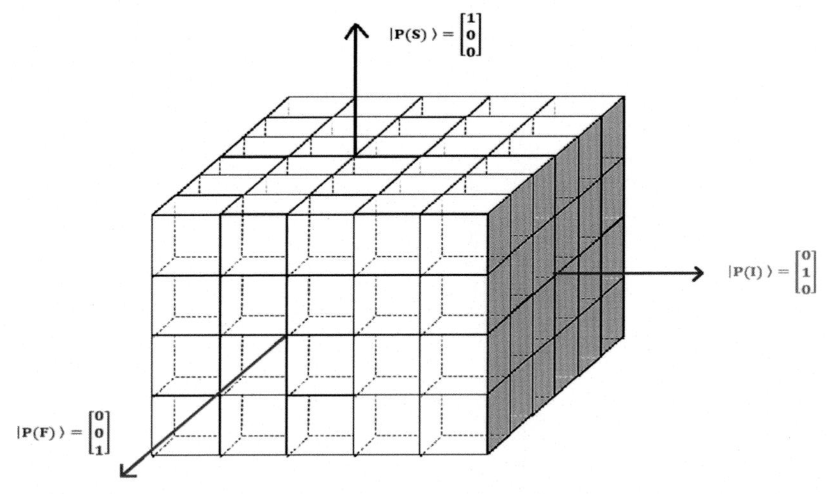

FIGURE 5.6 Three-dimensional grid or rectangular lattice for percolation analysis and relevant vectors as a phase space rather than physical space.

where there is complete information (probability) of the stateswhere $P(S) = |\langle P(S)|P\rangle|^2$, $P(L) = |\langle P(L)|P\rangle|^2$, and $P(L) = |\langle P(S)|P\rangle|^2$ so that the normalized values can be achieved as

$$\widetilde{P(S)} = \frac{|P(S)|P|^2}{\sqrt{|P(S)|P|^2 + |P(I)|P|^2 + |P(F)|P|^2}}$$

$$\widetilde{P(F)} = \frac{|P(F)|P|^2}{\sqrt{|P(S)|P|^2 + |P(I)|P|^2 + |P(F)|P|^2}}$$

$$\widetilde{P(I)} = \frac{|P(I)|P|^2}{\sqrt{|P(S)|P|^2 + |P(I)|P|^2 + |P(F)|P|^2}}$$

Let us focus on an $M \times N$ rectangular form of lattice for the preliminary definitions of the analysis

1. If a region D given that it is a region encompassing the cells in $M \times N$ $(D{:}D \in (M \times N))$, there is only one of those states as $P(S), P(F), P(I)$, and this region can be defined as controllable region symbolized as a set C (Fig. 5.7):

$$D : D \in (MXN) \rightarrow [\exists! P(S) \in (MXN)] \vee [\exists! P(I) \in (MXN)] \vee [\exists! P(F) \in (MXN)] \leftrightarrow D \in C.$$

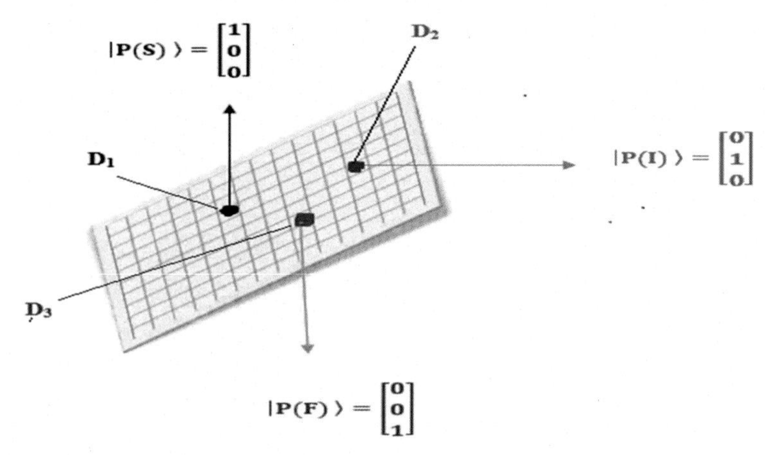

FIGURE 5.7 Illustration of some controllable states in a volume $M \times N \times R$.

2. If a region D given that it is a region encompassing the cells in $M \times N$ $(D{:}D \in (M \times N))$, there are only the intersection of two states of as $P(S), P(F), P(I)$, and this region can be defined as an uncertain region symbolized as a set U:

$$D{:}D \in (MXN) \rightarrow [\exists! P(S) \cap P(F) \in (MXN) \wedge [\exists! P(S) \cap P(I) \in (MXN)]$$
$$\wedge [\exists! P(I) \cap P(F) \in (MXN)] \leftrightarrow D \in U$$

so that we define the vectors as follows[2] (Fig. 5.8):

a. The uncertain region where the probability is $\alpha|P(S)\rangle + \beta|P(I)\rangle = |P\rangle$ and the resulting vector $\alpha|P(F)\rangle \otimes \beta|P(S)\rangle = |\text{Surprising}\rangle$ state
b. The uncertain region where the probability is $\alpha|P(S)\rangle + \beta|P(F)\rangle = |P\rangle$ and the resulting vector $\alpha|P(S)\rangle \otimes \delta|P(F)\rangle = |\text{Unexpected}\rangle$ state
c. The uncertain region where the probability is $\alpha|P(I)\rangle + \beta|P(S)\rangle = |P\rangle$ and the resulting vector $\beta|P(I)\rangle \otimes \alpha|P(S)\rangle = |\text{Dubious}\rangle$ state
d. The uncertain region where the probability is $\alpha|P(S)\rangle + \beta|P(I)\rangle = |P\rangle$ and the resulting vector $\alpha|P(S)\rangle \otimes \beta|P(I)\rangle = |\text{Risky}\rangle$ state
e. The uncertain region where the probability is $\alpha|P(F)\rangle + \beta|P(I)\rangle = |P\rangle$ and the resulting vector $\delta|P(F)\rangle \otimes \beta|P(I)\rangle = |\text{Ambiguous}\rangle$ state
f. The uncertain region where the probability is $\alpha|P(I)\rangle + \beta|P(F)\rangle = |P\rangle$ and the resulting vector $\beta|P(I)\rangle \otimes \delta|P(F)\rangle = |\text{Erroneous}\rangle$ state

---

2. The original quantum logic approach defined the meet (A ∧ B) of two events as the intersection of the two subspaces, and the join (A ∨ B) as the span of the two subspaces;. This makes sense when the events are compatible, but not when they are incompatible [4].

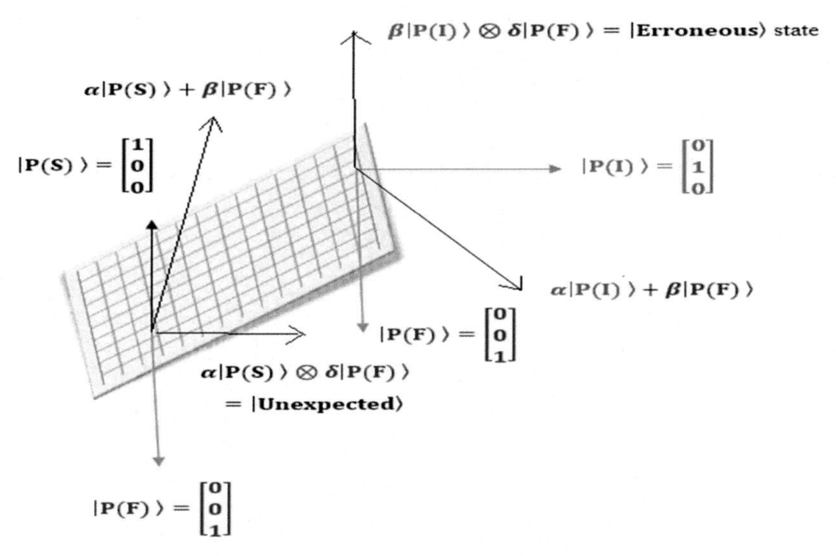

FIGURE 5.8 Illustration of some uncertain states in a volume M × N × R.

3. If a region D given that it is a region encompassing the cells in $M \times N$ $(D{:}D \in (M \times N))$, there are only the intersection of three states of as $P(S), P(F), P(I)$, and this region can be defined as a chaotic region symbolized as a set CH:

$$D{:}D \in (MXN) \rightarrow [\exists! P(S) \cap P(F) \cap P(I) \in (MXN)] \leftrightarrow D \in CH$$

Finally based on this model, the probability values as $(F)_{\text{real}}$, $P(T)_{\text{real}}$, and $P(I)_{\text{real}}$ and $P(F)_{\text{imaginary}}$, $P(T)_{\text{imaginary}}$, and $P(I)_{\text{imaginary}}$ are reduced into the doomsday calculation. In this analysis, our probability values are derived from the doomsday calculation. The doomsday argument (DA) is a probabilistic argument that purports to forecast the possible range of any future event based on an estimate of the total time passed so far. It is based on the idea of Copernican assumption assuming that there is nothing at all special about this moment [19]. Poundstone in [20] stated that:

> You may recognize this as common sense. If you met someone five days ago, it wouldn't be surprising for the affair to be over five days from now. It's too early for a tattoo or a deposit on a beach house for next summer. You may find this kind of estimation amusing or depressing or both. But the real question is, how accurate should we expect such estimates to be? Gott realized that you don't need fancy math to calculate that. All it takes is a diagram you can sketch on a napkin.

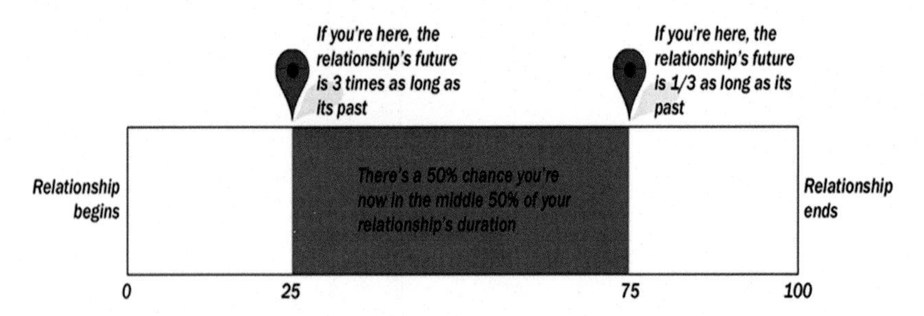

FIGURE 5.9 Boxplot.

Draw a horizontal bar representing your love affair's duration in time. Think of it as the scroll bar of a movie. The relationship's beginning is on the left, and its end is on the right. Since no one knows how long love will last, we cannot mark the bar in hours, days, or years. Instead, we will mark it in percentage points. The relationship's beginning is at 0%, and its end is at 100% (however long that is in real time). The present moment must fall somewhere between 0 and 100%, but we do not know where (Fig. 5.9)

> Still with me?
>
> I have shaded half the bar. It's the middle half, running from 25% to 75%. The present moment can be represented by a map pin ("You are here"). We'll assume it's equally likely to fall anywhere along the bar's length. That could be in the shaded part or the unshaded part. But because the shaded region is exactly 50% of the bar, we can say that the odds are 50:50 that the current moment falls within the shaded part.

Such an analysis is used in the statistics by using boxplot diagrams. An example of a boxplot is a method of depicting the distribution of data that is based on a five-number summary (the "minimum," the "first quartile (Q1)," the median, the third quartile (Q3), and the "maximum"). It may provide you with information about your outliers and the values they hold. It can also tell you whether or not your data are symmetrical, how closely your data are packed, and whether or not your data are skewed and in what direction (Fig. 5.10) [21].

At the end of the analysis, the median values in the boxplot are the used quantum decision-making model by the qutrit states.

5.5 Proposed solution

Many analyses regard the idea of potentiality to be crucial to the concept of risk. We use the term "risk" to describe situations in which bad occurrences are

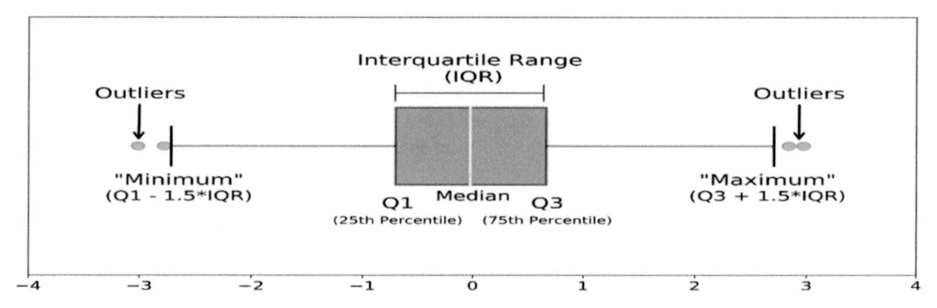

FIGURE 5.10 Different parts of a boxplot [21].

possible but do not occur. This idea is referred to by the term "possibility," which has been identified. In terms of their semantic link, the words "risk" and "possibility" might be regarded as a kind of hyponym for one another. Among the potential events that can happen, that is, the possibilities, some are wanted and others are unwanted. Finally, the term risk semantically refers to the probability of an unwanted event that may or may not occur, as in "the risk of a nuclear meltdown is one in a million." Probability is a measure of potentiality and uncertainty [21]. In this respect, we must define what kind of uncertainty is related to the concept of risk. To do this, we must classify the uncertain cases in terms of vectorial forms based on probabilities. The probabilistic forms that we use in this form are the neutrosophic definition of truth, indeterminacy, and falsity membership probability functions as base states. Such a classification is important because human cognition is situated; that is, it exhibits a vectorial form of behavior. In this respect, we first define real and imaginary probability functions based on real COVID-19 data. Then, we use probabilistic indeterminacy functions in real and imaginary forms to use as a criterion to understand the degree of risk in COVID-19 cases. We use the death rate probability function to assess this situation. From this perspective, if the death rate probability function exceeds the absolute difference between $|P(I)$ real—$P(I)$imaginary$|$, we label these zones as risk zones and conclude that there are some risky cases in uncertain situations in the COVID-19 cases. Second, we used real and hypothetical data to make a percolation model in terms of time to determine the critical day for the percolation of the disease. Finally, we use boxplots by assuming a doomsday argument to find the probability vector of uncertain situations.

5.6 Analysis

5.6.1 Analysis of the real and hypothetical data

The graph of $P(F)_{\text{real}}$, $P(T)_{\text{real}}$, and $P(I)_{\text{real}}$ and $P(F)_{\text{imaginary}}$, $P(T)_{\text{imaginary}}$, and $P(I)_{\text{imaginary}}$ values is shown in Fig. 5.11.

It is found that there is a significant difference between the death rate probability and the absolute difference between $\left|P(I)_{\text{real}} - P(I)_{\text{imaginary}}\right|$ in favor of death rates in terms of the Mann−Whitney U test (Fig. 5.12 and Table 5.2).

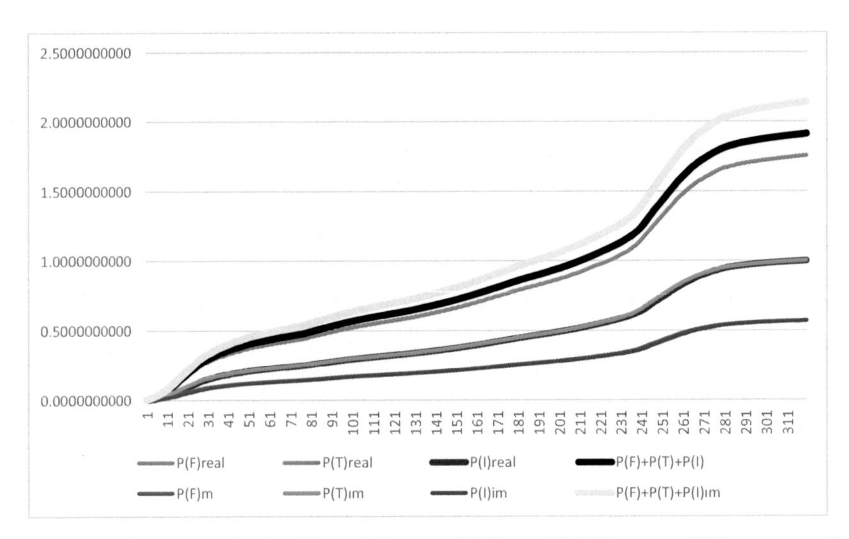

FIGURE 5.11 The graph of $P(F)_{\text{real}}$, $P(T)_{\text{real}}$, and $P(I)_{\text{real}}$ and $P(F)_{\text{imaginary}}$, $P(T)_{\text{imaginary}}$, and $P(I)_{\text{imaginary}}$ values.

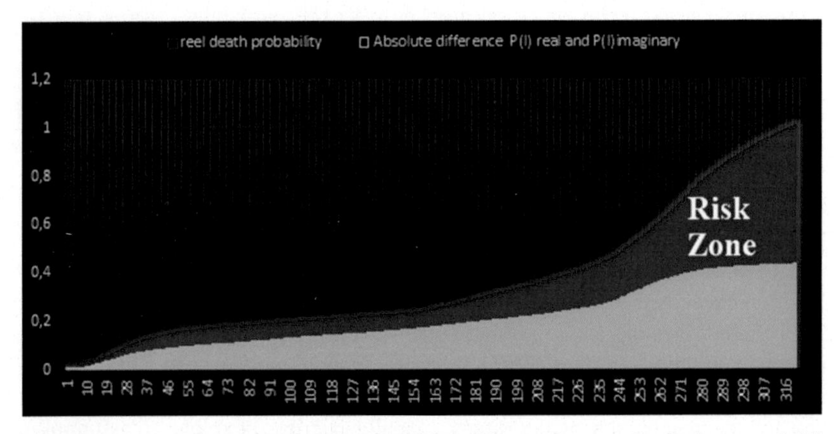

FIGURE 5.12 It is found that there is a significant difference between the death rate probability and the absolute difference between $|P(I)_{\text{real}} - P(I)_{\text{imaginary}}|$ in favor of death rates.

The general model can be found as follows by using the curve fitting tool in MATLAB:

$$f(x, y) = a + b \times \sin(m \times \pi \times x \times y) + c \times \exp(-(w \times y)^\wedge 2)$$

where x stands for $P(F)_{\text{real}}$, y stands for $P(T)_{\text{real}}$, z stands for $P(I)_{\text{real}}$, and w as weights that are the cumulative death rates ratio over the observed sum of

TABLE 5.2 Mann–Whitney U test result.

Variables	*N*	Mean rank	Sum of ranks	Test statistics	
Death rate	320	376,35	120431,00	Mann–Whitney U	33329,000
Difference	320	264,65	84689,00	Wilcoxon W	84689,000
Total	640			Z	−7,641
				Asymp. Sig. (two-tailed)	000

sample with the coefficients (with 95% confidence bounds) as follows with necessary coefficients fitting the curve of the data (Fig. 5.13);

$$a = -15.11(-5.591e+06,\ 5.591e+06); b = -0.293(-0.3417,\ -0.2444);$$

$$c = 15.77(-5.591e+06,\ 5.591e+06); m = 0.9661(0.9283,\ 1.004)$$

$$w = 0.00015(-28.03,\ 28.03)$$

Goodness of fit for this equation is SSE: 0.1106; R-square: 0.6875; adjusted R-square: 0.6835; RMSE: 0.01874

5.6.2 Curve fitting for the percolation values of the real and hypothesized data based on a modified version of electrical percolation in reverse micelles systems

Static percolation fitting for real values based on modified version of electrical percolation in reverse micelles systems[3]

$$f(x) = a^*(350-x)^\wedge(-b)$$

Coefficients (with 95% confidence bounds)[4]:

$$a1 = 1.004e+07 + i_{19}\epsilon(8.604e+06, 1.147e+07)$$

$$a2 = 0.6939 + i_{20}\epsilon(0.662, 0.7258)$$

3. Goodness of fit: SSE: 1.915e + 11 R-square: 0.9838 Adjusted R-square: 0.9837 RMSE: 2.458e + 04.
4. Goodness of fit: SSE: 1.754e + 12 R-square: 0.8525 Adjusted R-square: 0.8521 RMSE: 7.426e + 04.

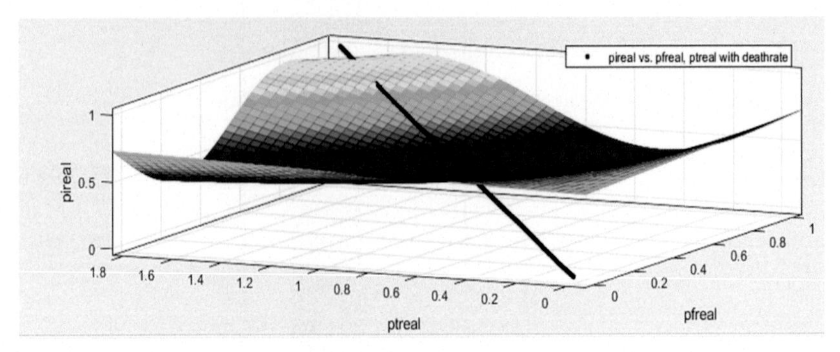

FIGURE 5.13 Curve fitting for $P(F)_{\text{real}}$ versus $P(T)_{\text{real}}$, $P(I)_{\text{real}}$ with the weights that are the cumulative death rates.

Static percolation fitting for hypothesized values based on a modified version of electrical percolation in reverse micelles systems[5]:

General model:

$$f(x) = a^*(350-x)^\wedge(-b)$$

Coefficients (with 95% confidence bounds) that can be given as neutrosophic numbers:

$$a = 1.454e + 07 + i_{21}\epsilon(1.2e + 07, 1.708e + 07)$$

$$b = 0.6531 + i_{22}\epsilon(0.6146, 0.6917)$$

5.7 Use cases

First, we create COVID-19 data based on real values and estimated R (reproduction number) for Turkey. Second, we create neutrosophic forms of real and hypothetical probabilistic data. As a third step, we apply the Mann–Whitney U test to infer whether there is a real danger or risk in our values. It should be noted that it is not a necessary step, and one can skip this step and only find a probability vector as well. We also make a percolation model for the real and hypothetical probability functions. Finally, we define the probability vector based on the median values of the boundaries of Q1 and Q3 in the boxplot so that we use the richer definition of uncertainty in the COVID-19 risk analysis (Fig. 5.14).

5. Goodness of fit: SSE: 7.869e + 12 R-square: 0.7852 Adjusted R-square: 0.7846 RMSE: 1.573e + 05.

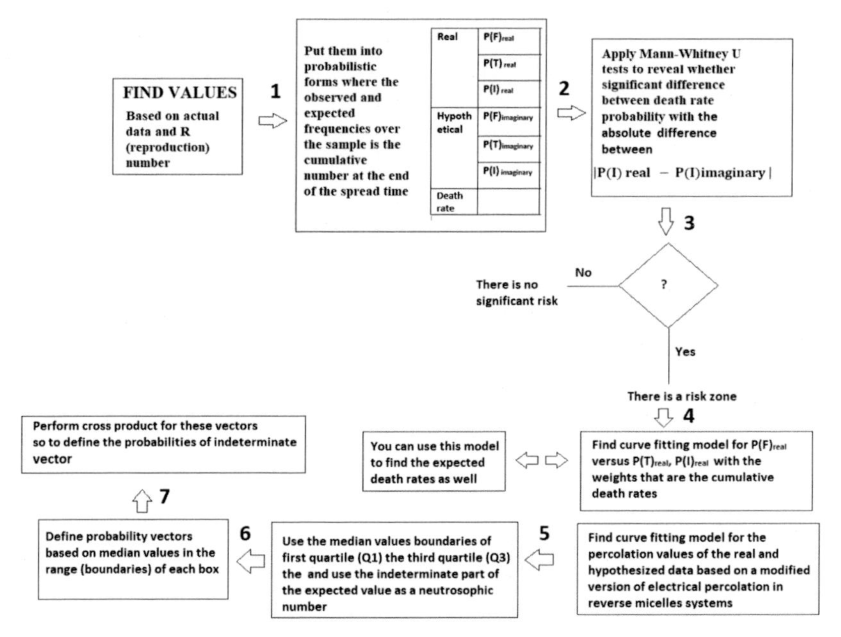

FIGURE 5.14 Strategy for the solution.

5.8 Applications

5.8.1 Analysis based on model proposition: epidemics employing neutrosophic form of percolation theory based on quantum decision-making model by the qutrit states

In the analysis, we assume that the median values between Q_1 and Q_3 boundaries are the most expected outcomes of $P(F)_{\text{real}}$, $P(T)_{\text{real}}$ and $P(I)_{\text{real}}$ and $P(F)_{\text{imaginary}}$, $P(T)_{\text{imaginary}}$, and $P(I)_{\text{imaginary}}$ distributions. We also take the $Q1$ and $Q3$ values depicted in Fig. 5.15 as the boundaries of the median values and use the indeterminate part of the expected value as a neutrosophic number.

Therefore, the probability vector can be defined as follows based on median values in the range (boundaries) of each box.

$$\alpha_r\left|P(S)\right\rangle + \beta_r\left|P(I)\right\rangle + \delta_r\left|P(F)\right\rangle = |P_{\text{real}}\rangle$$

where the coefficients are given as

$$\alpha_r = 0.418 + i_1 \in (0.292, 0.685)$$

$$\beta_r = 0.699 + i_2 \in (0.474, 1.145)$$

$$\delta_r = 0.416 + i_3 \in (0.292, 0.639)$$

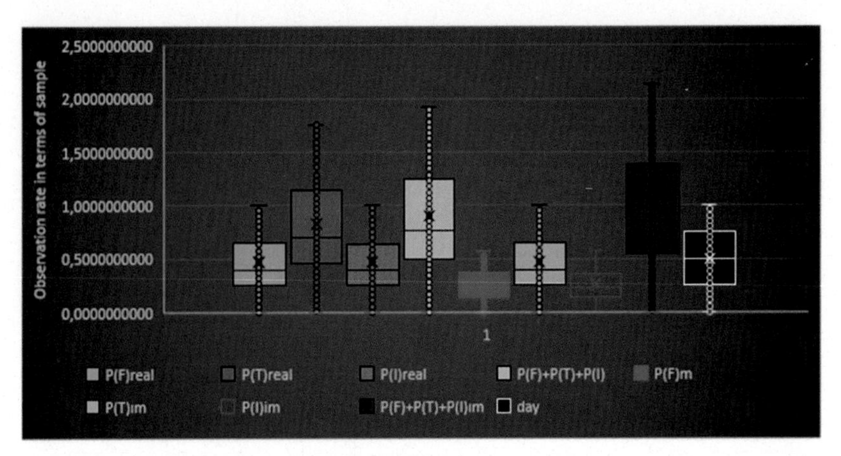

FIGURE 5.15 Boxplot graph of each probability distribution.

The imaginary probability vector can be defined as follows based on median values in the range (boundaries) of each box.

$$\alpha_i|P(S)\rangle + \beta_i|P(I)\rangle + \delta_i|P(F)\rangle = |P_{\text{imaginary}}\rangle$$

where the coefficients are given as

$$\alpha_i = 0.223 + i_4 \in (0.161, 0.378)$$

$$\beta_i = 0.418 + i_5 \in (0.262, 0.685)$$

$$\delta_i = 0.225 + i_6 \in (0.145, 0.363)$$

We can perform cross-product for these vectors as follows:

$$\begin{vmatrix} i & j & k \\ \alpha_r|\text{P(S)}\rangle & \beta_r|\text{P(I)}\rangle & \delta_r|\text{P(F)}\rangle \\ \alpha_i|\text{P(S)}\rangle & \beta_i|\text{P(I)}\rangle & \delta_i|\text{P(F)}\rangle \end{vmatrix} =$$

$$= i\begin{vmatrix} \beta_r|\text{P(I)}\rangle & \delta_r|\text{P(F)}\rangle \\ \beta_i|\text{P(I)}\rangle & \delta_i|\text{P(F)}\rangle \end{vmatrix} - j\begin{vmatrix} \alpha_r|\text{P(S)}\rangle & \delta_r|\text{P(F)}\rangle \\ \alpha_i|\text{P(S)}\rangle & \delta_i|\text{P(F)}\rangle \end{vmatrix} + k\begin{vmatrix} \alpha_r|\text{P(S)}\rangle & \beta_r|\text{P(I)}\rangle \\ \alpha_i|\text{P(S)}\rangle & \beta_i|\text{P(I)}\rangle \end{vmatrix}$$

$$= \left(\beta_r\delta_i|\text{P(I)}\rangle|\text{P(F)}\rangle - \delta_r\beta_i|\text{P(F)}\rangle|\text{P(I)}\rangle\right)\text{i}$$

$$- \left(\alpha_r\delta_i|\text{P(S)}\rangle|\text{P(F)}\rangle - \delta_r\alpha_i|\text{P(F)}|\text{P(S)}\rangle\right)\text{j} + \left(\alpha_r\beta_i|\text{P(S)}\rangle|\text{P(I)}\rangle - \beta_r\alpha_i|\text{P(I)}\rangle|\text{P(S)}\rangle\right)\text{k}$$

We can label the states to reduce the complexity of the notations based on the definitions in Fig. 5.3:

$$(((0.418 + i_1 \in (0.292, 0.685)^*(0.418 + i_5 \in (0.262, 0.685))|\text{Risky}\rangle -$$

$$(0.699 + i_2 \in (0.474.1.145))^*(0.223 + i_4 \in (0.161, 0.378)|\text{Dubious}\rangle)$$

Hence, we can find the results for each axis as follows:
For i axis:

$$\beta_r\delta_i|\text{Erroneous}\rangle - \delta_r\beta_i|\text{Ambiguous}\rangle = (0.699 + i_2 \in (0.474, 1.145))^*$$
$$(0.225 + i_6 \in (0.145, 0.363)) - (0.416 + i_3 \in (0.292, 0.639))$$
$$^*(0.418 + i_5 \in (0.262, 0.685))$$
$$= [(0.157 + I(0.449))]|\text{Erroneous}\rangle - [0.173 + \text{I}(0.476)]|\text{Ambiguos}\rangle$$

For j axis

$$(-\alpha_r\delta_i|\text{Unexpected}\rangle + \delta_r\alpha_i|\text{Surprising}\rangle)j = -(0.418 + i_1 \in (0.292, 0.685))$$
$$^*(0.225 + i_6 \in (0.145, 0.363)) + (0.416 + i_3 \in (0.292, 0.639))^*$$
$$(0.223 + i_4 \in (0.161, 0.378))$$
$$= -[(0.418 + \text{I}0.393)^*(0.225 + \text{I}(0.218))]|\text{Unexpected}\rangle + (0.416 + \text{I}0.347)^*$$
$$(0.223 + \text{I}0.217)|\text{Suprising}\rangle$$
$$= -[0.107 + \text{I}0.265]|\text{Unexpected}\rangle + (0.092 + \text{I}0.242)|\text{Suprising}\rangle$$

For k axis:

$$((\alpha_r\beta_i|\text{Risky}\rangle - \beta_r\alpha_i|\text{Dubious}\rangle) =$$
$$(((0.418 + i_1 \in (0.292, 0.685)^*(0.418 + i_5 \in (0.262, 0.685))|\text{Risky}\rangle$$
$$-(0.699 + i_2 \in (0.474.1.145))^*(0.223 + i_4 \in (0.161, 0.378)|\text{Dubious}\rangle)$$
$$= (0.418 + \text{I}(0.393))^*(0.418 + \text{I}(0.423))|\text{Risky}\rangle$$
$$-((0.699 + \text{I}0.671)^*(0.223 + I(0.217))|\text{Dubious}\rangle)$$
$$= (0.174 + \text{I}0.507)|\text{Risky}\rangle - (0.155 + \text{I}0.446)|\text{Dubious}\rangle$$

5.9 Discussion

In this paper, we try to make use of neutrosophic COVID-19 data and convert them into vector form to make semantically more explicit definition of the uncertain situation in the data. However, it should be noted that more rich semantic forms can also be created and used for the descriptive analysis of the COVID-19 data. Additionally, we focus on the Turkish sample as a whole, but we can focus on different regions or cities to get more detailed descriptions of the COVID-19 data.

5.10 Conclusions

In this study, we attempt to broaden the definition of risk by assuming the most probable conditions of the cases in the context of neutrosophic logic as T, F, and

I values. In the context of quantum decision-making theory, we refer to these three states as vectorial forms, and we discuss them in detail. Because of the non-commutative character of such states, we may generate a variety of distinct outputs, which allows us to extend the meaning of the risk in a larger area. Mostly, descriptive statistics and analysis are based on scalar data. Visualizations are also generally based on scalar data. However, sometimes, descriptive analysis for situations that are more related to cognitive and semantic characteristics might give more deep and detailed results than descriptive analysis based on scalar data. Therefore, we show that neutrosophic and vectorial forms of probability functions can be used for the analysis of COVID-19 spreads. Additionally, we show that percolation theory can be used in a neutrosophic form to model the spread of COVID-19. It is common practice to conduct studies on societal understandings of risk and safety without first establishing a clear understanding of the conceptual underpinnings of the "risk" phenomenon under investigation. Researchers who investigate how a given phenomenon (in this case, COVID-19) is connected with risk by society actors may find themselves in a dilemma about the true foundation of the connection [22]. The analysis presented enables clarification of these issues by proposing a mathematical semantic model for the descriptive analysis of COVID-19 spread.

The second contribution of this study is to combine the likelihood of an occurrence with the characterization of the cognitive states that correlate to that likelihood of an event. The idea of risk encompasses not only the likelihood of an event occurring, but also the cognitive processes characterized by many adjectives such as "erroneous," "dubious," "ambiguous," "hazy," "surprising," "unexpected," and "unexpected." The association between those two dimensions in the context of semantics and probability is not established, but we attempt to demonstrate how those states may be absorbed into each other in the context of semiotics and neutrosophic logic via the use of a mathematical model. Third, this study makes a significant addition by encapsulating percolation theory in the context of the analysis of COVID situations. In order to comprehend disease transmission patterns on temporal mobility networks, it is necessary to understand percolation theory. Despite the fact that we did not utilize the percolation theory as our primary tool, we constructed actual and hypothetical data to illustrate the potential danger zones in the setting of inadequate knowledge, in order to get feasible projection zones for the risky scenarios.

5.11 Outlook and future work

Generative adversarial networks can be used to predict the percolation of phase spaces of COVID-19 infections. Generic adversarial networks are based on the principle of training two networks: a generator network G that attempts to make a video and a discriminator network D that attempts to discern between "genuine-" and "fake-"produced data (Fig. 5.16). These networks may be trained against each other in a min−max game, in which the

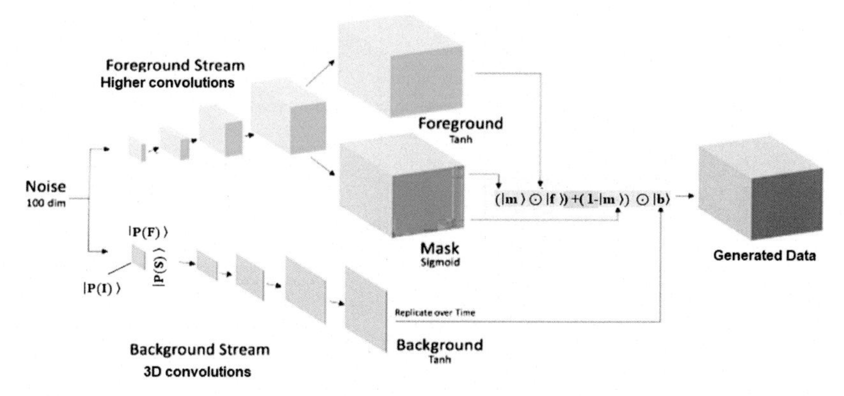

FIGURE 5.16 The modified version of the video generator network proposed in [23].

generator tries to mislead the discriminator as much as possible while the discriminator tries to figure out which samples are fake. The model learns without supervision to generate these pathways such that, when they are combined, the video looks real. Therefore, such a model can be used for predicting plausible futures of static data [23].

References

[1] F. Smarandache, B. Said, Neutrosophic theories in communication, Management and Information Technology, NOVA Science Publisher, 2020.

[2] J.M. Yearsleya, J.R. Busemeyer, Quantum cognition and decision theories: a tutorial, J. Math. Psychol. (2015).

[3] B.E. Baaquıe, Quantum Field Theory for Economics and Finance, Cambridge University Press, 2018.

[4] D. McMahon, Quantum Computing Explained, John Wiley & Sons, 2008.

[5] J.F. Kihlstrom, L. Park, Cognitive psychology: overview, reference module in neuroscience and biobehavioral psychology (2018). Available from: https://doi.org/10.1016/b978-0-12-809324-5.21702-1.

[6] https://www.washingtonpost.com/national/health-science/there-really-are-50-eskimo-words-for-snow/2013/01/14/e0e3f4e0-59a0-11e2-beee-6e38f5215402_story.html.

[7] J. Casti, X-Events: The Collapse of Everything, Harper Collins, 2013.

[8] H. Haoyu, D. Hengfang, W. Qi, G. Jianxi, Percolation of temporal hierarchical mobility networks during COVID-19. Phil. Trans. R. Soc. A. 380 (2022).

[9] F. Smarandache, Introduction to Neutrosophic Statistics, Sitech & Education Publishing, 2014.

[10] J.R. Busemeyer, Quantum models of cognition and decision, http://www.physics.smu.edu/web/seminars/seminar_archive/Spring2016/busemeyer.pdf. Retrieved from 24.10.21.

[11] J.R. Busemeyer, P.D. Bruza, Quantum Models of Cognition and Decision, Cambridge University Press, Cambridge, 2012.

[12] A. Łukasik, Quantum models of cognition and decision, Int. J. Parallel. Emergent. Distrib. Syst. 33 (3) (2018) 336–345. Available from: https://doi.org/10.1080/17445760.2017.1410547.

[13] S. Webb, If the universe is teeming with aliens...where is everybody? Fifty Solutions to the Fermi Paradox and the Problem of Extraterrestrial Life, Copernicus Books, New York, 2003.
[14] G.A. Landis, The Fermi Paradox: an approach based on percolation theory, J. Br. Interplanet. Soc. 51 (1998) 163–166. Available from: http://www.geoffreylandis.com/percolation.htp. Retrieved from 24.10.2021.
[15] M. Guettari, A. Aferni, Propagation Analysis of the Coronavirus Pandemic on the Light of the Percolation Theory [Online First], IntechOpen, 2021. Available from http://doi.org/10.5772/intechopen.97772, https://www.intechopen.com/online-first/76669.
[16] S.K. Köse, E. Demir, G. Aydoğdu, Estimation of the time dependent reproduction number of novel coronavirus (COVID 19) for Turkey in the late stage of the outbreak, Türkiye Klinikleri Biyoistatistik Derg. 13 (2021) 1.
[17] L. Böttcher, Epidemic processes. A dissertation submitted to attain the degree of doctor of sciences of eth Zurich (Dr. sc. ETH Zurich), Germany (2018).
[18] L. Böttcher, H.J. Herrmann, Computational Statistical Physics, Cambridge University Press, 2021.
[19] J.R. Gott, Time Travel in Einstein's Universe: The Physical Possibilities of Travel Through Time, Houghton Mifflin Harcourt, 2015.
[20] W. Poundstone, The doomsday calculation: how an equation that predicts the future is transforming everything we know about life and the universe, Little, Brown Spark (2019). London.
[21] Website: https://towardsdatascience.com/understanding-boxplots-5e2df7bcbd51.
[22] M. Boholm, The semantic field of risk, Saf. Sci. 92 (2017) 205–216.
[23] C. Vondrick, H. Pirsiavash, A. Torralba, Generating videos with scene dynamics, in: 29th Conference on Neural Information Processing Systems (NIPS), Barcelona, Spain, 2016. http://www.cs.columbia.edu/~vondrick/tinyvideo/paper.pdf (accessed 31.10.21).

Chapter 6

Use of neutrosophic statistics to analyze the measured data of diabetes

Usama Afzal[1] and Muhammad Aslam[2]

[1]School of Microelectronics, Tianjin University, Tianjin, P.R. China, [2]Department of Statistics, Faculty of Science, King Abdulaziz University, Jeddah, Saudi Arabia

6.1 Introduction and motivation

Diabetes is a long-lasting disease which is due to the increase in the sugar level in the blood of the body. Diabetes is expressed by a blue circle as shown in Fig. 6.1 [1].

As a person eats the meal, this meal (containing starch) reacts with water and produces glucose which increases the sugar level in the human body. The process is expressed in Fig. 6.2 [2].

There are two reasons for diabetes increase in the human body; first if the pancreas is not producing enough insulin and second if the cells of the human body are not responding properly to the insulin produced [3]. There are many symptoms of this disease like an increase in thirst, frequent urination, and an increase in appetite. Diabetes also becomes the reason for further different diseases [4]; like, it can damage human eyes, kidneys, and nerves. Generally, it is seen that diabetes is also a reason for high blood pressure [5]. There are three main kinds of diabetes [6], that is, type 1 diabetes, type 2 diabetes, and gestational diabetes, as shown in Fig. 6.3.

There are different treatments and precautions to control diabetes. For example, type 1 diabetes is treated by insulin injections. In this process, artificial insulin is injected into the human body to stop the diabetes incensement. Similarly, the treatment of type 2 diabetes includes a healthy diet, daily basis exercise, avoiding the use of tobacco, and controlling body weight [7]. Generally, it is seen that this type of diabetes is treated by different medications like insulin sensitizers containing or without insulin [6]. Similarly, gestational diabetes is treated after the baby's birth. Diabetes is a complex and long-term disease. So, numerous devices are used for the

Cognitive Intelligence with Neutrosophic Statistics in Bioinformatics.
DOI: https://doi.org/10.1016/B978-0-323-99456-9.00006-4

FIGURE 6.1 Universal symbol of diabetes.

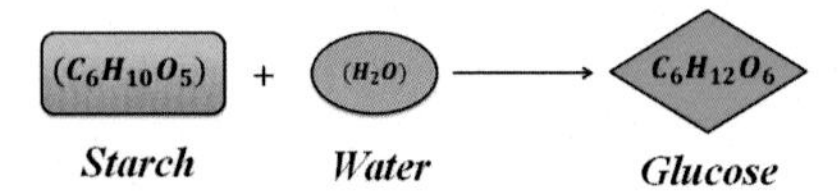

FIGURE 6.2 Conversion of starch into glucose.

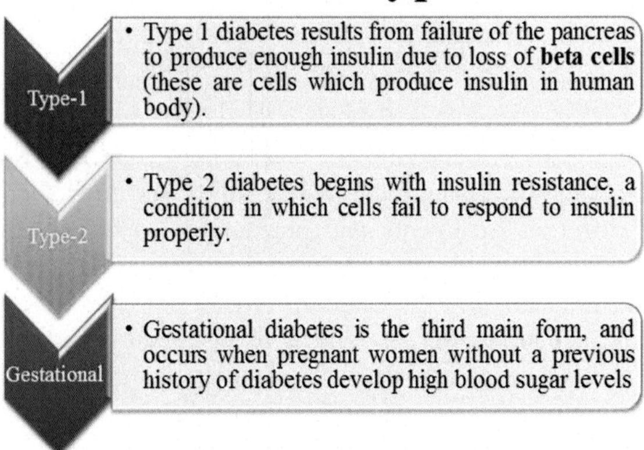

FIGURE 6.3 Types of diabetes.

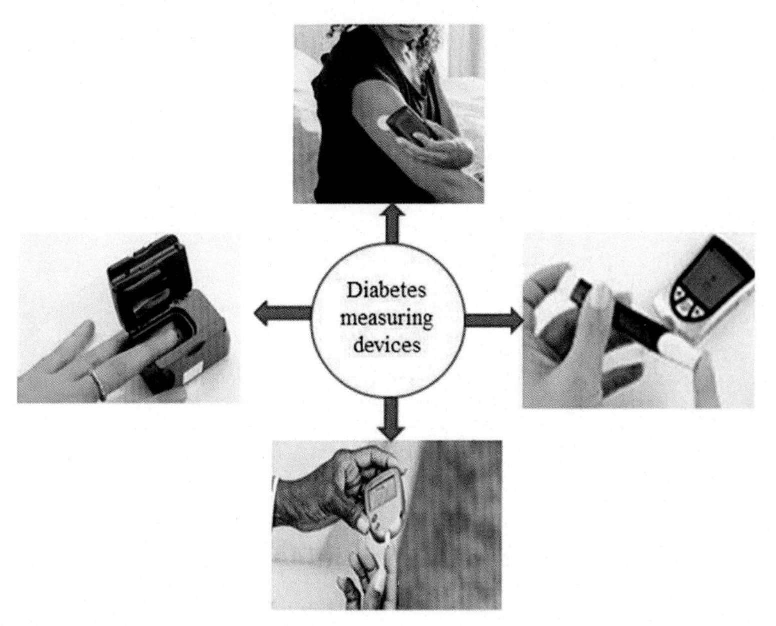

FIGURE 6.4 Diabetes measuring devices.

measurement of diabetes as shown in Fig. 6.4. There are two measuring units for diabetes [8], one is an international standard unit millimoles per liter (mmol/L), and another unit based on mass concentration is milligrams per deciliter (mg/dL).

The above figure shows different devices for measuring diabetes. Numerous measurements have been collected to study diabetes or sugar levels in the blood of humans. The main thing is not to collect the data, but the main focus is to analyze the data for getting useful information. Generally, it is seen that the analysis of data is performed by using different statistical approaches. Statistics is a significant branch of science used for the analysis of data [9] of a particular problem whether in table form or graphical form. It is an interdisciplinary field; its application can be seen in scientific research as well as others. There are a number of statistics approaches/logics for analyzing the data such as classical statistics (CS), fuzzy statistics (FS), and neutrosophic statistics (NS). Classical statistics only deals with determinate values, FS deals with only fuzzy sets, and NS deals with indeterminate value intervals. For expressing the difference between these approaches, we take an example of the statement. According to CS, a statement is only right or only wrong at a time. But in FS, the statement is between wrong and right, which means there is uncertainty. Similarly, in NS, the statement is right and wrong at the same time but with indeterminacy.

6.2 Background, definitions, and abbreviations

6.2.1 Abbreviations

- NS-Neutrosophic statistics
- FS-Fuzzy statistics
- CS-Classical statistics
- NF-Neutrosophic formula
- NM-Neutrosophic mean
- CM-Classical mean
- CAFC-Classical analysis flowchart
- NAFC-Neutrosophic analysis flowchart
- mg/dL-milligrams per deciliter
- mmol/L-millimoles per liter
- NV-Neutrosophic variable
- SL-Sugar levels

6.2.2 Background of statistical approaches

As we have already discussed, three approaches of statistics are in Section 6.1. Let us see bit of background of these approaches (but we will use CS and NS in this chapter for the analysis of diabetes measured data).

6.2.2.1 Neutrosophic statistics

Neutrosophic statistics is a more flexible statistics approach than CS and FS. The main purpose of this approach is to study and analyze the interval accurately with indeterminacy [10]. Neutrosophic statistics is more informative than classical statistics because CS only deals with determining values whereas NS deals with indeterminate values. This is a substantial benefit over the classical approach. Moreover, the neutrosophic technique is more helpful than classical techniques; see examples of ([11,12]; F [13]). Aslam [14,15] developed numerous statistical approaches/techniques under NS. The neutrosophic logic is also used for the analysis of indeterministic data. It is the generalization of fuzzy logic and is more reliable and informative than fuzzy logic [16]. Smarandache claimed it expressed the efficiency of neutrosophic logic over fuzzy logic [17]. Similarly, in 2014 Smarandache, F. proposed NS for the analysis of indeterminacy of data. Nowadays, NS has gotten importance for analysis and is used in different fields, such as applied sciences [18], medicine for diagnoses data measurement [19], astrophysics for measurement of earth speed data [20], humanistic [21], and material science to analyze the data [22].

6.2.2.2 Classical statistics

Classical statistics is generally known as frequentist statistics [23] because it consists of many ordinary formulas and approaches for the analysis of the value. As already mentioned in the above section, CS only analyzes the fixed-point values, that is, determined values, for example, the use of the classical least-square formula and classical mean formula, etc. [24]. As it deals only with fixed-point values, we work with interval values and fails in the analysis of interval values and can explain the problem properly. This is because the interval values consist of uncertainty and it is tough to deal with uncertainty through the CS. So, one can easily conclude that CS is unable to well analyze the data containing uncertainty [25].

6.2.2.3 Fuzzy statistics

A more flexible approach that deals with uncertainty is known as FS. In 1965, the fuzzy logic which is based on the fuzzy sets (that is why also called the fuzzy set theory) was developed by [26]. FS is used to analyze the data when there is uncertainty, that is, interval form. It is more flexible and accurate in dealing with the interval data having uncertainty. For understanding this concept, we take an example. In classical logic, if a number is a member of a set, then it will be "1" otherwise it will be "0", but in fuzzy, the number will be between "0" and "1", that is, [0, 1], simultaneously [27]. Different researchers have worked on the FS. For example, [28] proposed the statistical test using fuzzy logic. The purpose of FS is to use an indeterministic approach to analyze and observe the uncertainty in data. Similarly, the median analysis with the help of this statistical test was also proposed by [29]. Analysis of the interval data using signed-rank test was attempted by [30].

6.2.3 Definition of neutrosophic statistics

The neutrosophic formula for the neutrosophic variable (NV) A_N is written as:

$$A_{iN} = A_{iL} + A_{iU} I_N (i = 1, 2, 3, \ldots, n_N) \tag{6.1}$$

The NV has $A_N \in [A_L, A_U]$ interval with the size $n_N \in [n_{NL}, n_{NU}]$, and indeterminacy interval for this NV is $I_N \in [I_{NL}, I_{NU}]$.

For the above equation, $A_{iN} \in [A_{iL}, A_{iU}]$ has two parts: A_{iL} expressing the lower value which is under CS and $A_{iU} I_N$ is an upper part with indeterminacy $I_N \in [I_L, I_U]$. Similarly, the neutrosophic mean (NM) $\overline{A}_N \in [\overline{A}_L, \overline{A}_U]$ is defined as follows:

$$\overline{A}_N = \overline{A}_L - \overline{A}_U I_N I_N \in [I_L, I_U] \tag{6.2}$$

Here $\overline{A}_L = \sum_{i=1}^{nU} (A_{iL}/n_L)$ and $\overline{A}_U = \sum_{i=1}^{nU} (A_{iU}/n_U)$

Now, the neutrosophic analysis flowchart (NAFC) is shown in Fig. 6.5.

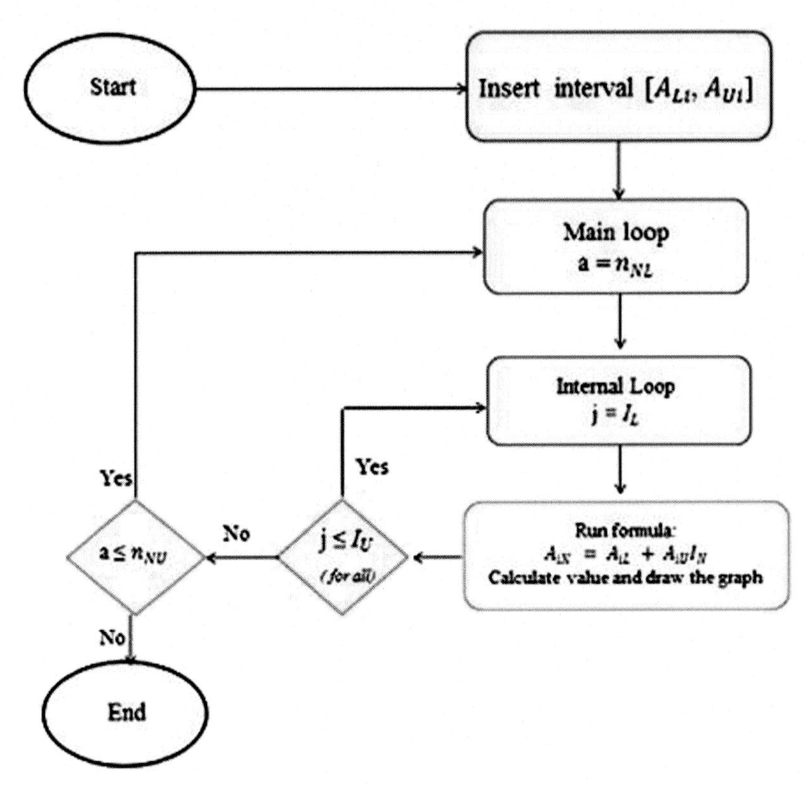

FIGURE 6.5 Neutrosophic analysis flowchart.

6.3 Literature review and state of the art

6.3.1 Previous work on diabetes

Many researchers have worked on the measurement and analysis of sugar levels in the blood of the human body. Urbina et al. [31] studied the fasting sugar level and insulin levels in the blood. They collected the data from about 216 children and young adults having the age from 13 to 27 years. A study of the correlation between fasting sugar level, insulin level, and echocardiographic LVM was performed by them. In 2001, [32] used a noninvasive method which was based on transmission line method, for measuring the sugar levels in the blood. They used millimeter wave noninvasively to study the change in sugar levels. In 2008, a health monitoring device based on a small needle to measure the sugar levels in the blood was developed by [33]. The device was based on the glucose enzyme sensor (GES). The sample of the blood was collected through a needle, and the GES was used to observe the sugar levels. Their device showed about 90% volumetric efficiency. Kumari and Mathana [34] proposed a novel chewing and swallowing (SCS) based on the diet method. They developed an algorithm to control diabetes and used an acoustic MEMS sensor to measure the sugar levels in the blood. They observed that about 50 patients controlled about 85% of diabetes by following their method. Yadav et al. [35] used the biosensor to measure and analyze diabetes. Generally, it is seen all previous researchers have used different measuring, but they all have used classical formulas to analyze data.

6.3.2 New proposed work

In this chapter, we have tried to develop a NS application in the analysis of data variance of sugar levels in the blood of the human body. A measurement of the sugar level in the blood has been performed with respect to the changes in age. Through this chapter, we have tried to analyze the variance of sugar level data, and for the first time, we have used NS for this purpose. Also, CS has been used to make a comparison between the results of both statistics types.

6.4 Problem

The purpose of this study is to focus on the application of NS in bio-information. We have used the neutrosophic formulas for the analysis of the changes in sugar levels with respect to age. We have collected the data of 150 people having diabetes issues aged from 45 to 60 years. The data have been collected through oral glucose tolerance tests. We have divided this collected data into two groups. group one, that is, "G1" data have been collected after 8 hours of fasting, and the second group, that is, "G2" after 2 hours of drinking glucose solution about 237 mm containing 75-g sugar. All data are collected at the normal body temperature. We are interested in the analysis of data variation with respect to the age of the patients. For this purpose, we have used classical and NS. Also, it is observed which one is more suitable to analyze the problem on the behalf of the method and formulas (Fig. 6.6).

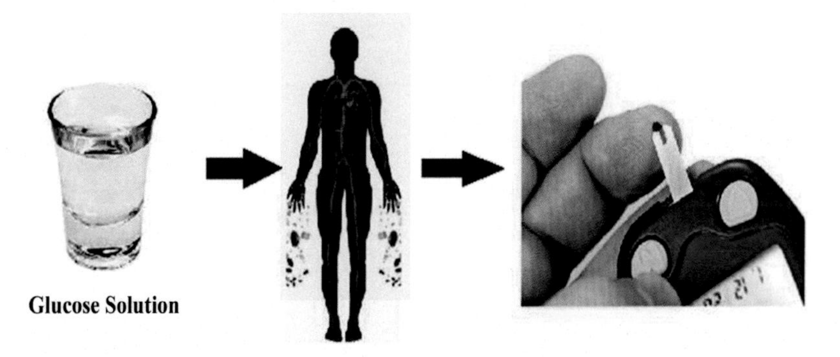

FIGURE 6.6 Diabetes measuring setup.

6.5 The proposed solution of the problem

For the understanding and analysis of the measured value, we have used CS and NS and compared their results. Their formulas and working procedure are as follows:

6.5.1 Classical statistics

As we have already described, CS only deals with the fixed-point value, that is, determined values. So, classical mean (CM) formula to convert these values into fixed-point values for CS analysis.

$$\overline{M}i = \frac{a_{Li} + a_{Ui}}{n}(i = 0, 1, 2, 3, 4\ldots N) \tag{6.3}$$

Here "$\overline{M}_i$" is the classical mean of the "ith" interval as well as a_{Li} and a_{Ui} are the lower and upper values of the interval, respectively. Also "n" is

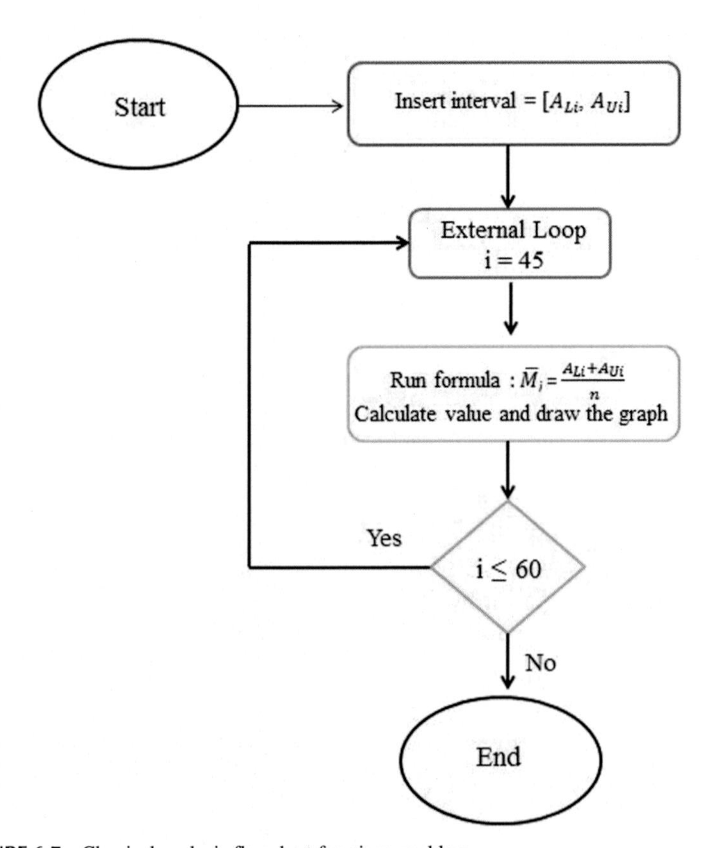

FIGURE 6.7 Classical analysis flowchart for given problem.

the number of values. For the following problem, $n = 2$ (as there are only two values of interval, that is, lower and upper) and "i" is taken as age so $i = \text{age} = 45, 46, 47, \ldots\ldots\ldots, 60$.

Now the **classical analysis flowchart** (CAFC) of the said problem is shown in Fig. 6.7.

6.5.2 Neutrosophic statistics

Our study belongs to the changes of sugar level in blood with respect to changes in age, so this factor can be written as the function of the age, that is, SL(age). Now, let SL(age)_N is measured interval value of the temperature, that is, $\text{SL(age)}_N \in [\text{SL(age)}_L, \text{SL(age)}_U]$; here SL(age)_L and SL(age)_U are the lower and upper values of the interval, respectively. The NF for temperature with indeterminacy $I_N \in [I_L, I_U]$ is written as follows:

$$\text{SL(age)}_N = \text{SL(age)}_L + \text{SL(age)}_U I_N; I_N \in [I_L, I_U] \quad (6.4)$$

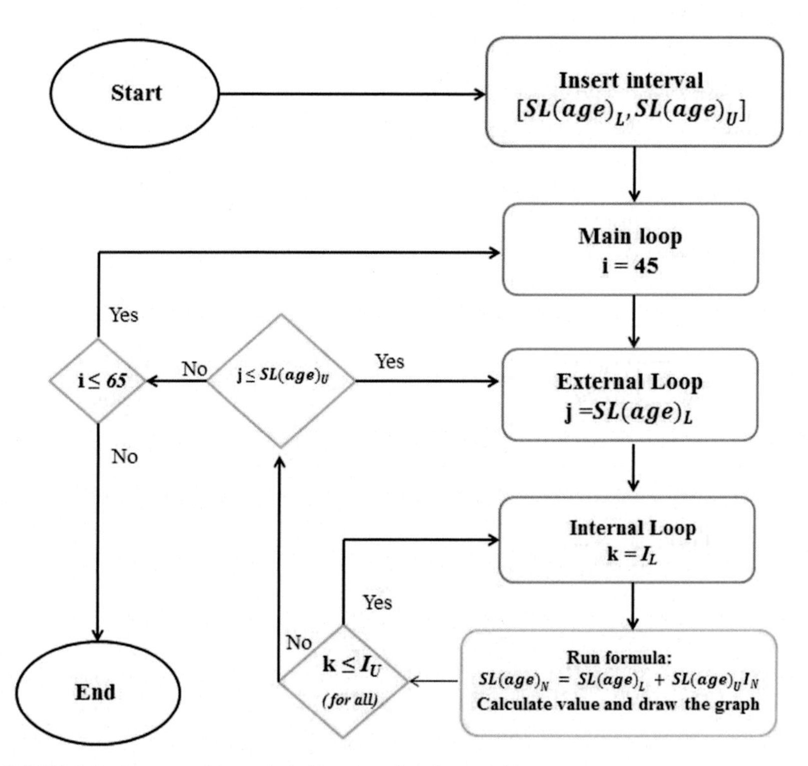

FIGURE 6.8 Neutrosophic analysis flowchart for given problem.

From the above NF of $\mathrm{SL(age)}_N \in [\mathrm{SL(age)}_L, \mathrm{SL(age)}_U]$ is an extension under the CS. The equation contains two parts, that is, $\mathrm{SL(age)}_L$ determined and $\mathrm{SL(age)}_U I_N$ indeterminate parts. Moreover, $I_N \in [I_L, I_U]$ is known as an indeterminacy interval. Also, the measured interval $\mathrm{SL(age)}_N \in [\mathrm{SL(age)}_L, \mathrm{SL(age)}_U]$ can be reduced to the classical or determined part if we choose $I_L = 0$, and I_U can find by using $(\mathrm{SL}(T)_U - \mathrm{SL}(T)_L)/\mathrm{SL}(T)_U$.

Now, the neutrosophic analysis flowchart (NAFC) of the said problem is shown in Fig. 6.8.

6.6 Analysis of data

The measured data of the sugar levels of patients with respect to age are shown in Table 6.1. This table expresses the data sugar level variance with respect to age from 45 to 60 years for both groups.

Table 6.1 shows the variation of the sugar level in the blood of the human body with respect to age (changing from 45 to 60 years). Also,

TABLE 6.1 Data of the sugar level in blood.

Age (years)	Data of "G1" (mg/dL)	Data of "G2" (mg/dL)
45	[159, 199]	[166, 206]
46	[150, 196]	[156, 202]
47	[139, 199]	[147, 207]
48	[142, 167]	[148, 173]
49	[152, 210]	[160, 218]
50	[143, 187]	[150, 194]
51	[151, 177]	[159, 185]
52	[140, 195]	[147, 207]
53	[154, 200]	[160, 206]
54	[142, 197]	[149, 204]
55	[150, 189]	[157, 196]
56	[160, 198]	[168, 206]
57	[162, 190]	[170, 198]
58	[146, 198]	[152, 204]
59	[149, 188]	[155, 194]
60	[177, 198]	[179, 205]

the average changes in sugar level are about (151−193 mg/dL) and (157.6−200.3 mg/dL).

6.7 Statistical conditions

We have used two types of statistical conditions, that is, CS and NS, for the data analysis of the said problem.

6.7.1 Classical analysis

For the classical analysis of the problem, we have used the above-said procedure as shown in Section 6.5.1. We have applied this procedure to all the measured values of the sugar level in blood as shown in Table 6.2.

6.7.2 Neutrosophic analysis

For the neutrosophic analysis of the problem, we have used the above-said procedure as expressed in Section 6.5.2. We have applied this procedure to

TABLE 6.2 Classical analysis of the data measured.

Age (years)	Data of "G1"mg/dL	Data of "G2"mg/dL
45	179	186
46	173	179
47	169	177
48	154.5	160.5
49	181	189
50	165	172
51	164	172
52	167.5	177
53	177	183
54	169.5	176.5
55	169.5	176.5
56	179	187
57	176	184
58	172	178
59	168.5	174.5
60	187.5	192

TABLE 6.3 Neutrosophic analysis of the data measured.

Age (years)	Data of "G1"mg/dL	Data of "G2"mg/dL
45	$159+199I_N$; $I_N \epsilon [0, 0.201]$	$166+206\ I_N$; $I_N \epsilon [0, 0.194]$
46	$150+196\ I_N$; $I_N \epsilon [0, 0.235]$	$156+202\ I_N$; $I_N \epsilon [0, 0.228]$
47	$139+199\ I_N$; $I_N \epsilon [0, 0.302]$	$147+207\ I_N$; $I_N \epsilon [0, 0.290]$
48	$142+167\ I_N$; $I_N \epsilon [0, 0.150]$	$148+173\ I_N$; $I_N \epsilon [0, 0.145]$
49	$152+210\ I_N$; $I_N \epsilon [0, 0.276]$	$160+218\ I_N$; $I_N \epsilon [0, 0.266]$
50	$143+187\ I_N$; $I_N \epsilon [0, 0.235]$	$150+194\ I_N$; $I_N \epsilon [0, 0.227]$
51	$151+177\ I_N$; $I_N \epsilon [0, 0.147]$	$159+185\ I_N$; $I_N \epsilon [0, 0.140]$
52	$140+195\ I_N$; $I_N \epsilon [0, 0.282]$	$147+207\ I_N$; $I_N \epsilon [0, 0.290]$
53	$154+200\ I_N$; $I_N \epsilon [0, 0.230]$	$160+206\ I_N$; $I_N \epsilon [0, 0.223]$
54	$142+197\ I_N$; $I_N \epsilon [0, 0.279]$	$149+204\ I_N$; $I_N \epsilon [0, 0.269]$
55	$150+189\ I_N$; $I_N \epsilon [0, 0.206]$	$157+196\ I_N$; $I_N \epsilon [0, 0.199]$
56	$160+198\ I_N$; $I_N \epsilon [0, 0.192]$	$168+206\ I_N$; $I_N \epsilon [0, 0.184]$
57	$162+190\ I_N$; $I_N \epsilon [0, 0.147]$	$170+198\ I_N$; $I_N \epsilon [0, 0.141]$
58	$146+198\ I_N$; $I_N \epsilon [0, 0.263]$	$152+204\ I_N$; $I_N \epsilon [0, 0.255]$
59	$149+188\ I_N$; $I_N \epsilon [0, 0.207]$	$155+194\ I_N$; $I_N \epsilon [0, 0.201]$
60	$177+198\ I_N$; $I_N \epsilon [0, 0.106]$	$179+205\ I_N$; $I_N \epsilon [0, 0.127]$

all the measured values of sugar levels in blood with their indeterminacy, respectively, as shown in Table 6.3.

6.8 Discussion

In this chapter, we have analyzed the variation of sugar levels in the human body with respect to variation of age. We have applied classical and NS to analyze the data. The graphs of the classical analysis are shown in Figs. 6.9 and 6.10, respectively. Similarly, the graphs of the neutrosophic analysis are shown in Figs. 6.11 and 6.12.

The above graphs are plotted to express the difference between the classical and neutrosophic statistical analyses of the measured data. From classical analysis graphs, it is seen that there are only fixed-point values that mean no variance or indeterminacy. The graph is not much flexible and only expresses the value of sugar level only at a specific point of age. One cannot find out more information from classical graphs. But on the other hand, neutrosophic analysis graphs are more informative as these are dealing with the

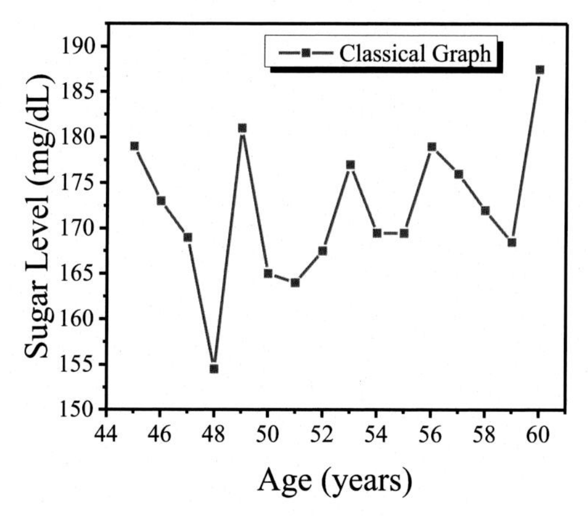

FIGURE 6.9 Classical plot of sugar level for G1.

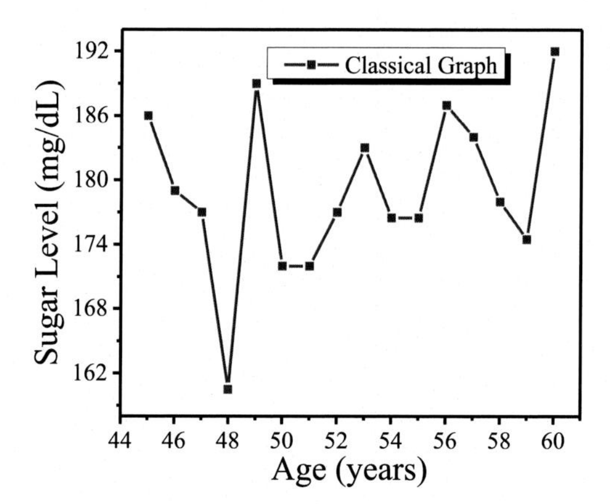

FIGURE 6.10 Classical plot of sugar level for G2.

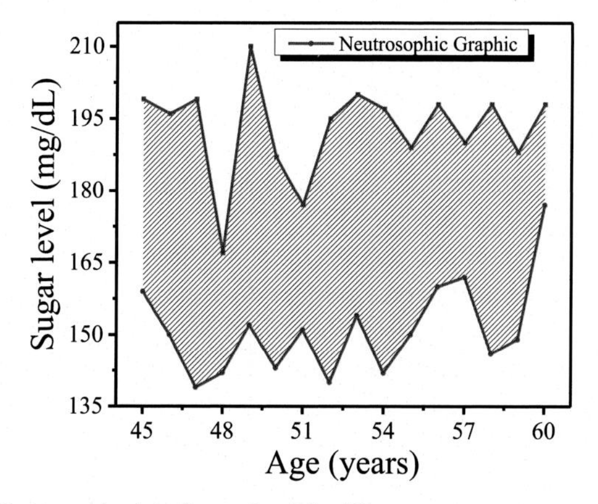

FIGURE 6.11 Neutrosophic plot of sugar level for G1.

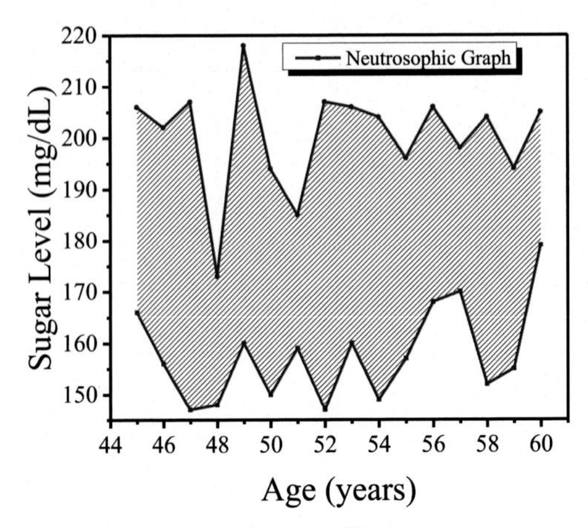

FIGURE 6.12 Neutrosophic plot of sugar level for G2.

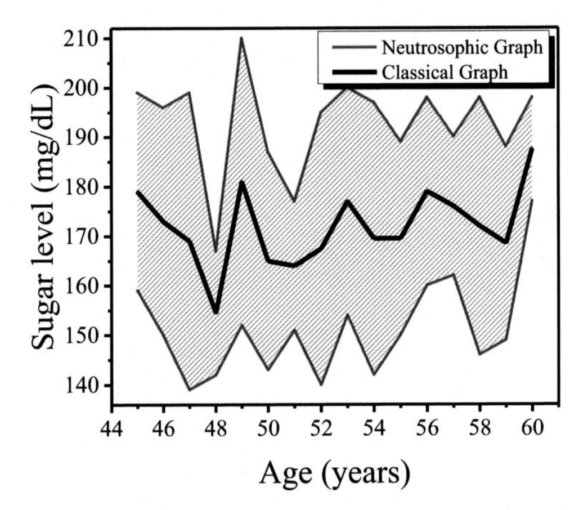

FIGURE 6.13 Combined graphs of neutrosophic and classical analyses for G1.

indeterminacy as well as the variance of data with respect to age. These are more flexible because one can find not only the value at a specific point but also find value at near points which are found in the indeterminacy range of that specific point [36]. Now, let us see the combined graphs of classical and neutrosophic analysis as shown in Figs. 6.13 and 6.14.

From the above graphs, it can be seen that the graph of classical analysis is less flexible and informative because this graph is drawn at fixed-point values; that is, there is no indeterminacy. Also, commonly researchers have used such type of graph (classical graph) as found in previous works. But the graph of neutrosophic analysis shows more flexibility. Also, this graph shows that the neutrosophic approach is a generalization of the classical

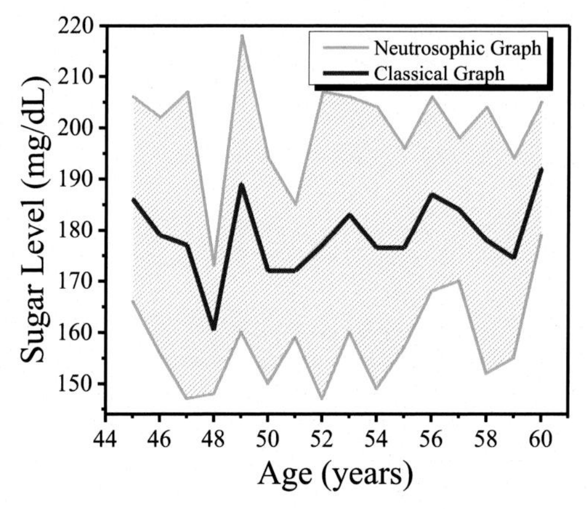

FIGURE 6.14 Combined graphs of neutrosophic and classical analyses for G2.

approach [37]; that is, one can also observe the classical graph through the neutrosophic approach as it is following between the middle of the neutrosophic graph [38–41].

6.9 Conclusion

In this chapter, we have tried to develop a neutrosophic statistical application in bio-information by analyzing the variance of sugar levels in the human body with respect to the variation of age. For the experiment, the sugar level of 150 diabetes patients has been measured by oral glucose tolerance tests, and the data have been divided into two groups. The variance analysis of sugar levels in the human body has been analyzed through neutrosophic and classical statistics. As a result, it is observed that classical analysis is not effective in explaining the variation of human sugar levels with respect to changes in age. But the neutrosophic analysis is more effective. It is concluded that NS is more effective, reliable, and much informative for studying the variance of the interval data with indeterminacy.

References

[1] T. Adebisi, Assessment of nutritional status of diabetic patients in Ogun State, Nigeria, Am. J. Hum. Ecol. 2 (4) (2013) 120–126.

[2] R.C. Ray, S. Behera, Bio-fuel agro-industrial production system in sweet sorghum, in: T.D. Pereira (Ed.), Sorghum: Cultivation, Varieties and Uses, *Nova Science Publishers*, *New York*, 2011, pp. 65–82.

[3] D. Sho, Greenspan's Basic & Clinical Endocrinology, McGraw-Hill Medical, 2011.

[4] R.R. Kalyani, M. Corriere, L. Ferrucci, Age-related and disease-related muscle loss: the effect of diabetes, obesity, and other diseases, Lancet Diabetes Endocrinol 2 (10) (2014) 819–829.

[5] J.K. Cruickshank, J.-C. Mbanya, R. Wilks, B. Balkau, N. McFarlane-Anderson, T. Forrester, Sick genes, sick individuals or sick populations with chronic disease? The emergence of diabetes and high blood pressure in African-origin populations, Int. J. Epidemiol. 30 (1) (2001) 111–117.
[6] W.H. Organization, Diabetes fact sheet N 312. October 2013. Archived from the original on, 26, 2013.
[7] H.A. Wani, S. Majid, M.S. Khan, A.A. Bhat, R.A. Wani, S.A. Bhat, et al., Scope of Honey in Diabetes and Metabolic Disorders Therapeutic Applications of Honey and its Phytochemicals, Springer, 2020, pp. 195–217.
[8] Abbottdiabetescare, Diabetes FAQs. From https://web.archive.org/web/20110706100159/http://www.abbottdiabetescare.com.au/diabetes-faq-measure-units.php, 2005.
[9] R.V. Hogg, J. McKean, A.T. Craig, Introduction to Mathematical Statistics, Pearson Education, 2005.
[10] M. Aslam, A study on skewness and kurtosis estimators of wind speed distribution under indeterminacy, Theor. Appl. Climatol. 143 (3) (2021) 1227–1234.
[11] J. Chen, J. Ye, S. Du, Scale effect and anisotropy analyzed for neutrosophic numbers of rock joint roughness coefficient based on neutrosophic statistics, Symmetry 9 (10) (2017) 208.
[12] J. Chen, J. Ye, S. Du, R. Yong, Expressions of rock joint roughness coefficient using neutrosophic interval statistical numbers, Symmetry 9 (7) (2017) 123.
[13] F. Smarandache, Introduction to Neutrosophic Statistics: Infinite Study, Romania-Educational Publisher, *Columbus, OH*, 2014.
[14] M. Aslam, Design of the Bartlett and Hartley tests for homogeneity of variances under indeterminacy environment, J. Taibah Univ. Sci. 14 (1) (2020) 6–10.
[15] M. Aslam, On detecting outliers in complex data using Dixon's test under neutrosophic statistics, J. King Saud. Univ.-Sci. 32 (3) (2020) 2005–2008.
[16] M. Aslam, Neutrosophic analysis of variance: application to university students, Complex. Intell. Syst. 5 (4) (2019) 403–407.
[17] F. Smarandache, Multispace & Multistructure. Neutrosophic Transdisciplinarity (100 Collected Papers of Science) (Vol. 4): Infinite Study, 2010.
[18] V. Christianto, R.N. Boyd, F. Smarandache, Three Possible Applications of Neutrosophic Logic in Fundamental and Applied Sciences, Infinite Study, 2020.
[19] J. Ye, Improved cosine similarity measures of simplified neutrosophic sets for medical diagnoses, Artif. Intell. Med. 63 (3) (2015) 171–179.
[20] M. Aslam, Enhanced statistical tests under indeterminacy with application to earth speed data, Earth Sci. Inform. (2021) 1–7.
[21] F. Smarandache, The Neutrosophic Research Method in Scientific and Humanistic Fields, 2010.
[22] U. Afzal, M. Aslam, A.H. Al-Marshadi, Analyzing Imprecise Graphene Foam Resistance Data, Materials Research Express, 2022.
[23] S.J. Press, Applied Multivariate Analysis: Using Bayesian and Frequentist Methods of Inference, Courier Corporation, 2005.
[24] M.J. Schervish, Theory of Statistics, Springer Science & Business Media, 2012.
[25] M. Aslam, N. Khan, M. Albassam, Control chart for failure-censored reliability tests under uncertainty environment, Symmetry 10 (12) (2018) 690.
[26] Z. Lotfi, Fuzzy sets, Inf. Control. 8 (3) (1965) 338–353.
[27] E.N. Nasibov, Fuzzy logic in statistical data analysis, in: M. Lovric (Ed.), International Encyclopedia of Statistical Science, Springer Berlin Heidelberg, Berlin, Heidelberg, 2011, pp. 558–563.

[28] P. Grzegorzewski, Testing statistical hypotheses with vague data, Fuzzy Sets Syst. 112 (3) (2000) 501–510.
[29] P. Grzegorzewski, k-sample median test for vague data, Int. J. Intell. Syst. 24 (5) (2009) 529–539.
[30] P. Grzegorzewski, M. Śpiewak, The sign test and the signed-rank test for interval-valued data, Int. J. Intell. Syst. 34 (9) (2019) 2122–2150.
[31] E.M. Urbina, S.S. Gidding, W. Bao, A. Elkasabany, G.S. Berenson, Association of fasting blood sugar level, insulin level, and obesity with left ventricular mass in healthy children and adolescents: The Bogalusa Heart Study, Am. heart J. 138 (1) (1999) 122–127.
[32] Y. Nikawa, D. Someya, Non-invasive measurement of blood sugar level by millimeter waves, in: Paper presented at the 2001 IEEE MTT-S International Microwave Sympsoium Digest (Cat. No. 01CH37157), 2001.
[33] K. Kawanaka, Y. Uetsuji, K. Tsuchiya, E. Nakamachi, Development of automatic blood extraction device with a micro-needle for blood-sugar level measurement, in: Paper presented at the Smart Structures, Devices, and Systems IV, 2008.
[34] S.K. Kumari, J. Mathana, Blood sugar level indication through chewing and swallowing from acoustic MEMS sensor and deep learning algorithm for diabetic management, J. Med. Syst. 43 (1) (2019) 1–9.
[35] A. Yadav, P. Sharan, A. Kumar, Surface plasmonic resonance based five layered structure-biosensor for sugar level measurement in human, Results in Optics, 1, 100002, 2020.
[36] U. Afzal, N. Ahmad, Q. Zafar, M. Aslam, Fabrication of a surface type humidity sensor based on methyl green thin film, with the analysis of capacitance and resistance through neutrosophic statistics, RSC Adv. 11 (61) (2021) 38674–38682.
[37] U. Afzal, H. Alrweili, N. Ahamd, M. Aslam, Neutrosophic statistical analysis of resistance depending on the temperature variance of conducting material, Sci. Rep. 11 (1) (2021) 23939. Available from: https://doi.org/10.1038/s41598-021-03347-z.
[38] U. Afzal, M. Aslam, K. Maryam, A.H. AL-Marshadi, F. Afzal, Fabrication and characterization of a highly sensitive and flexible tactile sensor based on indium zinc oxide (IZO) with imprecise data analysis, ACS Omega 7 (36) (2022) 32569–32576. Available from: https://doi.org/10.1021/acsomega.2c04156.
[39] U. Afzal, J. Afzal, M. Aslam, Analyzing the imprecise capacitance and resistance data of humidity sensors, Sens. Actuators: B. Chem. 367 (2022) 132092–132097. Available from: https://doi.org/10.1016/j.snb.2022.132092.
[40] U. Afzal, F. Afzal, K. Maryam, M. Aslam, Fabrication of flexible temperature sensors to explore indeterministic data analysis for robots as an application of Internet of Things, RSC Adv. 12 (2022) 17138–17145. Available from: https://doi.org/10.1039/d2ra03015b.
[41] U. Afzal, M. Aslam, F. Afzal, K. Maryam, N. Ahmad, Q. Zafar, Z. Farooq, Fabrication of a graphene-based sensor to detect the humidity and the temperature of a metal body with imprecise data analysis, RSC Adv. 12 (2022) 21297–21308. Available from: https://doi.org/10.1039/d2ra03474c.

Chapter 7

Analysis of changes in blood pressure of women during pregnancy through neutrosophic statistics

Usama Afzal[1] and Muhammad Aslam[2]
[1]School of Microelectronics, Tianjin University, Tianjin, P.R. China, [2]Department of Statistics, Faculty of Science, King Abdulaziz University, Jeddah, Saudi Arabia

7.1 Introduction and motivation

Pregnancy is a very important factor in life. It is also called gestation, because it is a period in which inside a woman one or more offspring develops [1]. Women get pregnant when she intimates with a man and the man's sperm fertilized the egg. The pregnancy of a woman can be divided into three trimesters each having approximately 3 months duration as shown in Fig. 7.1. In the first trimester, the sperm of a man fertilizes the ovary egg. This egg starts to move in the fallopian tube and at the end attaches to the uterus for developing itself into embryo and placenta [2]. Generally, it is seen that mostly natural death of an embryo, that is miscarriage, happens during this trimester. In the second or middle trimester, a woman can feel an embryo moment. About 90% or more babies can be survived outside the uterus. Sometimes, babies get some serious health issues like heartbeats, respiratory issues, and other development of disabilities [3].

To overcome such disabilities, prenatal care is very important [4]. It includes the physical examination, blood test, and other caring factors. During this critical period, different changes and weaknesses are felt by women in their bodies like vomiting, low blood pressure, and high blood pressure [2]. These changes are measured using a useful tool and careful analysis. Blood pressure is a very important factor during pregnancy. Sometimes, it becomes a heddle in the normal birth of babies. So, it should be properly measured and analyzed.

Cognitive Intelligence with Neutrosophic Statistics in Bioinformatics.
DOI: https://doi.org/10.1016/B978-0-323-99456-9.00010-6

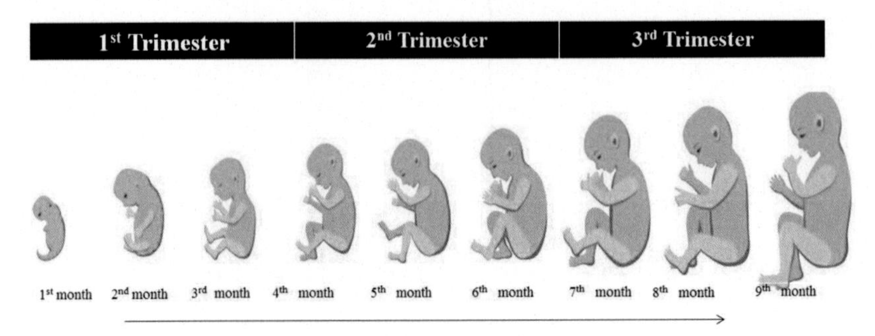

FIGURE 7.1 Embryo development cycle.

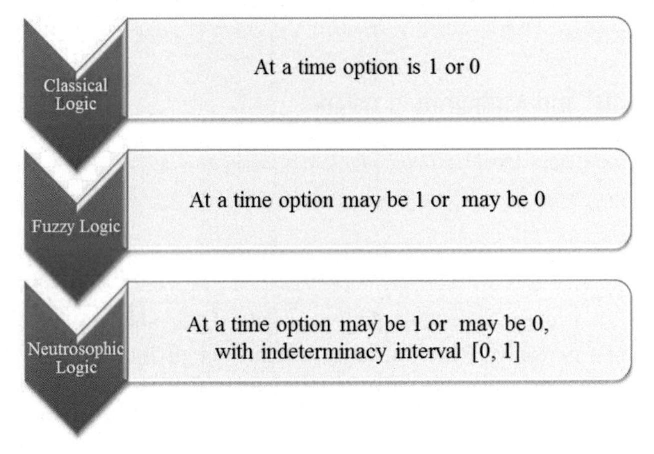

FIGURE 7.2 Example to show the difference between classical, fuzzy, and neutrosophic statistics.

For the analysis of data, statistics is the best option. In statistics, there are a number of logics for analyzing the data such as classical logic (CL), fuzzy logic (FL), and neutrosophic logic (NL). These logics have different conditions; such as, CL only deals with determined values, FL deals with only fuzzy set, and NL deals with indetermined value interval. Let us use an example to show the difference between above statistics approaches as shown in Fig. 7.2 which shows that how classical, fuzzy, and neutrosophic statistics analyze an option having "1" or "0."

Blood pressure is a very serious issue for women while giving baby birth. Sometimes women have low blood pressure issues and sometimes high blood pressure issues. Due to this, the analysis of blood pressure is very important. In this chapter, we have used neutrosophic statistics first time (neutrosophic formulas and graphs) to analyze the blood pressure variance with respect to the age of pregnant women.

7.2 Background, definitions, and abbreviations

Classical statistics (CS) contains ordinary approaches and formulas which can only analyze the determined value or fixed-point value, such as the classical mean formula and least square [5]. That is why it is also known as frequentist statistic (FS) [6]. For interval values, CS failed to analyze and explain the problem, because interval values contain uncertainty. So, one can easily conclude that CS is unable to well analyze the data containing uncertainty [7].

Similarly, fuzzy statistics (FS) is used to analyze the data when there is uncertainty, that is interval form. Fuzzy logic (fuzzy set theory) was proposed by Lotfi [8]. Fuzzy logic is more flexible than CL; like in CL, if a number is a member of a set, then it will be “1” otherwise it will be “0,” but in fuzzy the number will be between “0” and “1,” that is [0; 1], simultaneously [9]. Statistical test using the FL was introduced by Grzegorzewski [10]. The purpose of the fuzzy statistics is to use an indeterministic approach to analyze and observe the uncertainty in data. Similarly, the median analysis with the help of this statistical test was also proposed by Grzegorzewski [11]. Analysis of the interval data using signed-rank test was attempted by Grzegorzewski and Śpiewak [12].

The neutrosophic approach is used for the analysis of interval point values data having indeterminacy [13]. This is a substantial benefit over the classical approach because the CS approach only deals with fixed-point values data, having no indeterminacy. Moreover, the neutrosophic technique is more helpful than classical techniques; see examples of references [14–16]. Different statistical techniques were developed under neutrosophic statistics by Muhammad Aslam [17,18]. The NL is also used for the analysis of indeterministic data. It is the generalization of the FL and is more reliable and informative than FL [19]. Smarandache claimed it expressed the efficiency of NL over FL [20]. Similarly, in 2014 Smarandache proposed neutrosophic statistics (NS) for the analysis of indeterminacy of data. Nowadays, NS has got importance for analysis and is used in different fields, such as applied sciences [21], medicine for diagnoses of data measurement [22], astrophysics for measurement of earth speed data [23], humanistic [24], and material science [25].

7.2.1 Preliminaries

Let X_N be a neutrosophic variable having interval of variance $X_N \epsilon [X_L, X_U]$ having size $n_N \epsilon [n_{NL}, n_{NU}]$ with indeterminacy $I_N \epsilon [I_{NL}, I_{NU}]$, so the neutrosophic formula (NF) is written as follows [26]:

$$X_{iN} = X_{iL} + X_{iU} I_N (i = 1, 2, 3, \ldots, n_N) \tag{7.1}$$

X_{iN} $\in[X_{iL}, X_{IU}]$ has two parts: X_{iL} expressing the lower value which is under CS, and $X_{iU}I_N$ is a upper part with indeterminacy $I_N \in [I_L, I_U]$. Similarly, the neutrosophic mean (NM) $\overline{X}_N \in [\overline{X}_L, \overline{X}_U]$ is defined as follows:

$$\overline{X}_N = \overline{X}_L - \overline{X}_U I_N I_N \in [I_L, I_U] \tag{7.2}$$

Here $\overline{X}_L = \sum_{(i=1)}^{nU} \left(\frac{X_{iL}}{n_L}\right)$ and $\overline{X}_{\mathrm{U}} = \sum_{(i=1)}^{nU} \left(\frac{X_{iU}}{n_U}\right)$

Now, the **neutrosophic flowchart (NFC)** for solving a problem through neutrosophic statistics is shown in Fig. 7.3:

7.2.2 List of abbreviations

- NL: Neutrosophic logic
- FL: Fuzzy logic
- CL: Classical logic
- NS: Neutrosophic statistics
- FS: Fuzzy statistics
- CN: Classical statistics

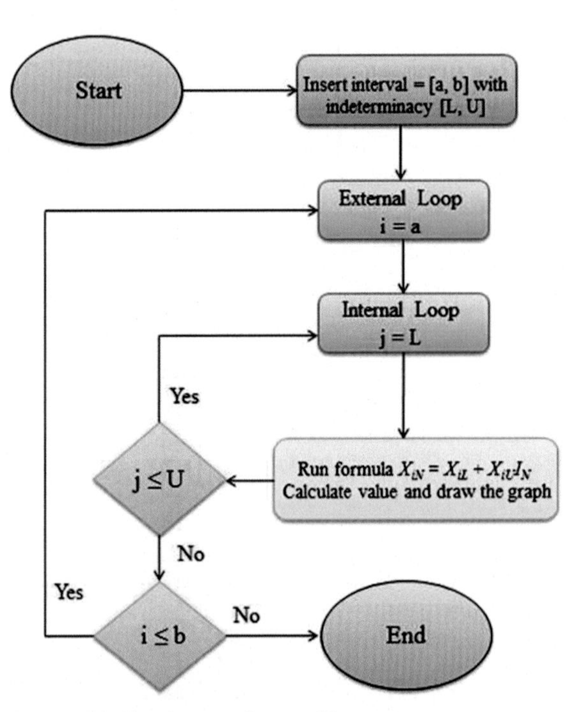

FIGURE 7.3 Neutrosophic flowchart to solve a problem.

- NF: Neutrosophic formula
- NM: Neutrosophic mean
- CM: Classical mean
- CSF: Classical flowchart
- NFC: Neutrosophic flowchart
- LBP: Low blood pressure
- HBP: High blood pressure

7.3 Literature review and state of the art

In 1976 Roberta and Christianson studied the blood pressure of women during the pregnancy with age and parity's inflection [27]. They collected the data of 6662 pregnant women, and for the analysis, they used mean (classical mean formula). From the result, they observed that nulliparas were having the problem of blood pressure more than the parous. M. D. D. Churchill et al. also studied blood pressure during the pregnancy period [28]. They used the blood pressure data of 209 nulliparous pregnant women, which was measured many times, that is, after 18 weeks, 28 weeks, and 36 weeks of gestation. They found that higher blood pressure at 28 and 36 weeks predicted low gestation, and at 18 weeks, gestation is unpredicted. M.D. J. Zhang et al., reported the blood pressure study in early stage of pregnancy by analyzing the data of 5167 women having singleton gestation [29]. They studied the blood pressure and pulse rate at baseline and mid-third trimester and found different rises for both, that is, 10 and 3 mmHg, respectively. Grindheim and Guro et al. studied the blood pressure changes and heart rate during pregnancy in 57 healthy women [30]. They measured the data by using the Dinamap and Finometer PRO at gestation periods 14–16, 22–24, 30–32, and 36 weeks and postpartum of 6 months. They statistically analyzed the data using mean formulation (Classical Formulation). In the end, they observed low blood pressure for 22–24 weeks.

This is an overview of the previous work based on blood pressure during pregnancy. From the above-mentioned references, it is observed that the measured data of blood pressure during pregnancy has been analyzed through classical formulas, graphs and tables, which are based on the fixed-point values. According to our best knowledge and literature review, we have not found any neutrosophic analysis of blood pressure. In this chapter, we have applied neutrosophic statistics to analyze the variance of blood pressure during pregnancy. Measurement of blood pressure changes has been attempted, and data are collected for the analysis. Through this chapter, we have tried to analyze the variance data measured directly from the patients using useful tools, and for the first time, we used neutrosophic statistics and also used classical statistics for this purpose.

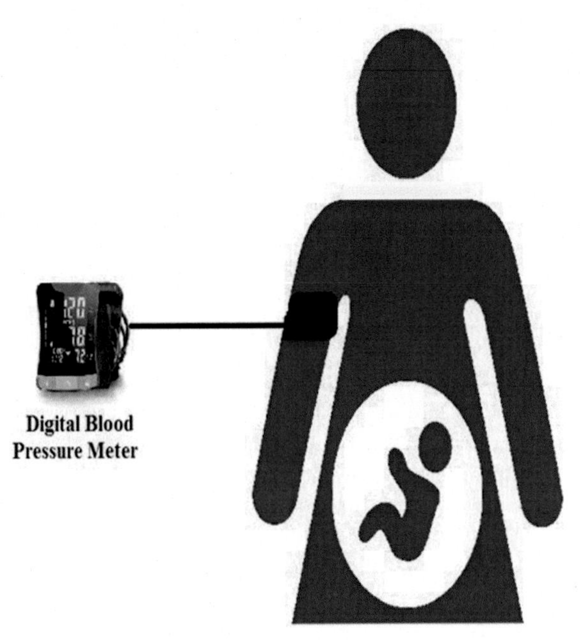

FIGURE 7.4 Setup to measure the blood pressure of pregnant women.

7.4 Problem

The purpose of this study is to highlight an application of neutrosophic statistics in bio-information. We have used the neutrosophic formulas for the analysis of the changes in blood pressure of pregnant women with respect to changes in their blood pressure. We have collected the data of 300 women aged 25 to 35 years, just one hour before baby delivery using the digital meter manually (all data have been collected with the consent of the patients) as shown in Fig. 7.4. We have observed two types of blood pressure issues, that are low blood pressure and high blood pressure. So, we have divided it into two groups: (A) having low blood pressure issues and (B) having high blood pressure issues. As we are interested in the analysis of data variation with respect to the age of pregnant women, we have used classical and neutrosophic statistics. Also, it is observed which one is more suitable to analyze the problem on the behalf of the method and formulas.

7.5 Proposed solution to the problem

For understanding and analyzing the measured value of both groups (A) and (B), we have used classical statistics and neutrosophic statistics and compared their results.

7.5.1 Classical statistics

As the measured data are showing the variance, CS only deals with the fixed-point value. So, classical mean (CM) formula to convert these values into fixed-point value for CS analysis is as follows:

$$\overline{M}_i = \frac{A_{Li} + A_{Ui}}{n} (i = 0, 1, 2, 3, 4 \ldots N) \quad (7.3)$$

Here, "$\overline{M}_i$" is classical mean of the "*i*th" interval, and A_{Li} and A_{Ui} are the lower and upper values of the interval, respectively. Also, "n" is the number of values. For the following problem, $n = 2$ (as there are only two values of interval, i.e., lower and upper) and "i" is taken as age so$i$ = age = 25, 26, 24 . . . 35.

Now the **classical flowchart** (CFC) of said problem is shown in Fig. 7.5:

7.5.2 Neutrosophic statistics

Our study belongs to the changes in low blood pressure and high blood pressure of women during the pregnancy with respect to change in age, so these factors can be written as the function of the age, that is, LBP(age) and HBP(age), respectively. Now, let LBP(age)$_N$ is measured interval value of the

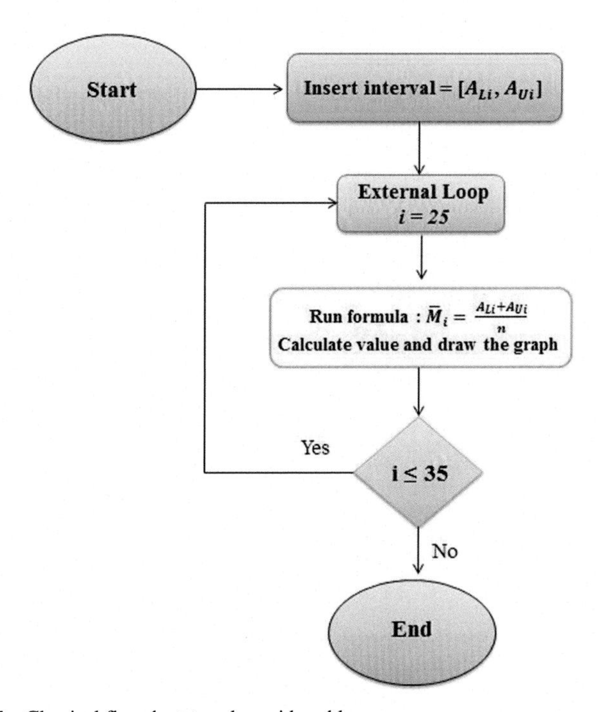

FIGURE 7.5 Classical flowchart to solve said problem.

low blood pressure, that is, $\text{LBP(age)}_N \in [\text{LBP(age)}_L, \text{LBP(age)}_U]$; here, LBP(age)_L and LBP(age)_U are the lower and upper values of the interval, respectively. The NF for low blood pressure with indeterminacy $I_N \in [I_L, I_U]$ is written as following:

$$\text{LBP(age)}_N = \text{LBP(age)}_L + \text{LBP(age)}_U I_N; I_N \in [I_L, \quad I_U] \tag{7.4}$$

From the above NF of $\text{LBP(age)}_N \in [\text{LBP(age)}_L, \text{LBP(age)}_U]$ is an extension under the classical statistics. The equation contains two parts, that is, LBP(age)_L determined part and $\text{LBP(age)}_U I_N$ indetermined part. Moreover, $I_N \in [I_L, I_U]$ is known as an indeterminacy interval. Also, the measured interval $\text{LBP(age)}_N \in [\text{LBP(age)}_L, \text{LBP(age)}_U]$ can be reduced to the classical or determined part if we choose $I_L = 0$, and I_U can be found by using $(\text{LBP(age)}_U - \text{LBP(age)}_L)/\text{LBP(age)}_U$. Similarly, for high blood pressure:

$$\text{HBP(age)}_N = \text{HBP(age)}_L + \text{HBP(age)}_U I_N; I_N \in [I_L, I_U] \tag{7.5}$$

Now, the **neutrosophic flowchart (NFC)** of said problem is shown in Fig. 7.6:

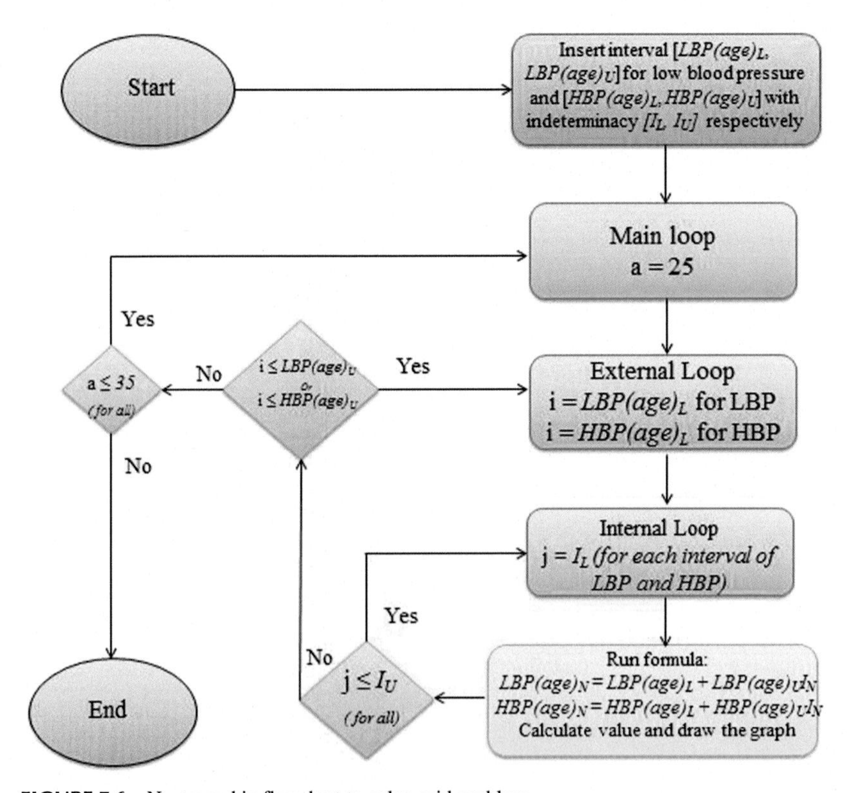

FIGURE 7.6 Neutrosophic flowchart to solve said problem.

7.6 Analysis of data

The measured data of the pregnant women of group (A) and group (B) are shown in Tables 7.1 and 7.2. Table 7.1 expresses the collected data of pregnant women having low blood pressure issues, and Table 7.2 expresses the collected data of pregnant women having high blood pressure.

The above tables show the variation of low blood pressure and high blood pressure with respect to age (changing from 25 to 35 years). Also, the average low blood pressure and average high blood pressure of pregnant women are (54.5/82.5–59.7/86.7 mmHg) and (85/136–90.5/142 mmHg), respectively.

7.7 Statistical conditions

We have used two types of the statistical conditions, that is, classical statistics and neutrosophic statistics, for the data analysis of the said problem.

7.7.1 Classical analysis

For the classical analysis of the problem, we have used the above-said procedure as shown in Section 7.5.1. We have applied this procedure, applied classical average formula, on both groups, that is, low blood pressure and high blood pressure, as shown in Tables 7.3 and 7.4, respectively.

TABLE 7.1 Measured data of group (A) "low blood pressure"

Age (years)	Low blood pressure (mmHg)	
	Diastolic	Systolic
25	[56, 59]	[83, 85]
26	[53, 58]	[80, 86]
27	[50, 57]	[87, 94]
28	[54, 58]	[83, 87]
29	[56, 64]	[82, 84]
30	[56, 60]	[79, 85]
31	[57, 61]	[78, 85]
32	[51, 58]	[80, 84]
33	[53, 59]	[85, 88]
34	[54, 62]	[81, 84]
35	[57, 61]	[88, 92]

TABLE 7.2 Measured data of group (B) "high blood pressure"

Age (years)	High blood pressure (mmHg)	
	Diastolic	Systolic
25	[82, 92]	[131, 139]
26	[84, 87]	[137, 143]
27	[87, 93]	[134, 137]
28	[83, 86]	[139, 146]
29	[86, 88]	[137, 140]
30	[85, 94]	[134, 143]
31	[88, 95]	[139, 141]
32	[86, 89]	[136, 140]
33	[89, 92]	[141, 143]
34	[81, 86]	[134, 142]
35	[83, 92]	[136, 147]

TABLE 7.3 Classical analysis of group (A) "low blood pressure"

Age (years)	Low blood pressure (mmHg)	
	Diastolic	Systolic
25	57.5	84
26	55.5	83
27	53.5	90.5
28	56	85
29	60	83
30	58	82
31	59	81.5
32	54.5	82
33	56	86.5
34	58	82.5
35	59	90

TABLE 7.4 Classical analysis of group (B) "high blood pressure"

Age (years)	Low blood pressure (mmHg)	
	Diastolic	Systolic
25	87	135
26	85.5	140
27	90	135.5
28	84.5	142.5
29	87	138.5
30	89.5	138.5
31	91.5	140
32	87.5	138
33	90.5	142
34	83.5	138
35	87.5	141.5

7.7.2 Neutrosophic analysis

For the neutrosophic analysis of the problem, we have used the above-said procedure as expressed in Section 7.5.2. We have applied this procedure on all the measured values of the low blood pressure and high blood pressure with their indeterminacy, respectively, as shown in Tables 7.5 and 7.6.

7.8 Discussion

In this chapter, we have analyzed the variation of blood pressure due to pregnancy in women. We have measured the data for two groups, that are women having low blood pressure issues (A) and women having high blood pressure issues (B). The plots/graphs of the classical analysis of groups (A) and (B) are shown in Figs. 7.7 and 7.8, respectively. Similarly, the plots/graphs of the neutrosophic analysis of groups (A) and (B) are shown in Figs. 7.9 and 7.10, respectively.

The above graphs are plotted to express the difference between the classical and neutrosophic statistical analyses of the measured data. From classical analysis, graphs are based on the classical average formula [26]. It is seen that for classical analysis there are only fixed-point values i.e. single value of low blood pressure and high blood pressure has been observed with respect to a specific age. That is why these graphs are not much flexible and

TABLE 7.5 Neutrosophic analysis of the data of group (A).

Age (years)	Low blood pressure (mmHg)	
	Diastolic	Systolic
25	$56+59I_N$; $I_N \in [0, 0.051]$	$85+83I_N$; $I_N \in [0, 0.024]$
26	$53+58I_N$; $I_N \in [0, 0.086]$	$80+86I_N$; $I_N \in [0, 0.070]$
27	$50+57I_N$; $I_N \in [0, 0.123]$	$87+94I_N$; $I_N \in [0, 0.074]$
28	$54+58I_N$; $I_N \in [0, 0.068]$	$83+87I_N$; $I_N \in [0, 0.046]$
29	$56+64I_N$; $I_N \in [0, 0.125]$	$82+84I_N$; $I_N \in [0, 0.024]$
30	$56+60I_N$; $I_N \in [0, 0.067]$	$79+85I_N$; $I_N \in [0, 0.071]$
31	$57+61I_N$; $I_N \in [0, 0.066]$	$78+85I_N$; $I_N \in [0, 0.082]$
32	$51+58I_N$; $I_N \in [0, 0.121]$	$80+84I_N$; $I_N \in [0, 0.048]$
33	$53+59I_N$; $I_N = [0, 0.102]$	$85+88I_N$; $I_N \in [0, 0.034]$
34	$54+62I_N$; $I_N \in [0, 0.129]$	$81+84I_N$; $I_N \in [0, 0.036]$
35	$57+61I_N$; $I_N \in [0, 0.066]$	$88+92I_N$; $I_N \in [0, 0.043]$

TABLE 7.6 Neutrosophic analysis of the data group (B).

Age (years)	High blood pressure (mmHg)	
	Diastolic	Systolic
25	$82+92I_N$; $I_N \in [0, 0.109]$	$131+139I_N$; $I_N \in [0, 0.058]$
26	$84+87I_N$; $I_N \in [0, 0.035]$	$137+143I_N$; $I_N \in [0, 0.042]$
27	$87+93I_N$; $I_N \in [0, 0.065]$	$134+137I_N$; $I_N \in [0, 0.022]$
28	$83+86I_N$; $I_N \in [0, 0.035]$	$139+146I_N$; $I_N \in [0, 0.048]$
29	$86+88I_N$; $I_N \in [0, 0.023]$	$137+140I_N$; $I_N \in [0, 0.022]$
30	$85+94I_N$; $I_N \in [0, 0.096]$	$134+143I_N$; $I_N \in [0, 0.063]$
31	$88+95I_N$; $I_N \in [0, 0.074]$	$139+141I_N$; $I_N \in [0, 0.014]$
32	$86+89I_N$; $I_N = [0, 0.034]$	$136+140I_N$; $I_N \in [0, 0.029]$
33	$89+92I_N$; $I_N = [0, 0.033]$	$141+143I_N$; $I_N \in [0, 0.014]$
34	$81+86I_N$; $I_N = [0, 0.058]$	$134+142I_N$; $I_N = [0, 0.056]$
35	$83+92I_N$; $I_N = [0, 0.098]$	$136+147I_N$; $I_N = [0, 0.075]$

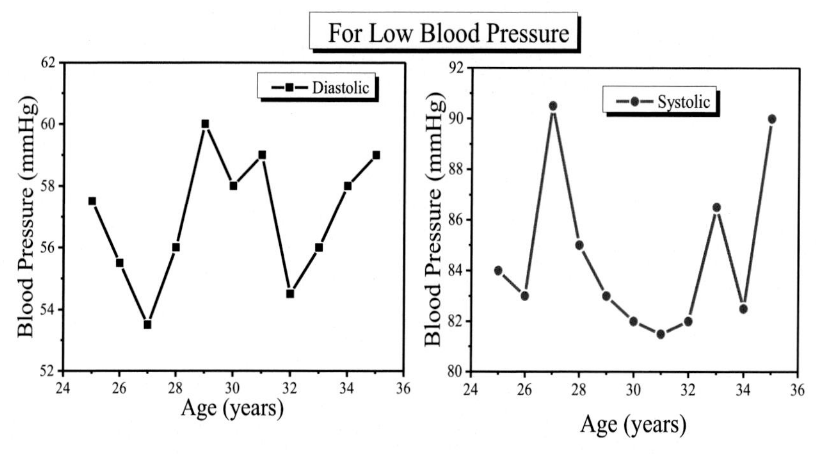

FIGURE 7.7 Classical graphs of diastolic and systolic limits of low blood pressure.

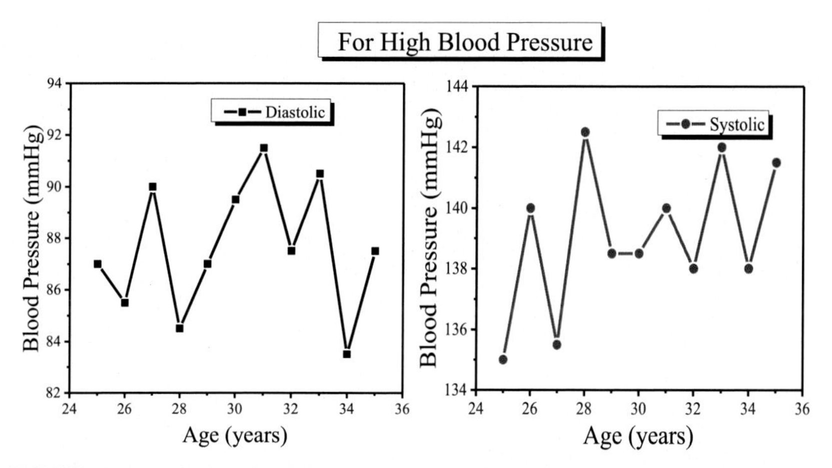

FIGURE 7.8 Classical graphs of diastolic and systolic limits of high blood pressure.

informative. But on the other hand, neutrosophic analysis graphs are more informative as these are dealing with the indeterminacy and the variance of data with respect to age. The formulations used by neutrosophic analysis are very effective and reliable [31]. Due to this, we have achieved more flexibility because one can find no value at a specific point but also find value at near points which are found in the indeterminacy range of that specific point. As a result, we can conclude that neutrosophic statistics is more effective, reliable, and informative to analyze the blood pressure variance with respect to change in age than classical statistics [32–35].

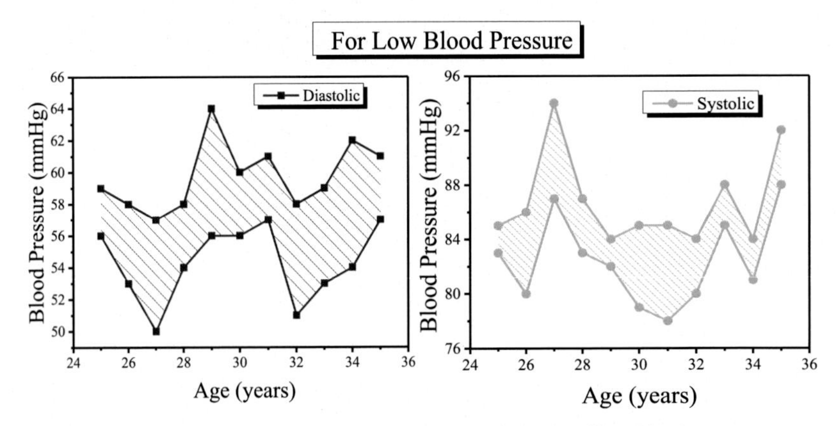

FIGURE 7.9 Neutrosophic graphs of diastolic and systolic limits of low blood pressure.

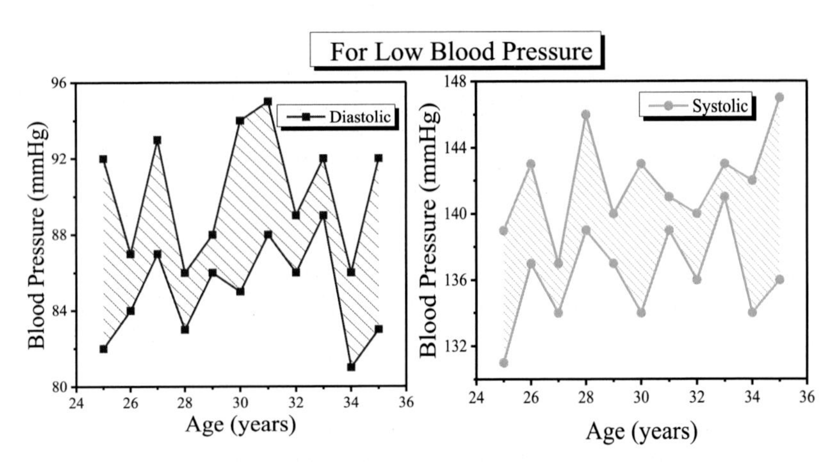

FIGURE 7.10 Neutrosophic graphs of diastolic and systolic limits of high blood pressure.

7.9 Conclusion

In this chapter, we have tried to develop a neutrosophic statistical application in bio-information by analyzing the blood pressure variance of pregnant women with respect to the variation of age. We have collected the data of 300 pregnant women aged 25 to 35 years and divided them into two groups (A) having low blood pressure and (B) high. Then, we analyze these measured data through the classical and neutrosophic statistics and observed the variation of blood pressure of pregnant women with respect to change in age. As a result, it is observed that classical analysis is not effective in explaining the variation of blood pressure with respect to change in age. But the neutrosophic analysis is more effective. It is concluded that neutrosophic statistics is more effective, reliable, and much informative in taking decisions than classical statistics.

References

[1] Sindhuja Murugaboopathi, Hephzibah Kirubamani, Awareness of complications of first trimester pregnancy, Indian J. Obstet. Gynecol. 7 (4 (Part-II)) (2019) 627–631.

[2] Tanvi Banker, An evaluation of rate of fatigue and sleep quality in pregnant women: a cross sectional study, Int. J. Sci. Healthc. Res. 6 (1) (2021).

[3] Casey Crump, An overview of adult health outcomes after preterm birth, Early Hum. Dev. 150 (2020) 105187.

[4] Hamza Ahmed Hassan Mohammed, Determination of Hemoglobin F Level in Normal Pregnant Women Referred to Dar Elber Specialized Medical Center in Khartoum City, Faculty of Graduate Studies and Scientific Research, Shendi University, 2018.

[5] M.J. Schervish, Theory of Statistics, Springer Science & Business Media, 2012.

[6] S.J. Press, Applied Multivariate Analysis, Using Bayesian and Frequentist Methods of Inference, Courier Corporation, 2005.

[7] M. Aslam, N. Khan, M. Albassam, Control chart for failure-censored reliability tests under uncertainty environment, Symmetry 10 (12) (2018) 690.

[8] Z. Lotfi, Fuzzy sets, Inf. Control. 8 (3) (1965) 338–353.

[9] E.N. Nasibov, Fuzzy logic in statistical data analysis, in: M. Lovric (Ed.), *International Encyclopedia of Statistical Science* (pp. 558–563), Springer Berlin Heidelberg, Berlin, Heidelberg, 2011.

[10] P. Grzegorzewski, Testing statistical hypotheses with vague data, Fuzzy Sets Syst. 112 (3) (2000) 501–510.

[11] P. Grzegorzewski, k-sample median test for vague data, Int. J. Intell. Syst. 24 (5) (2009) 529–539.

[12] P. Grzegorzewski, M. Śpiewak, The sign test and the signed-rank test for interval-valued data, Int. J. Intell. Syst. 34 (9) (2019) 2122–2150.

[13] M. Aslam, A study on skewness and kurtosis estimators of wind speed distribution under indeterminacy, Theor. Appl. Climatol. 143 (3) (2021) 1227–1234.

[14] J. Chen, J. Ye, S. Du, Scale effect and anisotropy analyzed for neutrosophic numbers of rock joint roughness coefficient based on neutrosophic statistics, Symmetry 9 (10) (2017) 208.

[15] J. Chen, J. Ye, S. Du, R. Yong, Expressions of rock joint roughness coefficient using neutrosophic interval statistical numbers, Symmetry 9 (7) (2017) 123.

[16] F. Smarandache, Introduction to Neutrosophic Statistics: Infinite Study, Romania-Educational Publisher, Columbus, OH, 2014.

[17] M. Aslam, Design of the Bartlett and Hartley tests for homogeneity of variances under indeterminacy environment, J. Taibah Univ. Sci. 14 (1) (2020) 6–10.

[18] M. Aslam, On detecting outliers in complex data using Dixon's test under neutrosophic statistics, J. King Saud. Univ.-Sci. 32 (3) (2020) 2005–2008.

[19] M. Aslam, Neutrosophic analysis of variance: application to university students, Complex. Intell. Syst. 5 (4) (2019) 403–407.

[20] Florentin Smarandache, Multispace & Multistructure. Neutrosophic Transdisciplinarity (100 Collected Papers of Sciences), IV, 2010.

[21] Victor Christianto, Robert N. Boyd, Florentin Smarandache, Three possible applications of neutrosophic logic in fundamental and applied sciences. Infinite Study, 2020.

[22] J. Ye, Improved cosine similarity measures of simplified neutrosophic sets for medical diagnoses, Artif. Intell. Med. 63 (3) (2015) 171–179.

[23] M. Aslam, Enhanced statistical tests under indeterminacy with application to earth speed data, Earth Sci. Inform. (2021) 1–7.

[24] Florentin Smarandache, The Neutrosophic Research Method in Scientific and Humanistic Fields, 2010.

[25] U. Afzal, M. Aslam, A.H. Al-Marshadi, Analyzing Imprecise Graphene Foam Resistance Data, Materials Research Express, 2022.

[26] U. Afzal, H. Alrweili, N. Ahamd, M. Aslam, Neutrosophic statistical analysis of resistance depending on the temperature variance of conducting material, Sci. Rep. 11 (1) (2021) 23939. Available from: https://doi.org/10.1038/s41598-021-03347-z.

[27] R.E. Christianson, Studies on blood pressure during pregnancy: I. influence of parity and age, Am. J. Obstet. Gynecol. 125 (4) (1976) 509–513.

[28] D. Churchill, I.J. Perry, D. Beevers, Ambulatory blood pressure in pregnancy and fetal growth, Lancet 349 (9044) (1997) 7–10.

[29] J. Zhang, J. Villar, W. Sun, M. Merialdi, H. Abdel-Aleem, M. Mathai, et al., Blood pressure dynamics during pregnancy and spontaneous preterm birth, Am. J. Obstet. Gynecol. 197 (2) (2007) 162–e161. –162. e166.

[30] G. Grindheim, M.-E. Estensen, E. Langesaeter, L.A. Rosseland, K. Toska, Changes in blood pressure during healthy pregnancy: a longitudinal cohort study, J. Hypertens. 30 (2) (2012) 342–350.

[31] U. Afzal, N. Ahmad, Q. Zafar, M. Aslam, Fabrication of a surface type humidity sensor based on methyl green thin film, with the analysis of capacitance and resistance through neutrosophic statistics, RSC Adv. 11 (61) (2021) 38674–38682.

[32] U. Afzal, A. Jamil, M. Aslam, Analyzing the imprecise capacitance and resistance data of humidity sensors, Sen. Actuators: B. Chem. 367 (2022)132092. Available from: https://doi.org/10.1016/j.snb.2022.132092.

[33] U. Afzal, F. Afzal, K. Maryam, M. Aslam, Fabrication of flexible temperature sensors to explore indeterministic data analysis for robots as an application of Internet of Things, RSC Adv. 12 (2022) 17138–17145. Available from: https://doi.org/10.1039/d2ra03015b.

[34] U. Afzal, M. Aslam, F. Afzal, K. Maryam, N. Ahmad, Q. Zafar, Z. Farooq, Fabrication of a graphene-based sensor to detect the humidity and the temperature of a metal body with imprecise data analysis, RSC Adv. 12 (2022) 21297–21308. Available from: https://doi.org/10.1039/d2ra03474c.

[35] U. Afzal, M. Aslam, K. Maryam, A.H. AL-Marshadi, F. Afzal, Fabrication and characterization of a highly sensitive and flexible tactile sensor based on indium zinc oxide (IZO) with imprecise data analysis, ACS Omega 7 (36) (2022) 32569–32576. Available from: https://doi.org/10.1021/acsomega.2c04156.

Chapter 8

Neutrosophic statistical analysis of changes in blood pressure, pulse rate and temperature of human body due to COVID-19

Usama Afzal[1] and Muhammad Aslam[2]
[1]*School of Microelectronics, Tianjin University, Tianjin, P.R. China,* [2]*Department of Statistics, Faculty of Science, King Abdulaziz University, Jeddah, Saudi Arabia*

8.1 Introduction and motivation

A novel disease coronavirus was identified in Wuhan, China, in 2019 [1]. WHO first nominated this disease as COVID-19 which stands for coronavirus disease 2019 [2]. This disease changes the normal body condition such as high temperature, pneumonia, and respiratory, which becomes a reason for death of the patient [3]. This disease is transformable, which generally spreads from person to person [4]. COVID-19 has badly affected all the fields of life, that is, economy [5], healthcare departments, and education [6]. The whole world is facing this pandemic. Still, there is not any approved medicine for COVID-19 by FDA, unfortunately. In March 2020, WHO published the first interim guidance to stop the spreading of COVID-19 named: "Responding to community spread of COVID-19."

Different measurements have been attempted to understand the nature of coronavirus and its effects on the body, such as the measurement of body temperature. Also, researchers are working on different technologies to control its spread. The analysis of these measured data is performed through statistics approaches. Statistics is a significant tool which is used for the analysis of data [7] of a particular problem whether in table form or graphical form. It is an interdisciplinary field; that is, its application can be seen in scientific research and others. Measured data of problems have two types, that is, categorical and numerical types, as shown in Fig. 8.1 with their subtypes.

Cognitive Intelligence with Neutrosophic Statistics in Bioinformatics.
DOI: https://doi.org/10.1016/B978-0-323-99456-9.00002-7

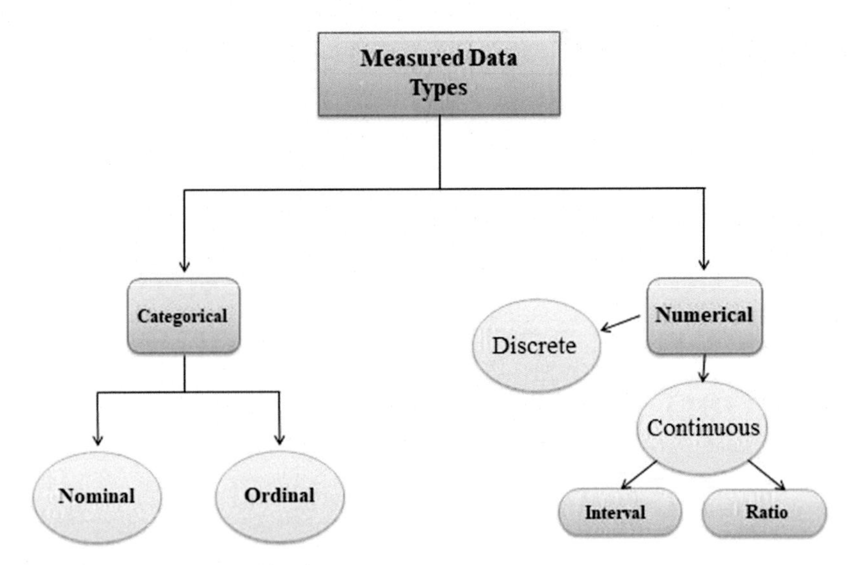

FIGURE 8.1 Data types with subtypes.

Categorical data are no quantitative type data and represent characteristics; that is, it represents data like gender, languages, profession. It has two subtypes, nominal and ordinal. Similarly, numerical data are a quantitative data type and represent numeric values. It has two subtypes, discrete, that is, fixed-point value, generally known as determining value/data, and continuous. Continuous further has two types, interval and ratio data, generally known as indeterminate data [7–9].

There are a number of approaches/logic for analyzing the data such as classical statistics (CS), fuzzy statistics (FS), and neutrosophic statistics (NS). Classical statistics only deals with determined values, fuzzy statistics deals with only fuzzy sets, and neutrosophic statistics deals with indeterminate intervals. Let us use an example to show the difference between the above statistics approaches as shown in Fig. 8.2 which shows that how classical, fuzzy, and neutrosophic statistics analyze true and false statements.

8.2 Background, definitions, and abbreviations

Classical statistics (CS) contains ordinary approaches and formulas which can only analyze the determined value or fixed-point value, such as the classical mean formula and least square. That is why it is also known as frequentist statistic (Press, 2005). But for interval values, classical statistics failed to analyze and explain the problem. But interval values contain uncertainty. So, one can easily conclude that classical statistics are unable to well analyze the data containing uncertainty.

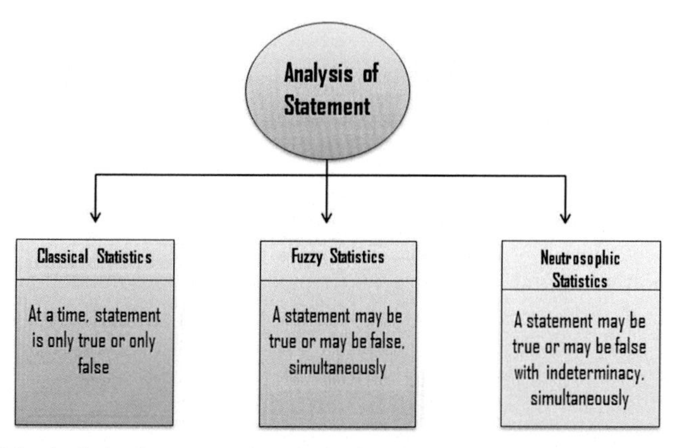

FIGURE 8.2 Analysis of statement through classical, fuzzy, and neutrosophic statistics.

Similarly, fuzzy statistics (FS) is used to analyze the data when there is uncertainty, that is, interval form. Fuzzy logic (fuzzy set theory) was proposed by Ref. [10]. Fuzzy logic is more flexible than classical logic; e.g., in classical logic, if a number is a member of a set, then it will be "1" otherwise it will be "0", but in fuzzy the number will be between "0" and "1", that is, [0; 1], simultaneously, [11]. Statistical test using the fuzzy logic was introduced by Przemysław [12]. The purpose of fuzzy statistics is to use an indeterministic approach to analyze and observe the uncertainty in data. Similarly, the median analysis with the help of this statistical test was also proposed by Przemysław [13]. Analysis of the interval data using signed-rank test was attempted by Ref. [14].

The neutrosophic approach is used for the analysis of interval point values data, that is, having indeterminacy [15]. This is a substantial benefit over the classical approach because the classical statistic approach only deals with fixed point values data, that is, having no indeterminacy. Moreover, the neutrosophic technique is more helpful than classical techniques; see examples in the following Refs. [16–18]. Different statistical techniques were developed under neutrosophic statistics by Aslam and Aslam [19,20]. The neutrosophic logic is also used for the analysis of indeterministic data. It is the generalization of fuzzy logic and is more reliable and informative than fuzzy logic [21]. Ref. [22] claimed it expressed the efficiency of neutrosophic logic over fuzzy logic. Similarly, in 2014 proposed neutrosophic statistics (NS) for analysis of the indeterminacy of data. Nowadays, NS has gotten importance for analysis and is used in different fields, such as applied sciences [23], medicine for diagnoses data measurement [24], astrophysics for measurement of earth speed data [25], humanistic [26], and material science [27].

8.2.1 Preliminaries

Let X_N be a neutrosophic variable having interval of variance $X_N \epsilon[X_L, X_U]$ having size $n_N \epsilon[n_{NL}, n_{NU}]$ with indeterminacy $I_N \epsilon[I_{NL}, I_{NU}]$, so the neutrosophic formula (NF) is written as follows [28]:

$$X_{iN} = X_{iL} + X_{iU} I_N (i = 1, 2, 3, \ldots, n_N) \tag{8.1}$$

$X_{iN} \epsilon[X_{iL}, X_{IU}]$ has two parts: X_{iL} expressing the lower value which is under classical statistics, and $X_{iU} I_N$ is a upper part with indeterminacy $I_N \in [I_L, I_U]$. Similarly, the neutrosophic mean (NM) $\overline{X}_N \in \left[\overline{X}_L, \overline{X}_U\right]$ is defined as follows:

$$\overline{X}_N = \overline{X}_L - \overline{X}_U I_N I_N \in [I_L, I_U] \tag{8.2}$$

Here $\overline{X}_L = \sum_{i=1}^{nU} (X_{iL}/n_L)$ and $\overline{X}_U = \sum_{i=1}^{nU} (X_{iU}/n_U)$

Now, the neutrosophic statistical algorithm (NSL) for computer programs can be written as:

Step 1: Start the program
Step 2: Start an external or main loop from $i = 1 \ to \ i < = n_N$.
Step 3: Run internal loop from $I_N = I_L \ to \ I_N = I_U$ (first for $i = 1$)
Step 4: Execute the formula $X_{iN} = X_{iL} + X_{iU} I_N$: calculate values
Step 5: Increment + + and go to step 3
Step 6: End internal loop
Step 7: Increment + + and go to step 2
Step 8: End external loop
Step 7: End the program

8.2.2 List of abbreviations

- CN: Classical statistics
- NF: Neutrosophic formula
- NM: Neutrosophic mean
- CM: Classical mean
- CSA: Classical statistical algorithm
- NSA: Neutrosophic statistical algorithm
- BP: Blood pressure
- PR: Pulse rate
- T: Temperature (human body)

8.3 Literature review and state of the art

In May 2020, George et al. provided a guide to representing the data and measurements of COVID-19 in form of benchmarking across countries. They discussed the important facts and measurements as well as some recommendations for accessing and analyzing the data [29]. A prediction model based on deep learning has been proposed by Ref. [30]. The implementation of this model is based on the TensorFlow for the prediction of corona cases by previously available data. But this model contains some limitations like it requires some data source. The first time the effects of physical distance on students due to COVID-19 in Indonesia were analyzed by Ref. [31]. They collected data from 248 students aged less than 18 years. They found that students were not able to carry out their study properly and especially their daily activities. Ref. [32] analyzed the effects of preexisting morbidities in patients with COVID-19. They used already existing data available on different sites and journals for systemic review and meta-analysis. They divided these morbidities into eight groups having 27,670 samples. A random or fixed-effect model was used to estimate the likelihood of deaths, and an assessment of publication bias was performed through the visual inspection of the funnel plot asymmetry and Egger's regression test. Ref. [33] developed an automatic algorithm to segment and measure lung infection due to COVID-19 with the help of CT scanning. They used about 275 CT scan datasets. Statistical measurements were used for measuring, F-measure, accuracy, precision, sensitivity, specificity, Jacquard, MCC, and Dice, which showed that the proposed method is more accurate, flexible, and efficient.

This is an overview of the previous work on COVID-19. Also, different research papers containing work on the measurement of body temperature due to COVID-19 have been reviewed as in Refs. [34–37]. In this chapter, we have applied neutrosophic statistics to analyze the variance of the data measured from COVID-19 effects on the human body. A measurement of the human body temperature, pulse rate, and blood pressure changes has been attempted, and data are collected for analysis. Through this chapter, we have tried to analyze the variance of COVID-19 data with respect to changes in age measured directly from the patients, and for the first time, we have used classical and neutrosophic statistics for this purpose. For finding the better one, we have compared the analysis method and results of both.

8.4 Problem

The purpose of this study is to highlight an application of neutrosophic statistics in bio-information. We have used the neutrosophic formulas for the analysis of the changes in body temperature, blood pressure, and pulse rate of COVID-19 patients as shown in Fig. 8.3. We have collected the data of one fifty (150) patients aged from 20 to 40 twice, that is, first when they suffered from COVID-19 and the other after a month of recovery time. The digital thermometer is used for measuring the body temperature and a digital blood pressure meter for blood pressure and pulse rate of the human body. The purpose of this work is to observe and analyze the indeterministic variation of the body temperature, blood pressure, and pulse rate with respect to

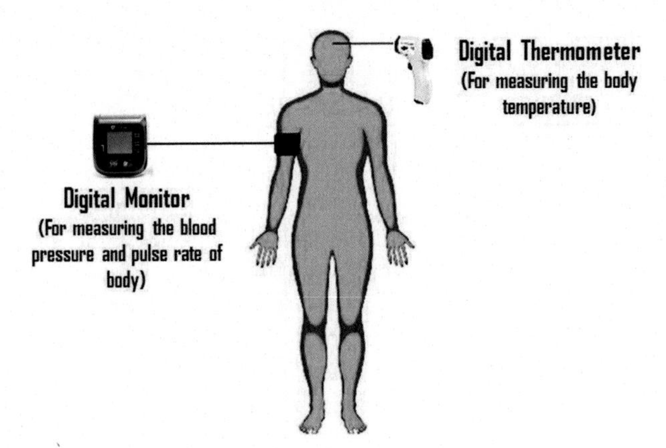

FIGURE 8.3 Setup to measure the data.

changes in age of the COVID-19-affected patients. So, we have used classical and neutrosophic approaches. We want to observe the best formula and method for analysis of the collected data.

8.5 Proposed solution to the problem

The proposed solution for the said problem is as follows:

8.5.1 Classical statistics

As the measured data are showing the variance, CS only deals with the fixed-point value, that is, determined values. So, the classical mean (CM) formula to convert these values into fixed-point values for CS analysis is as follows:

$$\overline{X}i = \frac{X_{Li} + X_{Ui}}{n} \quad (i = 0, 1, 2, 3, 4 \ldots N) \tag{8.3}$$

Here "$\overline{X}_i$" is the classical mean of the "ith" interval, and X_{Li} and X_{Ui} are the lower and upper values of the interval, respectively. Also, n is the number of values. For the following problem, $n = 2$ (as there are only two values of interval, i.e., lower and upper) and i is taken as age so $i = \text{age} = 20, 21, 22 \ldots 40$.

Now, the classical statistics algorithm of a said problem for a computer program can be written as:

Step 1: Start the program
Step 2: Run loop $i = 20$ *to* $i <= 40$

Step 3: Execute the formula: $\overline{X}i = \frac{X_{Li} + X_{Ui}}{n}$ and calculate values (the lower and upper values of interval with respect to age will be given by the user)
Step 4: Increment ++ and go to Step 2
Step 5: End the program

8.5.2 Neutrosophic statistics

Our study belongs to the change of temperature, blood pressure, and pulse rate with respect to change in age, so these factors can be written as the function of the age, that is, T(age), BP(age), and PR(age), respectively. Now, let $T(\text{age})_N$ is measured interval value of the temperature, that is,$T(\text{age})_N \in [T(\text{age})_L, T(\text{age})_U]$; here, $T(\text{age})_L$ and $T(\text{age})_U$ are the lower and upper value of the interval, respectively. The NF for temperature with indeterminacy $I_N \in [I_L, I_U]$ is written as follows:

$$T(\text{age})_N = T(\text{age})_L + T(\text{age})_U I_N; I_N \in [I_L, I_U] \tag{8.4}$$

From the above NF of $T(\text{age})_N \in [T(\text{age})_L, T(\text{age})_U]$ is an extension under the classical statistics. The equation contains two parts, that is, $T(\text{age})_L$ determined and $T(\text{age})_U I_N$ indeterminate parts. Moreover, $I_N \in [I_L, I_U]$ is known as an indeterminacy interval. Also, the measured interval $T(\text{age})_N \in [T(\text{age})_L, T(\text{age})_U]$ can be reduced to the classical or determined part if we choose $I_L = 0$ and I_U can be found by using $(R(T)_U - R(T)_L)/R(T)_U$. Similarly, for blood pressure and pulse rate as follows:

$$\text{BP}(\text{age})_N = \text{BP}(\text{age})_L + \text{BP}(\text{age})_U I_N; I_N \in [I_L, I_U] \tag{8.5}$$

$$\text{PR}(\text{age})_N = \text{PR}(\text{age})_L + \text{PR}(\text{age})_U I_N; I_N \in [I_L, I_U] \tag{8.6}$$

Now, the **neutrosophic statistical algorithm (NSL)** of the said problem for a computer program can be written as:

Step 1: Start the program
Step 2: Start an external or main loop from $i = 20$ *to* $i <= 40$.
Step 3: Run internal loop from $I_N = I_L = 0$ *to* $I_N = I_U$ (first for $i = 20$)
Step 4: Execute the formula for calculating values
$T(\text{age})_N = T(\text{age})_L + T(\text{age})_U I_N$ (for measuring the variance of body temperature)
$\text{BP}(\text{age})_N = \text{BP}(\text{age})_L + \text{BP}(\text{age})_U I_N$ (for measuring the variance of blood pressure)
$\text{PR}(\text{age})_N = \text{PR}(\text{age})_L + \text{PR}(\text{age})_U I_N$ (for measuring the variance of pulse rate)
Step 5: Increment ++ and go to Step 3
Step 6: End internal loop

Step 7: Increment + + and go to Step 2
Step 8: End external loop
Step 7: End the program

8.6 Analysis of data

The measured data of the patients are as shown in Tables 8.1 and 8.2. Table 8.1 expresses the data of patients collected when they suffered from

TABLE 8.1 Data measured when patients suffered from COVID-19.

Age (years)	Pulse rate (bpm)	Blood pressure (mmHg)		Temperature (F°)
		Diastolic	Systolic	
20	[66; 81]	[67; 76]	[101; 118]	[100.4; 105.1]
21	[61; 77]	[68; 77]	[102; 121]	[102.6; 104.2]
22	[64; 79]	[67; 74]	[112; 123]	[101.0; 106.6]
23	[64; 81]	[68; 76]	[102; 126]	[100.8; 105.0]
24	[62; 79]	[70; 77]	[103; 124]	[101.2; 104.9]
25	[67; 79]	[71; 76]	[100; 119]	[99.8; 106.9]
26	[63; 83]	[70; 77]	[99; 126]	[100.8; 105.2]
27	[64; 79]	[69; 77]	[104; 126]	[101.5; 104.6]
28	[63; 82]	[70; 76]	[101; 122]	[102.4; 105.3]
29	[66; 83]	[71; 77]	[101; 126]	[103.0; 105.7]
30	[62; 82]	[71; 78]	[110; 128]	[102.6; 104.3]
31	[70; 85]	[70; 73]	[109; 122]	[101.8; 105.2]
32	[62; 81]	[71; 77]	[108; 122]	[101.9; 105.8]
33	[65; 79]	[73; 79]	[113; 126]	[100.4; 106.1].
34	[67; 83]	[68; 72]	[105; 114]	[101.1; 105.9]
35	[66; 81]	[67; 75]	[107; 116]	[100.8; 104.5]
36	[62; 87]	[71; 79]	[112; 126]	[102.7; 103.4]
37	[70; 84]	[71; 75]	[111; 124]	[101.3; 103.7]
38	[71; 83]	[73; 76]	[113; 123]	[102.0; 103.6]
39	[64; 79]	[72; 77]	[111; 120]	[101.7; 103.4]
40	[59; 74]	[72; 79]	[112;119]	[100.1; 104.9]

TABLE 8.2 Data measured after recovery.

Age (years)	Pulse rate (bpm)	Blood pressure (mmHg)		Temperature (F°)
		Diastolic	Systolic	
20	[59; 74]	[74; 81]	[108; 125]	[98.2; 99.7]
21	[55; 72]	[74; 83]	[108; 127]	[97.6; 99.2]
22	[56; 71]	[75; 82]	[120; 131]	[98.0; 99.6]
23	[57; 74]	[75; 83]	[109; 131]	[97.5; 99.0]
24	[56; 73]	[76; 83]	[109; 130]	[97.2; 98.9]
25	[59; 71]	[79; 82]	[108; 127]	[97.3; 98.9]
26	[56; 76]	[76; 83]	[106; 131]	[96.8; 99.2]
27	[58; 73]	[77; 83]	[110; 130]	[97.5; 99.6]
28	[55; 74]	[78; 84]	[109; 130]	[96.4; 99.3]
29	[59; 76]	[76; 84]	[108; 133]	[99.0; 99.7]
30	[56; 76]	[77; 84]	[116; 134]	[98.2; 99.3]
31	[62; 77]	[78; 85]	[117; 130]	[96.8; 98.2]
32	[55; 74]	[76; 84]	[115; 129]	[96.9; 97.8]
33	[59; 73]	[79; 85]	[119; 130]	[97.4; 99.1].
34	[59; 75]	[76; 80]	[113; 122]	[97.1; 97.9]
35	[59; 74]	[74; 82]	[109; 123]	[96.8; 97.5]
36	[56; 81]	[77; 85]	[118; 132]	[96.7; 98.4]
37	[62; 76]	[79; 83]	[119; 132]	[97.3; 98.7]
38	[64; 76]	[80; 83]	[120; 130]	[97.0; 98.6]
39	[58; 73]	[78; 83]	[117; 126]	[96.7; 98.4]
40	[59; 74]	[80; 87]	[120;137]	[97.1; 98.9]

COVID-19, and Table 8.2 expresses the data collected after a month of recovery from COVID-19.

The above tables are showing the variation in human body pulse rate, blood pressure, and body temperature with respect to age (changing from 20 to 40 years). Also, the average changes in temperature, pulse rate, and blood pressure of patients suffering from COVID-19 are (101.7–104.3 F°), (71.5–79.78 bpm), and (66/110–74/121 mmHg), respectively, and the average changes in temperature, pulse rate and blood pressure of patient after

TABLE 8.3 Classical analysis of the data measured when patients suffered from COVID-19.

Age (years)	Pulse rate (bpm)	Blood pressure (mmHg)		Temperature (F°)
		Diastolic	Systolic	
20	73.5	71.5	109.5	102.7
21	69	72.5	111.5	103.4
22	71.5	70.5	117.5	103.8
23	72.5	72	113	102.9
24	70.5	73.5	113.5	103
25	73	73.5	109.5	103.3
26	73	73.5	11.5	103
27	71.5	73	114	103
28	72.5	73	111.5	103.8
29	74.5	74	113.5	104.3
30	72	74.5	119	103.4
31	77.5	71.5	115.5	103.5
32	71.5	74	115	103.8
33	72	76	118.5	103.2
34	75	70	109.5	103.5
35	73.5	71	109	102.6
36	74.5	75	119	103
37	77	73	117.5	102.4
38	77	74.5	118	102.8
39	71.5	74.5	115.5	102.5
40	66.5	75.5	120.5	102.5

recovery are (96.9−98.5 F°), (63.7−68.5 bpm), and (78/119.4−83.17/128 mmHg), respectively.

8.7 Statistical conditions

We have used two types of statistical conditions, that is, classical statistics and neutrosophic statistics, for the data analysis of the said problem.

8.7.1 Classical analysis

For the classical analysis of the problem, we have used the said procedure as shown in Section 8.5.1. We have applied the classical formula of average on measured values for classical analysis of variance of body temperature, pulse rate, and blood pressure with respect to change in age as shown in Tables 8.3 and 8.4.

TABLE 8.4 Classical analysis of the data measured after recovery.

Age (years)	Pulse rate (bpm)	Blood pressure (mmHg)		Temperature (F°)
		Diastolic	Systolic	
20	66.5	77.5	116.5	98.9
21	63.5	78.5	117.5	98.4
22	63.5	78.5	125.5	98.8
23	65.5	79	120	98.2
24	64.5	79.5	119.5	98
25	65	80.5	117.5	98.1
26	66	79.5	118.5	98
27	65.5	80	120	98.5
28	64.5	81	119.5	97.8
29	67.5	80	120.5	99.3
30	66	80.5	125	98.7
31	69.5	81.5	123.5	97.5
32	64.5	80	122	97.3
33	66	82	124.5	98.2
34	67	78	117.5	97.5
35	66.5	78	116	97.1
36	68.5	81	125	97.5
37	69	81	125.5	98
38	70	81.5	125	97.8
39	65.5	80.5	121.5	97.5
40	66.5	83.5	128.5	98

8.7.2 Neutrosophic analysis

For the neutrosophic analysis of the problem, we have used the above-said procedure as expressed in Section 8.5.2. We have applied this procedure to all the measured values of the body pulse rate, blood pressure, and body temperature with their indeterminacy, respectively, as shown in Tables 8.5 and 8.6.

TABLE 8.5 Neutrosophic analysis of the data measured when patients suffered from COVID-19.

Age (years)	Pulse rate (bpm)	Blood pressure (mmHg)		Temperature (F°)
		Diastolic	Systolic	
20	$66 + 81I_N;$ $I_N \in [0, 0.18]$	$67 + 76I_N;$ $I_N \in [0, 0.11]$	$101 + 118I_N;$ $I_N \in [0, 0.14]$	$100.4 + 105.1I_N;$ $I_N \in [0, 0.04]$
21	$61 + 77I_N;$ $I_N \in [0, 0.20]$	$68 + 77I_N;$ $I_N \in [0, 0.11]$	$102 + 121I_N;$ $I_N \in [0, 0.15]$	$102.6 + 104.2I_N;$ $I_N \in [0, 0.01]$
22	$64 + 79I_N;$ $I_N \in [0, 0.18]$	$67 + 74I_N;$ $I_N \in [0, 0.09]$	$112 + 123I_N;$ $I_N \in [0, 0.08]$	$101.0 + 106.6I_N;$ $I_N \in [0, 0.05]$
23	$64 + 81I_N;$ $I_N \in [0, 0.20]$	$68 + 76I_N;$ $I_N \in [0, 0.10]$	$102 + 126I_N;$ $I_N \in [0, 0.19]$	$100.8 + 105.0I_N;$ $I_N \in [0, 0.04]$
24	$62 + 79I_N;$ $I_N \in [0, 0.21]$	$70 + 77I_N;$ $I_N \in [0, 0.09]$	$103 + 124I_N;$ $I_N \in [0, 0.16]$	$101.2 + 104.9I_N;$ $I_N \in [0, 0.03]$
25	$67 + 79I_N;$ $I_N \in [0, 0.15]$	$71 + 76I_N;$ $I_N \in [0, 0.06]$	$100 + 119I_N;$ $I_N \in [0, 0.15]$	$99.8 + 106.9I_N;$ $I_N \in [0, 0.06]$
26	$63 + 83I_N;$ $I_N \in [0, 0.24]$	$70 + 77I_N;$ $I_N \in [0, 0.09]$	$99 + 126I_N;$ $I_N \in [0, 0.21]$	$100.8 + 105.2I_N;$ $I_N \in [0, 0.04]$
27	$64 + 79I_N;$ $I_N \in [0, 0.18]$	$69 + 77I_N;$ $I_N \in [0, 0.10]$	$104 + 126I_N;$ $I_N \in [0, 0.17]$	$101.5 + 104.6I_N;$ $I_N \in [0, 0.02]$
28	$63 + 82I_N;$ $I_N \in [0, 0.23]$	$70 + 76I_N;$ $I_N \in [0, 0.07]$	$101 + 122I_N;$ $I_N \in [0, 0.17]$	$102.4 + 105.3I_N;$ $I_N \in [0, 0.02]$
29	$66 + 83I_N;$ $I_N \in [0, 0.20]$	$71 + 77I_N;$ $I_N \in [0, 0.07]$	$101 + 126I_N$ $I_N \in [0, 0.19]$	$103.0 + 105.7I_N;$ $I_N \in [0, 0.02]$
30	$62 + 82I_N;$ $I_N \in [0, 0.24]$	$71 + 78I_N;$ $I_N \in [0, 0.08]$	$110 + 128I_N;$ $I_N \in [0, 0.14]$	$102.6 + 104.3I_N;$ $I_N \in [0, 0.01]$
31	$70 + 85I_N;$ $I_N \in [0, 0.17]$	$70 + 73I_N;$ $I_N \in [0, 0.04]$	$109 + 122I_N;$ $I_N \in [0, 0.10]$	$101.8 + 105.2I_N;$ $I_N \in [0, 0.03]$
32	$62 + 81I_N;$ $I_N \in [0, 0.23]$	$71 + 77I_N;$ $I_N \in [0, 0.07]$	$108 + 122I_N;$ $I_N \in [0, 0.11]$	$101.9 + 105.8I_N;$ $I_N \in [0, 0.03]$

(*Continued*)

TABLE 8.5 (Continued)

Age (years)	Pulse rate (bpm)	Blood pressure (mmHg)		Temperature (F°)
		Diastolic	Systolic	
33	$65+79I_N$; $I_N \in [0, 0.17]$	$73+79I_N$; $I_N \in [0, 0.07]$	$113+126I_N$; $I_N \in [0, 0.10]$	$100.4+106.1I_N$; $I_N \in [0, 0.05]$
34	$67+83I_N$; $I_N \in [0,0.19]$	$68+72I_N$; $I_N \in [0, 0.05]$	$105+114I_N$; $I_N \in [0, 0.07]$	$101.1+105.9I_N$; $I_N \in [0, 0.04]$
35	$66+81I_N$; $I_N \in [0, 0.18]$	$67+75I_N$; $I_N \in [0, 0.10]$	$107+116I_N$; $I_N \in [0, 0.07]$	$100.8+104.5I_N$; $I_N \in [0, 0.03]$
36	$62+87I_N$; $I_N \in [0, 0.28]$	$71+79I_N$; $I_N \in [0, 0.10]$	$112+126I_N$; $I_N \in [0, 0.11]$	$102.7+103.4I_N$; $I_N \in [0, 0.02]$
37	$70+84I_N$; $I_N \in [0, 0.16]$	$71+75I_N$; $I_N \in [0, 0.05]$	$111+124I_N$; $I_N \in [0, 0,10]$	$101.3+103.7I_N$; $I_N \in [0, 0.02]$
38	$71+83I_N$; $I_N \in [0, 0.14]$	$73+76I_N$; $I_N \in [0, 0.03]$	$113+123I_N$; $I_N \in [0, 0.08]$	$102.0+103.6I_N$; $I_N \in [0, 0.01]$
39	$64+79I_N$; $I_N \in [0, 0.18]$	$72+77I_N$; $I_N \in [0, 0.06]$	$111+120I_N$; $I_N \in [0, 0.07]$	$101.7+103.4I_N$; $I_N \in [0, 0.01]$
40	$59+74I_N$; $I_N \in [0, 0.20]$	$72+79I_N$; $I_N \in [0, 0.08]$	$112+119I_N$; $I_N \in [0, 0.05]$	$100.1+104.9I_N$; $I_N \in [0, 0.04]$

8.8 Discussion

In this chapter, we have analyzed the variation in the human body pulse rate, blood pressure, and temperature due to COVID-19. We have measured the data two times: first when they are affected by COVID-19 and second when they recovered. We have applied classical and neutrosophic statistics to analyze the data. The graphs of the classical analysis are shown in Figs. 8.4 and 8.5. Similarly, the graphs of neutrosophic analysis are shown in Figs. 8.6 and 8.7.

The above graphs are plotted to express the difference between the classical and neutrosophic statistical analyses of the measured data. From the classical graph, it is seen that it depends on fixed values of body temperature, pulse rate, and body blood pressure with respect to age variation; this is because the classical analysis has converted all intervals into fixed point by using the classical average formula [28] as can be seen in Tables 8.3 and 8.4. These graphs are not more flexible and only express single values of body temperature, blood pressure, and pulse rate at a specific age. One cannot find out more information from classical graphs [38]. But on the other

TABLE 8.6 Neutrosophic analysis of the data measured after recovery.

Age (years)	Pulse rate (bpm)	Blood pressure (mmHg)		Temperature (F°)
		Diastolic	Systolic	
20	$59+74I_N$; $I_N\in[0, 0.20]$	$74+81I_N$; $I_N\in[0, 0.08]$	$108+125I_N$; $I_N\in[0, 0.13]$	$98.2+99.7I_N$; $I_N\in[0, 0.01]$
21	$55+72I_N$; $I_N\in[0, 0.23]$	$74+83I_N$; $I_N\in[0, 0.10]$	$108+127I_N$; $I_N\in[0, 0.14]$	$97.6+99.2I_N$; $I_N\in[0, 0.01]$
22	$56+71I_N$; $I_N\in[0, 0.21]$	$75+82I_N$; $I_N\in[0, 0.08]$	$120+131I_N$; $I_N\in[0, 0.08]$	$98.0+99.6I_N$; $I_N\in[0, 0.01]$
23	$57+74I_N$; $I_N\in[0, 0.22]$	$75+83I_N$; $I_N\in[0, 0.09]$	$109+131I_N$; $I_N\in[0, 0.16]$	$97.5+99.0I_N$; $I_N\in[0, 0.01]$
24	$56+73I_N$; $I_N\in[0, 0.23]$	$76+83I_N$; $I_N\in[0, 0.08]$	$109+130I_N$; $I_N\in[0, 0.16]$	$97.2+98.9I_N$; $I_N\in[0, 0.01]$
25	$59+71I_N$; $I_N\in[0, 0.16]$	$79+82I_N$; $I_N\in[0, 0.03]$	$108+127I_N$; $I_N\in[0, 0.14]$	$97.3+98.9I_N$; $I_N\in[0, 0.01]$
26	$56+76I_N$; $I_N\in[0, 0.26]$	$76+83I_N$; $I_N\in[0, 0.08]$	$106+131I_N$; $I_N\in[0, 0.19]$	$96.8+99.2I_N$; $I_N\in[0, 0.02]$
27	$58+73I_N$; $I_N\in[0, 0.20]$	$77+83I_N$; $I_N\in[0, 0.07]$	$110+130I_N$; $I_N\in[0, 0.15]$	$97.5+99.6I_N$; $I_N\in[0, 0.02]$
28	$55+74I_N$; $I_N\in[0, 0.25]$	$78+84I_N$; $I_N\in[0, 0.07]$	$109+130I_N$; $I_N\in[0, 0.16]$	$96.4+99.3I_N$; $I_N\in[0, 0.02]$
29	$59+76I_N$; $I_N\in[0, 0.22]$	$76+84I_N$; $I_N\in[0, 0.09]$	$108+133I_N$; $I_N\in[0, 0.18]$	$99.0+99.7I_N$; $I_N\in[0, 0.07]$
30	$56+76I_N$; $I_N\in[0, 0.26]$	$77+84I_N$; $I_N\in[0, 0.08]$	$116+134I_N$; $I_N\in[0, 0.13]$	$98.2+99.3I_N$; $I_N\in[0, 0.01]$
31	$62+77I_N$; $I_N\in[0, 0.19]$	$78+85I_N$; $I_N\in[0, 0.08]$	$117+130I_N$; $I_N\in[0, 0.10]$	$96.8+98.2I_N$; $I_N\in[0, 0.01]$
32	$55+74I_N$; $I_N\in[0, 0.25]$	$76+84I_N$; $I_N\in[0, 0.09]$	$115+129I_N$; $I_N\in[0, 0.10]$	$96.9+97.8I_N$; $I_N\in[0, 0.09]$
33	$59+73I_N$; $I_N\in[0, 0.19]$	$79+85I_N$; $I_N\in[0, 0.07]$	$119+130I_N$; $I_N\in[0, 0.08]$	$97.4+99.1I_N$; $I_N\in[0, 0.01]$
34	$59+75I_N$; $I_N\in[0, 0.21]$	$76+80I_N$; $I_N\in[0, 0.05]$	$113+122I_N$; $I_N\in[0, 0.07]$	$97.1+97.9I_N$; $I_N\in[0, 0.008]$
35	$59+74I_N$; $I_N\in[0, 0.20]$	$74+82I_N$; $I_N\in[0, 0.09]$	$109+123I_N$; $I_N\in[0, 0.11]$	$96.8+97.5I_N$; $I_N\in[0, 0.007]$
36	$56+81I_N$; $I_N\in[0, 0.30]$	$77+85I_N$; $I_N\in[0, 0.09]$	$118+132I_N$; $I_N\in[0, 0.10]$	$96.7+98.4I_N$; $I_N\in[0, 0.01]$

(Continued)

TABLE 8.6 (Continued)

Age (years)	Pulse rate (bpm)	Blood pressure (mmHg)		Temperature (F°)
		Diastolic	Systolic	
37	$62 + 76I_N$; $I_N \in [0, 0.18]$	$79 + 83I_N$; $I_N \in [0, 0.04]$	$119 + 132I_N$; $I_N \in [0, 0.09]$	$97.3 + 98.7I_N$; $I_N \in [0, 0.01]$
38	$64 + 76I_N$; $I_N \in [0, 0.15]$	$80 + 83I_N$; $I_N \in [0, 0.03]$	$120 + 130I_N$; $I_N \in [0, 0.07]$	$97.0 + 98.6I_N$; $I_N \in [0, 0.01]$
39	$58 + 73I_N$; $I_N \in [0, 0.20]$	$78 + 83I_N$; $I_N \in [0, 0.06]$	$117 + 126I_N$; $I_N \in [0, 0.07]$	$96.7 + 98.4I_N$; $I_N \in [0, 0.01]$
40	$59 + 74I_N$; $I_N \in [0, 0.20]$	$80 + 87I_N$; $I_N \in [0, 0.08]$	$120 + 137I_N$; $I_N \in [0, 0.12]$	$97.1 + 98.9I_N$; $I_N \in [0, 0.01]$

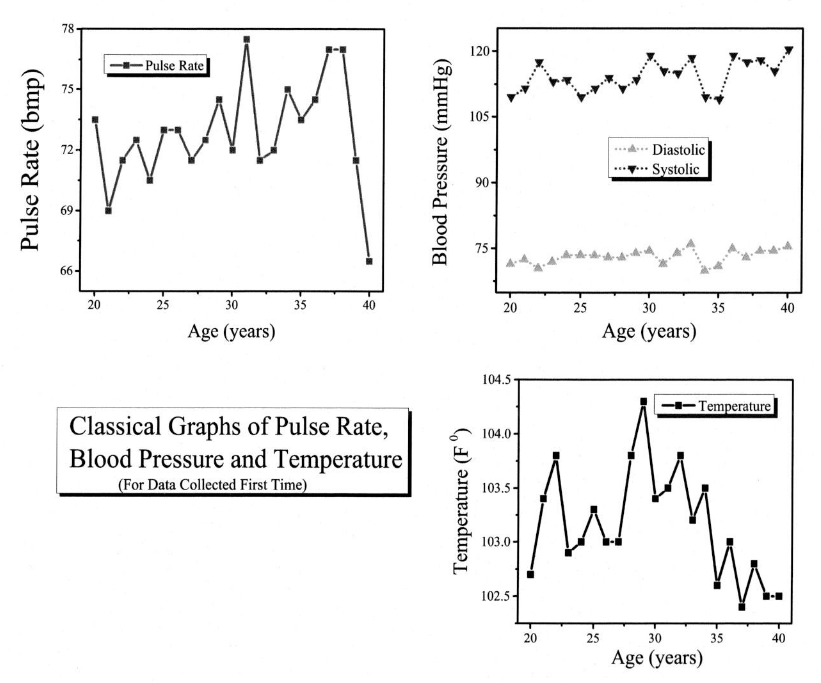

FIGURE 8.4 Classical plots of the data measured when patients suffered from COVID-19.

hand, neutrosophic analysis graphs are more informative as these are dealing with the indeterminacy as well as variance of data with respect to age. These are more flexible because one can find not only value at a specific point but

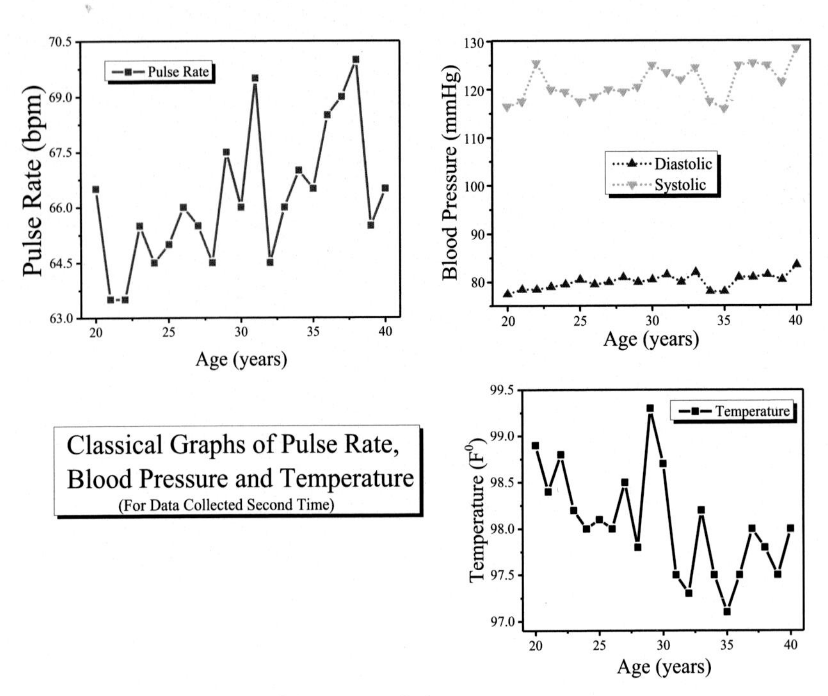

FIGURE 8.5 Classical plots of data measured after recovery.

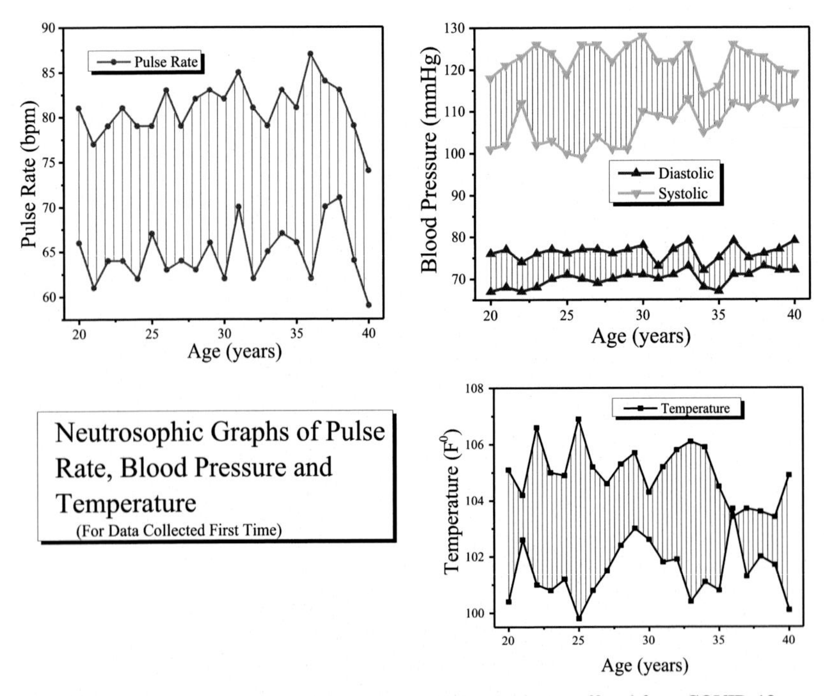

FIGURE 8.6 Neutrosophic plots of data measured when patients suffered from COVID-19.

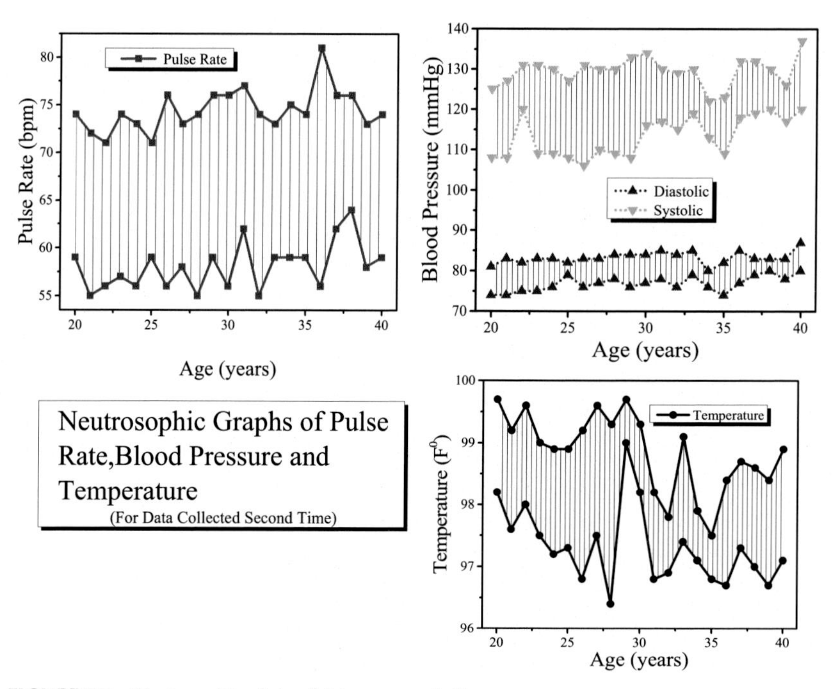

FIGURE 8.7 Neutrosophic plots of data measured after recovery.

also find value at near points which are found in the indeterminacy range of that specific point. In the literature review, we only found the classical graphs (single line or error bar graph) to express the data variation [39,40]. But these classical graphs are not suitable to express the data variation. That is why in the following work, we have moved to novel neutrosophic statistics approaches and graphs for understanding the data variation of body temperature, pulse rate, and blood pressure data with respect to change in age of the patients (for both measurements, that is, when they were suffered from COVID-19 and when they recovered from COVID-19), and we have found better result than the classical statistics approaches. In brief, we can conclude that neutrosophic statistics is more flexible, informative, and suitable for analyzing the said problem and much helpful in taking decision as compared to classical statistics [41,42].

8.9 Conclusions

In this chapter, we have tried to develop a neutrosophic statistical application in bio-information by analyzing the COVID-19 data which depends on the variance of body temperature, blood pressure, and pulse rate with respect to the change in the age of patients. We have the collected data with different digital tools of fifty hundred patients having age variations from 20 to 40 years. We have come to know that as the body temperature of patients due to COVID-19 increases which becomes a reason of increase in pulse rate and decrease in the body blood pressure. For observing this relation and collected data variation with respect to change in age, we have used the classical approach and the neutrosophic approach. As a result, we have gotten different graphs of classical and neutrosophic analyses. By comparing the method and obtained graphs, it is concluded that neutrosophic statistics is more effective, flexible, and informative to analyze a problem than classical statistics.

References

[1] C. Del Rio, P.N. Malani, 2019 novel coronavirus—important information for clinicians, JAMA 323 (11) (2020) 1039–1040.

[2] L.O. Gostin, COVID-19 reveals urgent need to strengthen the World Health Organization. Paper presented at the JAMA Health Forum, 2020.

[3] W. Alhazzani, M. Møller, Y. Arabi, M. Loeb, M. Gong, E. Fan, et al., Surviving Sepsis Campaign: guidelines on the management of critically ill adults with Coronavirus Disease 2019 (COVID-19), Intensive Care Med. (2020) 1–34.

[4] W.E. Wei, Z. Li, C.J. Chiew, S.E. Yong, M.P. Toh, V.J. Lee, Presymptomatic transmission of SARS-CoV-2—Singapore, January 23–March 16, 2020, Morb. Mortal. Wkly. Rep. 69 (14) (2020) 411.

[5] H.L. Feyisa, The World Economy at COVID-19 quarantine: contemporary review, Int. J. Econ. Financ. Manag. Sci. 8 (2) (2020) 63–74.

[6] S. Al-Samarrai, M. Gangwar, P. Gala, The impact of the COVID-19 pandemic on education financing, 2020.

[7] R.V. Hogg, J. McKean, A.T. Craig, Introduction to Mathematical Statistics, Pearson Education, 2005.

[8] H. Nassaji, Qualitative and Descriptive Research: Data Type Versus Data Analysis, Sage Publications Sage UK, London, England, 2015.

[9] Y. Zhang, A. Fu, C. Cai, P. Heng, Clustering categorical data. Paper presented at the Proceedings of the International Conference on Data Engineering, 2000.

[10] Z. Lotfi, Fuzzy sets, Inf. Control. 8 (3) (1965) 338–353.

[11] E.N. Nasibov, Fuzzy logic in statistical data analysis, in: M. Lovric (Ed.), International Encyclopedia of Statistical Science, Springer Berlin Heidelberg, Berlin, Heidelberg, 2011, pp. 558–563.

[12] P. Grzegorzewski, Testing statistical hypotheses with vague data, Fuzzy Sets Syst. 112 (3) (2000) 501–510.

[13] P. Grzegorzewski, k-sample median test for vague data, Int. J. Intell. Syst. 24 (5) (2009) 529–539.

[14] P. Grzegorzewski, M. Śpiewak, The sign test and the signed-rank test for interval-valued data, Int. J. Intell. Syst. 34 (9) (2019) 2122–2150.

[15] M. Aslam, A study on skewness and kurtosis estimators of wind speed distribution under indeterminacy, Theor. Appl. Climatol. 143 (3) (2021) 1227–1234.

[16] J. Chen, J. Ye, S. Du, Scale effect and anisotropy analyzed for neutrosophic numbers of rock joint roughness coefficient based on neutrosophic statistics, Symmetry 9 (10) (2017) 208.

[17] J. Chen, J. Ye, S. Du, R. Yong, Expressions of rock joint roughness coefficient using neutrosophic interval statistical numbers, Symmetry 9 (7) (2017) 123.

[18] F. Smarandache, Introduction to Neutrosophic Statistics: Infinite Study, Romania-Educational Publisher, Columbus, OH, 2014.

[19] M. Aslam, Design of the Bartlett and Hartley tests for homogeneity of variances under indeterminacy environment, J. Taibah Univ. Sci. 14 (1) (2020) 6–10.

[20] M. Aslam, On detecting outliers in complex data using Dixon's test under neutrosophic statistics, J. King Saud. Univ.-Sci. 32 (3) (2020) 2005–2008.

[21] M. Aslam, Neutrosophic analysis of variance: application to university students, Complex. Intell. Syst. 5 (4) (2019) 403–407.

[22] F. Smarandache, Multispace & Multistructure. Neutrosophic Transdisciplinarity (100 Collected Papers of Science) (Vol. 4), Infinite Study, 2010.

[23] V. Christianto, R.N. Boyd, F. Smarandache, Three Possible Applications of Neutrosophic Logic in Fundamental and Applied Sciences, Infinite Study, 2020.

[24] J. Ye, Improved cosine similarity measures of simplified neutrosophic sets for medical diagnoses, Artif. Intell. Med. 63 (3) (2015) 171–179.

[25] M. Aslam, Enhanced statistical tests under indeterminacy with application to earth speed data, Earth Sci. Inform. (2021) 1–7.

[26] F. Smarandache, The Neutrosophic Research Method in Scientific and Humanistic Fields, 2010.

[27] U. Afzal, M. Aslam, A.H. Al-Marshadi, Analyzing Imprecise Graphene Foam Resistance Data, Materials Research Express, 2022.

[28] U. Afzal, H. Alrweili, N. Ahamd, M. Aslam, Neutrosophic statistical analysis of resistance depending on the temperature variance of conducting material, Sci. Rep. 11 (1) (2021) 23939. Available from: https://doi.org/10.1038/s41598-021-03347-z.

[29] B. George, B. Verschuere, E. Wayenberg, B.L. Zaki, A guide to benchmarking COVID-19 performance data, Public. Adm. Rev. 80 (4) (2020) 696–700.

[30] M. Gupta, R. Jain, S. Taneja, G. Chaudhary, M. Khari, E. Verdú, Real-time measurement of the uncertain epidemiological appearances of COVID-19 infections, Appl. Soft Comput. 101 (2021) 107039.

[31] M. Ardan, F.F. Rahman, G.B. Geroda, The influence of physical distance to student anxiety on COVID-19, Indonesia, J. Crit. Rev. 7 (17) (2020) 1126–1132.

[32] M.M.A. Khan, M.N. Khan, M.G. Mustagir, J. Rana, M.S. Islam, M.I. Kabir, Effects of underlying morbidities on the occurrence of deaths in COVID-19 patients: a systematic review and meta-analysis, J. Glob. Health 10 (2) (2020).

[33] A. Oulefki, S. Agaian, T. Trongtirakul, A.K. Laouar, Automatic COVID-19 lung infected region segmentation and measurement using CT-scans images, Pattern Recognit. 114 (2021) 107747.

[34] S.V. Bhavani, E.S. Huang, P.A. Verhoef, M.M. Churpek, Novel temperature trajectory subphenotypes in COVID-19, Chest 158 (6) (2020) 2436.

[35] C. Dzien, W. Halder, H. Winner, M. Lechleitner, Covid-19 screening: are forehead temperature measurements during cold outdoor temperatures really helpful? Wien. Klinische Wochenschr. 133 (7) (2021) 331–335.

[36] E. Moisello, P. Malcovati, E. Bonizzoni, Thermal sensors for contactless temperature measurements, occupancy detection, and automatic operation of appliances during the COVID-19 pandemic: a review, Micromachines 12 (2) (2021) 148.
[37] F. Piccinini, G. Martinelli, A. Carbonaro, Reliability of body temperature measurements obtained with contactless infrared point thermometers commonly used during the COVID-19 pandemic, Sensors 21 (11) (2021) 3794.
[38] U. Afzal, N. Ahmad, Q. Zafar, M. Aslam, Fabrication of a surface type humidity sensor based on methyl green thin film, with the analysis of capacitance and resistance through neutrosophic statistics, RSC Adv. 11 (61) (2021) 38674–38682.
[39] U. Afzal, J. Afzal, M. Aslam, Analyzing the imprecise capacitance and resistance data of humidity sensors, Sens. Actuators: B. Chem. 367 (2022) 132092–132097. Available from: https://doi.org/10.1016/j.snb.2022.132092.
[40] U. Afzal, M. Aslam, K. Maryam, A.H. AL-Marshadi, F. Afzal, Fabrication and characterization of a highly sensitive and flexible tactile sensor based on indium zinc oxide (IZO) with imprecise data analysis, ACS Omega 7 (2022) 32569–32576. Available from: https://doi.org/10.1021/acsomega.2c04156.
[41] U. Afzal, F. Afzal, K. Maryam, M. Aslam, Fabrication of flexible temperature sensors to explore indeterministic data analysis for robots as an application of Internet of Things, RSC Adv. 12 (2022) 17138–17145. Available from: https://doi.org/10.1039/d2ra03015b.
[42] U. Afzal, M. Aslam, F. Afzal, K. Maryam, N. Ahmad, Q. Zafar, Z. Farooq, Fabrication of a graphene-based sensor to detect the humidity and the temperature of a metal body with imprecise data analysis, RSC Adv. 12 (2022) 21297–21308. Available from: https://doi.org/10.1039/d2ra03474c.

Chapter 9

A study of human respiration rate through neutrosophic statistics

Usama Afzal[1] and Muhammad Aslam[2]

[1]School of Microelectronics, Tianjin University, Tianjin, P.R. China, [2]Department of Statistics, Faculty of Science, King Abdulaziz University, Jeddah, Saudi Arabia

9.1 Introduction and motivation

Respiration is a biochemical process of the transformation of oxygen from the environment to the body of living things. In respiration, cells of an organism or tissue get oxygen and give out the carbon dioxide gas to the environment [1]. The inlet oxygen combines with glucose for energy purposes, and as a result, carbon is released from the body. The cell respiration formula can be written as follows in Fig. 9.1:

From Fig. 9.1, it can be seen that the reaction between glucose and oxygen produces carbon dioxide, water, and energy for the body. There are two types of respiration: first is aerobic and the other is anaerobic. Fig. 9.1 relation is called aerobic respiration because it uses the oxygen from the environment. Generally, humans and animals all depend on aerobic respiration. Similarly, anaerobic is the respiration type that does not depend on oxygen. Such type of respiration is found in bacteria and fungi [2] as shown in Fig. 9.2.

But the yeast respiration is not the same as anaerobic respiration [3] as shown in Fig. 9.3

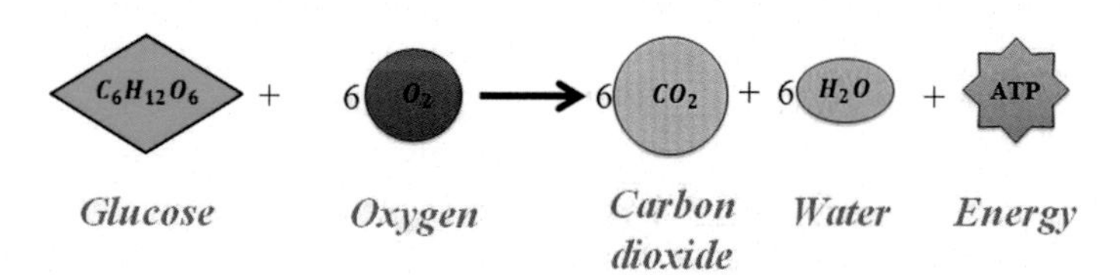

FIGURE 9.1 Human cell respiration formula.

Cognitive Intelligence with Neutrosophic Statistics in Bioinformatics.
DOI: https://doi.org/10.1016/B978-0-323-99456-9.00021-0

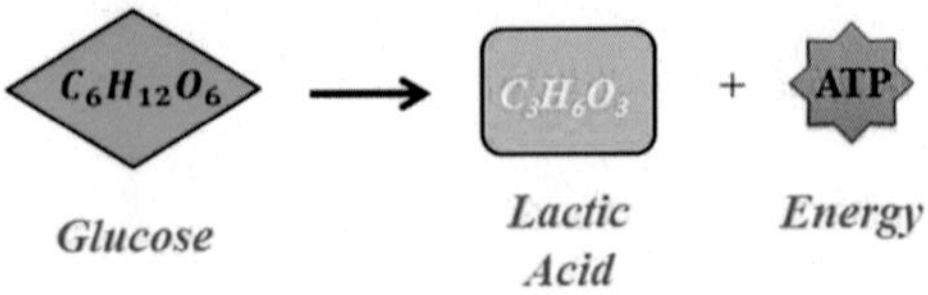

$C_6H_{12}O_6$ → $C_3H_6O_3$ + ATP

Glucose Lactic Acid Energy

FIGURE 9.2 Anaerobic respiration formula.

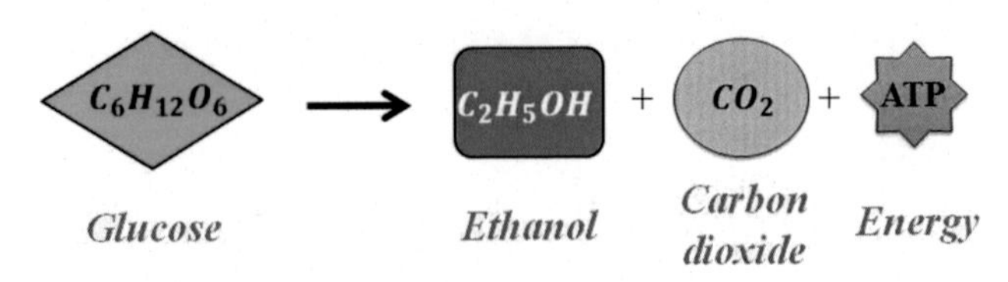

FIGURE 9.3 Yeast respiration formula.

As both animals including human beings and plants depend on respiration, different measurements have been attempted to understand the rate of respiration in human beings, animals, and plants through different methods. The analysis of these measured data is performed through statistics approaches. Statistics is a significant tool that is used for the analysis of data [4] of a particular problem whether in table form or graphical form. It is an interdisciplinary field as it is used in scientific research and other fields to collect, manipulate, and analyze data. There are a number of statistics approaches/logics for analyzing the data such as classical statistics (CS), fuzzy statistics (FS), and neutrosophic statistics (NS). Classical statistics only deals with determinate values, FS deals with only fuzzy sets, and NS deals with indeterminate value intervals.

9.2 Background, definitions, and abbreviations

9.2.1 Classical statistics

Classical statistics is generally known as frequentist statistics [5] because it consists of many ordinary formulas and approaches for the analysis of the value. As it is mentioned in Section 9.1, CS analyzes the fixed-point values, that is, without indetermined, for example, the use of the classical least-square formula and classical mean formula, etc., Schervish [6]. As it deals only with fixed-point value, when we deal with interval values it fails in the analysis of interval values and cannot explain the problem properly. This is because the interval values consist of uncertainty and it is tough to deal with uncertainty through the CS. So, one can easily conclude that CS is unable to well analyze the data containing uncertainty [7].

9.2.2 Fuzzy statistics

A more flexible approach that deals with uncertainty is known as FS. In 1965, the fuzzy logic which is based on the fuzzy sets (that is why also called the fuzzy set theory) was developed by [8]. FS is used to analyze the data when there is uncertainty, that is, interval form. It is more flexible and accurate in dealing with the interval data having uncertainty. For understanding this concept, we take an example. In classical logic, if a number is a member of a set, then it will be "1" otherwise it will be "0," but in fuzzy, the number will be between "0" and "1," that is, [0, 1], simultaneously [9]. Different researchers have worked on FS. For example, P. Grzegorzewski proposed the statistical test using the fuzzy logic [10]. The purpose of FS is to use an indeterministic approach to analyze and observe the uncertainty in data. Similarly, the median analysis with the help of this statistical test was also proposed by [11]. Analysis of the interval data using signed-rank test was attempted by [12].

9.2.3 Neutrosophic statistics

Neutrosophic statistics is a more flexible statistics approach than classical and FS. The main purpose of this approach is to study and analyze the interval accurately with indeterminacy [13]. Neutrosophic statistics is more informative than classical statistics because CS only deals with determined values but NS deals with indeterminate values. This is a substantial benefit over the classical approach. Moreover, the neutrosophic technique is more helpful than classical techniques; see examples of ([14,15]; F [16]). M. Aslam developed numerous statistical approaches/techniques under NS [17,18]. The neutrosophic logic is also used for the analysis of indeterministic data. It is the generalization of fuzzy logic and is more reliable and informative than fuzzy logic [19]. Smarandache claimed it expressed the efficiency of neutrosophic logic over fuzzy logic [20]. Similarly, in 2014 Smarandache, F. proposed NS for the analysis of indeterminacy of data. Nowadays, NS has gotten importance for analysis and is used in different fields, such as applied sciences [21], medicine for diagnoses data measurement [22], astrophysics for measurement of earth speed data [23], humanistic [24], and material science [25].

9.2.4 Difference between these statistics approach

For expressing the difference between these approaches, we take an example of the statement. According to CS, a statement is a time right or wrong. But in FS, the statement is between wrong and right, which means there is uncertainty. Similarly, in NS, the statement is right and wrong at the same time but with indeterminacy. For further differences, see Fig. 9.4.

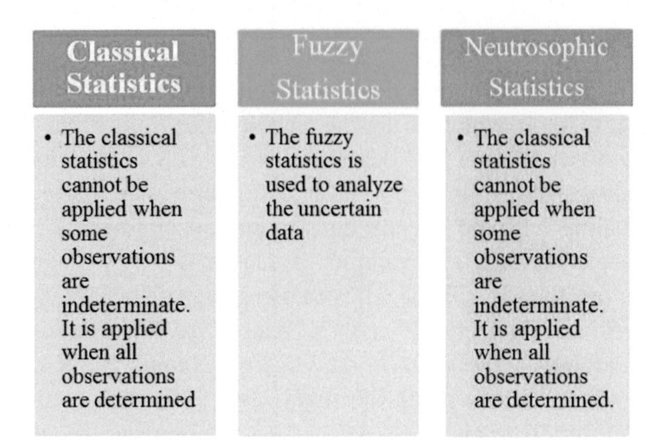

FIGURE 9.4 Difference between classical, fuzzy, and neutrosophic analyses.

9.2.5 Definition

The neutrosophic formula for the neutrosophic variable y_N is written as:

$$y_{iN} = y_{iL} + y_{iU} I_N (i = 1, 2, 3, \ldots, n_N) \quad (9.1)$$

The neutrosophic variable (NV) has $y_N \epsilon [y_L, y_U]$interval with the size $n_N \epsilon [n_{NL}, n_{NU}]$, and indeterminacy interval for this NV is $I_N \epsilon [I_{NL}, I_{NU}]$.

The above equation $y_{iN} \epsilon [y_{iL}, y_{IU}]$ has two parts: y_{iL} expressing the lower value which is under CS and $y_{iU} I_N$ is an upper part with indeterminacy $I_N \in [I_L, I_U]$. Similarly, the neutrosophic mean (NM) $\overline{y}_N \in [\overline{y}_L, \overline{y}_U]$ is defined as follows:

$$\overline{y}_N = \overline{y}_L - \overline{y}_U I_N \in [I_L, I_U] \quad (9.2)$$

Here $\overline{y}_L = \sum_{i=1}^{nU} (y_{iL}/n_L)$ and $\overline{y}_U = \sum_{i=1}^{nU} (y_{iU}/n_U)$

Now, the *neutrosophic statistical algorithm (NSL)* for computer programs can be written as:

Step 1: Start program
Step 2: Start an external or main loop from $i = 1$ to $i < = n_N$.
Step 3: Run an internal loop from $I_N = I_L$ to $I_N = I_U$ (first for i = 1)
Step 4: Execute the formula $y_{iN} = y_{iL} + y_{iU} I_N$: calculate values
Step 5: Increment + + and go to step 3
Step 6: End internal loop
Step 7: Increment + + and go to step 2
Step 8: End external loop
Step 7: End program

9.2.6 List of abbreviations

- NS: neutrosophic statistics
- FS: fuzzy statistics

- CN: classical statistics
- NF: neutrosophic formula
- NM: neutrosophic mean
- CM: classical mean
- CSA: classical statistical algorithm
- NSA: neutrosophic statistical algorithm
- RR: respiration rate
- rpm: respiration per minute

9.3 Literature review and state of the art

In October 2005, Venkatesh et al. used the ultra-wideband (UWB) system based on the impulse for the measurement and analysis of the respiration of the human body [26]. They used this UWB system to monitor the body's chest motion and then on the basis of the motion to estimate the respiration rate and pulse rate. They collected the experiment several times under different conditions and expressed the accurate measurement and analysis of it. The first contact-free measurement of respiration, as well as heartbeat and body movement during sleep, was measured by [27]. They used the high-resolution force sensor which was placed on bedposts. Similarly, the use of sound sensors for the measurement of heartbeats and respiration of the human body was explained by [28]. The human body sound was detected by the sound sensor, and then, it measured both heartbeats and respiration rate. Similarly, [29] used the remote compact sensors for the measurement of the human body's important signs, that is, respiration and pulse rates [29]. The sensor was based on electromagnetic waves with a 24 GHz frequency. They found an excellent result and analysis of the heartbeat and respiration. A 240 kHz ultrasonic sensor was used to measure the respiration rate [30]. [31] used digital temperature sensor (TMP100) with the 12C interface for measuring thermistor and chest expansion. He found the best measurement and data for analysis. Similarly, different devices have been used for the measurement of the respiration of the body like the seismo-cardiography based on wrist-worn accelerometer [32], Doppler radar technique [33], and much more. Similarly, Zhang, F. et al. proposed the diffraction model for finding human respiration rate through Wi-Fi devices [34]. All of these techniques have different measuring methods, but generally, all have used the classical methods for the analysis of their measured data.

In this chapter, we have applied NS to analyze the variance of the data measured of human body respiration rate. A measurement of the human body respiration rate changes with respect to age has been attempted, and data are collected for analysis. Through this chapter, we have tried to analyze the variance of respiration rate data, and for the first time, we have used NS for this purpose. Classical statistics also has been used for data analysis. For finding the better one, we have compared the analysis method and results of both.

9.4 Problem

The purpose of this study is to highlight an application of NS in bio-information. We have used the neutrosophic formulas for the analysis of the changes in respiration in the human body. We have collected the data of two hundred (200) people (all have normal health condition, which means no one has a breathing disorder) aged from 1 to 18, through the counting of breaths in 1 min. All data are then collected at normal body temperature with respect to age. The respiration process of the human body is shown in Fig. 9.5. The main goal of this work is to analyze the indeterministic variation of the human respiration rate with respect to changes in the age from 1 to 18 years. This data analysis is performed through the modified neutrosophic formula and classical formula. We want to observe the best formula and method for analyzing human respiration rate with changes in age.

9.5 The proposed solution of the problem

For the understanding and analysis of the measured value, we have used CS and NS and compared their results.

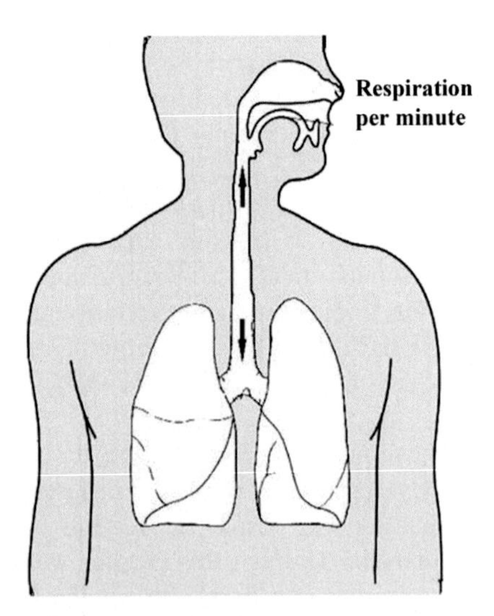

FIGURE 9.5 Human respiration measurement.

9.5.1 Classical statistics

As the measured data show the variance, CS only deals with the fixed-point value, that is, determined values. So, the classical mean (CM) formula to convert these values into fixed-point values for CS analysis is as follows:

$$\overline{A}i = \frac{a_{Li} + a_{Ui}}{n}(i = 0, 1, 2, 3, 4...N) \tag{9.3}$$

Here "$\overline{A}_i$" is the classical mean of the "ith" interval and a_{Li} and a_{Ui} are the lower and upper values of the interval, respectively. Also, n is the number of values. For the following problem, n = 2 (as there are only two values of interval, that is, lower and upper) and i is taken as age so $i = \text{age} = 1, 2, 3, ..., 18$

Now the *classical statistics algorithm (CSA)* of the said problem for a computer program can be written as:

Step 1: Start program
Step 2: Run loop $i = 1$ to $i <= 18$
Step 3: Execute formula: $\overline{A}i = \frac{a_{Li} + a_{Ui}}{n}$ and calculate values (the lower and upper values of interval with respect to age will be given by the user)
Step 4: Increment ++ and go to Step 2
Step 5: End program

9.5.2 Neutrosophic statistics

Our study belongs to the variation in human respiration rate with respect to change in age, so this factor can be written as the function of the age, that is, RR(age). Now, let $RR(age)_N$ is measured interval value of the respiration rate, that is, $RR(age)_N \in [RR(age)_L, RR(age)_U]$; here $RR(age)_L$ and $RR(age)_U$ are the lower and upper values of the interval, respectively. The NF for respiration rate with indeterminacy $I_N \in [I_L, I_U]$ is written as follows:

$$RR(age)_N = RR(age)_L + RR(age)_U I_N; I_N \in [I_L, \quad I_U] \tag{9.4}$$

From the above NF of $RR(age)_N \in [RR(age)_L, RR(age)_U]$ is an extension under the CS. The equation contains two parts, that is, $RR(age)_L$determined and $RR(age)_U I_N$ indeterminate parts. Moreover, $I_N \in [I_L, I_U]$ is known as an indeterminacy interval. Also, the measured interval $RR(age)_N \in [RR(age)_L, RR(age)_U]$ can be reduced to the classical or determined part if we choose $I_L = 0$, and I_U can find by using $(RR(T)_U - RR(T)_L)/RR(T)_U$.

Now, the *neutrosophic statistical algorithm (NSL)* of the said problem for a computer program can be written as:

Step 1: Start program
Step 2: Start an external or main loop from $i = 1$ to $i <= 18$.

Step 3: Run internal loop from $I_N = I_L = 0$ to $I_N = I_U$ (first for $i = 1$)
Step 4: Execute formula for calculating values
$RR(age)_N = RR(age)_L + RR(age)_U I_N$ (for measuring the variance of body respiration rate)
Step 5: Increment ++ and go to step 3
Step 6: End internal loop
Step 7: Increment ++ and go to step 2
Step 8: End external loop
Step 7: End program

9.6 Analysis of data

The measured data of the people are shown in Table 9.1. This table expresses the data of people with respect to changes in age from 1 to 18 years.

TABLE 9.1 Collected data on human respiration rate with respect to change in age.

Age (years)	Respiration rate (rpm)
1	[34, 39]
2	[26, 35]
3	[24, 34]
4	[25, 33]
5	[27, 30]
6	[20, 23]
7	[19, 25]
8	[15, 25]
9	[18, 26]
10	[16, 24]
11	[15, 19]
12	[16, 19]
13	[17, 20]
14	[15, 17]
15	[16, 18]
16	[16, 19]
17	[17, 20]
18	[13, 18]

Table 9.1 shows the variation of the respiration rate of the human body with respect to age (changing from 1 to 18 years). Also, the average change in respiration rate is about 20–25 rpm.

9.7 Statistical conditions

We have used two types of statistical conditions, that is, CS and NS, for the data analysis of the said problem.

9.7.1 Classical analysis

For the classical analysis of the problem, we have used the above-said procedure as shown in Section 9.5.1. We have applied the classical formula of average on measured values of the respiration rate for classical analysis as shown in Table 9.2.

TABLE 9.2 Classical analysis of the data measured.

Age (years)	Classical analysis
1	36.5
2	30.5
3	29
4	29
5	28.5
6	21.5
7	22
8	20
9	22
10	20
11	17
12	17.5
13	18.5
14	16
15	17
16	17.5
17	18.5
18	15.5

9.7.2 Neutrosophic analysis

For the neutrosophic analysis of the problem, we have used the above-said procedure as expressed in Section 9.5.2. We have applied the neutrosophic formula (depending on the indeterministic intervals for each data value changing with respect to the age of the people) to measured values of the respiration rate for neutrosophic analysis as shown in Table 9.3.

9.8 Discussion

In this chapter, we have analyzed the variation of the human respiration rate with respect to variation of age. We have applied classical and NS to analyze the data. The graphs of the classical and neutrosophic are as shown in Fig. 9.6 and Fig. 9.7, respectively.

The above graphs 9.6 and 9.7 are expressing the classical graph based on CS and neutrosophic graph based on NS, respectively. From the classical

TABLE 9.3 Neutrosophic analysis of the data.

Age (years)	Neutrosophic analysis
1	$34+39I_N$; $I_N \in [0, 0.128]$
2	$26+35I_N$; $I_N \in [0, 0.257]$
3	$24+34I_N$; $I_N \in [0, 0.294]$
4	$25+33I_N$; $I_N \in [0, 0.243]$
5	$27+30I_N$; $I_N \in [0, 0.100]$
6	$20+23I_N$; $I_N \in [0, 0.130]$
7	$19+25I_N$; $I_N \in [0, 0.240]$
8	$15+25I_N$; $I_N \in [0, 0.400]$
9	$18+26I_N$; $I_N \in [0, 0.308]$
10	$16+24I_N$; $I_N \in [0, 0.334]$
11	$15+19I_N$; $I_N \in [0, 0.211]$
12	$16+19I_N$; $I_N \in [0, 0.158]$
13	$17+20I_N$; $I_N \in [0, 0.150]$
14	$15+20I_N$; $I_N \in [0, 0.250]$
15	$16+18I_N$; $I_N \in [0, 0.112]$
16	$16+19I_N$; $I_N \in [0, 0.158]$
17	$17+20I_N$; $I_N \in [0, 0.150]$
18	$13+18I_N$; $I_N \in [0, 0.278]$

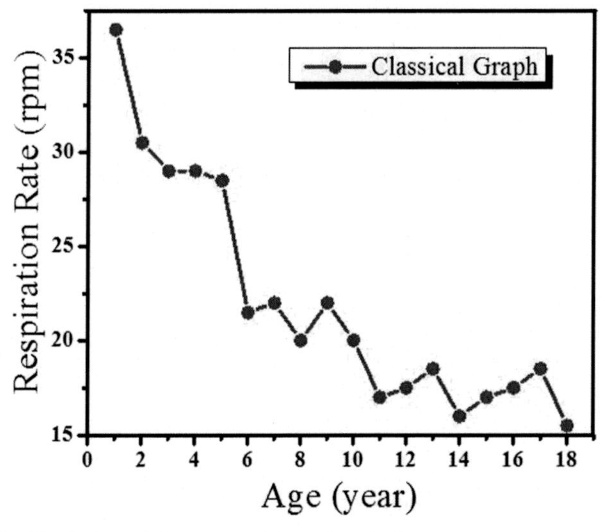

FIGURE 9.6 Classical graph for measured data of respiration.

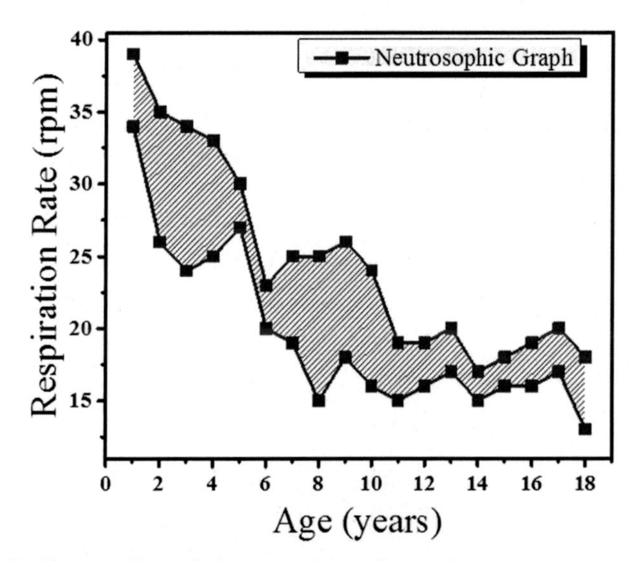

FIGURE 9.7 Neutrosophic graph for measured data of respiration.

graph, it is seen that it depends on fixed values of respiration rate as the classical analysis has converted all intervals into fixed-point values with respect to change in age [35]. That is why the graph is not much flexible and only expresses the value of respiration rate at only a specific point [36]. One cannot find out more information from classical graph. But on the other hand, the neutrosophic analysis graph is more informative and flexible [35] as this is dealing with the indeterminacy as well as the variance of respiration data with respect to change in age. For example, at 1 year age, the classical value of the respiration rate is 36.5 rpm (a single or fixed-point value), that is, RR (1 year) = 36.5 rpm. But on the other hand, neutrosophic analysis gives an equation $R(1\ \text{year}) = 34 + 39I_N$ with the indeterminacy interval $I_N \in [0, 0.128]$. According to neutrosophic analysis, the value of respiration rate of people aged 1 year lies between 34 and 39 by indeterminacy values.

9.8.1 Comparison section

For understanding the comparison of the classical and NS, we draw the combined graph as in Fig. 9.8:

From the above graph as shown in Fig. 9.8, it is seen that the human respiration rate has decreased with increase in the age of the people. Similarly, the graph also shows the comparison between classical and neutrosophic analyses. One can easily see that classical graphs are less flexible and informative to explain the human respiration rate with respect to change in age as these are drawn on fixed-point values. But neutrosophic graphs are enough flexible and informative to explain and conclude the problem [37,38]. Generally, it is seen that most researchers have used classical graphs in their works either in the form of single-line plot or in the form of error bar graph for expressing the data variation. But statistically, these are not effective plots as the error bar shows the error found in data, not variance. When a

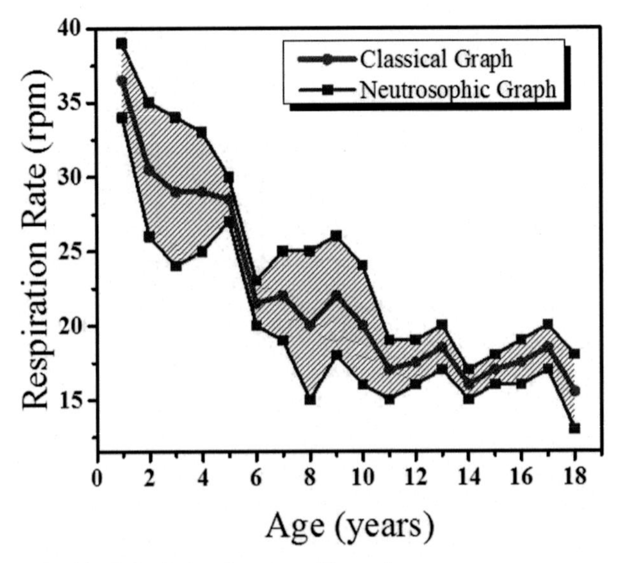

FIGURE 9.8 Combined classical and neutrosophic graphs.

researcher uses an error bar, it means that he is graphically expressing the data error (which may be due to personal error or data-collecting error) instead of data variance. But the neutrosophic approach is a remarkable approach for the analysis of the data variance which can easily run in any type of computer software (to execute the big data). Also, this graph shows that the neutrosophic approach is a generalization of the classical approach; that is, one can also observe the classical graph through the neutrosophic approach as it is following between the middle of the neutrosophic graph. As a result, it is observed that NS is more effective, reliable, and informative to analyze the human respiration rate with respect to change in age than CS [39,40].

9.9 Conclusion

In this chapter, we have tried to develop a neutrosophic statistical application in bio-information by analyzing the variation of the respiration rate of the human body with respect to the variation of age. For this purpose, we have examined the respiration rate of 200 people having age variations from 1 to 18 years and collected the data in conventional way, that is, respiration per minute (rpm) in intervals. It is observed that the respiration rate decreases with increases in the age of humans. As we are keenly interested in the analysis of data variation, the variance of respiration data has been analyzed through neutrosophic and CS. As a result, it is observed that classical analysis is not effective in explaining the variation of human respiration rate with respect to changes in age. But the neutrosophic analysis is more effective. It is concluded that NS is more effective, reliable, and much informative in taking decisions.

References

[1] Hinic-Frlog, S., Introductory Animal Physiology, 2019.

[2] W. Dilling, H. Cypionka, Aerobic respiration in sulfate-reducing bacteria, FEMS Microbiol. Lett. 71 (1–2) (1990) 123–127.

[3] S.-M. Fendt, U. Sauer, Transcriptional regulation of respiration in yeast metabolizing differently repressive carbon substrates, BMC Syst. Biol. 4 (1) (2010) 1–11.

[4] R.V. Hogg, J. McKean, A.T. Craig, Introduction to Mathematical Statistics, Pearson Education, 2005.

[5] Press, S.J., Applied Multivariate Analysis: Using Bayesian and Frequentist Methods of Inference, Courier Corporation, 2005.

[6] M.J. Schervish, Theory of Statistics, Springer Science & Business Media, 2012.

[7] M. Aslam, N. Khan, M. Albassam, Control chart for failure-censored reliability tests under uncertainty environment, Symmetry 10 (12) (2018) 690.

[8] Z. Lotfi, Fuzzy sets, Inf. Control. 8 (3) (1965) 338–353.

[9] E.N. Nasibov, Fuzzy logic in statistical data analysis, in: M. Lovric (Ed.), International Encyclopedia of Statistical Science, Springer Berlin Heidelberg, Berlin, Heidelberg, 2011, pp. 558–563.

[10] P. Grzegorzewski, Testing statistical hypotheses with vague data, Fuzzy Sets Syst. 112 (3) (2000) 501–510.

[11] P. Grzegorzewski, k-sample median test for vague data, Int. J. Intell. Syst. 24 (5) (2009) 529–539.

[12] P. Grzegorzewski, M. Śpiewak, The sign test and the signed-rank test for interval-valued data, Int. J. Intell. Syst. 34 (9) (2019) 2122–2150.

[13] M. Aslam, A study on skewness and kurtosis estimators of wind speed distribution under indeterminacy, Theor. Appl. Climatol. 143 (3) (2021) 1227–1234.

[14] J. Chen, J. Ye, S. Du, Scale effect and anisotropy analyzed for neutrosophic numbers of rock joint roughness coefficient based on neutrosophic statistics, Symmetry 9 (10) (2017) 208.

[15] J. Chen, J. Ye, S. Du, R. Yong, Expressions of rock joint roughness coefficient using neutrosophic interval statistical numbers, Symmetry 9 (7) (2017) 123.

[16] F. Smarandache, Introduction to Neutrosophic Statistics: Infinite Study, *Romania-Educational Publisher, Columbus, OH, USA*, 2014.

[17] M. Aslam, Design of the Bartlett and Hartley tests for homogeneity of variances under indeterminacy environment, J. Taibah Univ. Sci. 14 (1) (2020) 6–10.

[18] M. Aslam, On detecting outliers in complex data using Dixon's test under neutrosophic statistics, J. King Saud. Univ.-Sci. 32 (3) (2020) 2005–2008.

[19] M. Aslam, Neutrosophic analysis of variance: application to university students, Complex. Intell. Syst. 5 (4) (2019) 403–407.

[20] F. Smarandache, Multispace & Multistructure. Neutrosophic Transdisciplinarity (100 Collected Papers of Science), Infinite Study (2010).

[21] V. Christianto, R.N. Boyd, F. Smarandache, Three Possible Applications of Neutrosophic Logic in Fundamental and Applied Sciences, Infinite Study, 2020.

[22] J. Ye, Improved cosine similarity measures of simplified neutrosophic sets for medical diagnoses, Artif. Intell. Med. 63 (3) (2015) 171–179.

[23] M. Aslam, Enhanced statistical tests under indeterminacy with application to earth speed data, Earth Sci. Inform. (2021) 1–7.

[24] F. Smarandache, The Neutrosophic Research Method in Scientific and Humanistic Fields, 2010b.

[25] U. Afzal, M. Aslam, A.H. Al-Marshadi, Analyzing imprecise graphene foam resistance data, Mater. Res. Express (2022).

[26] S. Venkatesh, C.R. Anderson, N.V., Rivera, R.M. Buehrer, Implementation and analysis of respiration-rate estimation using impulse-based UWB, Paper presented at the MILCOM 2005-2005 IEEE Military Communications Conference, 2005.

[27] M. Brink, C.H. Müller, C. Schierz, Contact-free measurement of heart rate, respiration rate, and body movements during sleep, Behav. Res. Methods 38 (3) (2006) 511–521.

[28] H. Miwa, K. Sakai, Development of heart rate and respiration rate measurement system using body-sound, Paper presented at the 2009 9th International Conference on Information Technology and Applications in Biomedicine, 2009.

[29] J.H. Choi, D.K. Kim, A remote compact sensor for the real-time monitoring of human heartbeat and respiration rate, IEEE Trans. Biomed. Circ. Syst. 3 (3) (2009) 181–188.

[30] S.D. Min, J.K. Kim, H.S. Shin, Y.H. Yun, C.K. Lee, M. Lee, Noncontact respiration rate measurement system using an ultrasonic proximity sensor, IEEE Sens. J. 10 (11) (2010) 1732–1739.

[31] A. Agnihotri, Human body respiration measurement using digital temperature sensor with I2C interface, Int. J. Sci. Res. Publ. 3 (3) (2013) 1–8.
[32] M. Haescher, D.J. Matthies, J. Trimpop, B. Urban, A study on measuring heart-and respiration-rate via wrist-worn accelerometer-based seismocardiography (SCG) in comparison to commonly applied technologies, Paper presented at the Proceedings of the 2nd international Workshop on Sensor-based Activity Recognition and Interaction, 2015.
[33] J. Tu, T. Hwang, J. Lin, Respiration rate measurement under 1-D body motion using single continuous-wave Doppler radar vital sign detection system, IEEE Trans. Microw. Theory Tech. 64 (6) (2016) 1937–1946.
[34] F. Zhang, D. Zhang, J. Xiong, H. Wang, K. Niu, B. Jin, From fresnel diffraction model to fine-grained human respiration sensing with commodity wi-fi devices, in: Proceedings of the ACM on Interactive, Mobile, Wearable and Ubiquitous Technologies, 2018, vol. 2, iss (1), pp. 1–23
[35] U. Afzal, H. Alrweili, N. Ahamd, M. Aslam, Neutrosophic statistical analysis of resistance depending on the temperature variance of conducting material, Sci. Rep. 11 (1) (2021) 23939. Available from: https://doi.org/10.1038/s41598-021-03347-z.
[36] U. Afzal, N. Ahmad, Q. Zafar, M. Aslam, Fabrication of a surface type humidity sensor based on methyl green thin film, with the analysis of capacitance and resistance through neutrosophic statistics, RSC Adv. 11 (61) (2021) 38674–38682.
[37] U. Afzal, J. Afzal, M. Aslam, Analyzing the imprecise capacitance and resistance data of humidity sensors, Sens. Actuators: B. Chem. 367 (2022) 132092–132097. Available from: https://doi.org/10.1016/j.snb.2022.
[38] U. Afzal, F. Afzal, K. Maryam, M. Aslam, Fabrication of flexible temperature sensors to explore indeterministic data analysis for robots as an application of Internet, RSC Adv. 12 (2022) 17138–17145. Available from: https://doi.org/10.1039/d2ra03015b.
[39] U. Afzal, M. Aslam, F. Afzal, K. Maryam, N. Ahmad, Q. Zafar, Z. Farooq, Fabrication of a graphene-based sensor to detect the humidity and the temperature of a metal body with imprecise data analysis, RSC Adv. 12 (2022) 21297–21308. Available from: https://doi.org/10.1039/d2ra03474c.
[40] U. Afzal, M. Aslam, K. Maryam, A.H. AL-Marshadi, F. Afzal, Fabrication and characterization of a highly sensitive and flexible tactile sensor based on indium zinc oxide (IZO) with imprecise data analysis, ACS Omega 7 (36) (2022) 32569–32576. Available from: https://doi.org/10.1021/acsomega.2c04156.

Chapter 10

Neutrosophic diagnosis of rural women vulnerable to anemia

Nivetha Martin[1], U. Pavithra Krishnan[2], Florentin Smarandache[3] and Dragisa Stanujkic[4]

[1]Department of Mathematics, Arul Anandar College (Autonomous), Karumathur, Madurai, Tamil Nadu, India, [2]Department of Food Science and Technology, Arul Anandar College (Autonomous), Karumathur, Madurai, Tamil Nadu, India, [3]Mathematics, Physical and Natural Sciences Division, University of New Mexico, Gallup, NM, United States, [4]Technical Faculty in Bor, University of Belgrade, Bor, Serbia

10.1 Introduction and motivation

The domain of classical statistics dealing with deterministic observations of data plays a crucial role in drawing inferences from the collected research data. The ample availability of statistical testing methods smoothens the process of drawing inferences. Diagnostic tests are predominantly applied in the field of medicine and the related areas to diagnose especially the prevalence of diseases and disorders. Researchers of varied backgrounds have applied the classical diagnostic test for data analysis to conclude. To mention a few, Greiner et al. [1] analyzed veterinary data; Lalkhen et al. [2] examined the clinical data; and Leeflang et al. [3] studied the data on disease prevalence by making use of the diagnostic test in the classical sense.

The deterministic nature of data sets is always questionable, as the presence of uncertainty intervenes in the experimental scenario many times. To handle such uncertain instances, diagnostic test under a fuzzy environment was used by researchers to make inferences on disease data.

Hasima and Khan [4] used the fuzzy diagnostic test for identifying liver disease. A fuzzy logic-based medical expert system was developed by Jimmy et al. to diagnose chronic kidney disease. Sorina Zahana [5] applied score-based test under fuzzy logic and the pre-diagnosis test under fuzzy environment. Kim et al. [6] developed fuzzy rule-based system for liver diagnosis. Mumini et al. [7] constructed Genetic-Neuro-Fuzzy inferential model for diagnosing tuberculosis. Phelps and Hutson [8] applied the fuzzy gold standard method of a diagnostic test. Also, Castanho et al. [9],

Cognitive Intelligence with Neutrosophic Statistics in Bioinformatics.
DOI: https://doi.org/10.1016/B978-0-323-99456-9.00022-2

Smith and Slenning [10], and Bhise et al. [11] have applied fuzzy diagnostic tests to disease data.

In spite of the existing fuzzy diagnostic models, neutrosophic diagnostic tests are developed to handle the situations of indeterminacy, and the fuzzy logic is extended to neutrosophic logic to facilitate the management of indeterminacy. Neutrosophic sets are characterized by the values of truth, indeterminacy, and falsity. Neutrosophic statistics is an alternative to fuzzy statistics, and it helps in analyzing the data obtained from the population with fuzziness and indeterminacy. Smarandache [12] introduced neutrosophic statistics in which the statistical concepts were discussed under a neutrosophic environment.

Mumtaz Ali [13] has used neutrosophic recommender system for disease diagnosis. Basha et al. [14] have applied neutrosophic logic to construct a hybrid model for making inferences on X-rays. Muhammad Aslam [15] proposed a neutrosophic diagnostic test and a gold standard test. The test has been applied to identify the presence of diabetics. The proposed test has been validated by an example, but it has not been applied to research data as such to the best of our knowledge. The feasibility and the pragmatic nature of the neutrosophic diagnostic test developed by Muhammad Aslam [15] have made the authors apply the method to make the diagnostic study of anemic rural women with real-time data. Also, the neutrosophic diagnostic test was applied earlier only to diagnose the presence or the absence of diabetics, but in this paper, the neutrosophic gold standard diagnostic test was applied to diagnose the anemic and nonanemic persons and also the type of anemic present in the samples taken for study.

The paper is structured into many sections consisting of preliminaries, literature review, description of the study area, neutrosophic statistical analysis of the data, discussion of the results, conclusion, and future scope of the research work.

10.2 Background, definitions, and notations

The basic data representation and the preliminaries involved in the neutrosophic gold standard diagnostic test are presented in Tables 10.1 and 10.2.

10.3 Literature review and state of the art

The economic and social disparities have cleaved the populace into healthy and vulnerable groups in developing nations like India. The inadequate supply of nutritious foods and supplementary substances contributes to anemia, a serious health problem that affects the normal functioning of the human system, especially the womenfolk. Anemia is a hematological abnormality condition related to reduced oxygen transportation ability of the blood due to the reduction in red blood cells, packed cell volume, and hemoglobin. According to a statistical report of the World Health Organization, rural women of reproductive age are highly susceptible to anemia. Pregnant women face many health issues due to low levels of hemoglobin.

TABLE 10.1 Data representation under neutrosophic gold standard diagnostic test.

		Factual diagnosis				
		Disease +	**Disease −**	**Uncertainty positive**	**Uncertainty negative**	**Total**
Factual test reports	Disease +	$[O_L, O_U]$	$[M_L, M_U]$	$[P_L, P_U]$	$[p_L, p_U]$	$[O_L + M_L + P_L + p_L, O_U + M_U + P_U + p_U]$
	Disease -	$[V_L, V_U]$	$[W_L, W_U]$	$[Y_L, Y_U]$	$[y_L, y_U]$	$[V_L + W_L + Y_L + y_L, V_U + W_U + Y_U + y_U]$
	Uncertainty positive	$[d_L, d_U]$	$[e_L, e_U]$	$[G_L, G_U]$	$[g_L, g_U]$	$[d_L + e_L + G_L + g_L, d_U + e_U + G_U + g_U]$
	Uncertainty negative	$[k_L, k_U]$	$[q_L, q_U]$	$[Z_L, Z_U]$	$[z_L, z_U]$	$[k_L + q_L + Z_L + z_L, k_U + q_U + Z_U + z_U]$
	Total	$[O_L + V_L + d_L + k_L, O_U + V_U + d_U + k_U]$	$[M_L + W_L + e_L + q_L, M_U + W_U + e_U + q_U]$	$[P_L + Y_L + G_L + Z_L, P_U + Y_U + G_U + Z_U]$	$[p_L + y_L + g_L + z_L, p_U + y_U + g_U + z_U]$	N_N

TABLE 10.2 Basic terms of neutrosophic gold standard diagnostic test.

Terminology	Definition	Formula
Neutrosophic sensitivity	Percentage of right identification of the diseased persons in consensus with the test results	$\frac{[O_L, O_U]}{[O_L + V_L + d_L + k_L, O_U + V_U + d_U + k_U]}$
Neutrosophic practitioner sensitivity	Percentage of the persons being accepted by the practitioner to be diseased under uncertainty and with the test indicating the presence of disease	$\frac{[P_L, P_U]}{[P_L + Y_L + G_L + Z_L, P_U + Y_U + G_U + Z_U]}$
Neutrosophic test sensitivity	Percentage of the persons being accepted by the practitioner to be diseased and the tests indicating the presence of disease under uncertainty	$\frac{[d_L, dU]}{[O_L + V_L + d_L + k_L, O_U + V_U + d_U + k_U]}$
Neutrosophic practitioner-test sensitivity	Percentage of the diseased persons for which the practitioner and the test are under uncertainty	$\frac{[G_L, G_U]}{[P_L + Y_L + G_L + Z_L, P_U + Y_U + G_U + Z_U]}$
Neutrosophic specificity	Percentage of right identification of the non-diseased persons in consensus with the test results	$\frac{[W_L, W_U]}{[M_L + W_L + e_L + q_L, M_U + W_U + e_U + q_U]}$
Neutrosophic practitioner specificity	Percentage of the persons being accepted by the practitioner to be non-diseased under uncertainty and with the test indicating the absence of disease	$\frac{[y_L, y_U]}{[p_L + y_L + g_L + z_L, p_U + y_U + g_U + z_U]}$
Neutrosophic test specificity	Percentage of the persons being accepted by the practitioner to be non-diseased and the tests indicating the absence of disease under uncertainty	$\frac{[q_L, q_U]}{[M_L + W_L + e_L + q_L, M_U + W_U + e_U + q_U]}$
Neutrosophic practitioner-test specificity	Percentage of the non-diseased persons for which the practitioner and the test are under uncertainty	$\frac{[z_L, z_U]}{[p_L + y_L + g_L + z_L, p_U + y_U + g_U + z_U]}$
Neutrosophic practitioner positive predictiveValue	Percentage of diseased persons accepted by the practitioner under uncertainty	$\frac{[P_L, P_U]}{[O_L + M_L + P_L + p_L, O_U + M_U + P_U + p_U]}$
Neutrosophic test positive predictive value	Percentage of persons tested positive under uncertainty and are actually diseased	$\frac{[d_L, d_U]}{[d_L + e_L + G_L + g_L, d_U + e_U + G_U + g_U]}$

Premature births, low birth weight, postmartum hemorrhage are some of the most common consequences of anemic disorder. The prevalence of anemic is classified into severely anemic, moderately anemic, and less anemic based on the levels of hemoglobin.

Researchers have made a descriptive study on the prevalence of anemic among rural people of India, especially with a special focus on women. Bentley and Griffiths [16] have investigated the occurrences and causative factors of anemia among women in Andhra Pradesh, a state in South India. It was inferred that rural women are greatly affected in comparison to urban women, and especially, the factors contributing to such health burden are low income and low body mass index. Anu Rammohan [17] studied the impact of vegetarianism on the iron deficiency of women in India and discussed the dietary impacts. Maninder Kaur and Kochar [18] stated the causative reasons for anemia in rural women in Haryana, a state in North India. Sharadamani [19] conducted a research study on the people of Karaikudi and inferred that women are highly vulnerable to anemia. From the literature on rural women and anemia, it was inferred that the rural womenfolk of reproductive age are greatly suffering from this anemic disorder, and to the best of our knowledge, only descriptive study has been done so far using classical statistical tools. Henceforth, this paper attempts to make a diagnostic study on the presence or absence of anemia and the types of anemia using neutrosophic diagnostic tests.

10.4 Problem definition

To investigate the problem of diagnosis of anemia and the types of anemia of the rural womenfolk, the research study is conducted in Karumathur, a gram panchayat located in Chellampatti block in the Madurai district of Tamil Nadu state in India. The total geographical area is around 1630.3 hectares with a population of around 6737 and 1,902 housing settlements as per the 2011 census. The total male population is 3559, and the female population is 3178. The total literacy rate is 65.6%, and concerning females, it is 26.1%. The total working population is 55.1%. Agriculture and livestock rearing are the primary occupations of the people. The study area was economically and socially underdeveloped a few decades back with dominant practices of female infanticide and child marriage system, but presently, the establishment of schools and colleges has enlightened the people of this region. In spite of such developments, the practice of the girl's early marriage system still exists, and this has many consequential impacts on women health.

The subjects of the study are the women belonging to the age group of 18–40, and around 62.9% of the female population belongs to this group. The samples were chosen by the method of systematic sampling. This research study aims to categorize whether the subjects investigated are anemic or not. The hemoglobin levels are obtained from the subjects through clinical tests. If Hb levels are below the standard levels, then the subject is presumed to be anemic. But many times, there may arise certain discrepancies between investigation and the clinical test results. To handle such instances, the method of gold standard test under neutrosophic environment is used to make statistical analysis.

10.5 Analysis

Tables 10.3, 10.4, 10.5, and 10.6 represent the neutrosophic data

The results obtained by using the method of neutrosophic gold standard diagnostic test are presented in Table 10.7

10.6 Discussion

The neutrosophic sensitivity indeterminacy interval values from 61.58% to 63.39% under anemia represent that the woman has anemia, and certainly, the test results hold to be positive. The neutrosophic practitioner sensitivity indeterminacy interval values from 7.5% to 8.57% represent the percentage of women being accepted by the practitioner to be anemic under uncertainty and with the test indicating the presence of anemia. The neutrosophic test sensitivity indeterminacy interval values from 0.12% to 0.175% under anemia represent the percentage of women being accepted by the practitioner to be anemic and the tests indicating the presence of anemia under uncertainty. The neutrosophic practitioner-test sensitivity interval values from 32.5% to 34.28% under anemia represent the percentage of the anemic women for which the practitioner and the test are under uncertainty. In a similar fashion, the neutrosophic specificity indeterminacy interval values from 58.82% to 58.48% under anemia represent that the women have no anemia and the test will result negative.

The neutrosophic practitioner specificity indeterminacy interval values from 19.23% to 20.69% represent the percentage of women being accepted by the practitioner to be nonanemic under uncertainty and with the test indicating the absence of anemia. The neutrosophic test specificity indeterminacy interval values from 0.17% to 0.34% under anemia represent the percentage of women being accepted by the practitioner to be non-anemic, and the tests indicate the absence of anemia under uncertainty. The neutrosophic practitioner-test specificity interval values from 11.54% to 17.24% under anemia represent the percentage of the nonanemic women for which the practitioner and the test are under uncertainty. The neutrosophic practitioner positive predictive indeterminacy interval values from 0.149% to 0.15% represents the percentage of anemic women accepted by the practitioner under uncertainty, and the neutrosophic test positive predictive indeterminacy interval values from 6.45% to 8.82% represent the percentage of persons tested positive under uncertainty are anemic. Similarly, the same way of analysis shall be made for the category of mild, moderate, and severe anemic women.

TABLE 10.3 Neutrosophic data of anemia diagnosis.

		Factual diagnosis				
		Anemia +	**Anemia −**	**Uncertainty positive**	**Uncertainty negative**	**Total**
Factual test reports	Anemia +	[1050,1053]	[950,950]	[3,3]	[4,4]	[2007,2010]
	Anemia −	[650,650]	[1350,1355]	[3,4]	[5,6]	[2008,2015]
	Uncertainty positive	[2,3]	[3,4]	[12,13]	[14]	[31,34]
	Uncertainty negative	[3,4]	[4,8]	[17,20]	[3,5]	[27,37]
	Total	[1705,1710]	[2307,2317]	[35,40]	[26,29]	[4073,4096]

TABLE 10.4 Neutrosophic data of mild anemia diagnosis

		Factual diagnosis				
		Mild Anemia +	**Mild Anemia −**	**UncertaintyPositive**	**UncertaintyNegative**	**Total**
Factual test reports	Mild anemia +	[210,211]	[840,842]	[2,2]	[3,3]	[1055,1058]
	Mild anemia −	[130,130]	[520,522]	[2,4]	[3,5]	[655,661]
	Uncertainty positive	[2,2]	[2,3]	[7,8]	[10]	[21,23]
	Uncertainty negative	[2,4]	[3,5]	[12,15]	[2,4]	[19,28]
	Total	[344,347]	[1365,1372]	[23,29]	[18,22]	[1750,1770]

TABLE 10.5 Neutrosophic data of moderate anemia diagnosis.

		Factual diagnosis				
		Moderate anemia +	**Moderate anemia −**	**Uncertainty positive**	**Uncertainty negative**	**Total**
Factual test reports	Moderate anemia +	[420,421]	[630,632]	[2,3]	[3,3]	[1055,1059]
	Moderate anemia −	[260,260]	[390,392]	[1,3]	[4,5]	[655,660]
	Uncertainty positive	[1,2]	[1,3]	[10,13]	[10,11]	[22,29]
	Uncertainty negative	[2,3]	[3,4]	[11]	[2,3]	[18,21]
	Total	[683,686]	[1024,1031]	[24,30]	[19,22]	[1750,1769]

TABLE 10.6 Neutrosophic data of severe anemia diagnosis.

		Factual diagnosis				
		Severe anemia +	**Severe anemia -**	**Uncertainty positive**	**Uncertainty negative**	**Total**
Factual test reports	Severe anemia +	[315,316]	[735,737]	[2,2]	[1,1]	[1053,1056]
	Severe anemia −	[195,195]	[455,458]	[3,4]	[4,4]	[657,661]
	Uncertainty positive	[2,3]	[3,4]	[9,10]	[10]	[24,27]
	Uncertainty negative	[1,2]	[2,3]	[6,7]	[3,3]	[12,15]
	Total	[513,516]	[1195,1202]	[20,23]	[18]	[1746,1759]

TABLE 10.7 Results of neutrosophic gold standard diagnostic test.

Calculated values of neutrosophic statistical analysis	Anemia	Mild anemia	Moderate anemia	Severe anemia
Neutrosophic sensitivity	[61.58,63.39]	[61.05,60.81]	[61.49,61.37]	[61.4,61.2]
Neutrosophic practitioner sensitivity	[8.57,7.5]	[8.7,6.89]	[8.33,10]	[10,8.69]
Neutrosophic test sensitivity	[0.12,0.175]	[0.58,0.576]	[0.146,0.29]	[0.38,0.58]
Neutrosophic practitioner-test sensitivity	[34.28,32.5]	[30.43,27.58]	[41.67,43.33]	[45,43.5]
Neutrosophic specificity	[58.52,58.48]	[38.09,38.05]	[38.09,38.02]	[38.07.38.10]
Neutrosophic practitioner specificity	[19.23,20.69]	[16.67,22.72]	[21.05,22.73]	[22.22,22.22]
Neutrosophic test specificity	[0.17,0.34]	[0.22,0.36]	[0.29,0.233]	[0.167,0.24]
Neutrosophic practitioner-test specificity	[11.54,17.24]	[11.14,18.18]	[10.53,13.64]	[16.67,16.67]
Neutrosophic practitioner positive predictive value	[0.15,0.149]	[0.18,0.19]	[0.18,0.28]	[0.18,0.19]
Neutrosophic test positive predictive value	[6.45,8.82]	[9.52,8.69]	[4.55,6.89]	[0.33,11.11]

10.7 Conclusion

This research work applies the neutrosophic gold standard diagnostic test proposed to determine the presence and absence of anemic disease in rural women. The neutrosophic diagnostic test is more compatible, and it helps in analyzing the uncertain environment. It is inferred that rural women are greatly affected by anemic disease, and also, a majority of rural women are highly affected by moderate anemia as neutrosophic sensitivity of moderate anemia is higher in comparison with other types of anemia. The research work can be extended by exploring the factors contributing to anemic in rural women, and also, the vulnerability of urban women to anemic shall be compared concerning rural women.

References

[1]. M. Greiner, I.A. Gardner, Application of diagnostic tests in veterinary epidemiologic studies, Prev. Vet. Med. 45 (1–2) (2000) 43–59.

[2]. A.G. Lalkhen, A. McCluskey, Clinical tests: sensitivity and specificity, Cont. Educ. Anaesth. Crit. Care Pain. 8 (6) (2008) 221–223.

[3]. M.M. Leeflang, A.W. Rutjes, J.B. Reitsma, L. Hooft, P.M. Bossuyt, Variation of a test's sensitivity and specificity with disease prevalence, CMAJ 185 (11) (2013) E537–E544.

[4]. A. Hashmi, M.S. Khan, Diagnosis blood test for liver disease using fuzzy logic, Int. J. Sci.: Basic Appl. Res. (IJSBAR) 20 (1) (2015) 151–183.

[5]. S. Zahan, C. Michael, S. Nikolakeas, Fuzzy expert system for myocardial ischemia diagnosis, Fuzzy and Neuro-Fuzzy Systems in Medicine, CRC Press, 2017, pp. 211–242.

[6]. M.O. Omisore, O.W. Samuel, E.J. Atajeromavwo, A genetic-neuro-fuzzy inferential model for diagnosis of tuberculosis, Appl. Comput. Inform. 13 (1) (2017) 27–37.

[7]. Kim, J.W., & Oh, K.W., Development of fuzzy rule-based liver function test diagnosis system, in: Proceedings of the KOSOMBE Conference (Vol. 1992, No. 05, pp. 155–160), The Korea Society of Medical and Biological Engineering, 1992.

[8]. C.E. Phelps, A. Hutson, Estimating diagnostic test accuracy using a "fuzzy gold standard", Med. Decis. Mak. 15 (1) (1995) 44–57.

[9]. M.J. Castanho, L.C. Barros, A. Yamakami, L.L. Vendite, Fuzzy receiver operating characteristic curve: an option to evaluate diagnostic tests, IEEE Trans. Inf. Technol. Biomed. 11 (3) (2007) 244–250.

[10]. R.D. Smith, B.D. Slenning, Decision analysis: dealing with uncertainty in diagnostic testing, Prev. Vet. Med. 45 (1–2) (2000) 139–162.

[11]. V. Bhise, S.S. Rajan, D.F. Sittig, R.O. Morgan, P. Chaudhary, H. Singh, Defining and measuring diagnostic uncertainty in medicine: a systematic review, J. Gen. Intern. Med. 33 (1) (2018) 103–115.

[12]. Smarandache, F., Introduction to Neutrosophic Statistics, Infinite Study (2014). https://arxiv.org/pdf/1406.2000.

[13]. M. Ali, N.D. Thanh, N. Van Minh, A neutrosophic recommender system for medical diagnosis based on algebraic neutrosophic measures, Appl. Soft Comput. 71 (2018) 1054–1071.

[14]. S.H. Basha, A.M. Anter, A.E. Hassanien, A. Abdalla, Hybrid intelligent model for classifying chest X-ray images of COVID-19 patients using genetic algorithm and neutrosophic logic, Soft Comput. (2021) 1–16.

[15]. M. Aslam, O.H. Arif, R.A.K. Sherwani, New diagnosis test under the neutrosophic statistics: an application to diabetic patients, BioMed. Res. Int. 2020 (2020) 1–10.

[16]. M.E. Bentley, P.L. Griffiths, The burden of anemia among women in India, Eur. J. Clin. Nutr. 57 (2003) 52–60.

[17]. A. Rammohan., N. Awofeso., and M.-C. Robitaille, Addressing Female Iron-Deficiency Anaemia in India: Is Vegetarianism the Major Obstacle?, International Scholarly Research Network Public Health, pp. 1–9, 2011.
[18]. M. Kaur, G.K. Kochar, Burden of anaemia in rural and urban jat women in haryana state, India, Malaysian J. Nutr. 15 (2) (2009) 1–8.
[19]. Sharadamani, Iron deficiency anemia among rural population attending tertiary care teaching hospital, Trop. J. Pathol. Microbiol. 3 (1) (2017) 1–8.

Chapter 11

Various sampling inspection plans for cancer patients using gamma distribution under indeterminacy

Gadde Srinivasa Rao[1] **and Muhammad Aslam**[2]

[1]*Department of Mathematics and Statistics, University of Dodoma, Dodoma, Tanzania,*
[2]*Department of Statistics, Faculty of Science, King Abdulaziz University, Jeddah, Saudi Arabia*

11.1 Introduction

Cancer is one of the most dangerous and incurable diseases that necessitate the brachydactyly growth of cells intensifying the increase. It is a malevolent tumor that has abnormal cell growth with the potential to invade or spread to other parts of the human body. Once cancer spreads, small lampshades of cancer cell abandon the original tumor and progress to the other parts of the body; for more details, see [1]. The spread can either be direct, through blood, or the lymphatic system. Cancers can touch various organs, and each type of cancer has its unique characteristics. When mutations, which are changes in the sequence, occur, they cause the cells to forget how to stop dividing. After some time, the mass of cells becomes a tumor. Cancer of various types is known in relation to the location of cancers, namely cervical cancer, lung cancer, gynecological cancer, skin cancer, brain cancer, breast cancer, etc.; the specific type of cancer that is more common to an area or community than the other cancer is breast cancer. Recent advances in cancer biology have resulted in the need for increased statistical analysis of research data. Survival data analysis has been a very active research field for many decades now. Survival analysis or failure time data analysis is interested in the time from a defined time origin to the occurrence of a given event. In biomedicine, the typical example is the time from randomization to a given treatment until the event occurs leading to the observation of survival time for the patient involved. Usually, the objective to compare different treatment effects on the survival time while incorporating available information for each patient leads us to an indeterminacy statistical

Cognitive Intelligence with Neutrosophic Statistics in Bioinformatics.
DOI: https://doi.org/10.1016/B978-0-323-99456-9.00009-X

distribution. The present study aims at the survival time of cancer patients and follows to gamma distribution under indeterminacy.

Each year governments and organizations around the world fund thousands of clinical trials following the history of diseases and evaluating alternative treatments. Accurate analysis of the provided information is critical if the nature of the care for individuals is to be directly affected by the findings, also to save time and money. Most governments are applying the mechanism of test randomly to selected individuals to estimate the expected life or survival of the patients. To save the money and time, to test the survival time of the patient would be one quality control methodology, namely, acceptance sampling plans under indeterminacy. The oncologists are brainstorming to estimate the average survival time of the patients after attaching cancer due to their new method of treatment. In these situations, the oncologists are paying attention to test the null hypothesis that the average survival time of the patients is equal to the specified average survival time of the patients against the alternative hypothesis that the average survival time of the patients varies significantly. The null hypothesis could be rejected if the average survival time of the patients due to cancer, state acceptance number of patients, is more than or equal to the specified average survival time of the patients due to cancer.

More authors concentrate on studied sampling plans based on time-truncated life test for a variety of distributions. Some related articles can also be explored in Ref. [2–8]. The details about the acceptance sampling plans can be seen in Ref. [9,10]. The method of repetitive group acceptance sampling plan (RGASP) was first proposed by Ref. [11] for an attribute. The improvement of single acceptance sampling plan is known as repetitive sampling plan; for more details, explore [12–18].

The aforesaid authors concentrated on classical statistics to study the SSP and RSP, whereas some real-world applications related to cancer patient's lifetimes may not follow traditional statistical distributions. For those situations, recently more researchers are attracted by neutrosophic statistics. More details about the neutrosophic logics, their measure of determinacy, and indeterminacy are given in Ref. [19]. Numerous authors studied the neutrosophic logic for various real problems and showed its efficiency over fuzzy logic; for more details, refer Refs. [20–25]. The idea of neutrosophic statistics was given using the idea of neutrosophic logic, as given in Refs. [26–28]. The neutrosophic statistics gives information about the measure of determinacy and the measure of indeterminacy; see Ref. [29]. The neutrosophic statistics becomes classical statistics when no evidence is enrolled about the measure of indeterminacy. The acceptance sampling plans using the neutrosophic statistics are developed by Refs. [30–34].

The sampling plans in hand are based on traditional statistics, and a fuzzy environment could not give accounting data that is related to the measure of indeterminacy. Some works related to single sampling plan using a fuzzy approach can also be explored in Refs. [35–40]. More recently, Ref. [41]

developed testing average wind speed using sampling plan for Weibull distribution under indeterminacy, and Ref. [42] studied inspection plan for COVID-19 patients for Weibull distribution using repetitive sampling under indeterminacy. With the great motivation of the foreseen works, we propose the two sampling schemes such as single sampling and repetitive sampling plans for gamma distribution under indeterminacy to test the average survival time and remission time. By exploring the literature, there is no work on a time-truncated sampling plan for gamma distribution under indeterminacy. The present research work deals with two sampling schemes such as single sampling and repetitive sampling plans for gamma distribution under indeterminacy to test the average survival time and remission time. Very few works on sampling schemes are based on indeterminacy/neutrosophic statistics; thus this chapter targeted the development of two sampling schemes such as single sampling and repetitive sampling plans for gamma distribution under indeterminacy. Specifically, no attempt was made on repetitive sampling plans for any life distribution using indeterminacy.

Section 11.2 demonstrates the gamma distribution under an indeterminacy environment. Section 11.3 deals with the development of single sampling plan for gamma distribution under indeterminacy and also comparative study with existing traditional sampling plans. In Section 11.4, we propose the repetitive acceptance sampling plan for gamma distribution under indeterminacy and also give the comparative study and real example based on cancer data. In the end, concluding remarks, suggestions, and future research works are demonstrated in Section 11.5.

11.2 Gamma distribution for indeterminacy

Gamma distribution is based on two positive-valued parameters, namely, shape and scale or rate parameters. The gamma distribution (GD) is a generalization of exponential distribution and has more association with Erlang, normal, chi-square, beta, and some other distributions. This distribution has been used for modeling in various life sciences such as epidemiology, computational biology, medical sciences, biostatistics, and neuroscience. In recent years, more researchers studied the applications of gamma distribution in statistical quality control, reliability, queueing theory, survival analysis, and communication engineering; for more details, refer Ref. [43]. The present research is motivated by the idea of neutrosophic statistics given by Ref. [26] and extensive studies of Aslam from 2018 onward for various neutrosophic and indeterminacy probability distributions in different sampling and control chart schemes; some citations are given in the introduction section. Let us assume that $f(t_N)=f(t_L)+f(t_U)I_N; I_N \epsilon [I_L, I_U]$ be a neutrosophic probability density function (npdf) with determinate part $f(t_L)$, indeterminate part $f(t_U)I_N$, and indeterminacy period $I_N \epsilon [I_L, I_U]$. Note that $t_N \epsilon [t_L, t_U]$ be a neutrosophic random variable (nrv) follows the npdf. The npdf is the generalization of pdf under classical statistics. The proposed neutrosophic form of $f(t_N)\epsilon[f(t_L),f(t_U)]$ reduces to pdf under classical statistics when $I_L = 0$. Based on this information, the npdf of the GD is defined as follows

$$f(t_N) = \left\{(\theta^{\gamma}\Gamma\gamma)t_N^{\gamma-1}exp\left(-t_N/\theta\right)\right\} + \left\{(\theta^{\gamma}\Gamma\gamma)t_N^{\gamma-1}exp\left(-t_N/\theta\right)\right\}I_N; I_N\epsilon[I_L, I_U] \tag{11.1}$$

where γ and θ are shape and scale parameters, respectively. It is significant to note that the developed npdf of the GD is the oversimplification of pdf of the GD based on classical statistics. The neutrosophic form of the npdf of the GD reduces to the GD when $I_L = 0$. The neutrosophic cumulative distribution function (ncdf) of the GD is given by

$$F(t_N) = \left\{\frac{1}{\Gamma\gamma}\Gamma\left(\gamma, t_N/\theta\right)\right\} + \left\{\frac{1}{\Gamma\gamma}\Gamma\left(\gamma, t_N/\theta\right)\right\}I_N; I_N\epsilon[I_L, I_U] \tag{11.2}$$

where $\Gamma\left(\gamma, t_N/\theta\right)$ is the lower incomplete gamma function.

Average lifetime of neutrosophic GD is $\mu_{0N} = \gamma\theta + \gamma\theta I_N$. A product failure probability before the time t_{N0} is denoted as $p_N = F(t_N \leq t_{N0})$ and is conveyed below:

$$p_N = \left\{\frac{1}{\Gamma\gamma}\Gamma\left(\gamma, t_N/\theta\right)\right\} + \left\{\frac{1}{\Gamma\gamma}\Gamma\left(\gamma, t_N/\theta\right)\right\}I_N \tag{11.3}$$

Here, we express neutrosophic termination time t_{N0} a product of constant a and neutrosophic mean life μ_{0N}, i.e., $t_{N0}=a\mu_{0N}$. The scale parameter θ could be expressed in terms of neutrosophic mean μ_{0N}.

Therefore, Eq. (11.3) could be rewritten in terms of neutrosophic mean μ_{0N} as follows:

$$\begin{aligned} p_N &= \left\{\frac{1}{\Gamma\gamma}\Gamma\left(\gamma, a\mu_{0N}/\theta\right)\right\} + \left\{\frac{1}{\Gamma\gamma}\Gamma\left(\gamma, a\mu_{0N}/\theta\right)\right\}I_N \\ &= \left\{\frac{1}{\Gamma\gamma}\Gamma\left(\gamma, \frac{a\mu_{0N}}{\mu_N/\theta}\right)\right\} + \left\{\frac{1}{\Gamma\gamma}\Gamma\left(\gamma, \frac{a\mu_{0N}}{\mu_N/\theta}\right)\right\}I_N \\ &= \left\{\frac{1}{\Gamma\gamma}\Gamma\left(\gamma, a\theta/\frac{\mu_N}{\mu_{0N}}\right)\right\} + \left\{\frac{1}{\Gamma\gamma}\Gamma\left(\gamma, a\theta/\frac{\mu_N}{\mu_{0N}}\right)\right\}I_N \end{aligned} \tag{11.4}$$

where μ_N/μ_{0N} is the ratio of true average to stipulated average.

11.3 Single sampling plan under indeterminacy

The null and alternative hypotheses for the average traffic fatality are given below:

$$H_0{:}\mu_N = \mu_{0N} \text{ versus } H_1{:}\mu_N \neq \mu_{0N}$$

where μ_N is true average traffic fatality and μ_{0N} is the stipulated average traffic fatality. According to the reports received, the developed sampling plan is declared as follows:

Step-1: Choose a random sample of days n and document the average traffic fatality for these chosen locations. Stipulate the number of locations, say c, average traffic fatality μ_{0N}, and indeterminacy constraint $I_N \epsilon [I_L, I_U]$.

Step-2: Accept $H_0{:}\mu_N = \mu_{0N}$ if average traffic fatality in c locations is more than or equal to μ_{0N}, if not, reject $H_0{:}\mu_N = \mu_{0N}$.

The projected sampling scheme is based on three parameters n, c, and I_N, where $I_N \epsilon [I_L, I_U]$ is measured as the stipulated constraint and fix corresponding to the uncertainty intensity. The chance of accepting $H_0{:}\mu_N = \mu_{0N}$ is given by

$$L(p_N) = \sum_{i=0}^{c} \binom{n}{i} p_N^i (1-p_N)^{n-i} \tag{11.5}$$

where p_N is the probability of rejecting $H_0{:}\mu_N = \mu_{0N}$, and it is given in Eq. (11.4).

Suppose that α and β are type-I and type-II errors. The researcher is paying attention to concern the projected plan to examine $H_0{:}\mu_N = \mu_{0N}$ such that the chance of accepting $H_0{:}\mu_N = \mu_{0N}$ when it is true should be larger than $1-\alpha$ at μ/μ_0 and the chance of accepting $H_0{:}\mu_N = \mu_{0N}$ when it is wrong should be smaller than β at $\mu_N/\mu_{0N} = 1$. The plan constants to examine $H_0{:}\mu_N = \mu_{0N}$ will be determined in such a way that the below two inequalities are fulfilled.

$$L\left(p_{1N} | \mu_N/\mu_{0N}\right) \geq 1-\alpha \tag{11.6}$$

$$L\left(p_{2N} | \mu_N/\mu_{0N} = 1\right) \leq \beta \tag{11.7}$$

where p_{1N} and p_{2N} are defined by

$$p_{1N} = \left\{\frac{1}{\Gamma\gamma}\Gamma\left(\gamma, a\theta/\frac{\mu_N}{\mu_{0N}}\right)\right\} + \left\{\frac{1}{\Gamma\gamma}\Gamma\left(\gamma, a\theta/\frac{\mu_N}{\mu_{0N}}\right)\right\} I_N \tag{11.8}$$

and

$$p_{2N} = \left\{\frac{1}{\Gamma\gamma}\Gamma(\gamma, a\theta)\right\} + \left\{\frac{1}{\Gamma\gamma}\Gamma(\gamma, a\theta)\right\} I_N \tag{11.9}$$

The quantities of the plan constants n and c for $\beta = \{0.25, 0.10, 0.05\}$, $\alpha = 0.10$, $a = 0.5, 1.0$, and I_N are placed in Tables 11.1–11.7. Tables 11.1 and 11.2 are shown for the GD for $\gamma = 2$, Tables 11.3 and 11.4 for $\gamma = 2.5$, Tables 11.5 and 11.6 for $\gamma = 3$, and Table 11.7 for $\gamma = 1$ (exponential distribution). From these tables, we noticed the following few points.

1. When the values of a increases from 0.5 to 1.0 the value of n decreases.
2. It is observed that when the shape parameter increases from $\theta = 1 to \quad \theta = 3$ the values of n decrease when other parameters are fixed.
3. Further, it is observed that the indeterminacy parameter I_N also shows a considerable effect to minimizing the sample size.

TABLE 11.1 The SSP parameter when $\alpha = 0.10$; $\gamma = 2$; and $a = 0.50$.

β	$\frac{\mu_N}{\mu_{0N}}$	$I_U = 0.00$			$I_U = 0.02$			$I_U = 0.04$			$I_U = 0.05$		
		n	c	$L(p_1)$	n	c	$L(p_1)$	n	c	$L(p_1)$	n	c	$L(p_1)$
0.25	1.2	184	44	0.9016	179	45	0.9062	171	45	0.9050	167	45	0.9058
	1.3	97	22	0.9047	88	21	0.9029	84	21	0.9024	82	21	0.9032
	1.4	60	13	0.9080	62	14	0.9092	55	13	0.9022	57	14	0.9173
	1.5	44	9	0.9068	42	9	0.9055	40	9	0.9059	39	9	0.9069
	1.8	23	4	0.9066	22	4	0.9046	21	4	0.9038	20	4	0.9118
	2.0	19	3	0.9142	18	3	0.9150	17	3	0.9169	17	3	0.9105
0.10	1.2	316	73	0.9012	305	74	0.9039	291	74	0.9037	288	75	0.9062
	1.3	166	36	0.9059	154	35	0.9014	147	35	0.9003	147	36	0.9082
	1.4	103	21	0.9043	98	21	0.9048	98	22	0.9076	95	22	0.9136
	1.5	73	14	0.9029	74	15	0.9121	71	15	0.9076	69	15	0.9108
	1.8	43	7	0.9150	40	7	0.9233	39	7	0.9140	38	7	0.9147
	2.0	33	5	0.9280	32	5	0.9219	30	5	0.9263	29	5	0.9290
0.05	1.2	413	94	0.9005	398	95	0.9001	379	95	0.9034	374	96	0.9055
	1.3	212	45	0.9019	202	45	0.9013	196	46	0.9089	188	45	0.9023
	1.4	136	27	0.9030	129	27	0.9063	123	27	0.9059	120	27	0.9069
	1.5	97	18	0.9001	92	18	0.9025	92	19	0.9111	90	19	0.9097
	1.8	56	9	0.9265	49	8	0.9030	51	9	0.9237	50	9	0.9218
	2.0	42	6	0.9200	40	6	0.9191	38	6	0.9194	37	6	0.9201

TABLE 11.2 The SSP parameter when $\alpha = 0.10$; $\gamma = 2$; and $a = 1.00$.

β	$\frac{\mu_N}{\mu_{0N}}$	$I_U = 0.00$			$I_U = 0.02$			$I_U = 0.04$			$I_U = 0.05$		
		n	c	$L(p_1)$	n	c	$L(p_1)$	n	c	$L(p_1)$	n	c	$L(p_1)$
0.25	1.2	106	59	0.9097	97	56	0.9022	90	54	0.9021	90	55	0.9046
	1.3	52	28	0.9110	50	28	0.9121	50	29	0.9128	44	26	0.9041
	1.4	31	16	0.9007	35	19	0.9247	32	18	0.9190	28	16	0.9075
	1.5	22	11	0.9065	23	12	0.9153	22	12	0.9200	20	11	0.9065
	1.8	13	6	0.9323	13	6	0.9153	12	6	0.9328	10	5	0.9147
	2.0	10	4	0.9041	9	4	0.9269	9	4	0.9129	9	4	0.9051
0.10	1.2	175	95	0.9044	163	92	0.9021	157	92	0.9023	149	89	0.9012
	1.3	91	47	0.9002	85	46	0.9101	80	45	0.9093	75	43	0.9061
	1.4	56	28	0.9154	52	27	0.9103	50	27	0.9113	49	27	0.9130
	1.5	38	18	0.9006	38	19	0.9175	33	17	0.9003	34	18	0.9127
	1.8	19	8	0.9081	20	9	0.9289	20	9	0.9068	17	8	0.9133
	2.0	18	7	0.9241	15	6	0.9080	14	6	0.9203	14	6	0.9113
0.05	1.2	227	122	0.9041	213	119	0.9015	198	115	0.9013	196	116	0.9033
	1.3	115	59	0.9104	105	56	0.9044	101	56	0.9055	99	56	0.9079
	1.4	72	35	0.9012	65	33	0.9004	64	34	0.9116	61	33	0.9080
	1.5	53	25	0.9237	49	24	0.9178	47	24	0.9201	46	24	0.9224
	1.8	27	11	0.9110	26	11	0.9081	25	11	0.9068	22	10	0.9081
	2.0	21	8	0.9235	18	7	0.9038	17	7	0.9115	17	7	0.9008

TABLE 11.3 The SSP parameter when $\alpha = 0.10; \gamma = 2.5$; and $a = 0.50$.

β	$\frac{\mu_N}{\mu_{0N}}$	$I_U = 0.00$			$I_U = 0.02$			$I_U = 0.04$			$I_U = 0.05$		
		n	c	$L(p_1)$	n	c	$L(p_1)$	n	c	$L(p_1)$	n	c	$L(p_1)$
0.25	1.2	162	32	0.9033	158	33	0.9075	150	33	0.9066	142	32	0.9018
	1.3	81	15	0.9033	77	15	0.9003	73	15	0.9002	71	15	0.9012
	1.4	52	9	0.9042	54	10	0.9176	51	10	0.9191	50	10	0.9158
	1.5	37	6	0.9058	35	6	0.9056	38	7	0.9219	37	7	0.9219
	1.8	22	3	0.9235	21	3	0.9208	20	3	0.9190	19	3	0.9246
	2.0	17	2	0.9218	16	2	0.9222	15	2	0.9235	15	2	0.9180
0.10	1.2	279	53	0.9053	265	53	0.9013	251	53	0.9027	240	52	0.9004
	1.3	143	25	0.9018	135	25	0.9038	128	25	0.9032	125	25	0.9007
	1.4	92	15	0.9088	88	15	0.9012	83	15	0.9039	81	15	0.9022
	1.5	67	10	0.9017	63	10	0.9043	60	10	0.9009	58	10	0.9045
	1.8	40	5	0.9260	38	5	0.9240	36	5	0.9232	35	5	0.9233
	2.0	28	3	0.9159	27	3	0.9101	25	3	0.9150	25	3	0.9076
0.05	1.2	365	68	0.9021	345	68	0.9046	328	68	0.9004	319	68	0.9029
	1.3	187	32	0.9036	177	32	0.9032	168	32	0.9013	168	33	0.9083
	1.4	121	19	0.9020	114	19	0.9046	108	19	0.9041	105	19	0.9051
	1.5	89	13	0.9094	84	13	0.9100	80	13	0.9061	78	13	0.9050
	1.8	50	6	0.9240	48	6	0.9181	45	6	0.9207	44	6	0.9188
	2.0	39	4	0.9177	37	4	0.9158	35	4	0.9152	34	4	0.9153

TABLE 11.4 The SSP parameter when $\alpha = 0.10; \gamma = 2.5;$ and $a = 1.00$.

β	$\frac{\mu_N}{\mu_{0N}}$	$I_U = 0.00$			$I_U = 0.02$			$I_U = 0.04$			$I_U = 0.05$		
		n	c	$L(p_1)$	n	c	$L(p_1)$	n	c	$L(p_1)$	n	c	$L(p_1)$
0.25	1.2	85	46	0.9103	78	44	0.9069	75	44	0.9062	72	43	0.9016
	1.3	44	23	0.9214	37	20	0.9019	39	22	0.9123	38	22	0.9189
	1.4	26	13	0.9148	25	13	0.9126	24	13	0.9126	25	14	0.9279
	1.5	19	9	0.9106	18	9	0.9171	19	10	0.9299	17	9	0.9144
	1.8	10	4	0.9021	9	4	0.9240	9	4	0.9081	11	5	0.9200
	2.0	8	3	0.9196	8	3	0.9051	7	3	0.9311	7	3	0.9252
0.10	1.2	135	71	0.9009	133	73	0.9055	124	71	0.9067	120	70	0.9031
	1.3	70	35	0.9084	67	35	0.9101	64	35	0.9161	61	34	0.9123
	1.4	46	22	0.9204	44	22	0.9215	41	21	0.9002	38	20	0.9031
	1.5	31	14	0.9110	30	14	0.9021	30	15	0.9270	26	13	0.9006
	1.8	18	7	0.9218	15	6	0.9037	14	6	0.9148	14	6	0.9045
	2.0	14	5	0.9292	11	4	0.9071	13	5	0.9222	12	5	0.9411
0.05	1.2	177	92	0.9007	173	94	0.9082	161	91	0.9021	156	90	0.9026
	1.3	92	45	0.9003	86	44	0.9012	82	44	0.9104	77	42	0.9002
	1.4	56	26	0.9053	56	27	0.9016	51	26	0.9132	50	26	0.9123
	1.5	41	18	0.9084	37	17	0.9044	37	18	0.9207	37	18	0.9023
	1.8	21	8	0.9210	20	8	0.9219	19	8	0.9247	19	8	0.9137
	2.0	15	5	0.9036	17	6	0.9177	16	6	0.9231	15	6	0.9382

TABLE 11.5 The SSP parameter when $\alpha = 0.10; \gamma = 3$; and $a = 0.50$.

β	$\frac{\mu_N}{\mu_{0N}}$	$I_U = 0.00$			$I_U = 0.02$			$I_U = 0.04$			$I_U = 0.05$		
		n	c	$L(p_1)$	n	c	$L(p_1)$	n	c	$L(p_1)$	n	c	$L(p_1)$
0.25	1.2	151	25	0.909	137	24	0.9025	134	25	0.9094	131	25	0.9040
	1.3	78	12	0.914	74	12	0.9087	70	12	0.9062	68	12	0.9060
	1.4	50	7	0.907	47	7	0.9068	44	7	0.9090	43	7	0.9061
	1.5	38	5	0.919	36	5	0.9165	34	5	0.9151	33	5	0.9150
	1.8	20	2	0.916	19	2	0.9131	18	2	0.9110	17	2	0.9164
	2.0	20	2	0.955	19	2	0.9530	18	2	0.9517	17	2	0.9548
0.10	1.2	254	40	0.904	239	40	0.9036	231	41	0.9053	219	40	0.9026
	1.3	132	19	0.909	125	19	0.9035	118	19	0.9015	114	19	0.9052
	1.4	84	11	0.907	79	11	0.9060	75	11	0.9008	78	12	0.9170
	1.5	66	8	0.919	62	8	0.9189	58	8	0.9208	57	8	0.9155
	1.8	40	4	0.944	38	4	0.9407	36	4	0.9385	35	4	0.9378
	2.0	27	2	0.906	25	2	0.9077	24	2	0.9022	23	2	0.9045
0.05	1.2	330	51	0.904	311	51	0.9021	299	52	0.9031	285	51	0.9005
	1.3	172	24	0.905	162	24	0.9030	153	24	0.9002	148	24	0.9033
	1.4	111	14	0.903	104	14	0.9048	98	14	0.9037	101	15	0.9145
	1.5	86	10	0.917	74	9	0.9002	76	10	0.9158	74	10	0.9140
	1.8	46	4	0.909	43	4	0.9091	40	4	0.9115	39	4	0.9096
	2.0	39	3	0.928	36	3	0.9307	34	3	0.9290	33	3	0.9285

TABLE 11.6 The SSP parameter when $\alpha = 0.10; \gamma = 3;$ and $a = 1.00$.

β	$\frac{\mu_N}{\mu_{0N}}$	$I_U = 0.00$			$I_U = 0.02$			$I_U = 0.04$			$I_U = 0.05$		
		n	c	$L(p_1)$	n	c	$L(p_1)$	n	c	$L(p_1)$	n	c	$L(p_1)$
0.25	1.2	70	37	0.9090	67	37	0.9088	59	34	0.9015	61	36	0.9112
	1.3	34	17	0.9007	34	18	0.9164	31	17	0.9046	32	18	0.9133
	1.4	21	10	0.9043	20	10	0.9062	19	10	0.9110	22	12	0.9329
	1.5	18	8	0.9089	15	7	0.9010	14	7	0.9140	14	7	0.9018
	1.8	8	3	0.9071	10	4	0.9246	7	3	0.9182	7	3	0.9108
	2.0	6	2	0.9117	8	3	0.9412	7	3	0.9576	7	3	0.9534
0.10	1.2	115	59	0.9060	110	59	0.9055	100	56	0.9004	103	59	0.9083
	1.3	56	27	0.9018	57	29	0.9170	51	27	0.9058	50	27	0.9037
	1.4	39	18	0.9262	36	17	0.9020	32	16	0.9074	33	17	0.9187
	1.5	26	11	0.9006	27	12	0.9077	25	12	0.9274	23	11	0.9039
	1.8	14	5	0.9144	13	5	0.9229	13	5	0.9032	12	5	0.9253
	2.0	12	4	0.9386	9	3	0.9108	11	4	0.9341	11	4	0.9269
0.05	1.2	150	76	0.9069	145	77	0.9120	132	73	0.9010	129	73	0.9047
	1.3	74	35	0.9011	74	37	0.9179	69	36	0.9129	64	34	0.9029
	1.4	47	21	0.9120	43	20	0.9011	41	20	0.9027	40	20	0.9050
	1.5	36	15	0.9133	32	14	0.9080	32	15	0.9287	30	14	0.9036
	1.8	20	7	0.9264	19	7	0.9262	16	6	0.9021	17	7	0.9403
	2.0	13	4	0.9155	15	5	0.9354	12	4	0.9064	14	5	0.9329

TABLE 11.7 The SSP parameter when $\alpha = 0.10; \gamma = 1.0$; and $a = 0.50$.

β	$\frac{\mu_N}{\mu_{0N}}$	$I_U = 0.00$			$I_U = 0.02$			$I_U = 0.04$			$I_U = 0.05$		
		n	c	$L(p_1)$	n	c	$L(p_1)$	n	c	$L(p_1)$	n	c	$L(p_1)$
0.25	1.2	324	121	0.9026	313	121	0.9020	300	120	0.9014	290	118	0.9006
	1.3	162	59	0.9039	154	58	0.9003	146	57	0.9018	146	58	0.9046
	1.4	104	37	0.9084	98	36	0.9026	92	35	0.9014	93	36	0.9047
	1.5	72	25	0.9064	72	26	0.9141	67	25	0.9096	66	25	0.9081
	1.8	37	12	0.9085	36	12	0.9037	37	13	0.9215	34	12	0.9079
	2.0	29	9	0.9120	25	8	0.9016	27	9	0.9133	27	9	0.9042
0.10	1.2	–	–	–	–	–	–	–	–	–	–	–	–
	1.3	277	98	0.9017	267	98	0.9052	253	96	0.9011	246	95	0.9016
	1.4	175	60	0.9037	166	59	0.9035	163	60	0.9063	155	58	0.9023
	1.5	126	42	0.9084	116	40	0.9024	112	40	0.9033	110	40	0.9048
	1.8	65	20	0.9119	60	19	0.9023	58	19	0.9018	57	19	0.9022
	2.0	48	14	0.9082	49	15	0.9218	45	14	0.9044	44	14	0.9081
0.05	1.2	–	–	–	–	–	–	–	–	–	–	–	–
	1.3	363	127	0.9034	345	125	0.9014	333	125	0.9038	317	121	0.9001
	1.4	225	76	0.9019	217	76	0.9036	207	75	0.9014	203	75	0.9060
	1.5	162	53	0.9054	156	53	0.9088	148	52	0.9058	143	51	0.9011
	1.8	84	25	0.9019	81	25	0.9027	75	24	0.9005	76	25	0.9138
	2.0	64	18	0.9018	58	17	0.9016	56	17	0.9018	55	17	0.9025

Note that (–) represents parameters do not exist.

11.3.1 Comparative studies

The aim of this section is to study the effectiveness of the projected sampling plan with respect to sample size. The lesser the sample size is, more economical to examine the hypothesis about the average. Make a note that the developed sampling plan is the oversimplification of the plan based on traditional statistics if no uncertainty or indeterminacy is established while commemorating the average value. When $I_N = 0$, the developed sampling plan becomes the on-hand sampling plan. In Tables 11.1–11.7, the first spell of column, i.e., at $I_N = 0$, is the plan parameter of the traditional or existing sampling plans. From the results from the tables, we can conclude that the sample size is large in the traditional sampling plan as compared with the proposed sampling plan. For example, when $\alpha = 0.10, \beta = 0.10$, $\mu_N/\mu_{0N} = 1.5$, $\gamma = 2$, and $a = 0.5$ from Table 11.1, it can be seen that $n = 73$ from the plan under classical statistics and $n = 69$ for the projected sampling plan when $I_N = 0.05$. Furthermore, when $\gamma = 1$ the GD becomes an exponential distribution, and we constructed Table 11.7 for exponential distribution for comparison purpose. Table 11.7 depicts that GD shows less sample number as compared with exponential distribution. For example, when $\alpha = 0.10, \beta = 0.25$, $\mu_N/\mu_{0N} = 1.3$, $a = 0.5$, and $I_N = 0.04$ Table 11.7 shows that the sample size is 146 whereas proposed plan values are $n = 84$ for $\gamma = 2$, $n = 73$ for $\gamma = 2.5$, and $n = 70$ for $\gamma = 3$. From this study, it is concluded that the projected plan under indeterminacy is efficient over the existing sampling plan under traditional statistics with respect to sample size. Operating characteristic (OC) curve of plan of the GD when $\alpha = 0.10; \gamma = 2.5$; and $a = 0.50$ is depicted in Fig. 11.1. Therefore, the application of the proposed plan for testing the null hypothesis $H_0{:}\mu_N = \mu_{0N}$ demands a lesser sample as compared to the on-hand plan. The OC curve in Fig. 11.1 also shows the same performance. The researchers can apply the proposed plan under uncertainty to save time and money.

11.3.2 Applications for breast cancer data

The present section deals with the application of the proposed sampling plan for the GD under the indeterminacy obtained using a real example. Real data represent the survival times of 121 patients with breast cancer obtained from a large hospital in the period from 1929 to 1938; for more details, see Ref. [44]. For ready reference, the data are reported here.

Survival times: 0.3, 0.3, 4.0, 5.0, 5.6, 6.2, 6.3, 6.6, 6.8, 7.4, 7.5, 8.4, 8.4, 10.3, 11.0, 11.8, 12.2, 12.3, 13.5, 14.4, 14.4, 14.8, 15.5, 15.7, 16.2, 16.3, 16.5, 16.8, 17.2, 17.3, 17.5, 17.9, 19.8, 20.4, 20.9, 21.0, 21.0, 21.1, 23.0,

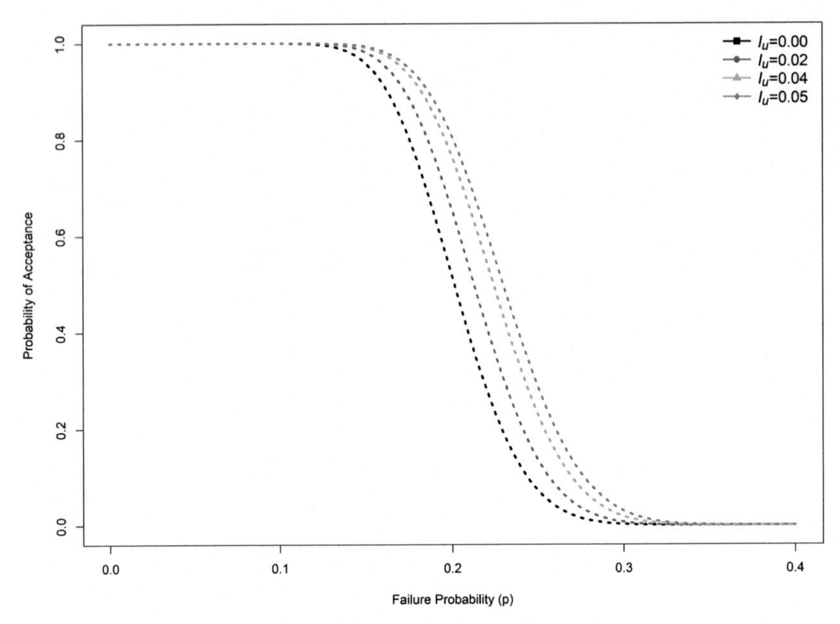

FIGURE 11.1 OC curve plan at different indeterminacy values.

23.4, 23.6, 24.0, 24.0, 27.9, 28.2, 29.1, 30.0, 31.0, 31.0, 32.0, 35.0, 35.0, 37.0, 37.0, 37.0, 38.0, 38.0, 38.0, 39.0, 39.0, 40.0, 40.0, 40.0, 41.0, 41.0, 41.0, 42.0, 43.0, 43.0, 43.0, 44.0, 45.0, 45.0, 46.0, 46.0, 47.0, 48.0, 49.0, 51.0, 51.0, 51.0, 52.0, 54.0, 55.0, 56.0, 57.0, 58.0, 59.0, 60.0, 60.0, 60.0, 61.0, 62.0, 65.0, 65.0, 67.0, 67.0, 68.0, 69.0, 78.0, 80.0, 83.0, 88.0, 89.0, 90.0, 93.0, 96.0, 103.0, 105.0, 109.0, 109.0, 111.0, 115.0, 117.0, 125.0, 126.0, 127.0, 129.0, 129.0, 139.0, and 154.0.

Survival times are very important in oncology for cancer patients. Due to the unpredictability and uncertainty, the survival times of breast cancer patient's data come from the statistical distribution under neutrosophic statistics. The oncologists are concerned to observe the average survival times under indeterminacy.

It is established that the survival time data come from the GD with shape parameter $\hat{\gamma} = 1.4959$ and scale parameter $\hat{\theta} = 30.9693$, and the maximum distance between the real-time data and the fitted GD is found from the Kolmogorov–Smirnov test statistic as 0.07618, and also, the p-value is 0.4837. The demonstration of the goodness of fit for the given model is shown in Fig. 11.2, the empirical and theoretical pdfs and Q-Q plots for the GD for the survival time data. The plan parameters for this shape parameter are shown in Table 11.8. For the proposed plan, the shape parameter is $\hat{\gamma}_N = (1 + 0.05) \times 1.4959 \approx 1.5707$ when $I_U = 0.05$. Suppose that oncologists are concerned to test $H_0{:}\mu_N = 48.6433$ with the aid of the proposed

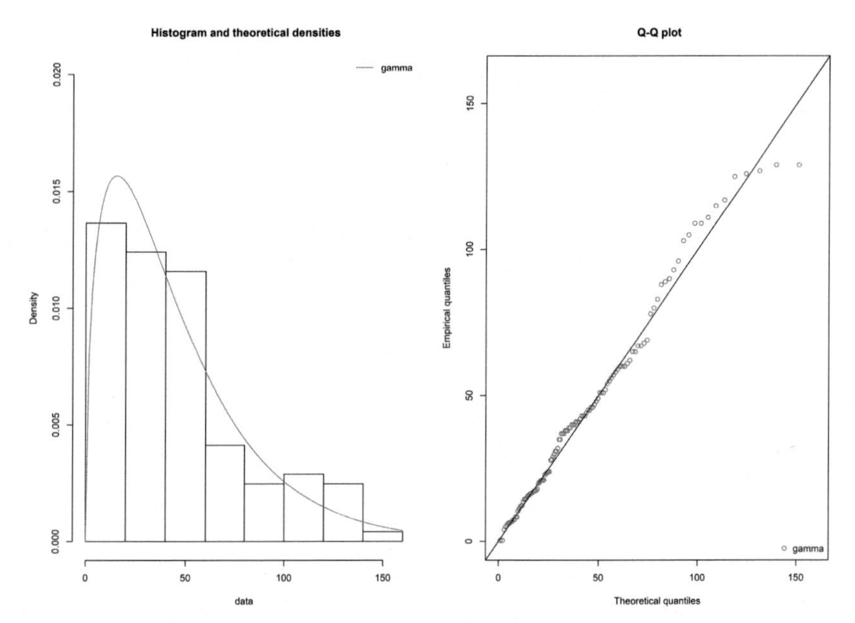

FIGURE 11.2 The empirical and theoretical pdf and Q-Q plots for the GD for breast cancer data.

sampling plan when $I_U = 0.05$, $\alpha = 0.10$, $\mu_N/\mu_{0N} = 1.5$, $a = 0.5$, and $\beta = 0.25$. From Table 11.8, it can be noted that $n = 48$ and $c = 14$. The projected sampling plan will be employed as: accept the null hypothesis $H_0{:}\mu_N = 48.6433$ if average survival time in 14 patients is more than or equal to 48.6433 survival time. From the data, it can be noted that the average survival time is greater than or equal to 48.6433 happened more than 14 times; therefore, the claim about the average survival time $H_0{:}\mu_N = 48.6433$ will be accepted. Using the real-world illustration, it is accomplished that the projected sampling will be obliged to check the average survival time of breast cancer patients.

11.4 Repetitive sampling plan under indeterminacy

The traditional repetitive acceptance sampling plan based on the truncated life test sampling scheme is developed by Refs. [45–47]. The step-by-step procedure to adopt the repetitive acceptance sampling plan is stated below:

Step 1: From a lot, choose a sample of size n. Conduct a life testing for these samples for a prespecified time say t_0. Indicate the average μ_{0N} and indeterminacy parameter $I_N \epsilon [I_L, I_U]$.

Step 2: Accept $H_0{:}\mu_N = \mu_{0N}$ if the average value in c_1 acceptance number is more than or equal to μ_{0N} (i.e., $\mu_{0N} \le c_1$). If average value in c_2 days is

TABLE 11.8 The SSP parameter when $\alpha = 0.10; \gamma = 1.4959$; and $a = 0.50$.

β	$\frac{\mu_N}{\mu_{0N}}$	$I_U = 0.00$			$I_U = 0.02$			$I_U = 0.04$			$I_U = 0.05$		
		n	c	$L(p_1)$	n	c	$L(p_1)$	n	c	$L(p_1)$	n	c	$L(p_1)$
0.25	1.2	234	69	0.9034	218	67	0.9005	209	67	0.9033	202	66	0.9004
	1.3	120	34	0.9039	115	34	0.9050	107	33	0.9041	105	33	0.9029
	1.4	73	20	0.9059	70	20	0.9061	71	21	0.9050	66	20	0.9045
	1.5	53	14	0.9085	51	14	0.9062	49	14	0.9056	48	14	0.9059
	1.8	29	7	0.9222	28	7	0.9193	27	7	0.9172	23	6	0.9015
	2.0	23	5	0.9104	22	5	0.9109	21	5	0.9126	20	5	0.9223
0.10	1.2	394	113	0.9011	378	113	0.9006	363	113	0.9001	349	111	0.9015
	1.3	201	55	0.9055	193	55	0.9037	182	54	0.9011	178	54	0.9041
	1.4	130	34	0.9094	121	33	0.9065	116	33	0.9076	114	33	0.9046
	1.5	92	23	0.9083	88	23	0.9101	81	22	0.9038	83	23	0.9074
	1.8	50	11	0.9103	48	11	0.9091	42	10	0.9000	41	10	0.9020
	2.0	39	8	0.9165	33	7	0.9033	35	8	0.9260	31	7	0.9030
0.05	1.2	515	146	0.9010	487	144	0.9004	474	146	0.9018	458	144	0.9016
	1.3	260	70	0.9046	246	69	0.9015	236	69	0.9022	228	68	0.9003
	1.4	168	43	0.9067	161	43	0.9071	151	42	0.9040	148	42	0.9037
	1.5	119	29	0.9053	114	29	0.9057	109	29	0.9090	107	29	0.9071
	1.8	61	13	0.9043	62	14	0.9189	56	13	0.9044	55	13	0.9027
	2.0	46	9	0.9021	44	9	0.9026	42	9	0.9046	41	9	0.9063

less than μ_0 (i.e., $\mu_{0N} > c_2$), then we reject $H_0{:}\mu_N = \mu_{0N}$ and terminate the test, where $c_1 \leq c_2$.

Step 3: Go to Step 1 and repeat the above procedure when $c_1 < \mu_{0N} \leq c_2$.

The developed repetitive acceptance sampling plan (RASP) based on above indeterminacy methodology consists of $n, c_1, c_2,$ and I_N, where $I_N \epsilon [I_L, I_U]$ is known as uncertainty level and it is predetermined. RASP is a generalization of single sampling plan under uncertainty studied in Section 11.3. The proposed RASP is reduced to a single sampling plan under uncertainty when $c_1 = c_2$. It is a tradition to assume that $t_0 = a\mu_0$ where a is the termination factor. The operating characteristic (OC) function can be obtained based on lot acceptance probability; for more details, refer [11], and it is defined as:

$$L(p_N) = \frac{P_a(p_N)}{P_a(p_N) + P_r(p_N)}; 0 < p_N < 1 \tag{11.10}$$

where $P_a(p_N)$ is the probability of accepting under $H_0{:}\mu_N = \mu_{0N}$ whereas $P_r(p_N)$ is the probability of rejecting at $H_0{:}\mu_N = \mu_{0N}$; these are obtained in the following expressions:

$$P_a(p_N) = \sum_{i=0}^{c_1} \binom{n}{i} p_N^i (1 - p_N)^{n-i} \tag{11.11}$$

and

$$P_r(p_N) = 1 - \sum_{i=0}^{c_2} \binom{n}{i} p_N^i (1 - p_N)^{n-i} \tag{11.12}$$

where p_N is the probability of unreliability, given in Eq. (11.3).

Using Eqs. (11.11) and (11.12), Eq. (11.10) becomes

$$L(p_N) = \frac{\sum_{i=0}^{c_1} \binom{n}{i} p_N^i (1 - p_N)^{n-i}}{\sum_{i=0}^{c_1} \binom{n}{i} p_N^i (1 - p_N)^{n-i} + 1 - \sum_{i=0}^{c_2} \binom{n}{i} p_N^i (1 - p_N)^{n-i}}; 0 < p_N < 1 \tag{11.13}$$

Suppose that α and β are type-I and type-II errors. The researcher is paying attention to concern the projected plan to examine $H_0{:}\mu_N = \mu_{0N}$ such that the chance of accepting $H_0{:}\mu_N = \mu_{0N}$ when it is true should be larger than $1 - \alpha$ at μ/μ_0 and the chance of accepting $H_0{:}\mu_N = \mu_{0N}$ when it is wrong should be smaller than β at $\mu_N/\mu_{0N} = 1$. In producer point of view, the probability of acceptance should be at least $1 - \alpha$ at acceptable quality level (AQL), p_{1N}. So, the producer demands the lot should be accepted at various levels of μ_N/μ_{0N}. Similarly, in consumer point of view the lot rejection probability should not exceed β at limiting quality level (LQL), p_{2N}. The plan constants to examine $H_0{:}\mu_N = \mu_{0N}$ will be determined in such a way that the below two inequalities are fulfilled simultaneously.

$$L(p_{1N}|\mu_N/\mu_{0N}) = \frac{\sum_{i=0}^{c_1}\binom{n}{i}p_{1N}{}^{i}(1-p_{1N})^{n-i}}{\sum_{i=0}^{c_1}\binom{n}{i}p_{1N}{}^{i}(1-p_{1N})^{n-i}+1-\sum_{i=0}^{c_2}\binom{n}{i}p_{1N}{}^{i}(1-p_{1N})^{n-i}} \geq 1-\alpha \tag{11.14}$$

$$L(p_{2N}|\mu_N/\mu_{0N}=1) = \frac{\sum_{i=0}^{c_1}\binom{n}{i}p_{2N}{}^{i}(1-p_{2N})^{n-i}}{\sum_{i=0}^{c_1}\binom{n}{i}p_{2N}{}^{i}(1-p_{2N})^{n-i}+1-\sum_{i=0}^{c_2}\binom{n}{i}p_{2N}{}^{i}(1-p_{2N})^{n-i}} \leq \beta \tag{11.15}$$

where p_{1N} and p_{2N} are defined in Eqs. (11.8) and (11.9), respectively.

The estimated designed parameters of the proposed plan should be minimizing the average sample number (ASN) at acceptable quality level. The ASN of the developed sampling scheme in terms of fraction defective (p_N) is given below:

$$ASN = \frac{n}{P_a(p_N)+P_r(p_N)} \tag{11.16}$$

Therefore, the design parameters for the proposed plan with minimum sample size will be obtained by solving the below optimization technique.

$$\begin{aligned}&\text{Minimize } ASN(p_{1N})\\&\text{subject to}\\&L(p_{1N})\geq 1-\alpha\\&L(p_{2N})\leq\beta\\&0\leq c_1\leq c_2\\&\text{where } n,c_1,c_2\in z\end{aligned} \tag{11.17}$$

The values of the designed parameters $n, c_1, and\ \ c_2$ for various values of $\beta = 0.25$, 0.10, 0.05, 0.01; $\alpha = 0.10$; $a = 0.5$ and 1.0; $\mu_N/\mu_{0N} = 1.1$, 1.2, 1.3, 1.4, 1.5, 1.8, 2.0; and $I_N = 0.0$, 0.02, 0.04, and 0.05 when shape parameter $\gamma = 2, 2.5, 3,$ and 1.0 are given in Tables 11.9–11.16. Tables 11.9–11.10 are shown for the GD for $\gamma = 2$, Tables 11.11–11.12 for $\gamma = 2.5$, Tables 11.13–11.14 for $\gamma = 3$, and Tables 11.15–11.16 for $\gamma = 1$ (exponential distribution). From these tables, we noticed the following few points.

1. When the values of a increases from 0.5 to 1.0, the value of ASN decreases.
2. It is observed that when the shape parameter increases from $\theta = 1\ to\ \theta = 3$ the values of ASN decrease when other parameters are fixed.

TABLE 11.9 The RSP plan parameter when $\alpha = 0.10; \gamma = 2;$ and $a = 0.50$.

β	$\frac{\mu_N}{\mu_{0N}}$	$I_U = 0.00$					$I_U = 0.02$					$I_U = 0.04$					$I_U = 0.05$				
		n	c_1	c_2	$L(p_1)$	ASN	n	c_1	c_2	$L(p_1)$	ASN	n	c_1	c_2	$L(p_1)$	ASN	n	c_1	c_2	$L(p_1)$	ASN
0.25	1.2	110	24	28	0.9023	156.30	61	12	18	0.9033	145.60	92	22	26	0.9038	135.48	94	23	27	0.9016	136.24
	1.3	41	7	11	0.9093	79.97	54	11	14	0.9025	78.01	47	10	13	0.9068	70.70	42	9	12	0.9030	65.20
	1.4	27	4	7	0.9076	49.50	40	8	10	0.9240	53.11	29	5	8	0.9089	49.34	24	4	7	0.9075	44.52
	1.5	19	2	5	0.9114	40.03	20	3	5	0.9003	30.96	19	3	5	0.9022	29.60	29	6	7	0.9017	33.60
	1.8	13	1	3	0.9314	22.87	15	2	3	0.9040	18.69	7	0	2	0.9095	16.61	11	1	3	0.9433	20.43
	2.0	8	0	2	0.9417	18.05	11	1	2	0.9004	14.24	7	0	2	0.9487	16.61	10	1	2	0.9062	13.11
0.1	1.2	138	28	35	0.9002	220.01	113	23	31	0.9030	215.38	125	28	35	0.9067	203.32	122	28	35	0.9087	199.65
	1.3	89	17	21	0.9049	118.44	70	13	18	0.9012	109.31	58	11	16	0.9010	98.81	61	12	17	0.9004	99.38
	1.4	41	6	10	0.9017	69.22	44	7	11	0.9018	69.41	37	6	10	0.9068	63.73	56	11	14	0.9041	70.78
	1.5	46	7	10	0.9172	60.73	39	6	9	0.9090	53.59	37	6	9	0.9120	51.26	20	2	6	0.9098	47.91
	1.8	11	0	3	0.9236	30.49	17	1	4	0.9197	30.42	16	1	4	0.9244	29.28	18	2	4	0.9120	25.01
	2.0	15	1	3	0.9374	22.97	24	3	4	0.9187	26.53	23	3	4	0.9164	25.37	14	1	3	0.9222	20.44
0.05	1.2	162	32	41	0.9046	262.47	159	33	42	0.9015	251.78	151	33	42	0.9097	244.44	148	33	42	0.9048	237.04
	1.3	79	13	20	0.9183	142.11	67	11	18	0.9020	128.53	68	12	19	0.9085	126.02	77	15	21	0.9008	117.41
	1.4	60	9	14	0.9086	89.13	52	8	13	0.9101	82.46	50	8	13	0.9027	77.99	44	7	12	0.9056	74.26
	1.5	51	7	11	0.9061	67.72	25	2	7	0.9002	61.98	41	6	10	0.9069	58.27	39	6	10	0.9246	58.14
	1.8	24	2	5	0.9317	35.94	24	2	5	0.9104	33.74	30	4	6	0.9241	35.32	16	1	4	0.9146	28.25
	2.0	24	2	4	0.9136	28.96	13	0	3	0.9039	24.98	12	0	3	0.9161	24.33	12	0	3	0.9065	23.43

TABLE 11.10 The RSP plan parameter when $\alpha = 0.10$; $\gamma = 2$; and $a = 1.00$.

β	$\frac{\mu_N}{\mu_{0N}}$	$I_U = 0.00$					$I_U = 0.02$					$I_U = 0.04$					$I_U = 0.05$				
		n	c_1	c_2	$L(p_1)$	ASN	n	c_1	c_2	$L(p_1)$	ASN	n	c_1	c_2	$L(p_1)$	ASN	n	c_1	c_2	$L(p_1)$	ASN
0.25	1.2	49	25	29	0.9014	80.95	47	25	29	0.9092	79.45	42	23	27	0.9065	74.27	38	21	25	0.9031	70.26
	1.3	32	16	18	0.9030	42.13	29	15	17	0.9006	38.94	33	18	20	0.9157	43.12	31	17	19	0.9021	40.54
	1.4	22	10	12	0.9022	29.93	9	3	6	0.9043	25.97	18	9	11	0.9262	26.73	18	9	11	0.9115	26.07
	1.5	22	11	11	0.9065	22.00	10	4	6	0.9176	17.71	16	8	9	0.9108	19.09	13	6	8	0.9323	20.57
	1.8	5	1	3	0.9406	11.95	5	1	3	0.9265	11.33	9	4	5	0.9522	11.65	10	5	5	0.9147	10.00
	2.0	3	0	2	0.9557	10.85	4	1	2	0.9113	6.02	6	2	3	0.9238	7.94	6	2	3	0.9170	7.83
0.1	1.2	76	38	44	0.9053	121.84	75	39	45	0.9052	118.94	67	36	42	0.9076	112.38	75	42	47	0.9012	107.73
	1.3	39	18	22	0.9042	59.56	43	21	25	0.9140	62.42	41	21	25	0.9283	61.81	32	16	20	0.9014	52.11
	1.4	23	9	13	0.9047	40.18	20	8	12	0.9118	38.61	30	15	17	0.9049	36.35	25	12	15	0.9155	35.99
	1.5	23	9	12	0.9099	31.42	13	4	8	0.9125	30.61	15	6	9	0.9200	25.79	15	6	9	0.9040	24.59
	1.8	7	1	4	0.9115	16.38	10	3	5	0.9043	13.93	12	4	6	0.9035	15.10	7	2	4	0.9166	11.86
	2.0	9	2	4	0.9021	12.03	11	4	5	0.9384	12.65	11	4	5	0.9231	12.33	8	2	4	0.9099	10.90
0.05	1.2	90	44	52	0.9081	147.04	74	37	45	0.9048	135.94	84	45	52	0.9079	129.83	65	34	42	0.9013	127.19
	1.3	48	21	27	0.9013	74.95	55	27	31	0.9000	69.63	33	15	21	0.9187	69.64	44	22	27	0.9094	65.09
	1.4	40	17	21	0.9044	50.59	30	13	17	0.9143	43.92	27	12	16	0.9121	41.03	28	13	17	0.9291	42.78
	1.5	18	6	10	0.9069	31.82	19	7	11	0.9288	33.15	23	9	13	0.9004	31.98	23	10	13	0.9062	29.76
	1.8	7	1	4	0.9115	16.38	9	2	5	0.9228	16.53	12	4	6	0.9035	15.10	11	3	6	0.9107	16.08
	2.0	9	2	4	0.9021	12.03	11	3	5	0.9205	13.56	10	3	5	0.9435	13.17	9	2	5	0.9516	14.44

TABLE 11.11 The RSP plan parameter when $\alpha = 0.10; \gamma = 2.5$; and $a = 0.50$.

β	$\frac{\mu_N}{\mu_{0N}}$	$I_U = 0.00$					$I_U = 0.02$					$I_U = 0.04$					$I_U = 0.05$				
		n	c_1	c_2	$L(p_1)$	ASN	n	c_1	c_2	$L(p_1)$	ASN	n	c_1	c_2	$L(p_1)$	ASN	n	c_1	c_2	$L(p_1)$	ASN
0.25	1.2	65	10	15	0.9038	132.95	78	14	18	0.9048	124.28	44	24	27	0.9020	63.29	38	21	24	0.9009	57.50
	1.3	52	8	11	0.9289	79.73	22	2	6	0.9079	67.26	21	10	13	0.9186	36.58	29	16	17	0.9030	32.98
	1.4	35	5	7	0.9301	49.23	29	4	6	0.9045	41.71	20	10	11	0.9025	23.07	17	8	10	0.9163	23.99
	1.5	21	2	4	0.9016	32.76	24	3	5	0.9380	36.31	11	5	6	0.9034	13.72	14	7	8	0.9405	17.08
	1.8	13	1	2	0.9176	17.13	22	3	3	0.9090	22.00	5	1	3	0.9496	10.88	3	0	2	0.9255	9.59
	2.0	17	2	2	0.9218	17.00	7	0	1	0.9015	10.43	4	1	2	0.9371	5.92	6	2	3	0.9536	7.90
0.1	1.2	99	15	22	0.9088	195.73	139	25	30	0.9008	188.02	48	24	30	0.9019	90.66	47	24	30	0.9058	90.10
	1.3	63	8	13	0.9005	105.27	48	6	11	0.9034	94.71	28	13	17	0.9202	47.40	32	16	19	0.9048	43.70
	1.4	42	5	8	0.9030	61.32	31	3	7	0.9137	62.33	32	16	17	0.9115	34.44	23	11	13	0.9121	28.81
	1.5	33	3	6	0.9026	49.02	31	3	6	0.9060	46.61	12	4	7	0.9108	21.19	16	7	9	0.9310	21.48
	1.8	18	1	3	0.9376	27.21	17	1	3	0.9375	25.79	10	2	5	0.9021	13.92	11	4	5	0.9077	12.27
	2.0	13	0	2	0.9193	20.99	13	0	2	0.9027	19.91	7	2	3	0.9169	8.35	7	2	3	0.9089	8.23
0.05	1.2	131	20	28	0.9010	223.43	124	20	28	0.9019	212.96	65	33	40	0.9180	109.68	68	36	42	0.9105	101.62
	1.3	81	11	16	0.9051	117.08	75	10	16	0.9033	117.61	38	18	22	0.9028	51.82	29	13	18	0.9120	51.68
	1.4	53	6	10	0.9089	76.34	50	6	10	0.9107	72.57	32	15	17	0.9012	36.14	22	9	13	0.9149	34.02
	1.5	38	3	7	0.9014	57.87	38	4	7	0.9046	51.43	16	6	9	0.9254	24.04	21	9	11	0.9002	24.43
	1.8	16	0	3	0.9145	31.21	32	3	5	0.9531	38.24	14	5	6	0.9032	14.96	8	2	4	0.9025	11.08
	2.0	25	1	3	0.9024	29.17	20	1	3	0.9473	26.26	8	2	4	0.9627	11.42	10	3	4	0.9015	10.79

TABLE 11.12 The RSP plan parameter when $\alpha = 0.10; \gamma = 2.5;$ and $a = 1.0.$

β	$\frac{\mu_N}{\mu_{0N}}$	$I_U = 0.00$					$I_U = 0.02$					$I_U = 0.04$					$I_U = 0.05$				
		n	c_1	c_2	$L(p_1)$	ASN	n	c_1	c_2	$L(p_1)$	ASN	n	c_1	c_2	$L(p_1)$	ASN	n	c_1	c_2	$L(p_1)$	ASN
0.25	1.2	43	21	25	0.9141	73.10	38	19	23	0.9005	66.77	44	24	27	0.9020	63.29	38	21	24	0.9009	57.50
	1.3	21	9	12	0.9056	36.52	18	8	11	0.9181	34.54	21	10	13	0.9186	36.58	29	16	17	0.9030	32.98
	1.4	12	4	7	0.9091	25.77	12	5	7	0.9120	19.81	20	10	11	0.9025	23.07	17	8	10	0.9163	23.99
	1.5	7	2	4	0.9036	14.05	11	4	6	0.9039	17.06	11	5	6	0.9034	13.72	14	7	8	0.9405	17.08
	1.8	12	5	5	0.9289	12.00	8	3	4	0.9506	10.33	5	1	3	0.9496	10.88	3	0	2	0.9255	9.59
	2.0	4	0	2	0.9072	8.37	8	3	4	0.9771	10.33	4	1	2	0.9371	5.92	6	2	3	0.9536	7.90
0.1	1.2	52	24	30	0.9019	96.44	48	23	29	0.9044	93.13	48	24	30	0.9019	90.66	47	24	30	0.9058	90.10
	1.3	35	15	19	0.9018	51.78	36	17	20	0.9037	47.74	28	13	17	0.9202	47.40	32	16	19	0.9048	43.70
	1.4	23	9	12	0.9045	32.58	20	8	11	0.9007	29.72	32	16	17	0.9115	34.44	23	11	13	0.9121	28.81
	1.5	18	7	9	0.9222	23.53	20	8	10	0.9029	24.19	12	4	7	0.9108	21.19	16	7	9	0.9310	21.48
	1.8	12	4	5	0.9191	13.57	15	6	6	0.9037	15.00	10	2	5	0.9021	13.92	11	4	5	0.9077	12.27
	2.0	8	2	3	0.9024	9.29	6	1	3	0.9490	10.02	7	2	3	0.9169	8.35	7	2	3	0.9089	8.23
0.05	1.2	69	32	39	0.9033	113.22	78	39	45	0.9047	109.96	65	33	40	0.9180	109.68	68	36	42	0.9105	101.62
	1.3	45	19	24	0.9010	61.81	37	16	21	0.9035	56.18	38	18	22	0.9028	51.82	29	13	18	0.9120	51.68
	1.4	27	10	14	0.9001	37.86	25	10	14	0.9342	38.66	32	15	17	0.9012	36.14	22	9	13	0.9149	34.02
	1.5	17	5	9	0.9136	28.16	19	7	10	0.9224	26.25	16	6	9	0.9254	24.04	21	9	11	0.9002	24.43
	1.8	11	3	5	0.9362	14.57	10	2	5	0.9278	15.06	14	5	6	0.9032	14.96	8	2	4	0.9025	11.08
	2.0	11	3	4	0.9109	11.99	8	1	4	0.9500	12.90	8	2	4	0.9627	11.42	10	3	4	0.9015	10.79

TABLE 11.13 The RSP plan parameter when $\alpha = 0.10; \gamma = 3.0$; and $a = 0.5$.

β	$\frac{\mu_N}{\mu_{0N}}$	$I_U = 0.00$					$I_U = 0.02$					$I_U = 0.04$					$I_U = 0.05$				
		n	c_1	c_2	$L(p_1)$	ASN	n	c_1	c_2	$L(p_1)$	ASN	n	c_1	c_2	$L(p_1)$	ASN	n	c_1	c_2	$L(p_1)$	ASN
0.25	1.2	69	9	13	0.9008	119.67	64	9	13	0.9127	113.96	42	5	10	0.9087	114.27	54	8	12	0.9009	99.70
	1.3	32	3	6	0.9012	60.54	30	3	6	0.9032	57.33	28	3	6	0.9084	54.44	27	3	6	0.9123	53.11
	1.4	31	3	5	0.9109	45.00	28	3	5	0.9262	42.05	26	3	5	0.9311	39.61	20	2	4	0.9107	32.75
	1.5	17	1	3	0.9338	30.90	16	1	3	0.9334	29.21	25	3	4	0.9182	30.05	15	1	3	0.9263	27.04
	1.8	20	2	2	0.9160	20.00	16	1	2	0.9317	20.17	14	1	2	0.9440	18.20	14	1	2	0.9388	18.00
	2.0	10	0	1	0.9175	13.95	8	0	1	0.9407	11.98	8	0	1	0.9319	11.70	9	0	1	0.9068	12.33
0.1	1.2	91	11	17	0.9035	171.62	98	13	19	0.9054	170.05	81	11	17	0.9007	153.57	90	13	19	0.9026	156.41
	1.3	58	6	10	0.9125	93.56	48	5	9	0.9073	83.43	51	6	10	0.9174	84.23	50	6	10	0.9112	81.40
	1.4	37	3	6	0.9087	58.14	35	3	6	0.9053	54.78	33	3	6	0.9043	51.76	32	3	6	0.9049	50.38
	1.5	24	1	4	0.9139	44.68	22	1	4	0.9234	42.71	32	3	5	0.9089	41.17	21	1	4	0.9048	38.53
	1.8	14	0	2	0.9331	24.58	27	2	3	0.9293	30.20	13	0	2	0.9203	21.87	13	0	2	0.9119	21.26
	2.0	20	1	2	0.9469	23.60	19	1	2	0.9446	22.34	13	0	2	0.9641	21.87	18	1	2	0.9376	20.88
0.05	1.2	129	16	23	0.9040	207.04	128	17	24	0.9019	198.65	114	16	23	0.9088	187.15	117	17	24	0.9042	183.91
	1.3	64	6	11	0.9019	105.46	83	10	14	0.9104	108.08	63	7	12	0.9076	98.65	68	8	13	0.9042	99.27
	1.4	50	4	8	0.9219	75.53	47	4	8	0.9217	71.31	38	3	7	0.9034	62.06	44	4	8	0.9070	64.58
	1.5	34	2	5	0.9031	49.71	32	2	5	0.9017	46.84	30	2	5	0.9031	44.27	26	1	5	0.9015	48.17
	1.8	27	1	3	0.9207	33.31	23	1	3	0.9444	30.60	23	1	3	0.9292	29.21	22	1	3	0.9326	28.27
	2.0	25	1	2	0.9027	27.27	16	0	2	0.9427	23.37	15	0	2	0.9427	22.01	17	0	2	0.9056	21.98

TABLE 11.14 The RSP plan parameter when $\alpha = 0.10; \gamma = 3.0$; and $a = 1.0$.

β	$\frac{\mu_N}{\mu_{0N}}$	$I_U = 0.00$					$I_U = 0.02$					$I_U = 0.04$					$I_U = 0.05$				
		n	c_1	c_2	$L(p_1)$	ASN	n	c_1	c_2	$L(p_1)$	ASN	n	c_1	c_2	$L(p_1)$	ASN	n	c_1	c_2	$L(p_1)$	ASN
0.25	1.2	39	19	22	0.9157	58.34	41	21	24	0.9197	60.09	29	15	18	0.9026	47.52	30	16	19	0.9108	48.85
	1.3	27	13	14	0.9092	30.91	24	12	13	0.9019	27.75	20	10	12	0.9318	29.07	20	10	12	0.9154	28.31
	1.4	16	7	8	0.9089	19.10	20	10	10	0.9062	20.00	10	4	6	0.9200	17.33	14	6	8	0.9068	20.06
	1.5	7	2	4	0.9372	14.26	5	1	3	0.9009	11.74	9	3	5	0.9069	14.43	14	7	7	0.9017	14.00
	1.8	7	2	3	0.9329	8.92	6	1	3	0.9369	10.20	4	1	2	0.9258	5.95	4	1	2	0.9192	5.87
	2.0	10	4	3	0.9019	8.82	4	0	2	0.9347	7.94	5	1	2	0.9102	6.28	7	3	3	0.9534	7.00
0.1	1.2	59	28	32	0.9013	80.62	48	23	28	0.9116	79.07	50	25	30	0.9006	77.05	53	28	32	0.9019	73.20
	1.3	32	13	17	0.9151	47.36	34	16	18	0.9006	40.49	26	12	15	0.9084	37.29	19	8	12	0.9207	38.99
	1.4	15	5	8	0.9098	25.53	16	6	9	0.9339	27.12	20	8	11	0.9127	27.66	15	5	9	0.9018	27.41
	1.5	9	2	5	0.9166	19.56	19	8	9	0.9184	21.01	6	1	4	0.9238	19.49	11	4	6	0.9051	15.71
	1.8	10	3	4	0.9303	11.54	9	2	4	0.9280	11.81	9	2	4	0.9065	11.19	6	1	3	0.9100	9.20
	2.0	8	2	3	0.9439	9.37	4	0	2	0.9347	7.94	5	1	2	0.9102	6.28	8	2	3	0.9066	8.78
0.05	1.2	60	27	33	0.9013	92.52	53	24	31	0.9001	93.24	55	27	33	0.9021	85.83	73	39	43	0.9007	88.38
	1.3	31	12	17	0.9295	53.04	28	11	16	0.9131	48.30	23	9	14	0.9053	44.81	27	12	16	0.9041	40.98
	1.4	19	6	10	0.9081	31.24	28	11	14	0.9009	33.36	18	7	10	0.9065	26.07	24	10	13	0.9028	29.67
	1.5	20	7	9	0.9010	23.35	20	6	10	0.9109	26.07	20	7	10	0.9002	23.83	14	4	8	0.9156	22.66
	1.8	9	2	4	0.9451	12.53	12	3	5	0.9264	13.96	6	1	3	0.9199	9.52	6	1	3	0.9100	9.20
	2.0	9	2	3	0.9110	9.92	6	0	3	0.9352	10.86	4	0	2	0.9198	7.38	8	2	3	0.9066	8.78

TABLE 11.15 The RSP plan parameter when $\alpha = 0.10; \gamma = 1.0$; and $a = 0.5$.

β	$\frac{\mu_N}{\mu_{0N}}$	$I_U = 0.00$					$I_U = 0.02$					$I_U = 0.04$					$I_U = 0.05$				
		n	c_1	c_2	$L(p_1)$	ASN	n	c_1	c_2	$L(p_1)$	ASN	n	c_1	c_2	$L(p_1)$	ASN	n	c_1	c_2	$L(p_1)$	ASN
0.25	1.2	160	56	63	0.9024	262.05	152	55	62	0.9013	251.30	113	41	49	0.9015	234.92	132	50	57	0.9003	228.62
	1.3	79	26	31	0.9034	130.81	71	24	29	0.9002	121.47	52	17	23	0.9019	119.98	81	30	34	0.9043	117.39
	1.4	61	20	23	0.9056	83.75	62	21	24	0.9019	83.72	60	21	24	0.9009	81.03	56	20	23	0.9069	77.32
	1.5	30	8	12	0.9002	60.05	45	15	17	0.9024	56.81	28	8	12	0.9018	56.75	31	10	13	0.9035	49.83
	1.8	15	3	6	0.9092	31.97	29	9	10	0.9030	33.01	21	6	8	0.9079	29.73	23	7	9	0.9266	32.10
	2.0	14	3	5	0.9020	22.27	21	6	7	0.9032	24.49	19	5	7	0.9218	26.78	22	7	8	0.9317	25.69
0.1	1.2	232	79	90	0.9016	389.40	216	76	87	0.9017	371.50	225	83	93	0.9028	354.81	190	70	81	0.9020	345.05
	1.3	114	36	44	0.9053	197.37	99	32	40	0.9041	184.16	93	31	39	0.9043	176.96	94	32	40	0.9067	176.32
	1.4	71	21	27	0.9015	119.16	81	26	31	0.9018	115.70	66	21	27	0.9081	112.61	68	22	28	0.9009	111.10
	1.5	64	19	23	0.9016	87.09	42	11	17	0.9004	86.67	47	14	19	0.9089	79.57	34	9	15	0.9082	83.05
	1.8	32	8	11	0.9053	44.72	20	4	8	0.9055	42.03	22	5	9	0.9232	43.52	19	4	8	0.9065	40.16
	2.0	20	4	7	0.9068	32.37	22	5	8	0.9292	34.51	14	2	6	0.9142	35.25	18	4	7	0.9191	30.19
0.05	1.2	292	99	112	0.9009	454.10	246	85	99	0.9012	436.24	289	106	118	0.9006	420.12	250	92	105	0.9013	406.48
	1.3	127	39	49	0.9024	227.31	117	37	47	0.9027	217.69	133	45	54	0.9028	208.30	153	54	62	0.9031	210.34
	1.4	81	23	31	0.9076	145.41	91	28	35	0.9050	137.49	85	27	34	0.9072	131.57	69	21	29	0.9029	130.26
	1.5	75	21	27	0.9022	106.02	63	18	24	0.9068	97.48	67	20	26	0.9028	96.86	63	19	25	0.9021	93.38
	1.8	38	9	13	0.9070	53.24	40	10	14	0.9093	53.72	39	10	14	0.9009	51.40	27	6	11	0.9210	50.94
	2.0	21	3	8	0.9130	45.21	30	7	10	0.9069	38.81	21	4	8	0.9077	36.13	24	5	9	0.9099	37.18

TABLE 11.16 The RSP plan parameter when $\alpha = 0.10; \gamma = 1.0$; and $a = 1.0$.

β	$\frac{\mu_N}{\mu_{0N}}$	$I_U = 0.00$					$I_U = 0.02$					$I_U = 0.04$					$I_U = 0.05$				
		n	c_1	c_2	$L(p_1)$	ASN	n	c_1	c_2	$L(p_1)$	ASN	n	c_1	c_2	$L(p_1)$	ASN	n	c_1	c_2	$L(p_1)$	ASN
0.25	1.2	132	78	82	0.9005	174.55	108	65	70	0.9036	164.07	97	60	65	0.9028	152.85	91	57	62	0.9013	146.69
	1.3	40	21	26	0.9031	87.26	58	34	37	0.9031	80.52	43	25	29	0.9031	75.96	44	26	30	0.9027	76.53
	1.4	37	20	23	0.9166	56.65	38	21	24	0.9026	56.08	30	17	20	0.9133	49.36	43	26	28	0.9143	54.36
	1.5	32	17	19	0.9017	41.59	29	16	18	0.9126	38.96	16	8	11	0.9085	33.79	26	15	17	0.9166	35.98
	1.8	18	9	10	0.9002	21.16	13	6	8	0.9156	20.53	15	8	9	0.9049	18.16	18	10	11	0.9252	21.30
	2.0	17	8	9	0.9078	19.57	16	8	9	0.9274	18.83	12	6	7	0.9115	14.74	10	5	6	0.9077	12.75
0.1	1.2	166	95	103	0.9025	251.81	131	76	85	0.9014	237.71	134	81	89	0.9004	220.93	145	90	97	0.9008	215.03
	1.3	65	34	41	0.9041	127.71	72	40	46	0.9041	118.66	77	45	50	0.9006	111.27	64	37	43	0.9056	110.92
	1.4	62	33	37	0.9090	83.37	44	23	28	0.9072	75.43	53	30	34	0.9178	75.24	42	23	28	0.9123	72.87
	1.5	40	20	24	0.9193	59.99	32	16	20	0.9035	51.89	40	22	25	0.9121	52.78	27	14	18	0.9143	48.74
	1.8	23	10	13	0.9111	31.90	18	8	11	0.9297	29.30	16	7	10	0.9092	26.17	14	6	9	0.9043	24.53
	2.0	14	5	8	0.9059	22.59	13	5	8	0.9382	23.56	9	3	6	0.9262	20.91	17	8	10	0.9253	21.77
0.05	1.2	154	85	97	0.9000	297.52	180	105	115	0.9006	276.14	165	99	109	0.9023	262.52	161	98	108	0.9027	257.45
	1.3	102	55	62	0.9017	146.94	83	45	53	0.9031	141.34	69	38	46	0.9027	133.05	88	51	58	0.9033	131.09
	1.4	63	32	38	0.9074	94.27	67	36	41	0.9025	89.26	65	36	41	0.9032	86.23	54	30	35	0.9006	77.84
	1.5	47	23	28	0.9200	69.25	31	14	20	0.9005	64.21	41	21	26	0.9095	61.28	41	22	26	0.9031	56.04
	1.8	22	9	13	0.9318	36.29	16	6	10	0.9039	30.81	17	7	11	0.9303	32.56	17	7	11	0.9163	30.71
	2.0	20	8	11	0.9254	27.18	15	5	9	0.9077	25.72	15	6	9	0.9166	22.58	19	8	11	0.9033	24.17

3. Further, it is observed that the indeterminacy parameter I_N also shows a considerable effect to minimizing the *ASN*.

11.4.1 Comparative studies

The aim of this section is to study the effectiveness of the projected RSP with respect to ASN. The lesser the ASN is, more economical to examine the hypothesis about the average. Make a note that the developed sampling plan is the oversimplification of the plan based on traditional statistics if no uncertainty or indeterminacy is established while commemorating the average value. When $I_N = 0$, the developed RSP becomes the on-hand sampling plan. In Tables 11.9–11.16 the first spell of column, i.e., at $I_N = 0$, is the plan parameter of the traditional or existing RSP. From the results from the tables, we can conclude that the ASN is large in traditional RSP as compared with proposed RSP. For example, when $\alpha = 0.10, \beta = 0.25$, $\mu_N/\mu_{0N} = 1.3$, $\gamma = 2$, and $a = 0.5$ from Table 11.9, it can be seen that $ASN = 79.97$ from the plan under classical statistics and $ASN = 65.30$ for the projected RSP when $I_N = 0.05$. Furthermore, when $\gamma = 1$ the GD becomes an exponential distribution, and we constructed Table 11.15 for exponential distribution for comparison purpose. Table 11.15 depicts that GD shows less ASN as compared with exponential distribution. For example, when $\alpha = 0.10, \beta = 0.10$, $\mu_N/\mu_{0N} = 1.4$, $a = 0.5$, and $I_N = 0.04$ Table 11.15 shows that the ASN is 112.61 whereas proposed plan values are $ASN = 63.73$ for $\gamma = 2$, ASN = 34.41 for $\gamma = 2.5$, and ASN = 51.76 for $\gamma = 3$. From this study, it is concluded that the projected plan under indeterminacy is efficient over the existing RSP under traditional statistics with respect to sample size. Operating characteristic (OC) curve of plan of the GD when $\alpha = 0.10, \beta = 0.10, \gamma = 3$, $\mu_N/\mu_{0N} = 1.3$, and $a = 0.5$ is depicted in Fig. 11.3. Therefore, the application of the proposed plan for testing the null hypothesis $H_0{:}\mu_N = \mu_{0N}$ demands a lesser ASN as compared to the on-hand plan. Moreover, a OC curve comparison between SSP and RSP is also displayed in Fig. 11.4. The OC curve in Fig. 11.4 also shows that RSP is superior to the SSP for the same specific parameters. The researchers advised as the proposed RSP under uncertainty is more economical to apply in medical sciences.

11.4.2 Applications for remission times of leukemia patients data

The present section deals with the application of the proposed sampling plan for the GD under the indeterminacy obtained using a real example. The dataset is taken from [48], and it constitutes the remission times, in weeks, for 35 leukemia patients. For ready reference, the data are given below.

Remission times (in weeks): 1, 3, 3, 6, 7, 7, 10, 12, 14, 15, 18, 19, 22, 26, 29, 34, 40, 1, 1, 2, 2, 3, 4, 5, 8, 8, 9, 11, 12, 14, 16, 18, 21, 31, and 44.

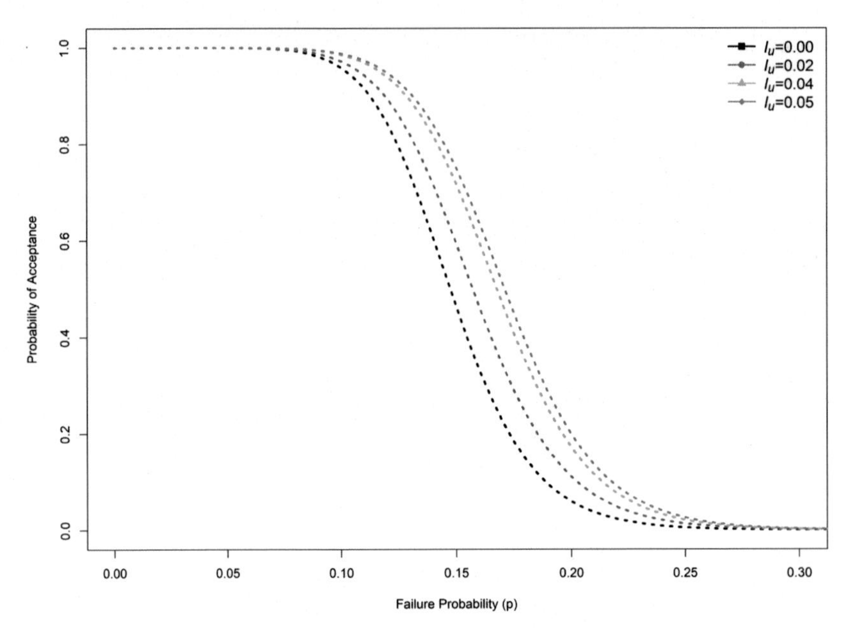

FIGURE 11.3 OC curve of RSP plan at different indeterminacy values.

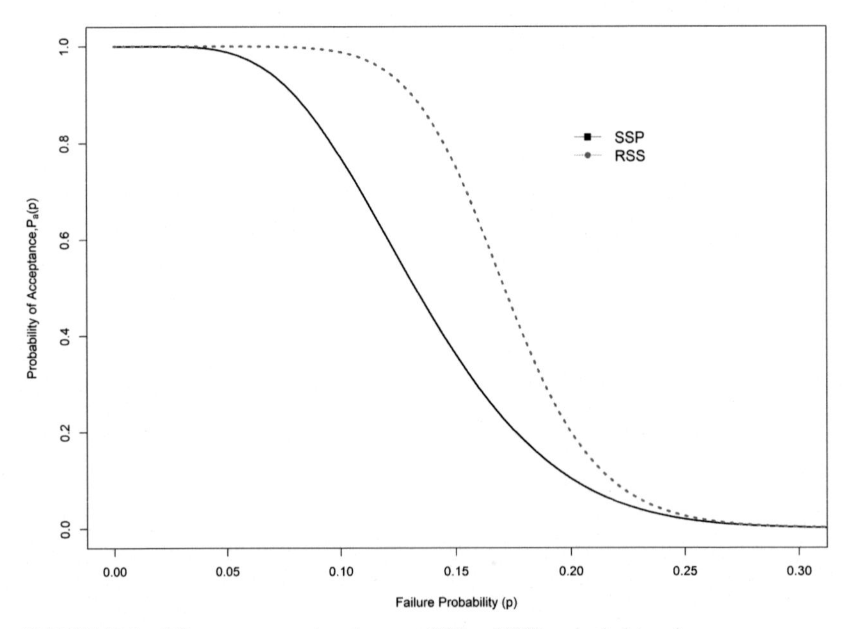

FIGURE 11.4 OC curves comparison between SSP and RSP under indeterminacy.

The remission time of leukemia patients is very crucial for oncologists to study the effectiveness of the medicine. The remission time of leukemia patients follows the statistical distribution under neutrosophic statistics due to unpredictability and uncertainty. The cancer specialists are anxious to observe the average remission times under indeterminacy.

It is evident that the remission time data follow the GD with shape parameter $\hat{\gamma} = 1.2863$ and scale parameter $\hat{\theta} = 10.5742$. The maximum distance between the real-time data and the fitted GD is found from the Kolmogorov–Smirnov test statistic as 0.0824 with p-value as 0.9713. The display of the goodness of fit for the given model is shown in Fig. 11.5, the empirical and theoretical cdfs and p-p plots for the GD for the remission time data. The plan parameters for this shape parameter are shown in Table 11.17. For the proposed plan, the shape parameter is $\hat{\gamma}_N = (1 + 0.04) \times 1.2863 \quad \approx 1.3378$ when $I_U = 0.04$. Suppose that a quality medical practitioner would like to use the proposed repetitive sampling plan for GD under indeterminacy to ensure the remission time of breast cancer patients is at least 20 weeks using the truncated life test for 10 weeks. Let the producer's risk be 10% at $\mu_N/\mu_{0N} = 1.5$ and the consumer's risk is 10%. From Table 11.17, with a = 0.50, $\beta = 0.10$, and $\alpha = 0.10$ for the repetitive sampling plan, it is found that the plan parameters are $n = 29$, $c_1 = 6$, $c_2 = 11$, and ASN = 66.07. Therefore, the plan could be implemented as follows: selecting a random sample of 29 leukemia patients from the arrived lot

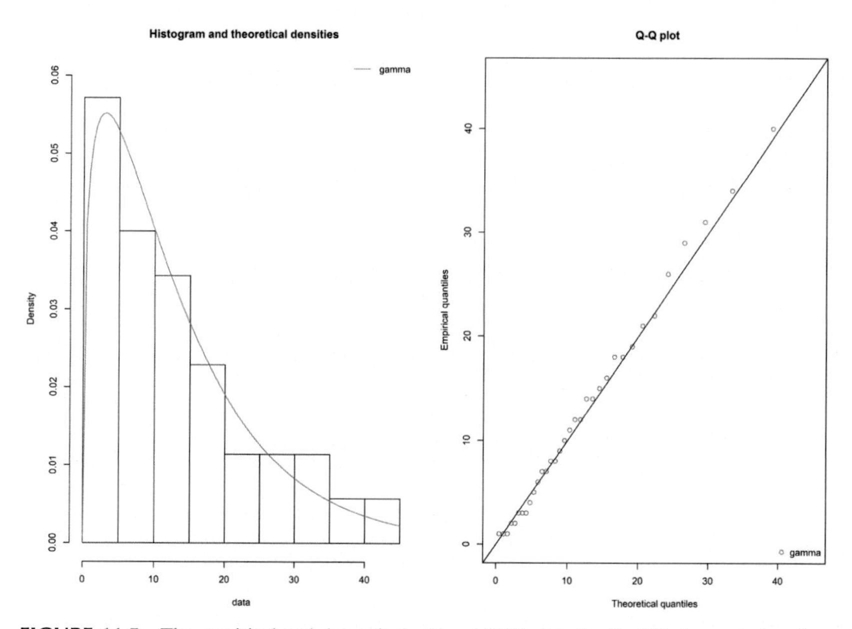

FIGURE 11.5 The empirical and theoretical pdf and Q-Q plots for the GD for remission time data.

TABLE 11.17 The RSP plan parameter when $\alpha = 0.10; \gamma = 1.2863$; and $a = 0.50$.

β	$\frac{\mu_N}{\mu_{0N}}$	$I_U = 0.00$					$I_U = 0.02$					$I_U = 0.04$					$I_U = 0.05$				
		n	c_1	c_2	$L(p_1)$	ASN	n	c_1	c_2	$L(p_1)$	ASN	n	c_1	c_2	$L(p_1)$	ASN	n	c_1	c_2	$L(p_1)$	ASN
0.25	1.2	133	40	46	0.9083	214.40	100	30	37	0.9047	200.10	115	37	43	0.9006	190.69	128	43	48	0.9014	186.70
	1.3	60	16	21	0.9064	111.24	68	20	24	0.9088	104.48	79	25	28	0.9012	103.12	87	29	31	0.9050	102.31
	1.4	34	8	12	0.9033	66.95	36	9	13	0.9046	67.55	52	16	18	0.9090	64.78	38	11	14	0.9045	58.31
	1.5	30	7	10	0.9056	49.34	29	7	10	0.9022	47.54	31	8	11	0.9097	49.08	24	6	9	0.9067	42.38
	1.8	19	4	6	0.9325	28.64	15	3	5	0.9201	24.13	15	3	5	0.9025	23.35	23	6	7	0.9009	26.52
	2.0	9	1	3	0.9156	17.72	14	3	4	0.9039	17.36	22	6	6	0.9088	22.00	19	5	6	0.9472	22.69
0.1	1.2	219	65	73	0.9011	313.46	156	46	56	0.9021	295.86	172	54	63	0.9025	282.54	163	52	61	0.9021	273.58
	1.3	85	22	29	0.9061	155.99	96	27	33	0.9025	147.51	79	22	29	0.9016	143.92	104	32	37	0.9011	141.06
	1.4	49	11	17	0.9121	102.43	44	10	16	0.9058	96.40	74	21	25	0.9050	96.90	69	20	24	0.9110	92.98
	1.5	44	10	14	0.9003	67.69	34	7	12	0.9014	68.26	29	6	11	0.9106	66.07	40	10	14	0.9002	61.66
	1.8	23	4	7	0.9116	36.42	22	4	7	0.9148	35.17	30	7	9	0.9125	36.79	25	5	8	0.9046	35.58
	2.0	12	1	4	0.9088	26.60	16	2	5	0.9049	27.49	21	4	6	0.9007	26.74	19	3	6	0.9137	28.66
0.05	1.2	225	64	76	0.9033	371.08	227	68	79	0.9008	348.15	196	60	72	0.9079	342.09	181	56	68	0.9022	327.36
	1.3	122	32	40	0.9010	185.42	117	32	40	0.9062	179.92	99	27	36	0.9053	176.09	88	24	33	0.9015	168.89
	1.4	66	15	22	0.9072	118.28	75	19	25	0.9067	111.79	72	19	25	0.9103	108.21	71	19	25	0.9043	105.21
	1.5	58	13	18	0.9019	83.22	52	12	17	0.9042	77.99	35	7	13	0.9016	76.18	49	12	17	0.9070	73.96
	1.8	22	3	7	0.9010	40.82	29	5	9	0.9143	43.96	28	5	9	0.9120	42.16	20	3	7	0.9004	37.14
	2.0	26	4	7	0.9043	34.47	25	4	7	0.9040	33.10	24	4	7	0.9052	31.83	29	6	8	0.9034	33.26

of patients and doing the truncated life test of remission time for 10 weeks. The proposed sampling plan will be implemented as: accept the null hypothesis $H_0{:}\mu_N = 14.1457$ if the average remission time of breast cancer patients in 20 weeks is less than 6 patients, but a lot of patients should be rejected as soon as the remission time of breast cancer patients exceeds 11 patients. Otherwise, the experiment would be repeated. Table 11.17 shows 22 patients before the average remission time of breast cancer patients of 14.1457. Therefore, the quality medical practitioner will reject the arrived lot of patients. Hence based on the real application, it is proficient that the developed sampling will be considerate to check the average remission time of leukemia patients.

11.5 Conclusions

An extensive analysis of two types of cancer data for gamma distribution based on indeterminacy using time-truncated single sampling and repetitive sampling plans are studied in this chapter. For known values of indeterminacy parameter, the plan quantities of the developed two sampling schemes parameters are obtained. Along with plan parameters, extensive tables are provided for ready reference to the researchers for the specified indeterminacy parametric values. The results show that when the values of constant multiplier increase from 0.5 to 1.0 the value of *ASN* decreases. Also, it is observed that if the shape parameter increases the values of *ASN* decrease. Furthermore, the indeterminacy parameter also shows a considerable effect to reduce the *ASN* values. The developed two sampling schemes are compared with the in-hand traditional statistic plans. The outcome depicts that the developed two sampling schemes under indeterminacy are more economical than the in-hand traditional sampling plans. In addition, between the proposed two sampling schemes under indeterminacy the repetitive sampling plan is more economical than the single sampling plan. It is also noticed that indeterminacy values play a vital role in sample size; when the indeterminacy values increase, then the average sample number is decreased. Thus the proposed two sampling schemes are thrifty to the researchers, especially in medical experimentation because medical experimentation needed more money and skilled professionals. Hence, the developed two sampling schemes under indeterminacy are recommended to be valid for testing the average number of patients due to cancer. The two real examples based on the breast cancer and leukemia patients for developed two sampling schemes under indeterminacy show an evidence. The proposed sampling plan could be adapted to other sciences and engineering fields for big data analytics. A further research can be established to extend our study to a multiple dependent state sampling plans and multiple dependent state repetitive sampling plans for various lifetime distributions.

References

[1] M. Dollinger, E. Rosenbaum, G. Cable, Understanding cancer, Everyone's Guide to Cancer Therapy, Andrews and McMeel, New York, 1991.

[2] R.R.L. Kantam, K. Rosaiah, G.S. Rao, Acceptance sampling based on life tests: Log-logistic model, J. Appl. Stat. 28 (1) (2001) 121–128.

[3] T.-R. Tsai, S.-J. Wu, Acceptance sampling based on truncated life tests for generalized Rayleigh distribution, J. Appl. Stat. 33 (6) (2006) 595–600.

[4] N. Balakrishnan, V. Leiva, J. López, Acceptance sampling plans from truncated life tests based on the generalized Birnbaum–Saunders distribution, Commun. Stat. -- Simul. Comput. 36 (3) (2007) 643–656.

[5] Y.L. Lio, T.-R. Tsai, S.-J. Wu, Acceptance sampling plans from truncated life tests based on the Birnbaum–Saunders distribution for percentiles, Commun. Stat. – Simul. Comput. 39 (1) (2009) 119–136.

[6] Y.L. Lio, T.-R. Tsai, S.-J. Wu, Acceptance sampling plans from truncated life tests based on the Burr type XII percentiles, J. Chin. Inst. Ind. Eng. 27 (4) (2010) 270–280.

[7] A. Al-Omari, S. Al-Hadhrami, Acceptance sampling plans based on truncated life tests for extended exponential distribution, Kuwait J. Sci. 45 (2018) 2.

[8] A.I. Al-Omari, Time truncated acceptance sampling plans for generalized inverted exponential distribution, Electron. J. Appl. Stat. Anal. 8 (1) (2015) 1–12.

[9] A. Yan, S. Liu, X. Dong, Variables two stage sampling plans based on the coefficient of variation, J. Adv. Mech. Design Systems Manuf. 10 (1) (2016) 1–12.

[10] C.-H. Yen, et al., A rectifying acceptance sampling plan based on the process capability index, Mathematics 8 (1) (2020) 141.

[11] R.E. Sherman, Design and evaluation of a repetitive group sampling plan, Technometrics 7 (1) (1965) 11–21.

[12] M. Aslam, et al., Decision rule of repetitive acceptance sampling plans assuring percentile life, Sci. Iran. 19 (3) (2012) 879–884.

[13] M. Aslam, Y.L. Lio, C.-H. Jun, *Repetitive acceptance sampling plans for burr type XII percentiles*. The, Int. J. Adv. Manuf. Technol. 68 (1) (2013) 495–507.

[14] M. Aslam, M. Azam, C.-H. Jun, Decision rule based on group sampling plan under the inverse gaussian distribution, Sequential Anal. 32 (1) (2013) 71–82.

[15] N. Singh, N. Singh, H. Kaur, A repetitive acceptance sampling plan for generalized inverted exponential distribution based on truncated life test, Int. J. Sci. Res. Math. Stat. Sci 5 (3) (2018) 58–64.

[16] A. Yan, S. Liu, Designing a repetitive group sampling plan for weibull distributed processes, Math. Probl. Eng. (2016) 5862071.

[17] M. Aslam, et al., Designing of a new monitoring t-chart using repetitive sampling, Inf. Sci. 269 (2014) 210–216.

[18] C.-H. Yen, C.-H. Chang, M. Aslam, Repetitive variable acceptance sampling plan for one-sided specification, J. Stat. Comput. Simul. 85 (6) (2015) 1102–1116.

[19] F. Smarandache, Neutrosophy. Neutrosophic Probability, Set, and Logic, ProQuest Information & Learning, 105, American Research Press, Ann Arbor, MI, 1998, pp. 118–123.

[20] F. Smarandache, H.E. Khalid, Neutrosophic precalculus and neutrosophic calculus, Infinite Study (2015).

[21] X. Peng, J. Dai, *Approaches to single-valued neutrosophic MADM based on MABAC*, TOPSIS and new similarity measure with score function, . Neural Comput. Appl. 29 (10) (2018) 939–954.

[22] M. Abdel-Basset, et al., Cosine similarity measures of bipolar neutrosophic set for diagnosis of bipolar disorder diseases, Artif. Intell. Med. 101 (2019) 101735.

[23] N.A. Nabeeh, et al., An integrated neutrosophic-topsis approach and its application to personnel selection: a new trend in brain processing and analysis, IEEE Access. 7 (2019) 29734–29744.

[24] J. Pratihar, et al., Transportation problem in neutrosophic environment, Neutrosophic Graph Theory and Algorithms, IGI Global, 2020, pp. 180–212.

[25] J. Pratihar, et al., Modified Vogel's approximation method for transportation problem under uncertain environment, Complex. Intell. Syst. (2020) 1–12.

[26] F. Smarandache, Introduction to neutrosophic statistics, Infinite Study, 2014.

[27] J. Chen, J. Ye, S. Du, Scale effect and anisotropy analyzed for neutrosophic numbers of rock joint roughness coefficient based on neutrosophic statistics, Symmetry 9 (10) (2017) 208.

[28] J. Chen, et al., Expressions of rock joint roughness coefficient using neutrosophic interval statistical numbers, Symmetry 9 (7) (2017) 123.

[29] M. Aslam, Introducing Kolmogorov–Smirnov Tests under Uncertainty: An Application to Radioactive Data, ACS Omega, 2019.

[30] M. Aslam, A new sampling plan using neutrosophic process loss consideration, Symmetry 10 (5) (2018) 132.

[31] M. Aslam, Design of sampling plan for exponential distribution under neutrosophic statistical interval method, IEEE Access (2018).

[32] M. Aslam, Design of sampling plan for exponential distribution under neutrosophic statistical interval method, IEEE Access. 6 (2018) 64153–64158.

[33] M. Aslam, A new attribute sampling plan using neutrosophic statistical interval method, Complex. Intell. Syst. (2019) 1–6.

[34] M. Aslam, et al., Time-truncated group plan under a weibull distribution based on neutrosophic statistics, Mathematics 7 (10) (2019) 905.

[35] E.B. Jamkhaneh, B. Sadeghpour-Gildeh, G. Yari, Important criteria of rectifying inspection for single sampling plan with fuzzy parameter, Int. J. Contemp. Math. Sci. 4 (36) (2009) 1791–1801.

[36] E.B. Jamkhaneh, B. Sadeghpour-Gildeh, G. Yari, Inspection error and its effects on single sampling plans with fuzzy parameters, Struct. Multidiscip. Optim. 43 (4) (2011) 555–560.

[37] B. Sadeghpour Gildeh, E. Baloui Jamkhaneh, G. Yari, Acceptance single sampling plan with fuzzy parameter, Iran. J. Fuzzy Syst. 8 (2) (2011) 47–55.

[38] R. Afshari, B. Sadeghpour Gildeh, Designing a multiple deferred state attribute sampling plan in a fuzzy environment, Am. J. Math. Manag. Sci. 36 (4) (2017) 328–345.

[39] X. Tong, Z. Wang, Fuzzy acceptance sampling plans for inspection of geospatial data with ambiguity in quality characteristics, Computers Geosci. 48 (2012) 256–266.

[40] G. Uma, K. Ramya, Impact of fuzzy logic on acceptance sampling plans – a review, Autom. Autonomous Syst. 7 (7) (2015) 181–185.

[41] M. Aslam, Testing average wind speed using sampling plan for Weibull distribution under indeterminacy, Sci. Rep. 11 (1) (2021) 7532.

[42] G.S. Rao, M. Aslam, Inspection plan for COVID-19 patients for Weibull distribution using repetitive sampling under indeterminacy, BMC Med. Res. Methodol. 21 (1) (2021) 229.

[43] H. Okagbue, M.O. Adamu, T.A. Anake, Approximations for the inverse cumulative distribution function of the gamma distribution used in wireless communication, Heliyon 6 (11) (2020) e05523.

[44] E.T. Lee, J. Wang, Statistical methods for survival data analysis, 476, John Wiley & Sons, 2003.

[45] S. Balamurali, C.-h Jun, Repetitive group sampling procedure for variables inspection, J. Appl. Stat. 33 (3) (2006) 327–338.

[46] M. Aslam, C.-H. Yen, C.-H. Jun, Variable repetitive group sampling plans with process loss consideration, J. Stat. Comput. Simul. 81 (11) (2011) 1417–1432.

[47] M. Aslam, et al., Developing a variables repetitive group sampling plan based on process capability index C pk with unknown mean and variance, J. Stat. Comput. Simul. 83 (8) (2013) 1507–1517.

[48] J.F. Lawless, Statistical models and methods for lifetime data, John Wiley & Sons, 2011, p. 362.

Chapter 12

Improved correlation coefficients in neutrosophic statistics for COVID patients using pentapartitioned neutrosophic sets

R. Radha[1], A. Stanis Arul Mary[1], Said Broumi[2,3], S. Jafari[4] and S.A. Edalatpanah[5]

[1]Department of Mathematics, Nirmala College for Women, Coimbatore, Tamil Nadu, India, [2]Laboratory of Information Processing, Faculty of Science, Ben M'Sik, University Hassan II, Casablanca, Morocco, [3]Regional Center for the Professions of Education and Training (C.R.M. E.F), Casablanca-Settat, Morocco, [4]Department of Mathematics, College of Vestsjaelland South, Slagelse, Denmark, [5]Department of Applied Mathematics, Ayandegan Institute of Higher Education, Tonekabon, Iran

12.1 Introduction

Fuzzy sets were introduced by Zadeh [1] in 1965 that permit the membership to perform valued within the interval [0,1], and set theory is an extension of classical pure mathematics. Fuzzy set helps to deal with the thought of uncertainty, unclearness, and impreciseness that is not attainable within the Cantorian set. As an Associate in Nursing extension of Zadeh's fuzzy set theory, intuitionistic fuzzy set (IFS) was introduced by Atanassov [2] in 1986 that consists of degree of membership and degree of nonmembership and lies within the interval of [0,1]. IFS theory was widely utilized in the areas of logic programming, decision-making issues, medical diagnosis, etc.

Florentin Smarandache [3] introduced the idea of neutrosophic set in 1995 that provides the information on neutral thought by introducing the new issue referred to as uncertainty within the set. Thus neutrosophic set was framed, and it includes the parts of truth membership function (T), indeterminacy membership function (I), and falsity membership function (F) severally. Neutrosophic sets deal with non-normal interval of $]-0\ 1+[$. Since

Cognitive Intelligence with Neutrosophic Statistics in Bioinformatics.
DOI: https://doi.org/10.1016/B978-0-323-99456-9.00017-9

neutrosophic set deals the indeterminateness effectively, it plays a very important role in several application areas embracing info technology, decision web, electronic database systems, diagnosis, multicriteria higher cognitive process issues, etc.

To method the unfinished data or imperfect data to unclearness, a brand new mathematical approach, that is, to deal the important world issues, Wang [4] introduced the idea of single-valued neutrosophic sets (SVNSs) that is additionally referred to as an extension of intuitionistic fuzzy sets and it became a really new hot analysis topic. Rama Malik et al. [5] projected the idea of pentapartitioned single-valued neutrosophic sets that relies on Belnap's five-valued logic and Smarandache's five numerical-valued logic. In PSVNS, indeterminacy is split into three functions referred to as "contradicition" (both true and false), "ignorance" (neither true nor false), and an "unknown" membership in order that PSVNS has five parts, T, C, U, F, and G, that additionally lies within the non-normal unit interval $]-0\ 1+[$. Further, R. Radha and A. Stanis Arul Mary [6,7] outlined a brand new hybrid model of pentapartitioned neutrosophic Pythagorean sets (PNPSs) and pentapartitioned neutrosophic topological spaces in 2021. Correlation coefficient may be an effective mathematical tool to live the strength of the link between two variables. A lot of researchers pay attention to the idea of varied correlation coefficients of the various sets like fuzzy set, IFS, SVNS, QSVNS. In 1999, D.A Chiang and N.P. Lin [8] projected the correlation of fuzzy sets underneath fuzzy setting. Later, D.H. Hong [9] outlined fuzzy measures for a coefficient of correlation of fuzzy numbers below Tw (the weakest t-norm) based mostly on fuzzy arithmetic operations.

Correlation coefficients play a very important role in several universal issues like multiple attribute cluster higher cognitive process, cluster analysis, pattern recognition, diagnosis, etc.; therefore several authors targeted the idea of shaping correlation coefficients to resolve the important world issues significantly using multicriteria decision-making strategies. Jun Ye [10] outlined the improved correlation coefficients of single-valued neutrosophic sets and interval-valued neutrosophic sets for multiple attribute higher cognitive process to beat the drawbacks of the correlation coefficients of SVNSs that are outlined in.

COVID-19 is a contagious respiratory disease caused by infection with the virus SARS-CoV-2. It usually spreads between people who are in close contact within six feet or two meters. The virus spreads through respiratory droplets released when someone breathes, coughs, sneezes, talks, or sings. These droplets can land in the mouth or nose of someone nearby or be inhaled. The virus can also spread if a person touches a surface or object with the virus on it and then touches another surface or object with the virus on it and then touches his or her mouth, nose, or eyes, although this is not considered to be the main way it spreads. The coronavirus pandemic has caused severe effects on human life and the global economy affecting all communities and individuals due to its rapid spreading, increasing the number of affected cases, and creating severe health issues and death cases worldwide. Spreading rapidly across borders, the coronavirus disease has created a global health crisis and caused numerous death cases all over the world. The survey is taken under COVID patients from pre-COVID, post-COVID, and during the COVID times and finds the most affected situation among the three cases. Finally, an illustrative example of COVID patients alternatives is given to demonstrate the application and effectiveness of the developed decision-making approaches under single-valued pentapartitioned neutrosophic and interval-valued pentapartitioned neutrosophic statistics environments.

In this paper, Section 12.2 gives some basic definitions of pentapartitioned neutrosophic sets and its complement, union, intersection, interval-valued pentapartitioned neutrosophic sets (IVPNS), improved correlation coefficient of pentapartitioned neutrosophic sets. In Section 12.3 and Section 12.4, we introduce the concept of improved correlation coefficient of pentapartitioned neutrosophic sets and also discuss some of its properties and decision-making method using the improved correlation coefficients of pentapartitioned neutrosophic sets. In Section 12.5, we introduce the concept of IVPNSs with some basic definitions and define improved correlation coefficients of IVPNS. Further, we have also discussed some of its properties and decision-making method using the improved correlation coefficient with interval-valued pentapartitioned neutrosophic environment. In Section 12.6 and Section 12.7, an illustrative example is given in the above-proposed correlation method particularly in multiple criteria decision-making problems. Section 12.8 concludes the paper. Several researchers are working on the treatment of COVID patients, and the different stages corresponding to the symptoms are also discussed with an example.

12.2 Preliminaries

12.2.1 Definition [3]

Let R be a universe. A neutrosophic set A on R can be defined as follows:

$$A = \{ < x, TR_A(x), IR_A(x), FR_A(x) > : x \in X\}$$

where $TR_A, IR_A, FR_A : U \rightarrow [0,1]$ and $0 \leq TR_A(x) + IR_A(x) + FR_A(x) \leq 3$

Here, $TR_A(x)$ is the degree of membership, $IR_A(x)$ is the degree of indeterminacy, and $FR_A(x)$ is the degree of nonmembership.

12.2.2 Definition [5]

Let P be a non-empty set. A pentapartitioned neutrosophic set A over P characterizes each element p in P a truth membership function TR_A, a contradiction membership function CR_A, an ignorance membership function GR_A, unknown membership function UR_A, and a false membership function FR_A, such that for each p in P

$$TR_A + CR_A + GR_A + UR_A + FR_A \leq 5$$

12.2.3 Definition [5]

The complement of a pentapartitioned neutrosophic set F on R is denoted by F^c and is defined as

$$F^c(\mathrm{x}) = \{ < x, FR_A(x), UR_A(x), 1 - GR_A(x), CR_A(x), TR_A(x) > : x \in R\}$$

12.2.4 Definition [5]

Let A $= < x, TR_A(x), CR_A(x), GR_A(x), UR_A(x), FR_A(x) >$ and B $= < x, TR_B(x), CR_B(x), GR_B(x), UR_B(x), FR_B(x) >$
are pentapartitioned neutrosophic sets. Then

$$\mathrm{A} \cup \mathrm{B} = < x, \max(TR_A(x), TR_B(x)), \max(CR_A(x), CR_B(x)), \min(GR_A(x), GR_B(x)), \min(UR_A(x), UR_B(x)), \min(F_A(x), F_B(x)), >$$

$$\mathrm{A} \cap \mathrm{B} = < x, \min(TR_A(x), TR_B(x)), \min(CR_A(x), CR_B(x)), \max(GR_A(x), GR_B(x)), \max(UR_A(x), UR_B(x)), \max(FR_A(x), FR_B(x)) >$$

12.3 Improved correlation coefficients

Based on the concept of correlation coefficient of PNSs, we have defined the improved correlation coefficients of PNSs in the following section.

12.3.1 Definition

Let P and Q be any two PNSs in the universe of discourse $R = \{r_1, r_2, r_3, \ldots, r_n\}$, then the improved correlation coefficient between P and Q is defined as follows

$$K(P, Q) = \frac{1}{5n}\sum_{k=1}^{n}[\mu_k(1 - \Delta\mathrm{TR}_k) + \varphi_k(1 - \Delta CR_k) + \delta_k(1 - \Delta GR_k) + \gamma_k(1 - \Delta UR_k) + \vartheta_k(1 - \Delta FR_k)] \quad (12.1)$$

where

$$\mu_k = \frac{5 - \Delta TR_k - \Delta TR_{\max}}{5 - \Delta TR_{\min} - \Delta TR_{\max}},$$

$$\varphi_k = \frac{5 - \Delta CR_k - \Delta CR_{\max}}{5 - \Delta CR_{\min} - \Delta CR_{\max}},$$

$$\delta_k = \frac{5 - \Delta GR_k - \Delta GR_{\max}}{5 - \Delta GR_{\min} - \Delta GR_{\max}},$$

$$\gamma_k = \frac{5 - \Delta UR_k - \Delta UR_{\max}}{5 - \Delta UR_{\min} - \Delta UR_{\max}},$$

$$\vartheta_k = \frac{5 - \Delta FR_k - \Delta FR_{\max}}{5 - \Delta FR_{\min} - \Delta FR_{\max}},$$

$$\Delta TR_K = \left|TR_P(r_k) - TR_Q(r_k)\right|,$$

$$\Delta CR_K = \left|CR_P(r_k) - CR_Q(r_k)\right|,$$

$$\Delta GR_K = \left|GR_P(r_k) - GR_Q(r_k)\right|,$$

$$\Delta UR_K = \left|UR_P(r_k) - UR_Q(r_k)\right|,$$

$$\Delta FR_K = \left|FR_P(r_k) - FR_Q(r_k)\right|,$$

$$\Delta TR_{\min} = \min_k\left|TR_P(r_k) - TR_Q(r_k)\right|,$$

$$\Delta CR_{\min} = \min_k\left|CR_P(r_k) - CR_Q(r_k)\right|,$$

$$\Delta GR_{\min} = \min_k\left|GR_P(r_k) - GR_Q(r_k)\right|,$$

$$\Delta UR_{\min} = \min_k\left|UR_P(r_k) - UR_Q(r_k)\right|,$$

$$\Delta FR_{\min} = \min_k\left|FR_P(r_k) - FR_Q(r_k)\right|,$$

$$\Delta TR_{\max} = \max_k\left|TR_P(r_k) - TR_Q(r_k)\right|,$$

$$\Delta CR_{\max} = \max_k\left|CR_P(r_k) - CR_Q(r_k)\right|,$$

$$\Delta GR_{\max} = \max_k\left|GR_P(r_k) - GR_Q(r_k)\right|,$$

$$\Delta UR_{\max} = \max_k\left|UR_P(r_k) - UR_Q(r_k)\right|,$$

$$\Delta FR_{\max} = \max_k\left|FR_P(r_k) - FR_Q(r_k))\right|,$$

For any $r_k \in R$ and $k = 1, 2, 3\ldots n$.

12.3.2 Theorem

For any two PNSs P and Q in the universe of discourse $R = \{r_1, r_2, r_3, \ldots, r_n\}$, the improved correlation coefficient $K(P, Q)$ satisfies the following properties.

1. $K(P, Q) = K(Q, P)$;
2. $0 \leq K(P, Q) \leq 1$;
3. $K(P, Q) = 1$ iff $P = Q$.

Proof

1. It is obvious and straightforward.
2. Here, $0 \leq \mu_k \leq 1$, $0 \leq \varphi_k \leq 1$, $0 \leq \delta_k \leq 1$, $0 \leq \gamma_k \leq 1$, $0 \leq \vartheta_k \leq 1$, $0 \leq 1 - \Delta TR_k \leq 1$, $0 \leq 1 - \Delta CR_k \leq 1$, $0 \leq 1 - \Delta GR_k \leq 1$, $0 \leq 1 - \Delta UR_k \leq 1$, $0 \leq 1 - \Delta FR_k \leq 1$. Therefore the following in equation satisfies
   $0 \leq \mu_k(1 - \Delta TR_k) + \varphi_k(1 - \Delta CR_k) + \delta_k(1 - \Delta GR_k) + \gamma_k(1 - \Delta UR_k) + \vartheta_k(1 - \Delta FR_k) \leq 5$. Hence we have $0 \leq K(P, Q) \leq 1$
3. If K (P, Q) = 1, then we get $\mu_k(1 - \Delta TR_k) + \varphi_k(1 - \Delta CR_k) + \delta_k(1 - \Delta GR_k) + \gamma_k(1 - \Delta UR_k) + \vartheta_k(1 - \Delta FR_k) = 5$. Since $0 \leq \mu_k (1 - \Delta TR_k) \leq 1$, $0 \leq \varphi_k(1 - \Delta CR_k) \leq 1$, $0 \leq \delta_k(1 - \Delta GR_k) \leq 1$, $0 \leq \gamma_k (1 - \Delta UR_k) \leq 1$ and $0 \leq \vartheta_k(1 - \Delta FR_k) \leq 1$, there are $\mu_k(1 - \Delta TR_k) = 1$, $\varphi_k(1 - \Delta CR_k) = 1$, $\delta_k(1 - \Delta GR_k) = 1$, $\gamma_k(1 - \Delta UR_k) = 1$ and $\vartheta_k (1 - \Delta FR_k) = 1$. Also, since $0 \leq \mu_k \leq 1$, $0 \leq \varphi_k \leq 1$, $0 \leq \delta_k \leq 1$, $0 \leq \gamma_k \leq 1$ and $0 \leq \vartheta_k \leq 1$, $0 \leq 1 - \Delta TR_k \leq 1$, $0 \leq 1 - \Delta CR_k \leq 1$, $0 \leq 1 - \Delta GR_k \leq 1$, $0 \leq 1 - \Delta UR_k \leq 1$, $0 \leq 1 - \Delta FR_k \leq 1$, we get $\mu_k = \varphi_k = \delta_k = \gamma_k = \vartheta_k = 1$ and $1 - \Delta TR_k = 1 - \Delta CR_k = 1 - \Delta GR_k = 1 - \Delta UR_k = 1 - \Delta FR_k = 1$. This implies $\Delta TR_k = \Delta TR_{\min} = \Delta TR_{\max} = 0$, $\Delta CR_k = \Delta CR_{\min} = \Delta CR_{\max} = 0$, $\Delta GR_k = \Delta GR_{\min} = \Delta GR_{\max} = 0$, $\Delta UR_k = \Delta UR_{\min} = \Delta UR_{\max} = 0$, $\Delta FR_k = \Delta FR_{\min} = \Delta FR_{\max} = 0$. Hence $TR_P(r_k) = TR_Q(r_k)$, $CR_P(r_k) = CR_Q(r_k)$, $GR_P(r_k) = GR_Q(r_k)$, $UR_P(r_k) = UR_Q(r_k)$, and $FR_P(r_k) = FR_Q(r_k)$ for any $r_k \in R$ and k = 1,2,3...n. Hence P = Q.

Conversely, assume that P = Q, this implies $TR_P(r_k) = TR_Q(r_k)$, $CR_P(r_k) = CR_Q(r_k)$, $GR_P(r_k) = GR_Q(r_k)$, $UR_P(r_k) = UR_Q(r_k)$, and $FR_P(r_k) = FR_Q(r_k)$ for any $r_k \in R$ and k = 1,2,3...n. Thus $\Delta TR_k = \Delta TR_{\min} = \Delta TR_{\max} = 0$, $\Delta CR_k = \Delta CR_{\min} = \Delta CR_{\max} = 0$, $\Delta GR_k = \Delta GR_{\min} = \Delta GR_{\max} = 0$, $\Delta UR_k = \Delta UR_{\min} = \Delta UR_{\max} = 0$, $\Delta FR_k = \Delta FR_{\min} = \Delta FR_{\max} = 0$. Hence we get $K(P, Q) = 1$.

The improved correlation coefficient formula which is defined is correct and also satisfies these properties in the above theorem. When we use any constant $\varepsilon > 3$ in the following expressions

$$\mu_k = \frac{\varepsilon - \Delta TR_k - \Delta TR_{\max}}{\varepsilon - \Delta TR_{\min} - \Delta TR_{\max}},$$

$$\varphi_k = \frac{\varepsilon - \Delta CR_k - \Delta CR_{\max}}{\varepsilon - \Delta CR_{\min} - \Delta CR_{\max}},$$

$$\delta_k = \frac{\varepsilon - \Delta GR_k - \Delta GR_{\max}}{\varepsilon - \Delta GR_{\min} - \Delta GR_{\max}},$$

$$\gamma_k = \frac{\varepsilon - \Delta UR_k - \Delta UR_{\max}}{\varepsilon - \Delta UR_{\min} - \Delta UR_{\max}},$$

$$\vartheta_k = \frac{\varepsilon - \Delta FR_k - \Delta FR_{\max}}{\varepsilon - \Delta FR_{\min} - \Delta FR_{\max}}$$

12.3.3 Example

Let $A = \{r, 0,0,0,0,0\}$ and $B = \{r, 0.4,0.2,0.5,0.1,0.2\}$ be any two PNSs in R. Therefore by Eq. (12.1), we get $K(A, B) = 0.871.56$. It shows that the above-defined improved correlation coefficient overcomes the disadvantages of the correlation coefficient.

In the following, we define a weighted correlation coefficient between PNSs since the differences in the elements are considered into account,

Let w_k be the weight of each element $r_k(k = 1,2...n)$, $w_k \in [0, 1]$, and $\sum_{k=1}^{n} w_k = 1$, then the weighted correlation coefficient between the PNSs A and B

$$K_w(A,B) = \frac{1}{5}\sum_{k=1}^{n} w_k[\mu_k(1 - \Delta TR_k) + \varphi_k(1 - \Delta CR_k) + \delta_k(1 - \Delta GR_k)$$
$$+ \gamma_k(1 - \Delta UR_k) + \vartheta_k(1 - \Delta FR_k)] \quad (12.2)$$

If $w = (1/n,1/n,1/n,...0.1/n)^T$, then Eq. (3.2) reduces to Eq. (3.1). $K_w(A, B)$ also satisfies the three properties in the above theorem.

12.3.4 Theorem

Let w_k be the weight for each element $r_k(\text{k} = 1,2,...n)$, $w_k \in [0, 1]$, and $\sum_{k=1}^{n} w_k = 1$, then the weighted correlation coefficient between the PNSs A and B which is denoted by $K_w(A, B)$ defined in Eq. (12.2) satisfies the following properties.

$$K_w(A,B) = K_w(B,A);$$

$$0 \leq K_w(A,B) \leq 1;$$

$$K_w(A,B) = 1 \text{ iff } A = B.$$

It is similar to proving the properties in Theorem 3.1.

12.4 Decision-making using the improved correlation coefficient of pentapartitioned neutrosophic sets

Multiple attribute decision-making (MADM) problems refer to make decisions when several attributes are involved in real-life problem. For example, one may buy a vehicle by analyzing the attributes which are given in terms of price, style, safety, comfort, etc.

Here, we consider a MADM problem with pentapartitioned neutrosophic Pythagorean information, and the characteristic of an alternative $A_i(i = 1,2,\ldots m)$ on an attribute $C_j(j = 1,2\ldots n)$ is represented by the following PNPSs:

$$A_i = \{(C_j, TR_{A_i}(C_j), CR_{A_i}(C_j), GR_{A_i}(C_j), UR_{A_i}(C_j), FR_{A_i}(C_j)\backslash C_j \in C, j = 1,2,\ldots n\}$$

where $TR_{A_i}(C_j), CR_{A_i}(C_j), GR_{A_i}(C_j), UR_{A_i}(C_j), FR_{A_i}(C_j) \in [0,1]$ and

$$0 \le TR^2_{A_j}(C_j) + CR^2_{A_j}(C_j) + GR^2_{A_j}(C_j) + UR^2_{A_j}(C_j) + FR^2_{A_j}(C_j) \le 3$$

$$\text{for } C_j \in C, j = 1,2,\ldots n \text{ and } I = 1,2,\ldots \text{m}.$$

To make it convenient, we are considering the following five functions $TR_{A_i}(C_j), CR_{A_i}(C_j), GR_{A_i}(C_j), UR_{A_i}(C_j), FR_{A_i}(C_j)$ in terms of pentapartitioned neutrosophic value (PNV)

$$d_{ij} = (t_{ij}, c_{ij}, g_{ij}, u_{ij}, f_{ij})(i = 1,2,\ldots m; j = 1,2\ldots n).$$

Here, the values of d_{ij} are usually derived from the evaluation of an alternative A_i with respect to a criteria C_j by the expert or decision-maker. Therefore we get a pentapartitioned neutrosophic decision matrix $D = (d_{ij})_{m\times n}$.

In the case of ideal alternative A^*, an ideal PNV can be defined by

$$d^*_j = (t^*_j, c^*_j, g^*_j, u^*_j, f^*_j) = (1,1,0,0,0)(j = 1,2\ldots n)\text{in the decision making method,}$$

Hence, the weighted correlation coefficient between an alternative A_i(i = 1,2…m) and the ideal alternative A^* is given by

$$K_w(A_i, A^*) = \frac{1}{5}\sum_{j=1}^{n} w_j[\mu_{ij}(1-\Delta t_{ij}) + \varphi_{ij}(1-\Delta c_{ij}) + \delta_{ij}(1-\Delta g_{ij})$$
$$+ \gamma_{ij}(1-\Delta u_{ij}) + \vartheta_{ij}(1-\Delta f_{ij})] \tag{12.3}$$

where

$$\mu_{ij} = \frac{5 - \Delta t_{ij} - \Delta t_{i\max}}{5 - \Delta t_{i\min} - \Delta t_{i\max}},$$

$$\varphi_{ij} = \frac{5 - \Delta c_{ij} - \Delta c_{i\max}}{5 - \Delta c_{i\min} - \Delta c_{i\max}},$$

$$\delta_{ij} = \frac{5 - \Delta g_{ij} - \Delta g_{i\max}}{5 - \Delta g_{i\min} - \Delta g_{i\max}},$$

$$\gamma_{ij} = \frac{5 - \Delta u_{ij} - \Delta u_{i\max}}{5 - \Delta u_{i\min} - \Delta u_{i\max}},$$

$$\vartheta_{ij} = \frac{5 - \Delta f_{ij} - \Delta f_{i\max}}{5 - \Delta f_{i\min} - \Delta f_{i\max}},$$

$$\Delta t_{ij} = \left| t_{ij} - t_j^* \right|,$$

$$\Delta c_{ij} = \left| c_{ij} - c_j^* \right|,$$

$$\Delta g_{ij} = \left| g_{ij} - g_j^* \right|,$$

$$\Delta u_{ij} = \left| u_{ij} - u_j^* \right|,$$

$$\Delta f_{ij} = \left| f_{ij} - f_j^* \right|,$$

$$\Delta t_{i\min} = \min_j \left| t_{ij} - t_j^* \right|,$$

$$\Delta g_{i\min} = \min_j \left| g_{ij} - g_j^* \right|,$$

$$\Delta c_{i\min} = \min_j \left| c_{ij} - c_j^* \right|,$$

$$\Delta u_{i\min} = \min_j \left| u_{ij} - u_j^* \right|,$$

$$\Delta f_{i\min} = \min_j \left| f_{ij} - f_j^* \right|,$$

$$\Delta t_{i\max} = \max_j \left| t_{ij} - t_j^* \right|,$$

$$\Delta c_{i\max} = \max_j \left| c_{ij} - c_j^* \right|,$$

$$\Delta g_{i\max} = \max_j \left| g_{ij} - g_j^* \right|,$$

$$\Delta u_{i\max} = \max_j \left| u_{ij} - u_j^* \right|,$$

$$\Delta f_{i\max} = \max_j \left| f_{ij} - f_j^* \right|,$$

For $i = 1,2\ldots m$ and $j = 1,2\ldots n$.

By using the above weighted correlation coefficient, we can derive the ranking order of all alternatives and we can choose the best one among those.

12.5 Interval-valued pentapartitioned neutrosophic sets

12.5.1 Definition

An IVPNS A in R is characterized by a truth membership function, a contradiction membership function, an unknown membership function, an ignorance membership function. For each point k in R, $TR_A = \left[TR_A^L(k), TR_A^U(k)\right]$, $CR_A = [CR_A^L(k), CR_A^U(k)]$, $GR_A = [GR_A^L(k), GR_A^U(k)]$, $UR_A = [UR_A^L(k), UR_A^U(k)]$ and $FR_A = [FR_A^L(k), FR_A^U(k)]$. Therefore an IVPNS can be denoted as

$$\begin{aligned} \mathrm{A} &= \{\mathrm{k}, TR_A^L(k), CR_A^L(k), GR_A^L(k), UR_A^L(k), FR_A^L(k) | k \in R\} \\ &= \{(\mathrm{k}, \left[TR_A^L(k), TR_A^U(k)\right], \left[CR_A^L(k), CR_A^U(k)\right], GR_A = \left[GR_A^L(k), GR_A^U(k)\right], \\ &\quad \left[UR_A^L(k), UR_A^U(k)\right], FR_A^L(k), FR_A^U(k) | k \in R)\} \end{aligned}$$

Then the sum of $TR_A(k), CR_A(k), GR_A(k), UR_A(k), FR_A(k)$ satisfies the condition

$$0 \le TR_A^L(k) + CR_A^L(k) + GR_A^L(k) + UR_A^L(k) + FR_A^L(k) \le 5.$$

The upper and lower ends of interval values are equal in an IVPNSs reduce to pentapartitioned neutrosophic sets.

12.5.2 Example

Let $R = \{a\}$ and $A = \{(a,\ [0.2,0.6],\ [0.3,0.8],\ [0.1,0.5],\ [0.1,0.8],\ [0.2,0.4]\}$ be IVPNS on R.

Then $\tau = \{0,1, A\}$ is a topology on R, and A is an IVPNS.

12.5.3 Definition

The complement of an IVPNS

$A = \{(k, \left[TR_A^L(k), TR_A^U(k)\right], \left[CR_A^L(k), CR_A^U(k)\right], \left[GR_A^L(k), GR_A^U(k)\right], \left[UR_A^L(k), UR_A^U(k)\right],$ $FR_A^L(k), FR_A^U(k) | k \in R)\}$is denoted by $A^C$and is defined as

$$\begin{aligned} A^C = \{(k, &\left[FR_A^L(k), FR_A^U(k)\right], \left[UR_A^L(k), UR_A^U(k)\right], \left[GR_A^L(k), GR_A^U(k)\right], \\ &\left[CR_A^L(k), UC(k)\right], TR_A^L(k), TR_A^U(k) \\ &| k \in R)\} \end{aligned}$$

12.5.4 Example

Let $R = \{a\}$ and $A = \{(a,$ [0.2,0.6], [0.3,0.8], [0.1,0.5], [0.1,0.8], [0.2,0.4]$\}$ be IVPNS on R. Then the complement of A on R is

$$A^C = \{(a, [0.2, 0.4], [0.1, 0.8], [0.1, 0.5], [0.3, 0.8], [0.2, 0.6]\}$$

12.5.5 Definition

An IVPNS A is contained in the other IVPNS B, $A \subseteq B$ if and only if $TR_A^L(k) \leq TR_B^U(k)$, $CR_A^L(k) \leq CR_B^U(k)$, $GR_A^L(k) \leq GR_B^U(k)$, $UR_A^L(k) \leq UR_B^U(k)$ and $FR_A^L(k) \leq FR_B^U(k)$ for any k in K.

12.5.6 Example

Let $R = \{a\}$ and $A = \{(a,$ [0.2,0.6], [0.3,0.8], [0.3,0.5], [0.3,0.9], [0.2,0.6]$\}$ and $B = \{(a,$ [0.3,0.7], [0.5,0.9], [0.2,0.5], [0.2,0.8], [0.1,0.6]$\}$ be IVPNS on R.

12.5.7 Definition

Two IVPNS A and B are equal, that is, $A = B$ if and only if $A \subseteq B$ and $B \subseteq A$.

12.5.8 Definition

Let $A = \{([TR_A^L, TR_A^U], [CR_A^L, CR_A^U], [GR_A^L, GR_A^U], [UR_A^L, UR_A^U], [FR_A^L, FR_A^U]\}$ and $B = \{[TR_B^L, TR_B^U], [CR_B^L, CR_B^U], [GR_B^L, GR_B^U], [UR_B^L, UR_B^U], [FR_B^L, FR_B^U]\}$ are two IVPNSs. Then, the union and intersection of two IVPNSs A and B are denoted by

$$A \cup B = \{[\max(TR_A^L, TR_B^L), \max(TR_A^U, TR_B^U)], [\max(CR_A^L, CR_B^L), \max(CR_A^U, CR_B^U)],$$

$$[\min(GR_A^L, GR_B^L), \min(GR_A^U, GR_B^U)], [\min(UR_A^L, UR_B^L), \min(UR_A^U, UR_B^U)],$$

$$[\min(FR_A^L, FR_B^L), \min(FR_A^U, FR_B^U)]$$

$$A \cap B = \{[\min(TR_A^L, TR_B^L), \min(TR_A^U, TR_B^U)], [\min(CR_A^L, CR_B^L), \min(CR_A^U, CR_B^U)],$$

$$[\max(GR_A^L, GR_B^L), \max(GR_A^U, GR_B^U)], [\max(UR_A^L, UR_B^L), \max(UR_A^U, UR_B^U)],$$

$$[\max(FR_A^L, FR_B^L), \max(FR_A^U, FR_B^U)]$$

12.5.9 Example

Let $R = \{a\}$ and $A = \{(a,$ [0.2,0.6], [0.3,0.8], [0.1,0.5], [0.1,0.8], [0.2,0.4]$\}$ and $B = \{(a,$ [0.1,0.7], [0.2,0.7], [0.2,0.5], [0.2,0.8], [0.1,0.6]$\}$ be IVPNS on R.

Then, the union and intersection of two IVPNS A and B are denoted by

$$A \cup B = \{(a, [0.2, 0.7], [0.3, 0.8], [0.1, 0.5], [0.1, 0.8], [0.1, 0.4]\}$$

$$A \cap B = \{(a, [0.1, 0.6], [0.2, 0.7], [0.1, 0.5], [0.3, 0.8], [0.2, 0.6]\}$$

12.5.9.1 Improved correlation coefficients of interval-valued pentapartitioned neutrosophic sets

In this section, we propose a correlation coefficient between IVPNSs as a generalization of the improved correlation coefficients of pentapartitioned neutrosophic sets.

12.5.10 Definition

Let P and Q be any two IVPNSs in the universe of discourse $R = \{r_1, r_2, r_3, \ldots, r_n\}$, then the improved correlation coefficient between P and Q is defined as follows

$$L(P, Q) = \frac{1}{10n}\sum_{k=1}^{n}[\mu_k^L(1 - \Delta TR_k^L) + \varphi_k^L(1 - \Delta CR_k^L) + \delta_k^L(1 - \Delta GR_k^L)$$

$$+ \gamma_k^L(1 - \Delta UR_k^L) + \vartheta_k^L(1 - \Delta FR_k^L) + \mu_k^U(1 - \Delta TR_k^U) + \varphi_k^U(1 - \Delta CR_k^U)$$

$$+ \delta_k^U(1 - \Delta GR_k^U) + \gamma_k^U(1 - \Delta UR_k^U) + \vartheta_k^U(1 - \Delta FR_k^U)] \quad (12.4)$$

where

$$\mu_k^L = \frac{5 - \Delta TR_k^L - \Delta TR_{\max}^L}{5 - \Delta TR_{\min}^L - \Delta TR_{\max}^L},$$

$$\varphi_k^L = \frac{5 - \Delta CR_k^L - \Delta CR_{\max}^L}{5 - \Delta CR_{\min}^L - \Delta CR_{\max}^L},$$

$$\delta_k^L = \frac{5 - \Delta GR_k^L - \Delta GR_{\max}^L}{5 - \Delta GR_{\min}^L - \Delta GR_{\max}^L},$$

$$\gamma_k^L = \frac{5 - \Delta UR_k^L - \Delta UR_{max}^L}{5 - \Delta UR_{min}^L - \Delta UR_{max}^L},$$

$$\vartheta_k^L = \frac{5 - \Delta FR_k^L - \Delta FR_{\max}^L}{5 - \Delta FR_{\min}^L - \Delta FR_{\max}^L},$$

$$\mu_k^U = \frac{5 - \Delta TR_k^U - \Delta TR_{\max}^U}{5 - \Delta TR_{\min}^L - \Delta TR_{\max}^L},$$

$$\varphi_k^U = \frac{5 - \Delta CR_k^U - \Delta CR_{\max}^U}{5 - \Delta CR_{\min}^U - \Delta CR_{\max}^U},$$

$$\delta_k^U = \frac{5 - \Delta GR_k^U - \Delta GR_{\max}^U}{5 - \Delta GR_{\min}^U - \Delta GR_{\max}^U},$$

$$\gamma_k^U = \frac{5 - \Delta UR_k^U - \Delta UR_{\max}^U}{5 - \Delta UR_{\min}^U - \Delta UR_{\max}^U},$$

$$\vartheta_k^U = \frac{5 - \Delta FR_k^U - \Delta FR_{\max}^U}{5 - \Delta FR_{\min}^U - \Delta FR_{\max}^U},$$

$$\Delta TR_{\min}^L = \min_k |TR_P^L(r_k) - TR_Q^L(r_k)|,$$

$$\Delta CR_{\min}^L = \min_k |CR_P^L(r_k) - CR_Q^L(r_k)|,$$

$$\Delta GR_{\min}^L = \min_k |GR_P^L(r_k) - GR_Q^L(r_k)|,$$

$$\Delta UR_{\min}^L = \min_k |UR_P^L(r_k) - UR_Q^L(r_k)|,$$

$$\Delta FR_{\min}^L = \min_k |FR_P^L(r_k) - FR_Q^L(r_k)|,$$

$$\Delta TR_{\min}^U = \min_k |TR_P^U(r_k) - TR_Q^U(r_k)|,$$

$$\Delta CR_{\min}^U = \min_k |CR_P^U(r_k) - CR_Q^U(r_k)|,$$

$$\Delta GR_{\min}^U = \min_k |GR_P^U(r_k) - GR_Q^U(r_k)|,$$

$$\Delta UR_{\min}^U = \min_k |UR_P^U(r_k) - UR_Q^U(r_k)|,$$

$$\Delta FR_{\min}^U = \min_k |FR_P^U(r_k) - FR_Q^U(r_k)|,$$

$$\Delta TR_{\max}^L = \max_k |TR_P^L(r_k) - TR_Q^L(r_k)|,$$

$$\Delta CR_{\max}^L = \max_k |CR_P^L(r_k) - CR_Q^L(r_k)|,$$

$$\Delta GR_{\max}^L = \max_k |GR_P^L(r_k) - GR_Q^L(r_k)|,$$

$$\Delta UR_{\max}^L = \max_k |UR_P^L(r_k) - UR_Q^L(r_k)|,$$

$$\Delta FR_{\max}^L = \max_k |FR_P^L(r_k) - FR_Q^L(r_k)|,$$

$$\Delta TR^U_{\max} = \max_k |TR^U_P(r_k) - TR^U_Q(r_k)|,$$

$$\Delta CR^U_{\max} = \max_k |CR^U_P(r_k) - CR^U_Q(r_k)|,$$

$$\Delta GR^U_{\max} = \max_k |GR^U_P(r_k) - GR^U_Q(r_k)|,$$

$$\Delta UR^U_{\max} = \max_k |UR^U_P(r_k) - UR^U_Q(r_k)|,$$

$$\Delta FR^U_{\max} = \max_k |FR^U_P(r_k) - FR^U_Q(r_k)|.$$

For any $r_k \in R$ and $k = 1, 2, 3\ldots n$.

12.5.11 Theorem

For any two IVPNSs P and Q in the universe of discourse $R = \{r_1, r_2, r_3, \ldots, r_n\}$, the improved correlation coefficient $L(P, Q)$ satisfies the following properties.

1. $L(P, Q) = L(Q, P)$;
2. $0 \leq L(P, Q) \leq 1$;
3. $L(P, Q) = 1$ iff $P = Q$.

In the following, we define a weighted correlation coefficient between PNSs since the differences in the elements are considered into account.

Let w_k be the weight of each element $r_k (k = 1,2\ldots n)$, $w_k \in [0, 1]$, and $\sum_{k=1}^{n} w_k = 1$, then the weighted correlation coefficient between the IVPNSs A and B

$$L(P, Q) = \frac{1}{10}\sum_{k=1}^{n} w_k[\mu^L_k(1 - \Delta TR^L_k) + \varphi^L_k(1 - \Delta CR^L_k) + \delta^L_k(1 - \Delta GR^L_k)$$

$$+ \gamma^L_k(1 - \Delta UR^L_k) + \vartheta^L_k(1 - \Delta FR^L_k) + \mu^U_k(1 - \Delta TR^U_k) + \varphi^U_k(1 - \Delta CR^U_k)$$

$$+ \delta^U_k(1 - \Delta GR^U_k) + \gamma^U_k(1 - \Delta UR^U_k) + \vartheta^U_k(1 - \Delta FR^U_k)] \tag{12.5}$$

If $w = (1/n, 1/n, 1/n, \ldots 0.1/n)^T$, then Eq. (12.5) reduces to Eq. (12.4). $K_w(A, B)$ also satisfies the three properties in the above theorem.

12.5.12 Theorem

Let w_k be the weight for each element $r_k (k = 1,2,\ldots n)$, $w_k \in [0, 1]$, and $\sum_{k=1}^{n} w_k = 1$, then the weighted correlation coefficient between the IVPNSs

A and B which is denoted by $L_w(A, B)$ defined in Eq. (3.2) satisfies the following properties.

$$L_w(A,B) = L_w(B,A);$$

$$0 \leq L_w(A,B) \leq 1;$$

$$L_w(A,B) = 1 \text{ iff } A = B.$$

It is similar to proving the properties in Theorem 3.1

12.6 Decision-making using the improved correlation coefficient of IVPNSs

Here, we consider a MADM problem with interval-valued pentapartitioned neutrosophic information, and the characteristic of an alternative $A_i(I = 1,2,...m)$ on an attribute $C_j(j = 1,2...n)$ is represented by the following IVPNSs:

$$A_i = \{(C_j, TR_{A_i}(C_j), CR_{A_i}(C_j), GR_{A_i}(C_j), UR_{A_i}(C_j), FR_{A_i}(C_j) | C_j \in C, j = 1,2,...n\}$$
$$= \{(C_j, [\inf TR_{A_i}(C_j), \sup TR_{A_i}(C_j)], [\inf CR_{A_i}(C_j), \sup CR_{A_i}(C_j)],$$

$$[\inf GR_{A_i}(C_j), \sup GR_{A_i}(C_j)], [\inf UR_{A_i}(C_j), \quad \sup UR_{A_i}(C_j)], [\inf FR_{A_i}(C_j),$$

$$\sup FR_{A_i}(C_j)]) | C_j \in C, j = 1,2,...n\}$$

where $TR_{A_i}(C_j), CR_{A_i}(C_j), GR_{A_i}(C_j), UR_{A_i}(C_j), FR_{A_i}(C_j) \in [0, 1]$ and $0 \leq TR_{A_i}(C_j) + CR_{A_i}(C_j) + GR_{A_i}(C_j) + UR_{A_i}(C_j) + FR_{A_i}(C_j) \leq 5$ for $C_j \in C, j=1,2,...n$

and $\mathrm{I} = 1,2,...m$.

To make it convenient, we are considering the following five functions

$TR_{A_i}(C_j) = [\inf TR_{A_i}(C_j), \sup TR_{A_i}(C_j)], CR_{A_i}(C_j) = [\inf CR_{A_i}(C_j), \sup CR_{A_i}(C_j)],$

$GR_{A_i}(C_j) = [\inf GR_{A_i}(C_j), \sup GR_{A_i}(C_j)], UR_{A_i}(C_j) = [\inf UR_{A_i}(C_j), \sup UR_{A_i}(C_j)],$

$FR_{A_i}(C_j) = \inf[GR_{A_i}(C_j), \sup GR_{A_i}(C_j)]$ in terms of interval valued pentapartitioned neutrosophic value (IVPNV)

$$d_{ij} = \left(\left[tr_{ij}^L, tr_{ij}^U\right], \left[cr_{ij}^L, cr_{ij}^U\right], \left[gr_{ij}^L, gr_{ij}^U\right], \left[ur_{ij}^L, ur_{ij}^U\right], \left[fr_{ij}^L, fr_{ij}^U\right]\right)$$
$$(i = 1,2,...m; j = 1,2...n).$$

Here, the values of d_{ij} are usually derived from the evaluation of an alternative A_i with respect to a criteria C_j by the expert or decision-maker.

Therefore we get an interval-valued pentapartitioned neutrosophic decision matrix $D = (d_{ij})_{m \times n}$.

In the case of an ideal alternative A^*, an ideal IVPNV can be defined by

$$d_{ij} = \left(\left[tr_{ij}^L, tr_{ij}^U\right], \left[cr_{ij}^L, cr_{ij}^U\right], \left[gr_{ij}^L, gr_{ij}^U\right], \left[ur_{ij}^L, ur_{ij}^U\right], \left[fr_{ij}^L, fr_{ij}^U\right]\right)$$

$= ([1, 1], [1, 1], [0, 0], [0, 0], [0, 0])(j = 1, 2 \ldots n)$ in the decision making method,

Hence, the weighted correlation coefficient between an alternative $A_i(i = 1,2\ldots m)$ and the ideal alternative A^* is given by

$$\begin{aligned} L_w(\mathrm{P}, \mathrm{Q}) = \frac{1}{10}\sum_{k=1}^{n} w_j \Big[& \mu_{ij}^L(1 - \Delta TR_{ij}^L) + \varphi_{ij}^L(1 - \Delta CR_{ij}^L) \\ & + \delta_{ij}^L\left(1 - \Delta GR_{ij}^L\right) + \gamma_{ij}^L\left(1 - \Delta UR_{ij}^L\right) + \vartheta_{ij}^L\left(1 - \Delta FR_{ij}^L\right) \\ & + \mu_{ij}^U\left(1 - \Delta TR_{ij}^U\right) + \varphi_{ij}^U(1 - \Delta CR_{ij}^U) \\ & + \delta_{ij}^U\left(1 - \Delta GR_{ij}^U\right) + \gamma_{ij}^U\left(1 - \Delta UR_{ij}^U\right) + \vartheta_{ij}^U\left(1 - \Delta FR_{ij}^U\right)\Big] \end{aligned} \tag{12.6}$$

where

$$\mu_{ij}^L = \frac{5 - \Delta tr_{ij}^L - \Delta tr_{i\max}^L}{5 - \Delta tr_{i\min}^L - \Delta tr_{i\max}^L},$$

$$\varphi_{ij}^L = \frac{5 - \Delta cr_{ij}^L - \Delta cr_{i\max}^L}{5 - \Delta cr_{i\min}^L - \Delta cr_{i\max}^L},$$

$$\delta_{ij}^L = \frac{5 - \Delta GR_{ij}^L - \Delta gr_{i\max}^L}{5 - \Delta gr_{i\min}^L - \Delta gr_{i\max}^L},$$

$$\gamma_{ij}^L = \frac{5 - \Delta ur_{ij}^L - \Delta ur_{i\max}^L}{5 - \Delta ur_{i\min}^L - \Delta ur_{i\max}^L},$$

$$\vartheta_{ij}^L = \frac{5 - \Delta fr_{ij}^L - \Delta fr_{i\max}^L}{5 - \Delta fr_{i\min}^L - \Delta fr_{i\max}^L},$$

$$\mu_{ij}^U = \frac{5 - \Delta tr_{ij}^U - \Delta tr_{i\max}^U}{5 - \Delta tr_{i\min}^U - \Delta tr_{i\max}^U},$$

$$\varphi_{ij}^U = \frac{5 - \Delta cr_{ij}^U - \Delta cr_{i\max}^U}{5 - \Delta cr_{i\min}^U - \Delta cr_{i\max}^U},$$

$$\delta_{ij}^{U} = \frac{5 - \Delta GR_{ij}^{U} - \Delta gr_{i\max}^{U}}{5 - \Delta gr_{i\min}^{U} - \Delta gr_{i\max}^{U}},$$

$$\gamma_{ij}^{U} = \frac{5 - \Delta ur_{ij}^{U} - \Delta ur_{i\max}^{U}}{5 - \Delta ur_{i\min}^{U} - \Delta ur_{i\max}^{U}},$$

$$\vartheta_{ij}^{U} = \frac{5 - \Delta fr_{ij}^{U} - \Delta fr_{i\max}^{U}}{5 - \Delta fr_{i\min}^{U} - \Delta fr_{i\max}^{U}},$$

$$\Delta tr_{ij}^{L} = |tr_{ij}^{L} - tr_{j}^{L^*}|,$$

$$\Delta cr_{ij}^{L} = |cr_{ij}^{L} - cr_{j}^{L^*}|,$$

$$\Delta gr_{ij}^{L} = |gr_{ij}^{L} - gr_{j}^{L^*}|,$$

$$\Delta ur_{ij}^{L} = |ur_{ij}^{L} - ur_{j}^{L^*}|,$$

$$\Delta fr_{ij}^{L} = |fr_{ij}^{L} - fr_{j}^{L^*}|,$$

$$\Delta tr_{ij}^{U} = |tr_{ij}^{U} - tr_{j}^{U^*}|,$$

$$\Delta cr_{ij}^{U} = |cr_{ij}^{U} - cr_{j}^{U^*}|,$$

$$\Delta gr_{ij}^{U} = |gr_{ij}^{U} - gr_{j}^{U^*}|,$$

$$\Delta ur_{ij}^{U} = |ur_{ij}^{U} - ur_{j}^{U^*}|,$$

$$\Delta fr_{ij}^{U} = |fr_{ij}^{U} - fr_{j}^{U^*}|,$$

$$\Delta tr_{i\min}^{L} = \min_j |tr_{ij}^{L} - tr_{j}^{L^*}|,$$

$$\Delta cr_{i\min}^{L} = \min_j |cr_{ij}^{L} - cr_{j}^{L^*}|,$$

$$\Delta gr_{imin}^{L} = min_j |gr_{ij}^{L} - gr_{j}^{L^*}|,$$

$$\Delta ur_{i\min}^{L} = \min_j |ur_{ij}^{L} - ur_{j}^{L^*}|,$$

$$\Delta fr_{i\min}^{L} = \min_j |fr_{ij}^{L} - fr_{j}^{L^*}|,$$

$$\Delta tr_{i\min}^{U} = \min_j |tr_{ij}^{U} - tr_{j}^{U^*}|,$$

$$\Delta cr^{U}_{i\min} = \min_j |cr^{U}_{ij} - cr^{U*}_{j}|,$$

$$\Delta gr^{U}_{i\min} = \min_j |gr^{U}_{ij} - gr^{U*}_{j}|,$$

$$\Delta ur^{U}_{i\min} = \min_j |ur^{U}_{ij} - ur^{U*}_{j}|,$$

$$\Delta fr^{U}_{i\min} = \min_j |fr^{U}_{ij} - fr^{U*}_{j}|,$$

$$\Delta tr^{L}_{i\max} = \max_j |tr^{L}_{ij} - tr^{L*}_{j}|,$$

$$\Delta cr^{L}_{i\max} = \max_j |cr^{L}_{ij} - cr^{L*}_{j}|,$$

$$\Delta gr^{L}_{i\max} = \max_j |gr^{L}_{ij} - gr^{L*}_{j}|,$$

$$\Delta ur^{L}_{i\max} = \max_j |ur^{L}_{ij} - ur^{L*}_{j}|,$$

$$\Delta fr^{L}_{i\max} = \max_j |fr^{L}_{ij} - fr^{L*}_{j}|,$$

$$\Delta tr^{U}_{i\max} = \max_j |tr^{U}_{ij} - tr^{U*}_{j}|,$$

$$\Delta cr^{U}_{i\max} = \max_j |cr^{U}_{ij} - cr^{U*}_{j}|,$$

$$\Delta gr^{U}_{i\max} = \max_j |gr^{U}_{ij} - gr^{U*}_{j}|,$$

$$\Delta ur^{U}_{i\max} = \max_j |ur^{U}_{ij} - ur^{U*}_{j}|,$$

$$\Delta fr^{U}_{i\max} = \max_j |fr^{U}_{ij} - fr^{U*}_{j}|,$$

For $I = 1,2,\ldots m$ and $j = 1,2,\ldots n$.

By using the above weighted correlation coefficient, we can derive the ranking order of all alternatives and we can choose the best one among those.

12.6.1 Example

This section deals with the example for the MADM problem with the given alternatives corresponding to the criteria alloted under pentapartitioned neutrosophic environment and interval-valued pentapartitioned neutrosophic environment.

12.7 Decision-making under pentapartitioned neutrosophic environment

For this example, the three potential alternatives are to be evaluated under the four different attributes. COVID-19 affects different people in different ways. Most infected people will develop mild to moderate illness and recover without hospitalization. The most common symptoms of COVID-19 are fever, cough, and tiredness. But there are many other possible signs and symptoms. The other symptoms are sore throat, vomiting, loss of taste or smell, diarrhea, etc. The three potential alternatives are A_1—during COVID, A_2—pre-COVID, and A_3—post-COVID, and the four different attributes are C_1—fever and cough, C_2—sore throat, C_3—shortness of breath, and C_4—tiredness. For the evaluation of an alternative A_1 with respect to an attribute C_1, it is obtained from the questionnaire of a domain expert. According to the attributes, we will derive the ranking order of all alternatives, and based on this ranking order, the customer will select the best one.

The weighted vector of the above attributes is given by $w = (0.2,0.35,0.25,0.2)$. Here, the alternatives are to be evaluated under the above four attributes by the form of IVPNSs. In general, the evaluation of an alternative A_i with respect to the attributes C_j $(i = 1,2,3, j = 1,2,3,4)$ will be done by the questionnaire of a domain expert. In particular, while asking the opinion about an alternative A_1 with respect to an attribute C_1, the possibility he (or) she says is that the statement true is 0.4, the statement both true and false is 0.3, the statement neither true nor false is 0.2, the statement false is 0.1, and the statement unknown is 0.4. It can be denoted in neutrosophic notation as $d_{11} = (0.4,0.3,\mathit{0.4},0.2,0.1)$.

$A_i \backslash C_j$	C_1	C_2	C_3	C_4
A_1	[0.4,0.3,0.4,0.2,0.1]	[0.5.0.4,0.5,0.3,0.2]	[0.4,0.1,0.4,0.1,0.1]	[0.6,0.2,0.6,0.3,0.2]
A_2	[0.4,0.2,0.6,0.1,0.2]	[0.3,0.3,0.5,0.2,0.1]	[0.1,0.4,0.2,0.3,0.2]	[0.5,0.3,0.6,0.1,0.1]
A_3	[0.3,0.4,0.4,0.3,0.4]	[0.5,0.1,0.6,0.2,0.1]	[0.4,0.5,0.4,0.3,0.2]	[0.3,0.2,0.5,0.2,0.2]

Then by using the proposed method, we will obtain the most desirable alternative. We can get the values of the correlation coefficient M_w (A_i, A^*) $(i = 1,2,3)$ by using Eq. (4.2).

Hence M_w $(A_1, A^*) = 0.586276$, M_w $(A_2, A^*) = 0.5640$, M_w $(A_3, A^*) = 0.56921$.

Thus ranking order of the three potential alternatives is $A1 > A3 > A2$. Therefore we can say that A1—during COVID patients are affected more by the symptoms of fever, cough, aches, difficulty in breathing, and chest pain. The decision-making method provided in this paper is more judicious and more vigorous.

12.7.1 Decision-making under interval-valued pentapartitioned neutrosophic environment

For this example, the three potential alternatives are to be evaluated under the four different attributes. The four potential alternatives are A_1—fever, A_2—cough and A_3—shortness of breathing, A_4—tiredness, and the different attributes are C_1—pre-COVID, C_2—during COVID, and C_3—post-COVID. For the evaluation of an alternative A_1 with respect to an attribute C_1, it is obtained from the questionnaire of a domain expert. According to the attributes, we will derive the ranking order of all alternatives, and based on this ranking order, the customer will select the best one.

The weighted vector of the above attributes is given by $w = (0.25,0.35,0.4)$. Here, the alternatives are to be evaluated under the above four attributes by the form of IVPNSs, In general, the evaluation of an alternative A_i with respect to the attributes C_j ($i = 1,2,3$, $j = 1,2,3,4$) will be done by the questionnaire of a domain expert. In particular, while asking the opinion about an alternative A_1 with respect to an attribute C_1, the possibility he (or) she says is that the statement true is [0.1,0.5], the statement both true and false is [0.3,0.6], the statement neither true nor false is [0.1,0.3], the statement false is [0.2,0.2], and the statement unknown is [0.4,0.6]. It can be denoted in neutrosophic notation as $d_{11} = ([0.1,0.5], [0.3,0.6], [0.4,0.6], [0.,0.3], [0.2,0.4])$.

$A_i \backslash C_j$	C_1	C_2	C_3
A_1	([0.1,0.5], [0.3,0.6], [0.4,0.6], [0.1,0.3], [0.2,0.4])	([0.3,0.5], [0.1,0.3], [0.2,0.3], [0.2,0.5], [0.1,0.5])	([0.3,0.6], [0.1,0.3], [0.2,0.3], [0.2,0.4], [0.3,0.7])
A_2	([0.3,0.4], [0.1,0.3], [0.1,0.7], [0.2,0.4], [0.3,0.5])	([0.1,0.4], [0.1,0.8], [0.2,0.4], [0.3,0.7], [0.1,0.7])	([0.1,0.8], [0.1,0.9], [0.2,0.7], [0.3,0.6], [0.3,0.4])
A_3	([0.1,0.4], [0.2,0.3], [0.4,0.5], [0.3,0.4], [0.2,0.3])	([0.4,0.8], [0.4,0.7], [0.7,0.4], [0.3,0.5], [0.2,0.3])	([0.1,0.3], [0.2,0.7], [0.2,0.6], [0.2,0.3], [0.1,0.6])
A_4	([0.1,0.7], [0.2,0.3], [0.1,0.5], [0.4,0.6], [0.2,0.4])	([0.5,0.6], [0.1,0.2], [0.3,0.5], [0.4,0.5], [0.3,04])	([0.4,0.5], [0.2,0.7], [0.2,0.4], [0.3,0.4], [0.1,0.2])

Then by using the proposed method, we will obtain the most desirable alternative. We can get the values of the correlation coefficient M_w (A_i, A^*) $(i = 1,2,3)$ by using Eq. (4.2).

Hence N_w $(A_1, A^*) = 0.52433$, N_w $(A_2, A^*) = 0.50036$, N_w $(A_3, A^*) =$ 0.*54386*. 0.53915.

Thus ranking order of the three potential alternatives is $A3 > A4 > A1 > A2$. Therefore we can say that A3—shortness of breathing problem effects more among COVID patients. The decision-making method provided in this paper is more judicious and more vigorous.

It is important to remember that most people who have COVID-19 recover quickly. But the potentially long-lasting problems from COVID-19 make it even more important to reduce the spread of COVID-19 by following the precautions. Precautions include wearing masks, social distancing, avoiding crowds, getting a vaccine when available, and keeping the hands clean.

12.8 Conclusion

In this paper, the concept of IVPNSs is introduced and its operations are studied with an example. We have outlined the improved correlation coefficient of PNSs and IVPNSs, and this is often applicable for a few cases once the correlation coefficient of PNSs is undefined (or) unmeaningful and additionally its properties are studied. Decision-making could be a process that plays a significant role in real-world issues. The major method in higher cognitive process is recognizing the matter (or) chance and deciding to deal with it. Here, we have mentioned the decision-making technique using the improved correlation of PNSs and IVPNSs, and significantly, an illustrative example is given in multiple attribute higher cognitive process issues that involves the many alternatives supported by varied criteria. Therefore our projected improved correlation of PNSs and IVPNSs helps to spot the foremost appropriate difference to the client supported on the given criteria. In the future, we can study the concept of improved correlation coefficients of pentapartitioned neutrosophic sets, and their properties are also studied with its application in medical diagnosis, clustering analysis, etc.

Funding

"This research received no external funding."

Acknowledgments

I would like to express my special thanks of gratitude to S.P. Rhea and R. Kathiresan for their guidance and constant support in completing the paper.

Conflicts of interest

"The authors declare no conflict of interest."

References

[1] L. Zadeh, Fuzzy sets, Inf. Control. 8 (1965) 87–96.
[2] K. Atanassov, Intuitionistic fuzzy sets, Fuzzy Sets Syst. 20 (1986) 87–96.

[3] F. Smarandache, A unifying field in logics, Neutrosophy: Neutrosophic Probability, Set and Logic, American Research Press, Rehoboth, 1998.

[4] H. Wang, F. Smarandache, Y.Q. Zhang, R. Sunderraman, Single valued neutrosophic sets, Multispace Multistruct. 4 (2010) 410–413.

[5] R. Malik, S. Pramanik, Pentapartitioned neutrosophic set and its properties, Neutrosophic Sets Syst. 36 (2020) 184–192.

[6] R. Radha, A. Stanis Arul Mary, Pentapartitioned neutrosophic pythagorean topological spaces, JXAT XIII (2021) 1–6.

[7] R. Radha, A. Stanis Arul Mary, Pentapartitioned neutrosophic pythagorean set, IRJASH 3 (2021) 62–82.

[8] D.A. Chiang, N.P. Lin, Correlation of fuzzy sets, Fuzzy Sets Syst. 102 (1999) 221–226.

[9] D.H. Hong, Fuzzy measures for a correlation coefficient of fuzzy numbers under Tw (the weakest tnorm)-based fuzzy arithmetic operations, Inf. Sci. 176 (2006) 150–160.

[10] J. Ye, Improved correlation coefficients of single valued neutrosophic sets and interval neutrosophic sets for multiple attribute decision making, J. Intell. Fuzzy Syst. 27 (2014) 2453–2462.

Chapter 13

Post-pandemic impact on the occupational shift of rural populace—a case study using neutrosophic comparison t-test

Nivetha Martin[1], S. Jegan Karuppiah[2], Florentin Smarandache[3], Rafael Rojas[4] and Maikel Yelandi Leyva Vazquez[5]

[1]Department of Mathematics, Arul Anandar College (Autonomous), Karumathur, Madurai, Tamil Nadu, India, [2]Department of Rural Development Science, Arul Anandar College (Autonomous), Karumathur, Madurai, Tamil Nadu, India, [3]Mathematics, Physical and Natural Sciences Division, University of New Mexico, Gallup, NM, United States, [4]Universidad Industrial de Santander, Bucaramanga, Santander, Colombia, [5]Universidad Regional Autónoma de los Andes (UNIANDES), Babahoyo, Los Ríos, Ecuador

13.1 Introduction and motivation

The statistical investigation is primarily used to determine the before and after impacts of the factors on various entities. Paired-comparison t-test is applied to find the before and after impacts in various scenarios; for instance, if a company has given training on production techniques, the paired-comparison t-test shall be used to determine the impact of training on the skill enrichment of the employees. Also, let us consider the case of school education, if the teacher introduces a new method of teaching, then the paired-comparison t-test shall be applied to find the impact of the new method on the knowledge enhancement of the students. In the medical field, to determine the impact of new drug treatment such a kind of comparison t-test is used to find the impact of the drug on the rate of recovery of the patient. So in general, the paired-comparison t-test is applied to find the before and after impacts, and it plays a vital role in making decisions on introducing new or retention of the new and old phenomena, respectively.

The classical paired-comparison t-test can be parametric or nonparametric depending on the number of observations, type of distribution, and the scale of data. Nonparametric tests equivalent to parametric t-tests are used extensively to

Cognitive Intelligence with Neutrosophic Statistics in Bioinformatics.
DOI: https://doi.org/10.1016/B978-0-323-99456-9.00020-9

make statistical investigations, in line with Wilcoxon signed-rank test, introduced by Frank Wilcoxon, which is a nonparametric test used as an alternative to parametric paired-comparison t-test. Bernard et al. have applied the classical Wilcoxon signed-rank test for cluster data [1]. Przemyslaw and Martyna [2] have used the test for interval-value data. The classical Wilcoxon signed-rank test is used to handle the data of deterministic nature, but in reality, the acquired data may not be completely deterministic. To investigate data consisting of uncertainty and impreciseness, fuzzy-based nonparametric tests were developed. Kahraman [3] discoursed the significances of the fuzzy nonparametric test. The classical Wilcoxon signed-rank test was discussed in the fuzzy environment to make inferences from vague data. Parchami [4] proposed and tested fuzzy hypotheses based on uncertain data sets. Taheri and Hesamian [5] presented the applications of the Wilcoxon signed-rank test in a fuzzy environment. As an extension of the fuzzy-based Wilcoxon signed-rank test, the neutrosophic-based Wilcoxon signed-rank test has started to gain more momentum amid the researchers as neutrosophic logic considers not only uncertainty but also takes into account indeterminacy. Guo et al. [6] have applied neutrosophic-based Wilcoxon signed-rank test to determine the performance of the proposed methods of lung segmentation in comparison with the earlier methods. Pattanayak et al. [7] have applied neutrosophic-based Wilcoxon signed-rank test to find the efficacy and consistency of the proposed forecasting methods. The brief literature on neutrosophic-based Wilcoxon signed-rank test shows the compatibility of neutrosophic logic over the fuzzy and crisp logic in determining the before and after impacts with quantitative data but not much with qualitative data.

Presently many extensive studies with a special focus on pandemic impacts on the economy, health, business, and so many others are ongoing, to mention a few, economic status of the nation, per capita income, gross domestic product (GDP) of the countries, immunity resistance of mankind, production capacity, and many others. The study on these aspects deals with quantitative data, but to study on the pandemic impacts on human emotions, social skills, the satisfaction rate of transitions in human lifestyles, purchasing behavior, and many of these kinds cannot be dealt completely with quantitative data as the impacts are more meaningful and realistic only if expressed in terms of qualitative data. Wilcoxon signed-rank test with neutrosophic quantitative data is dealt earlier by the researchers to make inferences on quantitative studies but not for qualitative studies. To bridge this gap, the Wilcoxon signed-rank test with neutrosophic quantification of qualitative data is used in this paper. As an instance to it, this research work intends to apply Wilcoxon signed-rank test for analyzing the qualitative data on determining the post-pandemic impact on the occupational shift of the rural populace. The paper is structured into many sections consisting of preliminaries, literature review, description of the study area, neutrosophic statistical analysis of the data, discussion of the results, conclusion, and future scope of the research work.

13.2 Literature review and state of the art

The pandemic has caused a lot of economic chaos, and it has especially affected the lives of several people with low annual income severely. The rural livelihood has been devastated greatly by the pandemic impacting their prime occupation which has made the switch to other occupations for the sake of survival. The occupational shifts of the rural populace have caused a lot of effects on their personal, social, and economic aspects, as switching to other occupations involves many hardships, and certainly, it is not an easy task. The satisfaction of transiting to other occupations might vary from person to person, and definitely, the economic constraints made them bind to the new workplace setting. The constrained accommodation and adaptation will result in various impacts, and such kinds of consequences of the occupational transition have to be studied intensively to determine the resultants of post-pandemic effects on occupational shifts. The personal satisfaction of the nature of the work, economic fulfillment, and social recognition before and after the occupational shift have to be studied.

Researchers have studied the pandemic impact on employment across the globe. Charlene and Sabrina [8] presented the pandemic impact of self-employment concerning the clustered samples based on gender, marital, and parental status. Hossain [9] discoursed the pandemic impact on employment and livelihood sustainability of the economically and socially downgraded people of Bangladesh. The Asia Foundation conducted a quick assessment of the pandemic impact on the employment status of the people of Nepal, especially those belonging to the middle cities [10]. The factors that obstruct economic hike were the key findings of the study. The International Labour Organization has stated in its report that around 2.3 crores of people have lost their jobs, and the Indian nation is not an exception to it. It has also published a report on the fall of the economy in G20 countries and on the recent trends and developments in the labor markets. The report of the Congressional Research Service [11] presents the fall of employment across the world and its implications on the per capita income, labor force, and various production sectors. Mohid Akther and Kamaraju [9] analyzed the crucial consequences of a pandemic on the surge of unemployment based on literature studies and secondary data. It was observed the unemployment rate was ebbing and dropping down during the first phase of the pandemic in India in urban and rural regions. The report published by Statista Research Development [12] presents the pandemic impact on rural and urban employment. The rural people suffered greatly to exercise their occupation due to continuous lockdown and improper compensation for the incurred economic losses. Yasotha [13] described the decimation of a pandemic on rural households and rural economies. The livelihood of several people was completely shattered, and it has caused the rural populace to migrate to urban places for survival. During the second phase of the pandemic, the people of rural regions started to step into alternative occupations, and these occupational shifts are considered as the alternative means of survival for the rural populace.

The paradigm shift in the occupation may not be much soothing to the rural populace personally as it may cause many consequences on personal, economic, and social factors. The focus on the qualitative analysis of the occupational shifts is very limited, and this research aims in determining the qualitative impacts of occupational shifts on the factors related to personal satisfaction, economic fulfillment, and social recognition using a neutrosophic case study. The triangular neutrosophic numbers are used to quantify the linguistic variables of Likert's five-point scale. Wilcoxon signed-rank test with triangular neutrosophic numbers is used to make a qualitative analysis of before and after impacts.

13.3 Problem definition

Karumathur is a grama panchayat chosen for the case study located in Chellampatti block of Madurai district of Tamil Nadu state in India. The area chosen for the case study is a rural region with a population of 6737 among which 55.1% are working. Agriculture, livestock rearing, and related activities are the primary occupations. The pandemic has caused the majority of the people to switch to other occupations for livelihood sustainability and their survival. The method of clustered sampling is used to group the samples under various clusters of gender, age, and marital status. A combination of the questionnaire and scheduled methods of data collection with triangular neutrosophic Likert's five-point scale is used in this study.

Very lowly satisfied	[(5.1,6.4,8.4);0.3,0.4,0.2]
Lowly satisfied	[(5.9,7.8,8.6);0.5,0.2,0.4]
No difference	[(6.3, 7.4, 8.1); 0.4,0.1,0.1]
Highly satisfied	[(5.8,7.8,8.8); 0.8,0.1,0.1]
Very highly satisfied	[(6.7,7.5,8.2; 0.9,0.1,0.1]

13.3.1 Hypothesis formulation

The research study aims in determining the impacts of occupational shifts on the personal satisfaction, economic fulfillment, and social recognition of the rural populace. The null hypotheses are formulated as follows:

H1: The personal satisfaction of the rural populace on the occupation has not been affected by the occupational shifts.

H2: The occupational shifts have not contributed to the economic fulfillment of the rural populace.

H3: The occupational shifts have not provided social recognition to the rural populace.

The Wilcoxon tests are performed to find the personal satisfaction on the occupation, economic fulfillment, and social recognition obtained by the occupation before and after the occupational impacts caused by the pandemic.

13.3.2 Discussion

Figs. 13.1A, 13.2A, and 13.3A represent the results of the Wilcoxon test using neutrosophic data, and Figs. 13.1B, 13.2B, and 13.3B represent the results of the Wilcoxon test using classical data. It is inferred from both cases that the P-value is less than .05 and the null hypothesis is rejected. The results very clearly show that the occupational shifts have caused considerate impacts on the economic fulfillment, social recognition, and personal satisfaction on the occupation with respect to the rural populace. This research

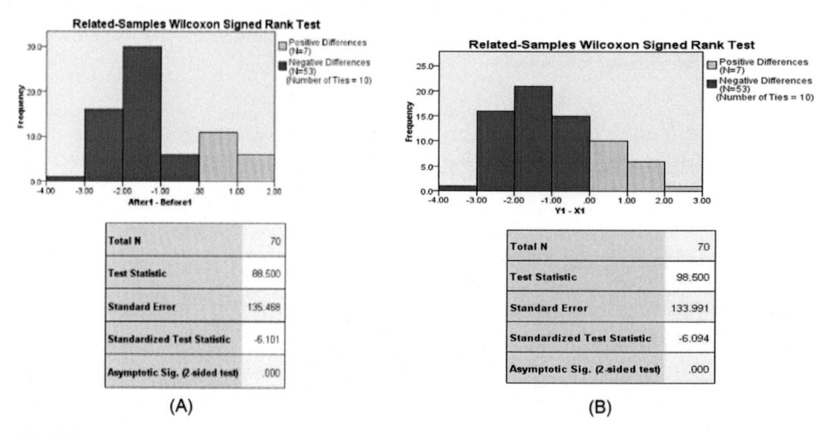

	(A)	(B)
Total N	70	70
Test Statistic	88.500	98.500
Standard Error	135.468	133.991
Standardized Test Statistic	-6.101	-6.094
Asymptotic Sig. (2-sided test)	.000	.000

(A) (B)

FIGURE 13.1 (A) Classical Wilcoxon result of economic fulfillment. (B) Neutrosophic Wilcoxon result of economic fulfillment.

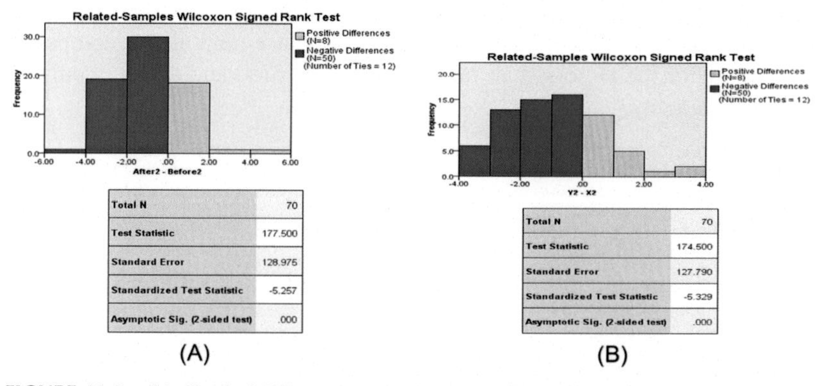

	(A)	(B)
Total N	70	70
Test Statistic	177.500	174.500
Standard Error	128.975	127.790
Standardized Test Statistic	-5.257	-5.329
Asymptotic Sig. (2-sided test)	.000	.000

(A) (B)

FIGURE 13.2 (A) Classical Wilcoxon result of social fulfillment. (B) Neutrosophic Wilcoxon result of social fulfillment.

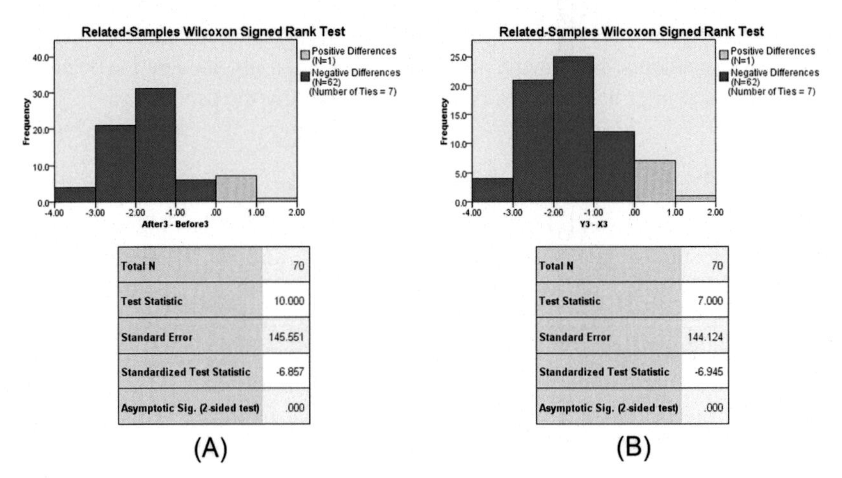

Total N	70
Test Statistic	10.000
Standard Error	145.551
Standardized Test Statistic	-6.857
Asymptotic Sig. (2-sided test)	.000

(A)

Total N	70
Test Statistic	7.000
Standard Error	144.124
Standardized Test Statistic	-6.945
Asymptotic Sig. (2-sided test)	.000

(B)

FIGURE 13.3 (A) Classical Wilcoxon result of personal fulfillment. (B) Neutrosophic Wilcoxon result of personal fulfillment.

work is a rudimentary study on the impacts of the occupational shifts, and this will help in the further investigation of other associated factors.

13.4 Conclusion and future work

This paper presents a neutrosophic qualitative study on the occupational impacts on the rural populace by considering three core factors such as personal satisfaction, economic fulfillment, and social recognition. The results are very evident on the transitional impacts caused by pandemic outbreaks on various facets of mankind. This work outlines the compatibility and consistency of the neutrosophic Wilcoxon signed-rank test with a case study in the context of handling qualitative data. This work shall be extended by applying neutrosophic statistics to explore the consequential impacts of occupational shifts on the urban populace, and also a comparative analysis of the factors shall be made between rural and urban populace for intensive study.

References

[1] B. Rosner, R.J. Glynn, M.-L.T. Lee, The Wilcoxon signed rank test for paired comparisons of clustered data, Biometrics 62 (1) (2006) 185–192.

[2] P. Grzegorzewski, M. Śpiewak, The sign test and the signed-rank test for interval-valued data, Int. J. Intell. Syst. 34 (2019) 2122–2150.

[3] C. Kahraman, C.E. Bozdag, D. Ruan, A.F. Ozok, Fuzzy sets approaches to statistical parametric and nonparametric tests, Int. J. Intell. Syst. 19 (2004) 1069–1087.

[4] A. Parchami, S.M. Taheri, B.S. Gildeh, M. Mashinchi, A simple but efficient approach for testing fuzzy hypotheses, J. Uncertain. Anal. Appl. 4 (1) (2016) 1–16.

[5] S.M. Taheri, G. Hesamian, A generalization of the Wilcoxon signed-rank test and its applications, Stat. Pap. 54 (2013) 457–470.
[6] Y. Guo, C. Zhou, H.P. Chan, A. Chughtai, J. Wei, L.M. Hadjiiski, et al., Automated iterative neutrosophic lung segmentation for image analysis in thoracic computed tomography, Med. Phys. 40 (8) (2013) 081912.
[7] R.M. Pattanayak, H.S. Behera, S. Panigrahi, A novel probabilistic intuitionistic fuzzy set based model for high order fuzzy time series forecasting, Eng. Appl. Artif. Intell. 99 (2021) 104136.
[8] C.M. Kalenkoski, S.W. Pabilonia, Initial impact of the COVID-19 pandemic on the employment and hours of self- employed coupled and single workers by gender and parental status, IZA – Inst. Labor. Econ. (2020) 1–46.
[9] M.I. Hossain, COVID-19 impacts on employment and livelihood of marginal people in Bangladesh: lessons learned and way forward, South. Asian Surv. 28 (1) (2021) 57–71.
[10] https://asiafoundation.org/wp-content/uploads/2021/04/Impact-of-the-Covid-19-Pandemic-on-Employment-in-Middle-order-Cities-of-Nepal.pdf.
[11] G. Falk, P.D. Romero, I.A. Nicchitta, E.C. Nyhof, Unemployment rates during the COVID-19 pandemic, Congress. Res. Serv. (2021) 1–29.
[12] https://www.statista.com/statistics/1142485/india-coronavirus-impact-on-rural-and-urban-employment-by-type/.
[13] Y. Margaret, Impact of COVID-19 on rural economy in India, J. Bus. Manage. 2 (1) (2020) 1–8.

Chapter 14

A novel approach of neutrosophic continuous probability distributions using AH-isometry with applications in medicine

Mohamed Bisher Zeina[1] and Mohammad Abobala[2]

[1]*Department of Mathematical Statistics, University of Aleppo, Aleppo, Syria,* [2]*Department of Mathematics, Tishreen University, Latakia, Syria*

14.1 Introduction

In 1995, Prof. Smarandache presented neutrosophic logic to study logical statements with its three parts; false, truth, and indeterminacy. From an algebraic point of view, Prof. Smarandache [1,2] extended real numbers set R to neutrosophic real numbers set $R(I) = R \cup \{I\}$ where I is indeterminacy and it fulfills $I^2 = I, 0 \cdot I = 0$[1].

Due to the generalization of sets and logic, many extensions to all branches of science have been made, including probability theory, statistics, reliability theory, queueing theory, artificial intelligence, data mining, algebra, linear algebra, mathematical analysis, complex analysis, differential equations, physics, philosophy, etc., [3–15].

Probability theory has many applications in the medical field. It is the base of reliability theory, survival analysis, risk theory, etc., [16]. Since life is full of indeterminacy, uncertainty, ambiguity, and unsureness, neutrosophic logic becomes a necessity to solve many lifetime problems especially in probability theory which is the science of randomness.

In this chapter, we are going to study several important continuous distributions, including uniform, exponential, beta, gamma, Rayleigh, normal, and chi-square. Also, we present a generalization of the definition of random variables presented in [15]; all of our studies will depend on AH-isometry presented by Mohammad Abobala and Ahmad Hatip in Ref. [17].

Cognitive Intelligence with Neutrosophic Statistics in Bioinformatics.
DOI: https://doi.org/10.1016/B978-0-323-99456-9.00014-3

14.2 Background, definitions, and notations

Definition 2.1

Let $R(I) = \{a + bI; a, b \in R\}$ where $I^2 = I, 0 \cdot I = 0$ are the neutrosophic field of reals. The one-dimensional isometry (AH-isometry) is defined as follows:

$$T: R(I) \rightarrow R \times R$$

$$T(a + bI) = (a, a + b)$$

Remark:

T is an algebraic isomorphism between two rings, and it has the following properties:

1. T is bijective.
2. T preserves addition and multiplication, that is,:

$$T[(a + bI) + (c + dI)] = T(a + bI) + T(c + dI)$$

And

$$T[(a + bI) \cdot (c + dI)] = T(a + bI) \cdot T(c + dI)$$

3. Since T is bijective, then it is invertible by:

$$T^{-1}: R \times R \rightarrow R(I)$$

$$T^{-1}(a, b) = a + (b - a)I$$

4. T preserves distances, that is,
   If $A = (x_1 + y_1 I, x_2 + y_2 I)$, $B = (x_3 + y_3 I, x_4 + y_4 I) \Rightarrow T(\|\overrightarrow{AB}\|) = \|T(\overrightarrow{AB})\|$
5. Two-dimensional isometry can be defined in the same way as follows:

$$T: R(I) \times R(I) \rightarrow R^2 \times R^2$$

$$T(a + bI, c + dI) = ((a, a + b), (c, c + d))$$

And:

$$T^{-1}: R^2 \times R^2 \rightarrow R(I) \times R(I)$$

$$T^{-1}((a, b), (c, d)) = (a + (b - a)I, c + (d - c)I)$$

Definition 2.2

We say that $a + bI \geq c + dI$ iff $a \geq c$ and $a + b \geq c + d$.

Definition 2.3

Let $f: R(I) \rightarrow R(I); f = f(X)$ and $X = x + yI \in R(I)$, then f is called a neutrosophic real function with one variable.

14.3 Literature review and state of the art

Patro and Smarandache [18] studied binomial and normal distributions and presented many examples and problems. Alhabib et al. [19] presented some other probability distributions, including uniform, exponential, and Poisson distributions assuming that the parameters of the mentioned distributions have indeterminacy; that is, it is of the form $a+bI$. Alhasan and Smarandache [20] studied the Weibull distribution and a class of Weibull family of distributions assuming also that its parameters are indeterminant parameters and presented some expected applications. Sherwani et al. [21] studied neutrosophic beta distributions, and its properties also presented some applications. Zeina and Hatip [15] defined a neutrosophic form of random variables noted by X_N where $X_N = X + I$ and I is indeterminacy and studied its characteristic values and presented some examples and applications. F. Smarandache extended and generalized the neutrosophic probability concept and multivariate probability to plithogenic probability in Ref. [22]. Many applications of neutrosophic probability theory were studied like reliability theory [6], queueing theory [12,13,23], information fusion and identification [24], etc.

The main problem in previous studies is neutrosophic probability theory.

14.4 Neutrosophic random variables

In this section, we are going to generalize the definition of random variables presented by Zeina et al. [15] using AH-isometry. This generalization is mathematically stronger and more general than previous definitions because it is supported and built with algebraic structures.

Definition 3.1:

We will define neutrosophic random variable X_N as a real function:

$$X_N:\Omega_N \to R(I);\Omega_N = \Omega_1 \times \Omega_2(I)$$

with:

$$X_N = X_1 + X_2 I, I^2 = I, 0 \cdot I = 0$$

where X_1, X_2 are random variables defined on Ω_1, Ω_2, respectively, and taking values in R.

Using AH-isometry, we can transform X_N from $R(I)$ to R^2 as follows:

$$T(X_N) = (X_1, X_1 + X_2)$$

which means neutrosophic random variable corresponds to classical joint random vector of variables $X_1, X_1 + X_2$ where $X_1 + X_2$ is the convolution of X_1, X_2.

Theorem 3.1:

Characteristics of neutrosophic random variable X_N are as follows:

1. $E(X_N) = E(X_1) + E(X_2)I$
2. $V(X_N) = V(X_1) + [V(X_1 + X_2) - V(X_1)]I$
3. $\sigma(X_N) = \sigma(X_1) + [\sigma(X_1 + X_2) - \sigma(X_1)]I$

Proof:

1. $T[E(X_N)] = T[E(X_1 + X_2 I)] = (E(X_1), E(X_1 + X_2)) = (E(X_1), E(X_1) + E(X_2)) \Rightarrow$ $E(X_N) = T^{-1}(E(X_1), E(X_1) + E(X_2)) = E(X_1) + E(X_2)I$
2. $T\left[E\left(X_N^2\right)\right] = T\left[E(X_1 + X_2 I)^2\right] = T\left[E\left(X_1^2 + (2X_1X_2 + X_2^2)I\right)\right] = \left(E\left(X_1^2\right),\right.$ $\left.E(X_1 + X_2)^2\right) \Rightarrow E\left(X_N^2\right) = E\left(X_1^2\right) + \left[E(X_1 + X_2)^2 - E\left(X_1^2\right)\right]I$

$$\begin{aligned} T[V(X_N)] &= T\left[E\left(X_N^2\right) - (E(X_N))^2\right] = \left(E\left(X_1^2\right), E(X_1 + X_2)^2\right) - \left((E(X_1))^2, \right) \\ &= \left(E\left(X_1^2\right) - (E(X_1))^2, E(X_1 + X_2)^2 - (E(X_1) + E(X_2))^2\right) \\ &= (V(X_1), V(X_1 + X_2)) \Rightarrow V(X_N) = T^{-1}(V(X_1), V(X_1 + X_2)) \\ &= V(X_1) + [V(X_1 + X_2) - V(X_1)]I \end{aligned}$$

Notice that if X_1, X_2 are independent, then:

$$V(X_N) = V(X_1) + V(X_2)I$$

3. $\sigma(X_N) = \sqrt{V(X_N)} = (V(X_N))^{\frac{1}{2}}$

$$T(\sigma(X_N)) = (V(X_1), V(X_1 + X_2))^{\left(\frac{1}{2}, \frac{1}{2}\right)} = (\sigma(X_1), \sigma(X_1 + X_2))$$

$$\Rightarrow \sigma(X_N) = T^{-1}(\sigma(X_1), \sigma(X_1 + X_2)) = \sigma(X_1) + [\sigma(X_1 + X_2) - \sigma(X_1)]I$$

14.5 Some neutrosophic probability density functions

First, we are going to present and derive definitions of neutrosophic gamma and beta functions which are very useful in studying many probability density functions.

14.5.1 Neutrosophic gamma function

We can find the value of gamma function at neutrosophic point $a_N = a_1 + a_2 I$ using the formula:

$$\Gamma(a_N) = \Gamma(a_1) + I\{\Gamma(a_1 + a_2) - \Gamma(a_1)\}$$

where:

$$\Gamma(a) = \int_0^\infty x^{a-1}e^{-x}dx; a > 0$$

Proof:

Let $f(x) = x^{a_N - 1}e^{-x}$, so:

$$T(f(x)) = \left(x^{a_1-1}e^{-x}, x^{a_1+a_2-1}e^{-x}\right)$$

$$\left(\int_0^\infty x^{a_1-1}e^{-x}dx, \int_0^\infty x^{a_1+a_2-1}e^{-x}dx\right) = (\Gamma(a_1), \Gamma(a_1 + a_2))$$

Taking T^{-1} yields to the proof.

Remark:

Neutrosophic gamma function $\Gamma(a_N)$ is defined when a_N, that is, $a_1 > 0, a_1 + a_2 > 0$.

Example:

$\Gamma(I)$ is undefined because $I = 0 + 1 \cdot I$ and $0 \not> 0$.

$$\Gamma(0.5 + 2I) = \Gamma(0.5) + I\{\Gamma(2.5) - \Gamma(0.5)\} = \sqrt{\pi} + I\left(1.5 \cdot 0.5 \cdot \sqrt{\pi} - \sqrt{\pi}\right)$$

Remark:

Since $T[\Gamma(m + nI + 1)] = (\Gamma(m + 1), \Gamma(m + n + 1)) = (m!, (m + n)!); m, n \in \mathbb{N}$.
So $\Gamma(m + nI + 1) = (m + nI)! = m! + I\{(m + n)! - m!\}$
And it is the formal form of the factorial function.

14.5.2 Neutrosophic beta function

We can find the value of beta function at neutrosophic points $a_N = a_1 + a_2 I$, $b_N = b_1 + b_2 I$ using the formula:

$$\beta(a_N, b_N) = \int_0^1 x^{a_N-1}(1-x)^{b_N-1}dx = \beta(a_1, b_1) + \{\beta(a_1 + a_2, b_1 + b_2) - \beta(a_1, b_1)\}I$$

where:

$$\beta(a, b) = \int_0^1 x^{a-1}(1-x)^{b-1}dx; a, b > 0$$

Proof:

Let $f(x) = x^{a_N - 1}(1-x)^{b_N - 1}$,

so:

$$T[f(x)] = \left(x^{a_1-1}(1-x)^{b_1-1}, x^{a_1+a_2-1}(1-x)^{b_1+b_2-1}\right)$$

$$\left(\int_0^1 x^{a_1-1}(1-x)^{b_1-1}dx, \int_0^1 x^{a_1+a_2-1}(1-x)^{b_1+b_2-1}dx\right) = (\beta(a_1,b_1), \beta(a_1+a_2, b_1+b_2))$$

So:

$$\beta(a_N, b_N) = \beta(a_1, b_1) + I\{\beta(a_1+a_2, b_1+b_2) - \beta(a_1, b_1)\}$$

We let it as an exercise to the reader to prove that:

$$\beta(a_N, b_N) = \frac{\Gamma(a_N)\Gamma(b_N)}{\Gamma(a_N + b_N)}$$

14.5.3 Neutrosophic exponential distribution

In classical probability theory, we say that the continuous random variable X follows exponential distribution with parameter $\lambda > 0$ if its pdf is given by:

$$f(x) = \lambda e^{-\lambda x}; x > 0$$

And we have:

$$E(X) = \frac{1}{\lambda}$$

$$V(X) = \frac{1}{\lambda^2}$$

Substituting $\lambda_N = \lambda_1 + \lambda_2 I, x_N = x_1 + x_2 I$ where I is indeterminacy we get:

$$f(x_N) = \lambda_N e^{-\lambda_N x_N}; x_N > 0$$

$$\begin{aligned} T[f(x_N)] &= T\left[\lambda_N e^{-\lambda_N x_N}\right] = T\left[(\lambda_1 + \lambda_2 I)e^{-(\lambda_1+\lambda_2 I)(x_1+x_2 I)}\right] \\ &= T[(\lambda_1 + \lambda_2 I)]T\left[e^{-(\lambda_1+\lambda_2 I)(x_1+x_2 I)}\right] = T[(\lambda_1 + \lambda_2 I)]\left[e^{-T(\lambda_1+\lambda_2 I)T(x_1+x_2 I)}\right] \\ &= (\lambda_1, \lambda_1 + \lambda_2)e^{-(\lambda_1, \lambda_1+\lambda_2)(x_1, x_1+x_2)} = \left(\lambda_1 e^{-\lambda_1 x_1}, (\lambda_1 + \lambda_2)e^{-(\lambda_1+\lambda_2)(x_1+x_2)}\right) \end{aligned}$$

Notice that the isometric image of the neutrosophic exponential distribution is two exponential distributions, first one is with parameter λ_1 and the second one is with parameter $\lambda_1 + \lambda_2$.

So:

$$f(x_N) = T^{-1}\left(\lambda_1 e^{-\lambda_1 x_1}, (\lambda_1 + \lambda_2)e^{-(\lambda_1+\lambda_2)(x_1+x_2)}\right)$$
$$= \lambda_1 e^{-\lambda_1 x_1} + I\left((\lambda_1 + \lambda_2)e^{-(\lambda_1+\lambda_2)(x_1+x_2)} - \lambda_1 e^{-\lambda_1 x_1}\right)$$

$$F(x_N) = \int_0^{x_N} f(t_N)dt_N$$

Since $T[f(t_N)] = \left(\lambda_1 e^{-\lambda_1 t_1}, (\lambda_1 + \lambda_2)e^{-(\lambda_1+\lambda_2)(t_1+t_2)}\right)$ and $\left(\int_0^{x_1} \lambda_1 e^{-\lambda_1 t_1} dt_1, \int_0^{x_1+x_2} (\lambda_1 + \lambda_2)e^{-(\lambda_1+\lambda_2)(t_1+t_2)} d(t_1 + t_2)\right) = \left(1 - e^{-\lambda_1 x_1}, 1 - e^{-(\lambda_1+\lambda_2)(x_1+x_2)}\right)$

so:

$$F(x_N) = 1 - e^{-\lambda_1 x_1} + I\left\{e^{-\lambda_1 x_1} - e^{-(\lambda_1+\lambda_2)(x_1+x_2)}\right\}$$

$$E(X_N) = \int_0^{\infty} x_N f(x_N)dx_N$$

We have:

$$T[x_N f(x_N)] = \left[x_1 \lambda_1 e^{-\lambda_1 x_1}, (x_1 + x_2)(\lambda_1 + \lambda_2)e^{-(\lambda_1+\lambda_2)(x_1+x_2)}\right]$$

Integrating from 0 to ∞ yields:

$$T[E(X_N)] = \left(\frac{1}{\lambda_1}, \frac{1}{\lambda_1 + \lambda_2}\right)$$

So:

$$E(X_N) = T^{-1}\left(\frac{1}{\lambda_1}, \frac{1}{\lambda_1 + \lambda_2}\right) = \frac{1}{\lambda_1} + I\left(\frac{1}{\lambda_1 + \lambda_2} - \frac{1}{\lambda_1}\right)$$

In the same way, we can find that:

$$E\left(X_N^2\right) = \frac{2}{\lambda_1^2} + I\left[\frac{2}{(\lambda_1+\lambda_2)^2} - \frac{2}{\lambda_1^2}\right]$$

So:

$$\mathrm{Var}(X_N) = \frac{2}{\lambda_1^2} + I\left[\frac{2}{(\lambda_1+\lambda_2)^2} - \frac{2}{\lambda_1^2}\right] - \left[\frac{1}{\lambda_1} + I\left(\frac{1}{\lambda_1+\lambda_2} - \frac{1}{\lambda_1}\right)\right]^2$$
$$= \frac{1}{\lambda_1^2} + I\left[\frac{1}{(\lambda_1+\lambda_2)^2} - \frac{1}{\lambda_1^2}\right]$$

$$\sigma(X_N)=\sqrt{\frac{1}{\lambda_1^2}+I\left[\frac{1}{(\lambda_1+\lambda_2)^2}-\frac{1}{\lambda_1^2}\right]}=\frac{1}{\lambda_1}+I\left(\frac{1}{\lambda_1+\lambda_2}-\frac{1}{\lambda_1}\right)$$

14.5.4 Neutrosophic uniform distribution

In classical probability theory, we say that the continuous random variable X follows a uniform distribution with parameters $a,b\in\mathbb{R}$ if its pdf is given by:

$$f(x)=\frac{1}{b-a};a<x<b$$

And we have:

$$E(X)=\frac{a+b}{2}$$

$$V(X)=\frac{(b-a)^2}{12}$$

Substituting $a_N=a_1+a_2I, b_N=b_1+b_2I, x_N=x_1+x_2I$ where I is indeterminacy we get:

$f(x_N)=\frac{1}{b_N-a_N};a_N<x_N<b_N$ (under the condition that b_N-a_N is invertible, that is, $b_1-a_1\neq 0$ and $(b_1+b_2)-(a_1+a_2)\neq 0$)

By direct computing, we get:

$$f(x_N)=\frac{1}{b_1+b_2I-a_1-a_2I}=\frac{1}{(b_1-a_1)+(b_2-a_2)I}=x+Iy$$

which results in:

$$x=\frac{1}{b_1-a_1},y=-\frac{a_2-b_2}{(b_1-a_1)(b_1-a_1+b_2-a_2)}$$

So:

$$f(x_N)=\frac{1}{b_1-a_1}-\frac{a_2-b_2}{(b_1-a_1)(b_1-a_1+b_2-a_2)}I;a_N<x_N<b_N$$

Using AH-isometry:

$$T[f(x_N)]=\frac{T(1)}{T(b_N)-T(a_N)}=\frac{(1,1)}{(b_1,b_1+b_2)-(a_1,a_1+a_2)}$$

$$=\left(\frac{1}{b_1-a_1},\frac{1}{(b_1-a_1+b_2-a_2)}\right)$$

Notice that the isometric image of the neutrosophic uniform distribution is two uniform distributions, first one is with parameters a_1, b_1 and the second one is with parameter $b_1 + b_2, a_1 + a_2$.

$$\Rightarrow f(x_N) = T^{-1}\left(\frac{1}{b_1 - a_1}, \frac{1}{(b_1 - a_1 + b_2 - a_2)}\right)$$

$$= \frac{1}{b_1 - a_1} + \left[\frac{1}{(b_1 - a_1 + b_2 - a_2)} - \frac{1}{b_1 - a_1}\right] I$$

$$= \frac{1}{b_1 - a_1} - \frac{a_2 - b_2}{(b_1 - a_1)(b_1 - a_1 + b_2 - a_2)} I$$

Same results!

$$E(X_N) = \frac{a_N + b_N}{2} = \frac{a_1 + a_2 I + b_1 + b_2 I}{2} = \frac{a_1 + b_1}{2} + \frac{a_2 + b_2}{2} I$$

Using AH-isometry:

$$E(X_N) = \int_{a_N}^{b_N} x_N f(x_N) dx_N$$

$$T[E(X_N)] = T\left[\int_{a_N}^{b_N} x_N f(x_N) dx_N\right] = \left(\int_{a_1}^{b_1} \frac{x_1}{b_1 - a_1} dx_1, \int_{a_1+a_2}^{b_1+b_2} \frac{x_1 + x_2}{b_1 - a_1 + b_2 - a_2} d(x_1 + x_2)\right)$$

$$= \left(\frac{a_1 + b_1}{2}, \frac{a_1 + b_1 + a_2 + b_2}{2}\right) \Rightarrow E(X_N) = T^{-1}\left(\frac{a_1 + b_1}{2}, \frac{a_1 + b_1 + a_2 + b_2}{2}\right)$$

$$= \frac{a_1 + b_1}{2} + \left(\frac{a_1 + b_1 + a_2 + b_2}{2} - \frac{a_1 + b_1}{2}\right) I = \frac{a_1 + b_1}{2} + \frac{a_2 + b_2}{2} I$$

Let us compute $E(X_N^2)$ using AH-isometry:
Notice that $x_N^2 = x_1^2 + (2x_1x_2 + x_2^2) I$

$$T\left[E\left(X_N^2\right)\right] = \left(\int_{a_1}^{b_1} \frac{x_1^2}{b_1 - a_1} dx_1, \int_{a_1+a_2}^{b_1+b_2} \frac{x_1^2 + 2x_1x_2 + x_2^2}{b_1 - a_1 + b_2 - a_2} d(x_1 + x_2)\right)$$

$$= \left(\frac{x_1^3}{3(b_1 - a_1)}\Big|_{a_1}^{b_1}, \frac{(x_1 + x_2)^3}{3(b_1 - a_1 + b_2 - a_2)}\Big|_{a_1 + a_2}^{b_1 + b_2}\right)$$

$$= \left(\frac{b_1^2 + b_1 a_1 + a_1^2}{3}, \frac{(b_1 + b_2)^2 + (b_1 + b_2)(a_1 + a_2) + (a_1 + a_2)^2}{3}\right) \Rightarrow$$

$$E\left(X_N^2\right) = T^{-1}\left(\frac{b_1^2 + b_1a_1 + a_1^2}{3}, \frac{(b_1+b_2)^2 + (b_1 + b_2)(a_1 + a_2) + (a_1+a_2)^2}{3}\right)$$

$$= \frac{b_1^2 + b_1a_1 + a_1^2}{3}$$

$$+ \left[\frac{(b_1+b_2)^2 + (b_1 + b_2)(a_1 + a_2) + (a_1+a_2)^2}{3} - \frac{b_1^2 + b_1a_1 + a_1^2}{3}\right]I$$

$$T[V(X_N)] = T\left[E\left(X_N^2\right)\right] - T[E(X_N)]^2$$

$$= \left(\frac{b_1^2 + b_1a_1 + a_1^2}{3}, \frac{(b_1+b_2)^2 + (b_1 + b_2)(a_1 + a_2) + (a_1+a_2)^2}{3}\right)$$

$$- \left(\left(\frac{a_1+b_1}{2}\right)^2, \left(\frac{a_1+b_1+a_2+b_2}{2}\right)^2\right) = \left(\frac{b_1^2 + b_1a_1 + a_1^2}{3} - \left(\frac{a_1+b_1}{2}\right)^2,\right.$$

$$\left.\frac{(b_1+b_2)^2 + (b_1 + b_2)(a_1 + a_2) + (a_1+a_2)^2}{3} - \left(\frac{a_1+b_1+a_2+b_2}{2}\right)^2\right)$$

$$= \left(\frac{(b_1-a_1)^2}{12}, \frac{(b_1-a_1+b_2-a_2)^2}{12}\right) \Rightarrow$$

$$V(X_N) = T^{-1}\left(\frac{(b_1-a_1)^2}{12}, \frac{(b_1-a_1+b_2-a_2)^2}{12}\right)$$

$$= \frac{(b_1-a_1)^2}{12} + \left[\frac{(b_1-a_1+b_2-a_2)^2}{12} - \frac{(b_1-a_1)^2}{12}\right]I$$

$$\sigma(X_N) = \frac{b_1-a_1}{\sqrt{12}} + \left[\frac{b_1-a_1+b_2-a_2}{\sqrt{12}} - \frac{b_1-a_1}{\sqrt{12}}\right]I$$

The same results can be found by direct computing.

14.5.5 Neutrosophic χ^2 distribution

In classical probability theory, we say that the continuous random variable X follows χ^2 distribution with ν degrees of freedom if its pdf is given by:

$$f(x) = \frac{1}{2^{\frac{\nu}{2}}\Gamma\left(\frac{\nu}{2}\right)} e^{-\frac{x}{2}} x^{\frac{\nu}{2}-1}; x > 0$$

Transforming this pdf into neutrosophic yields:

$$f(x_N) = \frac{1}{2^{\frac{\nu_N}{2}}\Gamma\left(\frac{\nu_N}{2}\right)} e^{-\frac{x_N}{2}} x_N^{\frac{\nu_N}{2}-1}; x > 0$$

where $\Gamma(.)$ here is the neutrosophic gamma function defined in 5.1.

Transforming to R^2 using the AH-isometry yields:

$$T[f(x_N)] = \left(\frac{1}{2^{\frac{\nu_1}{2}}\Gamma\left(\frac{\nu_1}{2}\right)} e^{-\frac{x_1}{2}} x_1^{\frac{\nu_1}{2}-1}, \frac{1}{2^{\frac{\nu_1}{2}+\frac{\nu_2}{2}}\Gamma\left(\frac{\nu_1}{2}+\frac{\nu_2}{2}\right)} e^{-\frac{x_1}{2}-\frac{x_2}{2}} (x_1+x_2)^{\frac{\nu_1}{2}+\frac{\nu_2}{2}-1} \right)$$

Notice that the isometric image of the neutrosophic χ^2 distribution is two χ^2distributions, first one is with parameter ν_1 and the second one is with parameter $\nu_1 + \nu_2$.

So:

$$f(x_N) = \frac{1}{2^{\frac{\nu_1}{2}}\Gamma\left(\frac{\nu_1}{2}\right)} e^{-\frac{x_1}{2}} x_1^{\frac{\nu_1}{2}-1}$$

$$+ I \left\{ \frac{1}{2^{\frac{\nu_1}{2}+\frac{\nu_2}{2}}\Gamma\left(\frac{\nu_1}{2}+\frac{\nu_2}{2}\right)} e^{-\frac{x_1}{2}-\frac{x_2}{2}} (x_1+x_2)^{\frac{\nu_1}{2}+\frac{\nu_2}{2}-1} - \frac{1}{2^{\frac{\nu_1}{2}}\Gamma\left(\frac{\nu_1}{2}\right)} e^{-\frac{x_1}{2}} x_1^{\frac{\nu_1}{2}-1} \right\}$$

Let us calculate $E(X_N), \mathrm{Var}(X_N)$:

$$T(x_N f(x_N)) = \left(x_1 \frac{1}{2^{\frac{\nu_1}{2}}\Gamma\left(\frac{\nu_1}{2}\right)} e^{-\frac{x_1}{2}} x_1^{\frac{\nu_1}{2}-1}, (x_1+x_2) \frac{1}{2^{\frac{\nu_1}{2}+\frac{\nu_2}{2}}\Gamma\left(\frac{\nu_1}{2}+\frac{\nu_2}{2}\right)} e^{-\frac{x_1}{2}-\frac{x_2}{2}} (x_1+x_2)^{\frac{\nu_1}{2}+\frac{\nu_2}{2}-1} \right)$$

So:

$$\left(\int_0^\infty x_1 \frac{1}{2^{\frac{\nu_1}{2}}\Gamma\left(\frac{\nu_1}{2}\right)} e^{-\frac{x_1}{2}} x_1^{\frac{\nu_1}{2}-1} dx_1, \int_0^\infty (x_1+x_2) \frac{1}{2^{\frac{\nu_1}{2}+\frac{\nu_2}{2}}\Gamma\left(\frac{\nu_1}{2}+\frac{\nu_2}{2}\right)} e^{-\frac{x_1}{2}-\frac{x_2}{2}} (x_1+x_2)^{\frac{\nu_1}{2}+\frac{\nu_2}{2}-1} d(x_1+x_2) \right)$$

$$= (\nu_1, \nu_1+\nu_2)$$

So $E(X_N) = T^{-1}(\nu_1, \nu_1+\nu_2) = \nu_1 + I(\nu_1+\nu_2-\nu_1) = \nu_1 + I\nu_2$

To calculate $E(X_N^2)$, we need to integrate the quantity $x_N^2 f(x_N)$ neutrosophically, so let us go back to isometry:

$$T\left(x_N^2 f(x_N)\right) =$$

$$\left(x_1^2 \frac{1}{2^{\frac{\nu_1}{2}}\Gamma\left(\frac{\nu_1}{2}\right)} e^{-\frac{x_1}{2}} x_1^{\frac{\nu_1}{2}-1}, (x_1+x_2)^2 \frac{1}{2^{\frac{\nu_1}{2}+\frac{\nu_2}{2}}\Gamma\left(\frac{\nu_1}{2}+\frac{\nu_2}{2}\right)} e^{-\frac{x_1}{2}-\frac{x_2}{2}} (x_1+x_2)^{\frac{\nu_1}{2}+\frac{\nu_2}{2}-1} \right)$$

So:

$$\left(\int_0^\infty x_1^2 \frac{1}{2^{\frac{\nu_1}{2}}\Gamma\left(\frac{\nu_1}{2}\right)} e^{-\frac{x_1}{2}} x_1^{\frac{\nu_1}{2}-1} dx_1, \int_0^\infty (x_1+x_2)^2 \frac{1}{2^{\frac{\nu_1}{2}+\frac{\nu_2}{2}}\Gamma\left(\frac{\nu_1}{2}+\frac{\nu_2}{2}\right)} e^{-\frac{x_1}{2}-\frac{x_2}{2}} (x_1+x_2)^{\frac{\nu_1}{2}+\frac{\nu_2}{2}-1} d(x_1+x_2) \right)$$

$$= \left(2\nu_1 + \nu_1^2, 2(\nu_1+\nu_2) + (\nu_1+\nu_2)^2\right)$$

So $E(X_N^2) = T^{-1}(2\nu_1 + \nu_1^2, 2(\nu_1 + \nu_2) + (\nu_1 + \nu_2)^2) = 2\nu_1 + \nu_1^2 + I(2(\nu_1 + \nu_2) + (\nu_1 + \nu_2)^2 - 2\nu_1 + \nu_1^2)$

Since $V(X_N) = E(X_N^2) - [E(X_N)]^2$, so: $T[V(X_N)] = T[E(X_N^2)] - T[E(X_N)]^2 = (2\nu_1 + \nu_1^2, 2(\nu_1 + \nu_2) + (\nu_1 + \nu_2)^2) - (\nu_1^2, (\nu_1 + \nu_2)^2) = (2\nu_1, 2(\nu_1 + \nu_2))$

So:

$$V(X_N) = T^{-1}(2\nu_1, 2(\nu_1 + \nu_2)) = 2\nu_1 + I[2(\nu_1 + \nu_2) - 2\nu_1] = 2\nu_1 + 2I\nu_2$$

Also, we can find that:

$$\sigma(X_N) = \sqrt{2\nu_1 + 2I\nu_2} = \sqrt{2\nu_1} + I\left\{\sqrt{2(\nu_1 + \nu_2)} - \sqrt{2\nu_1}\right\}$$

14.5.6 Neutrosophic normal distribution

In classical probability theory, we say that the continuous random variable X follows normal distribution or sometimes we call it Gaussian distribution with parameters μ, σ^2 if its pdf is given by:

$$f(x) = \frac{1}{\sigma\sqrt{2\pi}} e^{-\frac{1}{2}\left(\frac{x-\mu}{\sigma}\right)^2}; x \in \mathbb{R}, \mu \in \mathbb{R}, \sigma > 0$$

In neutrosophic theory, parameters μ, σ^2 and the variable x are neutrosophic numbers, that is,

$$f(x_N) = \frac{1}{\sigma_N\sqrt{2\pi}} e^{-\frac{1}{2}\left(\frac{x_N-\mu_N}{\sigma_N}\right)^2}; x_N \in \mathbb{R}(I), \mu_N \in \mathbb{R}(I), \sigma_N > 0$$

where $\sigma_N = \sigma_1 + \sigma_2 I, \mu_N = \mu_1 + \mu_2 I, x_N = x_1 + x_2 I$

Mapping to R^2 yields:

$$T(f(x_N)) = \left(\frac{1}{\sigma_1\sqrt{2\pi}} e^{-\frac{1}{2}\left(\frac{x_1-\mu_1}{\sigma_1}\right)^2}, \frac{1}{(\sigma_1 + \sigma_2)\sqrt{2\pi}} e^{-\frac{1}{2}\left(\frac{x_1+x_2-\mu_1-\mu_2}{\sigma_1+\sigma_2}\right)^2}\right)$$

Notice that the isometric image of the neutrosophic normal distribution is two normal distributions, first one is with parameters μ_1, σ_1^2 and the second one is with parameter $\mu_1 + \mu_2, (\sigma_1 + \sigma_2)^2$.

So:

$$f(x_N) = \frac{1}{\sigma_1\sqrt{2\pi}} e^{-\frac{1}{2}\left(\frac{x_1-\mu_1}{\sigma_1}\right)^2} + I\left[\frac{1}{(\sigma_1 + \sigma_2)\sqrt{2\pi}} e^{-\frac{1}{2}\left(\frac{x_1+x_2-\mu_1-\mu_2}{\sigma_1+\sigma_2}\right)^2} - \frac{1}{\sigma_1\sqrt{2\pi}} e^{-\frac{1}{2}\left(\frac{x_1-\mu_1}{\sigma_1}\right)^2}\right]$$

$$T(x_N f(x_N)) = \left(x_1 \frac{1}{\sigma_1\sqrt{2\pi}} e^{-\frac{1}{2}\left(\frac{x_1-\mu_1}{\sigma_1}\right)^2}, (x_1 + x_2)\frac{1}{(\sigma_1 + \sigma_2)\sqrt{2\pi}} e^{-\frac{1}{2}\left(\frac{x_1+x_2-\mu_1-\mu_2}{\sigma_1+\sigma_2}\right)^2}\right)$$

And:

$$\left(\int_{-\infty}^{+\infty} x_1 \frac{1}{\sigma_1\sqrt{2\pi}} e^{-\frac{1}{2}\left(\frac{x_1-\mu_1}{\sigma_1}\right)^2} dx_1, \int_{-\infty}^{+\infty} (x_1+x_2)\frac{1}{(\sigma_1+\sigma_2)\sqrt{2\pi}} e^{-\frac{1}{2}\left(\frac{x_1+x_2-\mu_1-\mu_2}{\sigma_1+\sigma_2}\right)^2} d(x_1+x_2)\right)$$
$$= (\mu_1, \mu_1+\mu_2)$$

So:

$$E(X_N) = T^{-1}(\mu_1, \mu_1+\mu_2) = \mu_1 + \mu_2 I$$

Also, we can find that:

$$T\left[E\left(X_N^2\right)\right] = \left(\mu_1^2+\sigma_1^2, (\mu_1+\mu_2)^2 + (\sigma_1+\sigma_2)^2\right)$$

So:

$$E\left(X_N^2\right) = \mu_1^2+\sigma_1^2 + I\left[(\mu_1+\mu_2)^2 + (\sigma_1+\sigma_2)^2 - \mu_1^2 + \sigma_1^2\right]$$

$$T(V(X_N)) = T\left[E\left(X_N^2\right)\right] - T[E(X_N)]^2 = \left(\mu_1^2+\sigma_1^2, (\mu_1+\mu_2)^2 + (\sigma_1+\sigma_2)^2\right)$$
$$- (\mu_1, \mu_1+\mu_2)^2 = \left(\sigma_1^2, (\sigma_1+\sigma_2)^2\right)$$

Hence:

$$V(X_N) = \sigma_1^2 + I\left[(\sigma_1+\sigma_2)^2 - \sigma_1^2\right]$$

Also:

$$\sigma(X_N) = \sigma_1 + I\sigma_2$$

Remark 1:

When $\mu_N = 0, \sigma_N = 1$, we get neutrosophic standard normal distribution with probability density function:

$$f(x_N) = \frac{1}{\sqrt{2\pi}} e^{-\frac{x_N{}^2}{2}}$$
$$= \frac{1}{\sqrt{2\pi}} e^{-\frac{x_1^2}{2}} + I\left[\frac{1}{\sqrt{2\pi}} e^{-\frac{(x_1+x_2)^2}{2}} - \frac{1}{\sqrt{2\pi}} e^{-\frac{x_1^2}{2}}\right]$$

Remark 2:

If X_N follows neutrosophic normal distribution with parameters μ_N, σ_N^2, then $Z_N = \frac{X_N - \mu_N}{\sigma_N}$ follows neutrosophic standard normal distribution.

Remark 3:

In previous works, there was no definition for standard normal distribution since neutrosophy was defined only for parameters and in standard normal distributions we have no parameters!

Remark 4:

Let $Z_N \sim N(0,1)$; to calculate the probability $P(Z_N < a_N)$ where $a_N = a_1 + a_2 I$, we can use classical standard normal distribution tables since:

$$P(Z_N < a_N) = \int_{-\infty}^{a_N} f(z_N)dz_N$$

$$T(f(z_N)) = \left(\frac{1}{\sqrt{2\pi}}e^{-\frac{z_1^2}{2}}, \frac{1}{\sqrt{2\pi}}e^{-\frac{(z_1+z_2)^2}{2}}\right)$$

And $\left(\int_{-\infty}^{a_1} \frac{1}{\sqrt{2\pi}}e^{-\frac{z_1^2}{2}}dz_1, \int_{-\infty}^{a_1+a_2} \frac{1}{\sqrt{2\pi}}e^{-\frac{(z_1+z_2)^2}{2}}d(z_1+z_2)\right) = (\Phi(a_1), \Phi(a_1+a_2))$

So:

$$P(Z_N < a_N) = T^{-1}(\Phi(a_1), \Phi(a_1+a_2)) = \Phi(a_1) + I[\Phi(a_1+a_2) - \Phi(a_1)] = \Phi_N(a_N)$$

14.5.7 Fundamental theorem of neutrosophic probability density functions

Let X_N be a neutrosophic continuous random variable with neutrosophic vector of parameters $\Theta_N = \Theta_1 + \Theta_2 I$, then:

1. $T(f(x_N;\Theta_N))$ turns into two classical probability density functions, first is with parameters vector Θ_1 and the second is with parameters vector $\Theta_1 + \Theta_2$.
2. $f(x_N;\Theta_N) = f(x_1;\Theta_1) + I\{f(x_1+x_2;\Theta_1+\Theta_2) - f(x_1;\Theta_1)\}$
3. $F(x_N;\Theta_N) = F(x_1;\Theta_1) + I\{F(x_1+x_2;\Theta_1+\Theta_2) - F(x_1;\Theta_1)\}$

where $X_N = X_1 + X_2 I, \Theta_N = \Theta_1 + \Theta_2 I, I$ is indeterminacy.

Proof:

1. Using two-dimensional AH-isometry, we can find:

$$T[f(x_N;\Theta_N)] = f[T(x_N); T(\Theta_N)] = f[(x_1, x_1+x_2); (\Theta_1, \Theta_1+\Theta_2)]$$
$$= (f(x_1;\Theta_1), f(x_1+x_2;\Theta_1+\Theta_2))$$

2. Using two-dimensional AH-isometry, we can find:

$$T[f(x_N;\Theta_N)] = f[T(x_N); T(\Theta_N)] = f[(x_1, x_1+x_2); (\Theta_1, \Theta_1+\Theta_2)]$$
$$= (f(x_1;\Theta_1), f(x_1+x_2;\Theta_1+\Theta_2))$$

which yields by taking T^{-1} to both sides:

$$f(x_N;\Theta_N) = f(x_1;\Theta_1) + I\{f(x_1+x_2;\Theta_1+\Theta_2) - f(x_1;\Theta_1)\}$$

3. Has become clear and straightforward.

Example 1:

In this example, we are going to find the neutrosophic form of Rayleigh probability density function using the previous theorem and then using AH-isometry and compare the results.

Solution:

In classical probability theory, we say that X follows Rayleigh distribution with parameter $a > 0$ if its probability density function is:

$$f(x;a) = \frac{2}{a} x e^{-\frac{x^2}{a}}; a > 0$$

1. Using theorem 5.7

$$f(x_N;a_N) = f(x_1;a_1) + I\{f(x_1 + x_2;a_1 + a_2) - f(x_1;a_1)\}$$
$$= \frac{2}{a_1} x_1 e^{-\frac{x_1^2}{a_1}} + I\left\{\frac{2}{a_1 + a_2}(x_1 + x_2) e^{-\frac{(x_1+x_2)^2}{a_1+a_2}} - \frac{2}{a_1} x_1 e^{-\frac{x_1^2}{a_1}}\right\}$$

2. Using AH-isometry:

$$f(x_N) = \frac{2}{a_N} x_N e^{-\frac{x_N^2}{a_N}}$$

$$f(x_N) = \frac{2}{a_1 + a_2 I}(x_1 + x_2 I) e^{-\frac{(x_1+x_2 I)^2}{a_1+a_2 I}}$$

$$T[f(x_N)] = \frac{T(2)}{T(a_1 + a_2 I)} T(x_1 + x_2 I) T\left(e^{-\frac{(x_1+x_2 I)^2}{a_1+a_2 I}}\right)$$

$$= \frac{(2,2)}{(a_1, a_1 + a_2)}(x_1, x_1 + x_2) e^{\left(-\frac{x_1^2}{a_1}, -\frac{(x_1+x_2)^2}{a_1+a_2}\right)}$$

$$= \left(\frac{2}{a_1} x_1 e^{-\frac{x_1^2}{a_1}}, \frac{2}{a_1 + a_2}(x_1 + x_2) e^{-\frac{(x_1+x_2)^2}{a_1+a_2}}\right)$$

So:

$$f(x_N) = \frac{2}{a_1} x_1 e^{-\frac{x_1^2}{a_1}} + I\left\{\frac{2}{a_1 + a_2}(x_1 + x_2) e^{-\frac{(x_1+x_2)^2}{a_1+a_2}} - \frac{2}{a_1} x_1 e^{-\frac{x_1^2}{a_1}}\right\}$$

Example 2:

Beta distribution.

Solution:

In classical probability theory, we say that X follows beta distribution with parameters $a, b > 0$ if its probability density function is:

$$f(x;a,b) = \frac{1}{\beta(a,b)} x^{a-1}(1-x)^{b-1}; 0 < x < 1$$

In neutrosophic probability theory:

$$f(x_N;a_N,b_N)=\frac{1}{\beta(a_N,b_N)}x_N^{a_N-1}(1-x_N)^{b_N-1};0<x<1$$

Using fundamental theorem:

$$f(x_N;a_N,b_N)=\frac{1}{\beta(a_1,b_1)}x_1^{a_1-1}(1-x_1)^{b_1-1}$$
$$+I\left\{\frac{1}{\beta(a_1+a_2,b_1+b_2)}(x_1+x_2)^{a_1+a_2-1}(1-x_1-x_2)^{b_1+b_2-1}-\frac{1}{\beta(a_1,b_1)}x_1^{a_1-1}(1-x_1)^{b_1-1}\right\}$$

It is left as an exercise to the reader to prove this result.

14.6 Applications and numerical examples

Here, we will represent many solved problems and examples based on neutrosophic probability density functions:

Example 6.1

Suppose that patients infected with COVID-19 have survival time following neutrosophic exponential distribution with parameter $\lambda_N=\frac{1}{15}+I$. Medical studies show that if a patient takes more than 30 days to survive then he will die; what is the probability that an infected person with COVID-19 will die?! And what is the mean survival time?

Solution

Suppose that X_N describes survival time of infected people with COVID-19, then its pdf is:

$$f(x_N)=\lambda_1e^{-\lambda_1x_1}+I\left((\lambda_1+\lambda_2)e^{-(\lambda_1+\lambda_2)(x_1+x_2)}-\lambda_1e^{-\lambda_1x_1}\right)$$

We have $\lambda_N=\lambda_1+\lambda_2I=\frac{1}{15}+I$ so $\lambda_1=\frac{1}{15},\lambda_2=1$ so:

$$f(x_N)=\frac{1}{15}e^{-\frac{x_1}{15}}+I\left(\frac{16}{15}e^{-\left(\frac{16}{15}\right)(x_1+x_2)}-\frac{1}{15}e^{-\frac{x_1}{15}}\right)$$

$$F(x_N)=P(X_N\le x_N)=1-e^{-\frac{x_1}{15}}+I\left\{e^{-\frac{x_1}{15}}-e^{-\frac{15}{16}(x_1+x_2)}\right\}$$

Probability that an infected person with COVID-19 will die is $P(X_N>30)=1-P(X_N\le 30)=1-F(30)=1-e^{-\frac{30}{15}}+I\left\{e^{-\frac{30}{15}}-e^{-\frac{15}{16}30}\right\}=$ $0.865+0.135I$

The element I describes indeterminacy, uncertainty, and inaccuracy in the probability of the patient to die because λ_N is inaccurate.

Mean survival time is $E(X_N)=\frac{1}{\lambda_1}+I\left(\frac{1}{\lambda_1+\lambda_2}-\frac{1}{\lambda_1}\right)=15+I\left(\frac{15}{16}-15\right)=$ $15-14.06I$

Example 6.2

Suppose that in a normally distributed sample of glucose measures of 100 diabetic people, the mean of glucose was $130 + 10I$ mg/dL with standard deviation equal to 15 + I mg/dL.

1. What is the proportion of people with glucose less than 150 mg/dL?
2. What is the proportion of people with glucose higher than 250 mg/dL?
3. What is the proportion of people with glucose between 150 and 250 mg/dL?
4. What is number of people with glucose higher than 150 mg/dL?

Solution

1. Let X_N be the neutrosophic random variable that describes glucose measure, then $X_N \sim N\left(130 + 10I, (15+I)^2\right) \Leftrightarrow X_N \sim N(130 + 10I, 225 + 31I)$. Proportion of people with glucose less than 150 mg/dL is equal to the probability $P(X_N < 150)$ so:

$$P(X_N < 150) = P\left(Z_N < \frac{150 - 130 - 10I}{15 + I}\right) = P\left(Z_N < \frac{20 - 10I}{15 + I}\right)$$

$$= P\left(Z_N < 1.33 + \frac{-150 - 20}{15 \times 16}I\right) = P(Z_N < 1.33 - 0.71I)$$

$$= \Phi(1.33) + I[\Phi(0.62) - \Phi(1.33)] = 0.908 + I(0.732 - 0.908)$$

$$= 0.908 - 0.176I$$

2. Proportion of people with glucose higher than 250 mg/dL is $P(X_N > 250) = 1 - P(X_N \leq 250) = 1 - P\left(Z_N \leq \frac{250 - 130 - 10I}{15 + I}\right) = 1 - P\left(Z_N \leq \frac{120 - 10I}{15 + I}\right) = 1 - P\left(Z_N \leq 8 + \frac{-150 - 120}{15 \times 16}I\right) = 1 - P(Z_N \leq 8 - 1.125I) = 1 - [\Phi(8) + I(\Phi(6.875) - \Phi(8))] = 1 - [1 + I(1 - 1)] = 0$
3. Proportion of people with glucose between 150 and 250 mg/dL is $P(150 < X_N < 250) = \Phi_N\left(\frac{250 - 130 - 10I}{15 + I}\right) - \Phi_N\left(\frac{150 - 130 - 10I}{15 + I}\right) = 1 - (0.908 - 0.176I) = 0.092 + 0.176I$
4. Number of people with glucose higher than 150 mg/dL is $100 \times P(X_N > 150) = 100 \times (1 - 0.908 + 0.176I) = 100 \times (0.092 + 0.176I) = 9.2 + 17.6I$

Example 6.3

CRP of a random sample follows χ^2 distribution with $24 + I$ degrees of freedom.

1. What is the percentage of people with slightly elevated CRP? (a person with CRP between 3 and 10 is considered slightly elevated).
2. What is the maximum CRP of the first ordered 25% of the sample?

Solution

Let $X_N \sim \chi^2(24+I)$ be the random variable that describes the CRP of the sample.

We have $\nu_N = 24 + I$ so $T(\nu_N) = (24, 25)$ and $T\left(\frac{\nu_N}{2}\right) = (12, 12.5)$.

1. We want to find the percentage of people who are with CRP between 3 and 10 where $T(3) = (3, 3)$ and $T(10) = (10, 10)$.

   Percentage of people with slightly elevated CRP is $P(3 \leq X_N \leq 10) = \int_3^{10} \frac{1}{2^{\frac{24+I}{2}}\Gamma\left(\frac{24+I}{2}\right)} e^{-\frac{x_N}{2}} x_N^{\frac{24+I}{2}-1} dx_N$ and:

$$T\left(\int_3^{10} \frac{1}{2^{12}\Gamma(12)} e^{-\frac{x_1}{2}} x_1^{11} dx_1\right)$$

$$= \left(\int_3^{10} \frac{1}{2^{12}\Gamma(12)} e^{-\frac{x_1}{2}} x_1^{11} dx_1, \int_3^{10} \frac{1}{2^{12.5}\Gamma(12.5)} e^{-\frac{x_1}{2}-\frac{x_2}{2}} (x_1 + x_2)^{11.5} d(x_1 + x_2)\right)$$

$$= (0.005453, 0.003347)$$

   Hence, the percentage is $T^{-1}(0.005453, 0.003347) = 0.005453 - 0.002106I$.

2. Maximum CRP of the first ordered 25% of the sample satisfies the equation $\int_0^{\mathrm{CRP_{Max}}} f(x_N) dx_N = 0.25$

   or:

$$\left(\int_0^{\mathrm{CRP_{Max}}} \frac{1}{2^{12}\Gamma(12)} e^{-\frac{x_1}{2}} x_1^{11} dx_1, \int_0^{\mathrm{CRP_{Max}}} \frac{1}{2^{12.5}\Gamma(12.5)} e^{-\frac{x_1}{2}-\frac{x_2}{2}} (x_1 + x_2)^{11.5} d(x_1 + x_2)\right)$$

$$= (0.25, 0.25)$$

   Solving numerically, we get:

$$T(\mathrm{CRP_{Max}}) = (8.4384, 8.8680)$$

   So:

$$\mathrm{CRP_{Max}} = 8.4384 + 0.4296I$$

14.7 Discussion

The formal form of probability density functions is presented in this chapter. We achieved this formal form and proved it by transmitting from $R(I)$ space to $R \times R$ space using AH-isometry.

The form of neutrosophic probability distribution function in $R \times R$ was:

$$(f(x_1;\Theta_1), f(x_1+x_2;\Theta_1+\Theta_2))$$

where $\Theta_N = \Theta_1 + \Theta_2 I$ is the parameters vector, and its form in $R(I)$ space is:

$$f(x_N;\Theta_N) = f(x_1;\Theta_1) + I\{f(x_1+x_2;\Theta_1+\Theta_2) - f(x_1;\Theta_1)\}$$

We have applied this theorem to find the form of two distributions, Rayleigh distribution and beta distribution, and then, we compared the results with direct computing.

Many applications in the medical field were presented, and numerical examples were successfully solved. The main advantage of this chapter is that it presents general form of neutrosophic probability density functions allowing readers to write the form of any neutrosophic probability density function with easy computations. This chapter opens the way to all the applications of probability theory in many fields including medical fields.

14.8 Conclusions

In this chapter, many probability density functions have been studied and written in its formal form. Also, the properties and characteristics of theirs were discussed. Several numerical examples and applications in the medical field were also presented and solved. General theorem of the probability distribution function was presented and proved.

14.9 Outlook and future work

After the theorems and definitions presented in this chapter, we can now redefine many existing definitions in its formal form, especially the goodness-of-fit tests. So we can know the distribution of any presented data and take our decisions based on the fitted distribution. We are looking forward to study goodness-of-fit tests in the neutrosophic environment and use it in many applications, and we also define some statistical tests and models that are helpful in many real-life fields.

References

[1] F. Smarandache, Neutrosophy. Neutrosophic Probability, Set, and Logic, Amer. Res. Press, United States, 1998.

[2] F. Smarandache, Generalization of the intuitionistic fuzzy set to the neutrosophic set, Int. Conf. Granul. Comput. (2006) 8–42.

[3] A. AL-Nafee, S. Broumi, F. Smarandache, Neutrosophic soft bitopological spaces, Int. J. Neutrosophic Sci. 14 (1) (2021) 47–56.

[4] A.A. Abd El-Khalek, A.T. Khalil, M.A. Abo El-Soud, I. Yasser, A robust machine learning algorithm for cosmic galaxy images classification using neutrosophic score features, Neutrosophic Sets Syst. 42 (2021) 79–101.

[5] M. Abobala, M. Ibrahim, An introduction to refined neutrosophic number theory, Neutrosophic Sets Syst. 45 (2021) 40–53.
[6] K. Alhasan, A. Salama, F. Smarandache, Introduction to neutrosophic reliability theory, Int. J. Neutrosophic Sci. 15 (1) (2021) 52–61.
[7] R. Ali, A short note on the solution of n-refined neutrosophic linear diophantine equations, Int. J. Neutrosophic Sci. 15 (1) (2021) 43–51.
[8] S. Banitalebi, A. Borzooei, Neutrosophic special dominating set in neutrosophic graphs, Neutrosophic Sets Syst. 45 (2021) 26–39.
[9] S. Debnath, Application of intuitionistic neutrosophic soft sets in decision making based on game theory, Int. J. Neutrosophic Sci. 14 (2) (2021) 83–97.
[10] F. Smarandache, Indeterminacy in neutrosophic theories and their applications, Int. J. Neutrosophic Sci. 15 (2) (2021) 89–97.
[11] I. Shahzadi, M. Aslam, H. Aslam, Neutrosophic statistical analysis of income of YouTube channels, Neutrosophic Sets Syst. 39 (2021) 101–106.
[12] M.B. Zeina, Neutrosophic event-based queueing model, Int. J. Neutrosophic Sci. 6 (1) (2020) 48–55.
[13] M.B. Zeina, Erlang service queueing model with neutrosophic parameters, Int. J. Neutrosophic Sci. 6 (2) (2020) 106–112.
[14] M. Abobala, Neutrosophic real inner product spaces, Neutrosophic Sets Syst. 43 (2021) 225–246.
[15] M.B. Zeina, A. Hatip, Neutrosophic random variables, Neutrosophic Sets Syst. 39 (2021) 44–52.
[16] New Science Theory, [Online]. Available from: https://www.new-science-theory.com/probability-science.php.
[17] M. Abobala, A. Hatip, An algebraic approach to neutrosophic euclidean geometry, Neutrosophic Sets Syst. 43 (2021) 114–123.
[18] S.K. Patro, F. Smarandache, The neutrosophic statistical distribution - more problems, more solutions, Neutrosophic Sets Syst. 12 (2016) 73–79.
[19] R. Alhabib, M. Ranna, H. Farah, A. Salama, Some neutrosophic probability distributions, Neutrosophic Sets Syst. 22 (2018) 30–38.
[20] K. Alhasan, F. Smarandache, Neutrosophic Weibull distribution and neutrosophic family Weibull distribution, Neutrosophic Sets Syst. 28 (2019) 191–199.
[21] R. Sherwani, M. Naeem, M. Aslam, M. Raza, M. Abid, S. Abbas, Neutrosophic beta distribution with properties and applications, Neutrosophic Sets Syst. 41 (2021) 209–214.
[22] F. Smarandache, Plithogenic probability & statistics are generalizations of multivariate probability & statistics, Neutrosophic Sets Syst. 43 (2021) 280–289.
[23] H. Rashad, M. Mohamed, Neutrosophic theory and its application in various queueing models: case studies, Neutrosophic Sets Syst. 42 (2021) 117–135.
[24] F. Smarandache, N. Abbas, Y. Chibani, B. Hadjadji, Z.A. Omar, PCR5 and neutrosophic probability in target identification, Neutrosophic Sets Syst. 16 (2017) 76–79.

Chapter 15

Monitoring COVID-19 cases under uncertainty

Nasrullah Khan[1], Muhammad Aslam[2] and Ushna Liaquat[3]
[1]*Department of Statistics and Auctorial Science, University of the Punjab, Lahore, Pakistan,*
[2]*Department of Statistics, Faculty of Science, King Abdulaziz University, Jeddah, Saudi Arabia,*
[3]*Dental Section, Faisalabad Medical University, Faisalabad, Pakistan*

15.1 Introduction

Coronavirus is a zoonotic disease that transformed from animals to the human beings. The coronavirus spread out from Wuhan, China, in 2019. The COVID-19 affected the economy and health of more than 250 countries in the world. The confirmed number of cases in China was reported to be more than 90,000 according to March 10, 2020. To date, Italy, United States, and Iran are among the most affected countries in the world. The number of deaths due to COVID-19 is more than 3500 in China, more than 500 in Italy, and more than 300 in Iran. The first case of the COVID-19 patient was detected in Pakistan on February 26, 2020; see Ref. [1]. The study on the COVID-19 is among the hottest area of research nowadays. The second wave and new form of COVID-19 hit several countries of the world. Hui et al. [2] studied the threat of COVID-19 on people's health. Khan et al. [3] discussed the risk due to COVID-19 and discussed the reasons for the spread of the coronavirus. Chen et al. [4] presented a work on the genome structure of the coronavirus. Noreen et al. [5] stated that the high population density of cities and illiteracy are the major causes of the spread of COVID-19 disease. Due to the nonserious attitude of the mass public toward this virus, the government declared SOPs of social distancing and frequent handwashing when the number of confirmed cases increased manifold. Almost 20% of the world population were forced to lockdown when the confirmed cases increased exponentially throughout the globe. Most of the countries were successful in controlling the rapid spread of COVID-19 cases by adopting drastic measures regarding the closure of educational institutions, public offices, shopping malls, national and international transportation, social and economic activities, etc. According to the data dashboard by Johns Hopkins University CSSE COVID-19 data,

Cognitive Intelligence with Neutrosophic Statistics in Bioinformatics.
DOI: https://doi.org/10.1016/B978-0-323-99456-9.00015-5

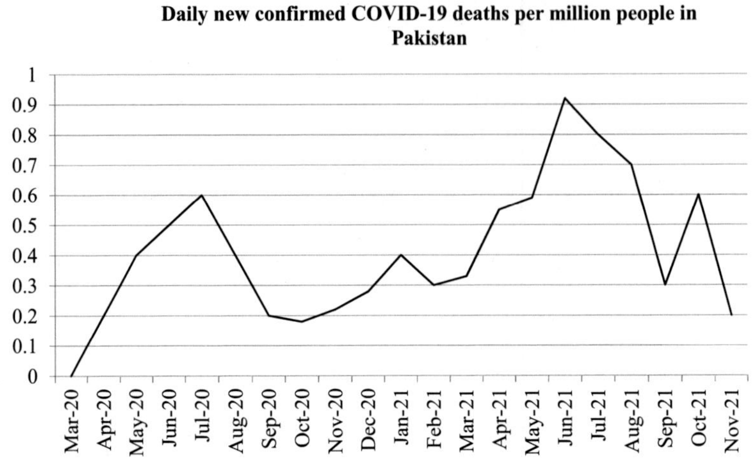

Daily new confirmed COVID-19 deaths per million people in Pakistan

FIGURE 15.1 Daily new confirmed cases of deaths per million in Pakistan.

June–July 2020 and May–June 2021 faced the highest number of deaths in Pakistan. Fig. 15.1 shows a daily new number of confirmed deaths in Pakistan from March 1, 2020, to November 9, 2021.

Vaccination is also treated as useless activity by the people of Pakistan as only 22.18% get vaccinated as the global vaccination is 39.83% for fully vaccinated population (WHO Health Emergency Dashboard). On the contrary, Khan et al. [6] showed concerns over the routine pediatric vaccination in Pakistan during the COVID-19 pandemic. The authorities should adopt aggressive measures to get optimal coverage of vaccination through telehealth, walk-through vaccination, involvement of community pharmacies particularly in rural areas where still strong barriers exist like price, illiteracy, and hesitancy. Perveen et al. [1] examined different factors relating to impeding the COVID-19 vaccination among the public of Pakistan. About 3.5 to 4.5 million deaths can be stopped using the effective vaccination program throughout the world [7].

Perveen et al. [1] presented the details of vaccines approved or authorized to vaccinate people as given in Table 15.1.

A control chart is an important tool that can be applied for the monitoring of the death rate due to COVID-19. The notion of control chart was presented by Walter A. Shewhart during the 1920s in which the collected observations are plotted with two control limits known as the upper control limit (UCL) and the lower control limit (LCL). When the plotted observations fall between these two limits, then the state of the process is declared as an in-control process, otherwise the process is declared as out of control [8]. Ahmad et al. [9] developed a control chart scheme for monitoring the quality of coal. In health science, several control chart procedures have been presented during the last few decades. Liang et al. [10] presented the control

TABLE 15.1 Details of vaccine approved in Pakistan.

No.	Name	Vaccine type	Primary developer	Country of origin
1	AstraZeneca (AZD1222): also known as Vaxzevria and Covishield	Adenovirus vaccine	BARDA, OWS	UK
2	CoronaVac	Inactivated vaccine (formalin with alum adjuvant)	Sinovac	China
3	BBIBP-Corv	Inactivated vaccine	Beijing Institute of Biological Products; China National Pharmaceutical Group (Sinopharm)	China
4	Convidecia (Ad5-nCoV)	Recombinant vaccine (adenovirus type 5 vector)	CanSino Biologics	China
5	Sputnik V	Recombinant adenovirus vaccine (rAd26 and rAd5)	Gamaleya Research Institute, Acellena Contract Drug Research and Development	Russia

chart scheme to diagnose concrete bridges based on the EWMA control chart and reliability analysis. Simpson et al. [11] developed high-quality healthcare information proficiency. The control chart can be an effective tool to monitor the average number of deaths due to coronavirus. The control charts indicate clearly on which day or month the minimum or the maximum number of cases are reported and how many deaths occurred from the confirmed COVID-19 cases. The control charts have been widely applied in monitoring healthcare issues. Tennant et al. [12] discussed the monitoring of patients using the control chart. Neuburger et al. [13] used a control chart to monitor clinical performance. Vazquez-Montes et al. [14] used a chart for watching severe episodes of patients. Aslam et al. [15] presented a control chart for monitoring blood glucose levels. Dey et al. [16] presented excellent work on COVID-19.

By exploring the literature and best of our knowledge, there is no work on a control chart for monitoring the reported cases due to COVID-19 when uncertainty about the average number of cases is presented. In this chapter, we will introduce a c control chart for monitoring the average number of reported cases in Pakistan. It is expected that the proposed control chart will monitor the reported cases effectively in the presence of uncertainty.

15.2 Materials and methods

The purpose of the proposed control chart is to monitor the cases reported due to COVID-19 in December 2020. Let there be uncertainty about the reported cases. Suppose that $\overline{c}_N \epsilon [\overline{c}_L, \overline{c}_U]$ be neutrosophic average of reported cases, where $\overline{c}_L$ be the lower value of the average and $\overline{c}_U$ be the upper value of the neutrosophic average. The neutrosophic form of $\overline{c}_N \epsilon [\overline{c}_L, \overline{c}_U]$ can be written as

$$\overline{c}_N = \overline{c}_L + \overline{c}_U I_N; I_N \epsilon [I_L, I_U] \quad (15.1)$$

Note that the first part $\overline{c}_L$ denotes the determined part and $\overline{c}_U I_N$ denotes the indeterminate or uncertain part of the neutrosophic form and $I_N \epsilon [I_L, I_U]$ be the measure of indeterminacy associated with an average number of reported cases. The proposed control chart reduces to the traditional c chart when $I_L = 0$. The control limits of the neutrosophic c_N chart are given by

$$LCL_N = \overline{c}_N - k\sqrt{\overline{c}_N}; \overline{c}_N \epsilon [\overline{c}_L, \overline{c}_U] \quad (15.2)$$

$$UCL_N = \overline{c}_N + k\sqrt{\overline{c}_N}; \overline{c}_N \epsilon [\overline{c}_L, \overline{c}_U] \quad (15.3)$$

Note that k presents the control limits coefficient.

15.3 Design of neutrosophic control chart

Let LCL_N be the lower control limit and UCL_N be the upper control limit for the neutrosophic statistic, and then, the pair of these control limits can be developed as

$$LCL_N = \overline{c}_N - k\sqrt{\overline{c}_N} \quad (15.4)$$

$$UCL_N = \overline{c}_N + k\sqrt{\overline{c}_N} \quad (15.5)$$

$$\overline{c}_N \epsilon [\overline{c}_L, \overline{c}_U] \quad (15.6)$$

where I_N is the indeterminacy in deaths due to COVID-19 and taken as 10% and 20%, respectively. The lower and upper control limits for the indeterminacy of 10% and 20% in deaths for different countries are shown in Tables 15.2 and 15.3, respectively. Figs. 15.2 and 15.3 are constructed for the number of deaths in different countries using 10% and 20% indeterminacy, respectively.

TABLE 15.2 Upper and lower control limits for 10% indeterminacy in deaths in different countries.

Country	$\overline{c}$	$\overline{c}_N \epsilon [\overline{c}_L, \overline{c}_U]$	LCL_N	UCL_N
Pakistan	31.48325	[30.922154,32.044353]	[14,15]	[47,49]
India	510.9581	[508.697701,513.218578]	[441,445]	[576,581]
Afghanistan	7.219512	[6.950821,7.488204]	[0,0]	[14,15]
Iran	121.0756	[119.975287,122.175974]	[87,89]	[152,155]
Italy	155.2189	[153.973015,156.464753]	[116,118]	[191,193]
Saudi Arabia	24.84314	[24.344708,25.341566]	[9,10]	[39,40]

TABLE 15.3 **The upper and lower control limits for 20% indeterminacy in deaths in different countries.**

Country	$\bar{c}$	$\bar{c}_N \epsilon [\bar{c}_L, \bar{c}_U]$	LCL_N	UCL_N
Pakistan	31.48325	[30.361055,32.605452]	[13,15]	[46,49]
India	510.9581	[506.437263,515.479016]	[438,447]	[573,583]
Afghanistan	7.219512	[6.682129,7.756895]	[−1,0]	[14,16]
Iran	121.0756	[118.874943,123.276318]	[86,89]	[151,156]
Italy	155.2189	[152.727147,157.710622]	[115,120]	[189,195]
Saudi Arabia	24.84314	[23.846279,25.839995]	[9,10]	[38,41]

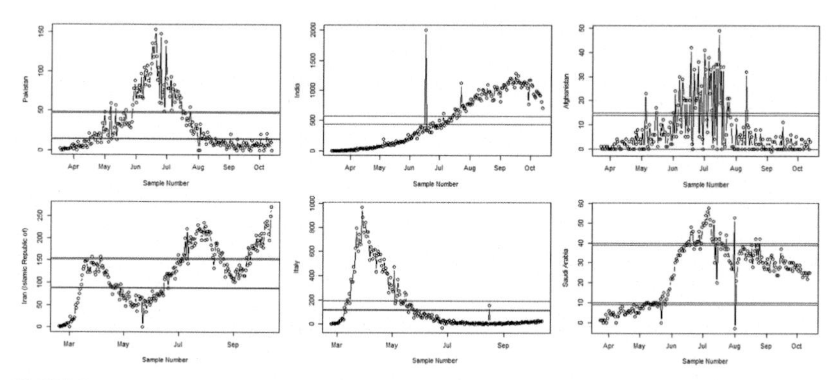

FIGURE 15.2 10% indeterminacy in deaths in different countries.

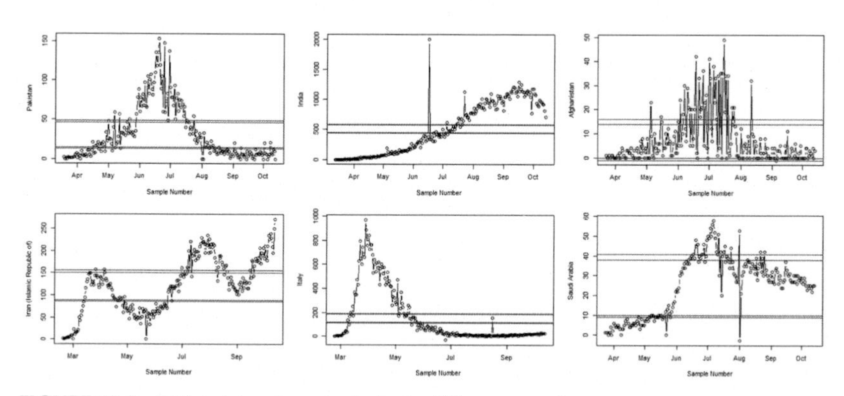

FIGURE 15.3 20% indeterminacy in deaths in different countries.

15.4 Results

In this section, we will discuss the applications of the proposed control chart for monitoring the reported cases in December 2020 in Pakistan. The control charts will be presented for the given the prefixed value of $I_N \epsilon [I_L, I_U]$. The data having indeterminacy $I_U = 0.7\%$ are shown in Table 15.4. For example, for $I_N \epsilon [0, 0.7\%]$, the values of $\overline{c}_N = [\overline{c}_L, \overline{c}_U]$ are as follows $\overline{c}_N = [2690.45, 2711.24]$. Suppose that for the real data, the specified average run length (ARL) is 370. The control limits for $I_U = 0.7\%$ are calculated as follows

$$LCL_N = [2690.45, 2711.24] - 3\sqrt{[2690.45, 2711.24]} = [2535, 2555]$$

$$UCL_N = [2690.45, 2711.24] + 3\sqrt{[2690.45, 2711.24]} = [2846, 2867]$$

The control limits for $I_U = 0.7\%$ and $k = 3.00$ are shown in Table 15.4. Fig. 15.4 shows the proposed control chart for $I_U = 0.7\%$. The control chart figures for any value of I_U can be made similarly.

TABLE 15.4 Control limits for reported cases when $I_U = 0.7\%$.

Date	Cases	LCL_L	UCL_L	LCL_U	UCL_U
11/1/2020	1123	2535	2846	2555	2867
11/2/2020	1167	2535	2846	2555	2867
11/3/2020	1313	2535	2846	2555	2867
11/4/2020	1302	2535	2846	2555	2867
11/5/2020	1376	2535	2846	2555	2867
11/6/2020	1502	2535	2846	2555	2867
11/7/2020	1436	2535	2846	2555	2867
11/8/2020	1650	2535	2846	2555	2867
11/9/2020	1637	2535	2846	2555	2867
11/10/2020	1708	2535	2846	2555	2867
11/11/2020	1808	2535	2846	2555	2867
11/12/2020	2304	2535	2846	2555	2867
11/13/2020	2165	2535	2846	2555	2867
11/14/2020	2443	2535	2846	2555	2867
11/15/2020	2128	2535	2846	2555	2867
11/16/2020	2050	2535	2846	2555	2867
11/17/2020	2298	2535	2846	2555	2867
11/18/2020	2547	2535	2846	2555	2867
11/19/2020	2738	2535	2846	2555	2867
11/20/2020	2843	2535	2846	2555	2867
11/21/2020	2665	2535	2846	2555	2867
11/22/2020	2756	2535	2846	2555	2867
11/23/2020	2954	2535	2846	2555	2867
11/24/2020	3009	2535	2846	2555	2867
11/25/2020	3306	2535	2846	2555	2867
11/26/2020	3113	2535	2846	2555	2867
11/27/2020	3045	2535	2846	2555	2867
11/28/2020	2829	2535	2846	2555	2867
11/29/2020	2839	2535	2846	2555	2867
11/30/2020	2458	2535	2846	2555	2867

(Continued)

TABLE 15.4 (Continued)

Date	Cases	LCL_L	UCL_L	LCL_U	UCL_U
12/1/2020	2829	2535	2846	2555	2867
12/2/2020	3499	2535	2846	2555	2867
12/3/2020	3262	2535	2846	2555	2867
12/4/2020	3119	2535	2846	2555	2867
12/5/2020	3308	2535	2846	2555	2867
12/6/2020	3795	2535	2846	2555	2867
12/7/2020	2885	2535	2846	2555	2867
12/8/2020	2963	2535	2846	2555	2867
12/9/2020	3138	2535	2846	2555	2867
12/10/2020	3047	2535	2846	2555	2867
12/11/2020	2729	2535	2846	2555	2867
12/12/2020	3369	2535	2846	2555	2867
12/13/2020	2362	2535	2846	2555	2867
12/14/2020	2459	2535	2846	2555	2867
12/15/2020	2731	2535	2846	2555	2867

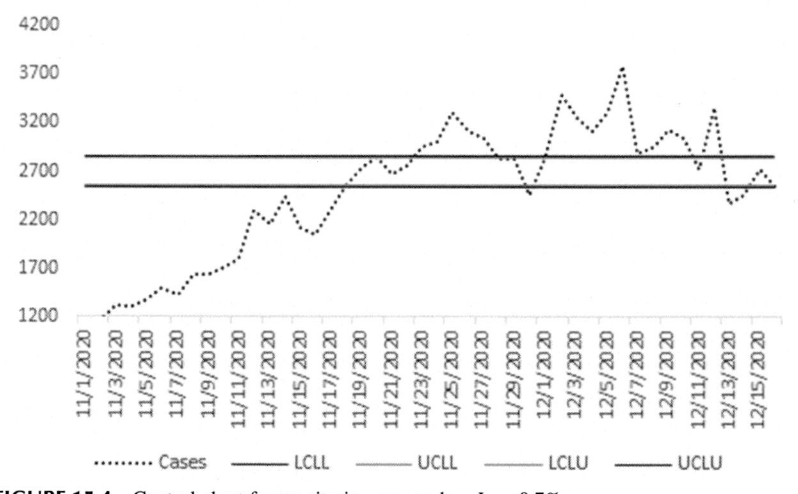

FIGURE 15.4 Control chart for monitoring cases when $I_U = 0.7\%$.

15.5 Discussions

From Fig. 15.4, it can be seen that the number of reported cases is beyond the control limits. The figure clearly indicates that the Government of Pakistan should take a series of actions to bring back the number of cases to an in-control state. In addition, it can be seen from Fig. 15.4 that several points are within the indeterminate intervals. From the study, it is quite clear that the proposed control chart gives information about the state of the process and about the points which are in indeterminate limits and need attention. Table 15.5 has been generated for Pakistan data with 50% indeterminacy and two pairs of control limits. Likewise, Table 15.6 has been generated for Pakistan data with 30% indeterminacy and two pairs of control limits. Table 15.7 has been generated for two limits with 20% indeterminacy for Pakistan. Figs. 15.5–15.7 are given for the plotting of control limits with 50%, 30%, and 20% indeterminacy for Pakistan, respectively.

TABLE 15.5 Control limits for reported cases when $I_U = 0.5\%$.

Date	Cases	LCL_L	UCL_L	LCL_U	UCL_U
11/1/2020	1123	2519	2831	2571	2883
11/2/2020	1167	2519	2831	2571	2883
11/3/2020	1313	2519	2831	2571	2883
11/4/2020	1302	2519	2831	2571	2883
11/5/2020	1376	2519	2831	2571	2883
11/6/2020	1502	2519	2831	2571	2883
11/7/2020	1436	2519	2831	2571	2883
11/8/2020	1650	2519	2831	2571	2883
11/9/2020	1637	2519	2831	2571	2883
11/10/2020	1708	2519	2831	2571	2883
11/11/2020	1808	2519	2831	2571	2883
11/12/2020	2304	2519	2831	2571	2883
11/13/2020	2165	2519	2831	2571	2883
11/14/2020	2443	2519	2831	2571	2883
11/15/2020	2128	2519	2831	2571	2883
11/16/2020	2050	2519	2831	2571	2883

(Continued)

TABLE 15.5 (Continued)

Date	Cases	LCL_L	UCL_L	LCL_U	UCL_U
11/17/2020	2298	2519	2831	2571	2883
11/18/2020	2547	2519	2831	2571	2883
11/19/2020	2738	2519	2831	2571	2883
11/20/2020	2843	2519	2831	2571	2883
11/21/2020	2665	2519	2831	2571	2883
11/22/2020	2756	2519	2831	2571	2883
11/23/2020	2954	2519	2831	2571	2883
11/24/2020	3009	2519	2831	2571	2883
11/25/2020	3306	2519	2831	2571	2883
11/26/2020	3113	2519	2831	2571	2883
11/27/2020	3045	2519	2831	2571	2883
11/28/2020	2829	2519	2831	2571	2883
11/29/2020	2839	2519	2831	2571	2883
11/30/2020	2458	2519	2831	2571	2883
12/1/2020	2829	2519	2831	2571	2883
12/2/2020	3499	2519	2831	2571	2883
12/3/2020	3262	2519	2831	2571	2883
12/4/2020	3119	2519	2831	2571	2883
12/5/2020	3308	2519	2831	2571	2883
12/6/2020	3795	2519	2831	2571	2883
12/7/2020	2885	2519	2831	2571	2883
12/8/2020	2963	2519	2831	2571	2883
12/9/2020	3138	2519	2831	2571	2883
12/10/2020	3047	2519	2831	2571	2883
12/11/2020	2729	2519	2831	2571	2883
12/12/2020	3369	2519	2831	2571	2883
12/13/2020	2362	2519	2831	2571	2883
12/14/2020	2459	2519	2831	2571	2883
12/15/2020	2731	2519	2831	2571	2883

TABLE 15.6 Control limits for reported cases when $I_U = 0.3\%$.

Date	Cases	LCL_L	UCL_L	LCL_U	UCL_U
11/1/2020	1123	2529	2841	2561	2872
11/2/2020	1167	2529	2841	2561	2872
11/3/2020	1313	2529	2841	2561	2872
11/4/2020	1302	2529	2841	2561	2872
11/5/2020	1376	2529	2841	2561	2872
11/6/2020	1502	2529	2841	2561	2872
11/7/2020	1436	2529	2841	2561	2872
11/8/2020	1650	2529	2841	2561	2872
11/9/2020	1637	2529	2841	2561	2872
11/10/2020	1708	2529	2841	2561	2872
11/11/2020	1808	2529	2841	2561	2872
11/12/2020	2304	2529	2841	2561	2872
11/13/2020	2165	2529	2841	2561	2872
11/14/2020	2443	2529	2841	2561	2872
11/15/2020	2128	2529	2841	2561	2872
11/16/2020	2050	2529	2841	2561	2872
11/17/2020	2298	2529	2841	2561	2872
11/18/2020	2547	2529	2841	2561	2872
11/19/2020	2738	2529	2841	2561	2872
11/20/2020	2843	2529	2841	2561	2872
11/21/2020	2665	2529	2841	2561	2872
11/22/2020	2756	2529	2841	2561	2872
11/23/2020	2954	2529	2841	2561	2872
11/24/2020	3009	2529	2841	2561	2872
11/25/2020	3306	2529	2841	2561	2872
11/26/2020	3113	2529	2841	2561	2872
11/27/2020	3045	2529	2841	2561	2872
11/28/2020	2829	2529	2841	2561	2872
11/29/2020	2839	2529	2841	2561	2872
11/30/2020	2458	2529	2841	2561	2872
12/1/2020	2829	2529	2841	2561	2872

(Continued)

TABLE 15.6 (Continued)

Date	Cases	LCL_L	UCL_L	LCL_U	UCL_U
12/2/2020	3499	2529	2841	2561	2872
12/3/2020	3262	2529	2841	2561	2872
12/4/2020	3119	2529	2841	2561	2872
12/5/2020	3308	2529	2841	2561	2872
12/6/2020	3795	2529	2841	2561	2872
12/7/2020	2885	2529	2841	2561	2872
12/8/2020	2963	2529	2841	2561	2872
12/9/2020	3138	2529	2841	2561	2872
12/10/2020	3047	2529	2841	2561	2872
12/11/2020	2729	2529	2841	2561	2872
12/12/2020	3369	2529	2841	2561	2872
12/13/2020	2362	2529	2841	2561	2872
12/14/2020	2459	2529	2841	2561	2872
12/15/2020	2731	2529	2841	2561	2872

TABLE 15.7 Control limits for reported cases when $I_U = 0.2\%$.

Date	Cases	LCL_L	UCL_L	LCL_U	UCL_U
11/1/2020	1123	2535	2846	2555	2867
11/2/2020	1167	2535	2846	2555	2867
11/3/2020	1313	2535	2846	2555	2867
11/4/2020	1302	2535	2846	2555	2867
11/5/2020	1376	2535	2846	2555	2867
11/6/2020	1502	2535	2846	2555	2867
11/7/2020	1436	2535	2846	2555	2867
11/8/2020	1650	2535	2846	2555	2867
11/9/2020	1637	2535	2846	2555	2867
11/10/2020	1708	2535	2846	2555	2867
11/11/2020	1808	2535	2846	2555	2867
11/12/2020	2304	2535	2846	2555	2867

(Continued)

TABLE 15.7 (Continued)

Date	Cases	LCL_L	UCL_L	LCL_U	UCL_U
11/13/2020	2165	2535	2846	2555	2867
11/14/2020	2443	2535	2846	2555	2867
11/15/2020	2128	2535	2846	2555	2867
11/16/2020	2050	2535	2846	2555	2867
11/17/2020	2298	2535	2846	2555	2867
11/18/2020	2547	2535	2846	2555	2867
11/19/2020	2738	2535	2846	2555	2867
11/20/2020	2843	2535	2846	2555	2867
11/21/2020	2665	2535	2846	2555	2867
11/22/2020	2756	2535	2846	2555	2867
11/23/2020	2954	2535	2846	2555	2867
11/24/2020	3009	2535	2846	2555	2867
11/25/2020	3306	2535	2846	2555	2867
11/26/2020	3113	2535	2846	2555	2867
11/27/2020	3045	2535	2846	2555	2867
11/28/2020	2829	2535	2846	2555	2867
11/29/2020	2839	2535	2846	2555	2867
11/30/2020	2458	2535	2846	2555	2867
12/1/2020	2829	2535	2846	2555	2867
12/2/2020	3499	2535	2846	2555	2867
12/3/2020	3262	2535	2846	2555	2867
12/4/2020	3119	2535	2846	2555	2867
12/5/2020	3308	2535	2846	2555	2867
12/6/2020	3795	2535	2846	2555	2867
12/7/2020	2885	2535	2846	2555	2867
12/8/2020	2963	2535	2846	2555	2867
12/9/2020	3138	2535	2846	2555	2867
12/10/2020	3047	2535	2846	2555	2867
12/11/2020	2729	2535	2846	2555	2867
12/12/2020	3369	2535	2846	2555	2867
12/13/2020	2362	2535	2846	2555	2867
12/14/2020	2459	2535	2846	2555	2867
12/15/2020	2731	2535	2846	2555	2867

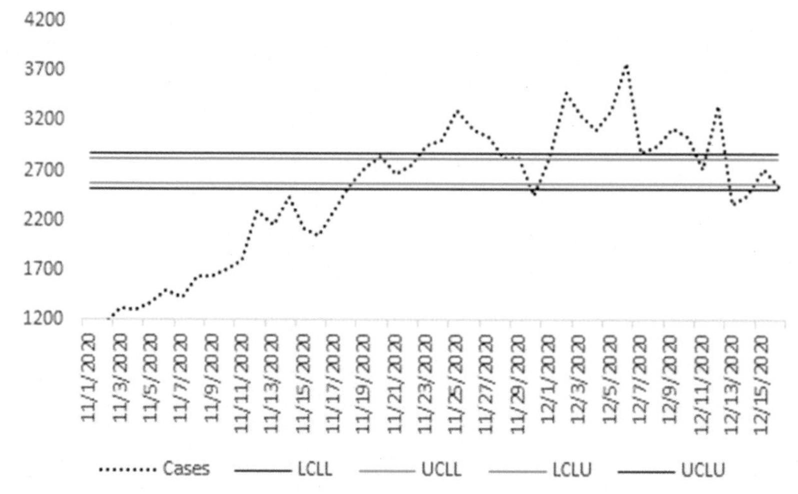

FIGURE 15.5 Plot of control limits with 50% indeterminacy for Pakistan.

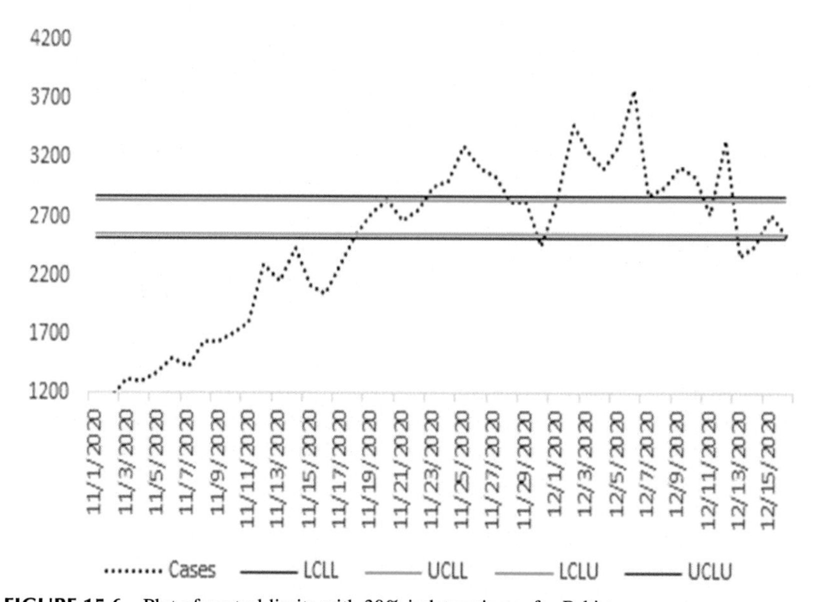

FIGURE 15.6 Plot of control limits with 30% indeterminacy for Pakistan.

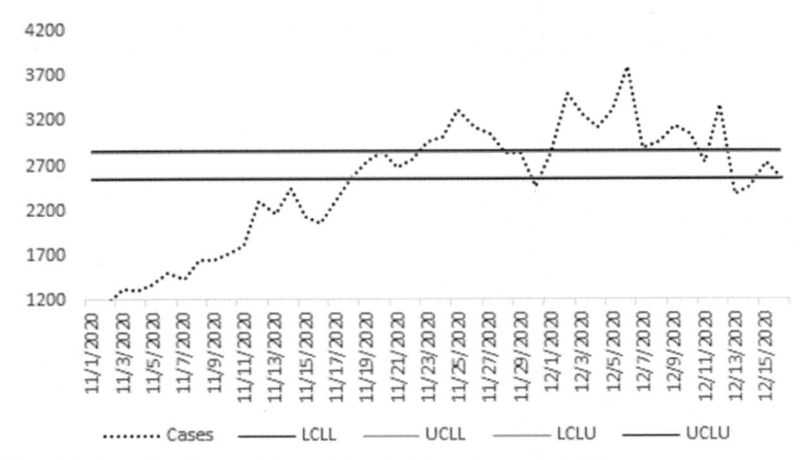

FIGURE 15.7 Plot of control limits with 20% indeterminacy for Pakistan.

15.6 Conclusions

In this paper, the *c* control chart under indeterminacy was proposed to monitor the cases reported due to COVID-19 in Pakistan. The application of the proposed control chart was given using real data of COVID-19 reported cases in December 2020. From this study, it can be noted that the proposed chart clearly indicates that the reported cases were out of control in Pakistan due to COVID-19. The proposed control chart indicated that several points are within the indeterminate intervals. Based on this information, it is recommended that the practitioners can apply the control chart for monitoring the reported cases under uncertainty due to COVID-19. The proposed control chart using a double sampling scheme can be studied in future research.

References

[1] S. Perveen, M. Akram, A. Nasar, A. Arshad-Ayaz, A. Naseem, Vaccination-hesitancy and vaccination-inequality as challenges in Pakistan's COVID-19 response, J. Commun. Psychol. (2021). Available from: https://doi.org/10.1002/jcop.22652.

[2] D.S. Hui, E. I Azhar, T.A. Madani, F. Ntoumi, R. Kock, O. Dar, et al., The continuing 2019-nCoV epidemic threat of novel coronaviruses to global health—the latest 2019 novel coronavirus outbreak in Wuhan, China, Int. J. Infect. Dis. 91 (2020) 264–266.

[3] S. Khan, R. Siddique, A. Ali, M. Xue, G. Nabi, Novel coronavirus, poor quarantine, and the risk of pandemic, J. Hospital Infect. (2020).

[4] Y. Chen, Q. Liu, D. Guo, Emerging coronaviruses: genome structure, replication, and pathogenesis, J. Med. Virolo. (2020).

[5] N. Noreen, S. Dil, S. Niazi, I. Naveed, N. Khan, F. Khan, et al., COVID 19 pandemic & Pakistan; limitations and gaps, Global Biosecurity 1 (4) (2020).

[6] A. Khan, A. Bibi, K. Sheraz Khan, A. Raza Butt, H.A. Alvi, A. Zahra Naqvi, et al., Routine pediatric vaccination in pakistan during COVID-19: how can healthcare professionals help? Front. Pediatrics 8 (859) (2020). Available from: https://doi.org/10.3389/fped.2020.613433.

[7] E. Robertson, K.S. Reeve, C.L. Niedzwiedz, J. Moore, M. Blake, M. Green, et al., Predictors of COVID-19 vaccine hesitancy in the UK household longitudinal study, Brain Behav. Immun. 94 (2021) 41–50. Available from: https://doi.org/10.1016/j.bbi.2021.03.008.

[8] D.C. Montgomery, Introduction to Statistical Quality Control, John Wiley & Sons, 2020.

[9] L. Ahmad, M. Aslam, C.-H. Jun, Coal quality monitoring with improved control charts, Eur. J. Sci. Res. 125 (2) (2014) 427–434.

[10] Z. Liang, J. Cao, J. Zhou, A statistical method for health diagnosis of concrete bridge based on EWMA control chart and reliability analysis (2010).

[11] D. Simpson, T. Roberts, C. Walker, K.D. Cooper, F. O'Brien, Using statistical process control (SPC) chart techniques to support data quality and information proficiency: the underpinning structure of high-quality health care, Qual. Prim. Care 13 (1) (2005) 37–44.

[12] R. Tennant, M.A. Mohammed, J.J. Coleman, U. Martin, Monitoring patients using control charts: a systematic review, Int. J. Qual. Health Care 19 (4) (2007) 187–194.

[13] J. Neuburger, K. Walker, C. Sherlaw-Johnson, J. van der Meulen, D.A. Cromwell, Comparison of control charts for monitoring clinical performance using binary data, BMJ Qual. Saf. 26 (11) (2017) 919–928.

[14] M.D. Vazquez-Montes, R. Stevens, R. Perera, K. Saunders, J.R. Geddes, Control charts for monitoring mood stability as a predictor of severe episodes in patients with bipolar disorder, Int. J. Bipolar Disord. 6 (1) (2018) 7.

[15] M. Aslam, G.S. Rao, N. Khan, F.A. Al-Abbasi, EWMA control chart using repetitive sampling for monitoring blood glucose levels in type-II diabetes patients, Symmetry 11 (1) (2019) 57.

[16] S.K. Dey, M.M. Rahman, U.R. Siddiqi, A. Howlader, Analyzing the epidemiological outbreak of COVID-19: a visual exploratory data analysis (EDA) approach, J. Med. Virol. (2020).

Chapter 16

Neutrosophic statistical analysis of hybrid work model of medical professionals

A. Aleeswari[1], Nivetha Martin[2], Florentin Smarandache[3] and Darjan Karabasevic[4]

[1]Department of Management, PSNA College of Engineering and Technology, Dindigul, Tamil Nadu, India, [2]Department of Mathematics, Arul Anandar College (Autonomous), Karumathur, Madurai, Tamil Nadu, India, [3]Mathematics, Physical and Natural Sciences Division, University of New Mexico, Gallup, NM, United States, [4]University Business Academy, Novi Sad, Serbia

16.1 Introduction and motivation

A working environment in a developing nation like India is characterized by a company with employees bound to company and labor legislation that fixes the period of working. The quite common problem of every working professional is to set a balance between work and life. The disability in establishing a work–life balance has caused professionals to quit their job. The nature of work is the deciding factor in the working style of the professionals. Work from home (WFH) is a quite common work culture for IT professionals; as many of the companies in India are administered by foreign nations, few Indian-based companies also follow this working system. But this kind of working culture may not be apposite to all other professionals especially to the medical field. The medical professionals are comprised of doctors, nurses, emergency workers, medical technicians, pharmacists, well-trained medical representatives, and all other people who are related to the field. The persons of this profession have to be in personal touch with the common people to care for their needs.

The pandemic has caused a lot of chaos in the normal functioning of mankind and intervened in their work–life aspects. The main motive behind switching to online work mode is to bridge the employees with work by avoiding production losses. The decision-makers of the company integrated an online working system into their operative framework and incorporated various strategies to make working at home more feasible. Unfortunately, COVID-19 has also set new norms in the working style of medical professionals. The pandemic situation had drawn

Cognitive Intelligence with Neutrosophic Statistics in Bioinformatics.
DOI: https://doi.org/10.1016/B978-0-323-99456-9.00023-4

them to a new working environment by refraining them from visiting hospitals and engaging with the patients directly. The conventional working hours are changed by the venture of online platforms, and the practice of work from home had made erratic deviations and huge impacts on the work–life balance. The pandemic has caused the development of a new working style that is routed to a hybrid work model (HWM) for medical professionals. The frontline medical workers subjected themselves to the welfare of COVID patients for the entire pandemic period with unconditional service, but they exercised the hybrid work model to attend to the needs of other patients. HWM is characterized by both online and offline working modes by the medical professionals which have streamlined their working pattern especially the working style of the medical representatives who play a vital role in promoting and selling pharmaceutical products and equipment to hospitals, clinics, and medical colleges.

Medical representatives must establish good networking and robust connectivity with other medical professionals by making regular visits to demonstrate products and equipment to the customers. The outbreak of pandemics has alienated them from meeting their customers and has shrunken the opportunities for widening their network. HWM in comparison with WFH is more convincing to the medical representatives in exercising their work as the former working models provide space for both direct and virtual engagements. The impacts of WFH are studied by many researchers as they were highly inquisitive to determine the consequences of this new normal working pattern on health, social aspects, and productivity. Oakman et al. [1] reviewed the health impacts of work from home and discussed the role of gender and made a few recommendations for enhancing employer's health, and Xia et al. [2] discussed the impacts of work at home on the welfare of the users of an office workstation. Delaney [3] stated the healthcare needs that are to be taken care of during pandemic times. The online mode of working burdened the employees as they faced many challenges in accommodating virtual offices and meetings. This kind of switching paralyzed the social skills of the employees. The alienation of the employees by social distancing from the workplace and colleagues diminished the social well-being of the employees. The manifestations of the new normal in working style are statistically studied to find consequential impacts of work from home on the health and social aspects of the professionals, and one of the major findings of this impact study is the challenges of work–life demarcations.

The hybrid working model is gaining momentum after the postpandemic period, and now it is getting institutionalized in some workplaces. It is predicted that many companies will move to HWM in the coming years for creating an effective workplace with positivity for better productivity. With such background, this research work aims in finding the before and after impacts of networking and marketing through a hybrid work model practiced by medical representatives using neutrosophic statistics. This is indeed a new initiative made in this paper, as statistical investigation on the efficacy of HWM has not been made so far in a neutrosophic environment. The magnitude of empirical exploration of HWM is less in comparison with that of the theoretical studies. The need for such statistical study is required for further critical examination of the dynamics of HWM. Neutrosophic paired t-test is the best choice to determine the before and after impacts. The choice of neutrosophic representation of data is very significant as it eliminates the intervention of indeterminacy at times of making inferences.

16.2 Background, definitions, and notations

This section presents the underlying concepts and the steps involved in the neutrosophic paired t-test. For further references, the readers shall refer to the neutrosophic approach discussed by Albassam and Aslam [4].

The set of paired neutrosophic observations of size $n_N \in [n_L, n_U]$ considered for the research study is of the form $Y_{1N} = Y_{IL} + Y_{1U} I_{1N}$; $I_{1N} \in [I_{1L}, I_{1U}]$ and $Y_{2N} = Y_{2L} + Y_{2U} I_{2N}$; $I_{2N} \in [I_{2L}, I_{2U}]$. The determined observations are represented by Y_{1L}, Y_{2L}, and the indeterminate observations are denoted by $Y_{1U} I_{1N}$, $Y_{2U} I_{2N}$. The indeterminacy intervals related to neutrosophic random variables $Y_{1N} \in [Y_{1L}, Y_{I\mathrm{U}}]$ and $X_{2N} \in [X_{2L}, X_{2U}]$ are of the form $I_{1N} \in [I_{1L}, I_{1U}]$, $I_{2N} \in [I_{2L}, I_{2U}]$. The differences between $Y_{1N} \in [Y_{1L}, Y_{1U}]$ are represented by a neutrosophic variable D_N where $D_N \in [D_L, D_U]$

The sum of the square of deviation of $D_N \in [D_L, D_U]$ from $\overline{D}_N \in [\overline{D}_L, \overline{D}_U]$ is given by

$$S_{DN}{}^2 = \frac{\sum_{j=1}^{n_N} \left(D_i - \overline{D}_{iN}\right)^2}{n_N - 1} \tag{16.1}$$

The neutrosophic variance is denoted by $S_{DN}{}^2$ where $S_{DN}{}^2 \in [S_{DL}{}^2, S_{DU}{}^2]$

The neutrosophic standard deviation is denoted by S_{DN}, where $S_{DN} \in [S_{DL}, S_{DU}]$ is defined by

$$S_{DN} = \frac{\sqrt{\sum_{j=1}^{n_N} \left(D_i - \overline{D}_{iN}\right)^2}}{n_N - 1} \tag{16.2}$$

The neutrosophic average of the neutrosophic variable $D_N \in [D_L, D_U]$ is given as follows

$$\sum_{j=1}^{n_N} \left(D_i - \overline{D}_{iN}\right)^2 = \begin{bmatrix} \min\begin{pmatrix} (D_{Lj} + D_{Uj} I_L)(\overline{D}_L + \overline{D}_U I_L), (D_{Lj} + D_{Uj} I_L)(\overline{D}_L + \overline{D}_U I_U) \\ (D_{Lj} + D_{Uj} I_L)(\overline{D}_L + \overline{D}_U I_L), (D_{Lj} + D_{Uj} I_L)(\overline{D}_L + \overline{D}_U I_U) \end{pmatrix} \\ \max\begin{pmatrix} (D_{Lj} + D_{Uj} I_L)(\overline{D}_L + \overline{D}_U I_L), (D_{Lj} + D_{Uj} I_L)(\overline{D}_L + \overline{D}_U I_U) \\ (D_{Lj} + D_{Uj9} I_L)(\overline{D}_L + \overline{D}_U I_L), (D_{Li} + D_{Ui} I_L)(\overline{D}_L + \overline{D}_U I_U) \end{pmatrix} \end{bmatrix} \tag{16.3}$$

The neutrosophic paired t-test aims in determining whether the mean difference between the paired data is significantly different from zero. Let μ_{DN}

represents the mean differences of the paired data, H_{0N} represents the neutrosophic null hypothesis, and H_{1N} represents the neutrosophic alternate hypothesis

$$\begin{cases} H_{0N}: \mu_{DN=D_{0N}} \\ H_{1N}: \mu_{DN \neq D_{0N}} \end{cases}$$

The test statistic is given by

$$t_N = \frac{\overline{D}_N - \overline{D}_{ON}}{S_{DN}/n\sqrt{n}} \tag{16.4}$$

The neutrosophic test is given by

$$t_N = t_N + t_U I_{tN} \tag{16.5}$$

The test statistic of the paired t-test under classical statistics is represented by t_L, and the neutrosophic indeterminate part is denoted by $t_U I_{tN}$, where $I_{tN} \epsilon [I_{tL}, I_{tU}]$. If $I_{tL} = 0$, then neutrosophic paired t-test gets reduced to the classical paired t-test.

The steps involved are

1. Formulation of null and alternate neutrosophic hypothesis
2. Fixation of the level of significance α
3. Calculation of the values of neutrosophic t-test static $t_N \epsilon [t_L, t_U]$
4. Find the critical values and the degrees of freedom $\gamma_N = n_N - 1$
   Reject the null hypothesis if the critical value is less than the calculated value, and accept the null hypothesis if the calculated value is less than the critical value.

16.3 Literature review and state of the art

The parametric paired t-test of classical statistics is commonly applied to find the before and after impacts from the deterministic data sets. Mohammad Javad Koohsari [5], Charlene et al. [6], and a few other researchers have used classically paired t-test in making inferences on pandemic impacts on the well-being of the employees. The statistical processing of imprecise and uncertain data is very difficult, and this has paved the way for the introduction of fuzzy statistics. Zadeh [7] introduced fuzzy logic, and it has several statistical applications. The fuzzy logic deals with truth and falsity values, and it cannot be applied to indeterminate data. This is the point of evolution of neutrosophic statistics. Smarandache [8] introduced the theory of neutrosophy and defined neutrosophic sets with three components of truth, indeterminate, and falsity membership values. Neutrosophic statistics is widely applied to make inferences from indeterminate data. Albassam and Aslam [4] applied neutrosophic paired t-test to make inferences from clinical data, and it was observed that the neutrosophic approach of paired t-test was highly accommodative and more feasible in comparison with fuzzy and classical paired t-tests. To the best of our knowledge, the neutrosophic paired t-test is not used to determine the before and after impacts of the hybrid work model on product marketing and sales for medical professionals. This paper aims in finding the consequential impacts on work productivity of medical professionals using a hybrid work model based on the approach of neutrosophic paired t-test discussed by Albassam and Aslam [4].

16.4 Application to the work productivity of medical professionals using hybrid work model

To determine the consequential impacts of the hybrid work model on product sales and marketing, neutrosophic paired t-test is applied. The two factors such as networking with customers and product sales are considered to determine work productivity. Networking with the customers here refers to the acquisition of new customers. The data on networking and product sales by the medical professionals for a period of 2 months [1 January–1 March 2021] before and during [1 September–1 November 2021] lockdown are considered for investigation. The neutrosophic percentage of customer acquisition and sales was collected from 30 medical professionals belonging to the Madurai district through telephonic interviews, and they are presented as follows in Table 16.1

Based on the calculations as stated in Section 16.2, the values of standard deviation and t-test statistic are obtained as [2.69, 3.08] and [− 1.15,3.91],

TABLE 16.1 Neutrosophic data of networking and sales of medical representatives.

Medical representatives	Percentage of networking before pandemic	Percentage of networking after pandemic	Percentage of sales before pandemic	Percentage of sales after pandemic
1	[35.2,35.6]	[37.3,37.4]	[36.2,36.6]	[34.2,34.6]
2	[45.3,46.4]	[42.3,42.5]	[47.3,47.5]	[35.3,35.5]
3	[27.3,27.8]	[30.5,30.6]	[37.3,37.9]	[26.5,26.9]
4	[19.5, 19.7]	[20.3,20.4]	[29.6, 29.7]	[30.7, 30.9]
5	[17,17]	[17.2,17.4]	[22.5,22.6]	[20.4,20.7]
6	[20.3,20.5]	[21.3,21.5]	[30.3,30.5]	[31.4,31.6]
7	[16.5,16.8]	[15.3,15.4]	[18.5,18.8]	[16.7,16.8]
8	[32.5,32.8]	[32.6,32.7]	[35.5,35.8]	[30.5,30.8]
9	[45.3,45.5]	[46.3,46.5]	[47.3,47.5]	[46.6,46.7]
10	[50.2,50.4]	[50.4,50.6]	[43.2,43.4]	[44.2,44.5]

(*Continued*)

TABLE 16.1 (Continued)

Medical representatives	Percentage of networking before pandemic	Percentage of networking after pandemic	Percentage of sales before pandemic	Percentage of sales after pandemic
11	[41.3,41.6]	[42.1,42.3]	[39.3,39.6]	[34.3,34.7]
12	[62.3,62.7]	[63.1,63.4]	[43.3,43.7]	[43.3,43.7]
13	[54.2,54.3]	[55.4,55.6]	[44.2,44.4]	[43.2,43.5]
14	[42.3,42.6]	[43.3,43.4]	[52.3,52.7]	[51.3,51.9]
15	[20.3,20.5]	[22.2,22.4]	[24.3,24.6]	[22.3,22.6]
16	[16.5,16.8]	[21.3,21.5]	[26.6,26.9]	[25.6,25.7]
17	[27.3,27.8]	[15.3,15.4]	[32.3,32.9]	[31.3,31.6]
18	[19.5, 19.7]	[21.3,21.4]	[29.5, 29.9]	[28.3,28.5]
19	[18,18.3]	[19.2,19.3]	[19,19.3]	[16,16.3]
20	[64,64.3]	[65.2,65.3]	[63,63.3]	[62.4,62.5]
21	[45.3,45.4]	[46.3,46.5]	[47.3,47.4]	[40.2,40.3]
22	[50.2,50.4]	[50.3,50.5]	[54.2,54.6]	[53.4,53.7]
23	[41.3,41.6]	[42.3,42.6]	[42.3,42.7]	[41.3,41.5]
24	[29,29.4]	[30.3,30.4]	[43.3,43.8]	[42.5,42.6]
25	[27.3,27.8]	[28.3,28.8]	[44.2,44.7]	[42.5,42.6]
26	[19.5, 19.7]	[21.3,21.5]	[23.5, 23.7]	[20.4,20.6]
27	[20.3,20.5]	[21.3,21.4]	[15.3,15.7]	[14.3,14.7]
28	[16.5,16.8]	[17.2,17.4]	[19.5,19.9]	[20.4,20.5]
29	[45.3,45.4]	[46.3,46.5]	[46.3,46.5]	[44.3,44.5]
30	[55.3,55.4]	[56.3,56.4]	[52.7,52.9]	[50.2,50.3]

respectively. It is observed that the test statistic value is more than the critical value at 5% level of significance for 29 degrees of freedom, and henceforth, the null hypothesis is rejected.

16.5 Discussions

The test statistic value indicates the rejection of the null hypothesis stating that there are no significant differences in the percentage of networking and sales promotion by exercising conventional working models and hybrid working models. The result explicates that the hybrid working model has created considerate impacts on networking and sales promotion. On comparing the results with the classical paired t-test, it is observed that the p-value is less than the test significance ($\alpha = 0.05$), and hence, the null hypothesis is rejected. In both neutrosophic and classical data, the null hypothesis is rejected.

This research work paves way for further investigation on the factors that contribute to such significant differences in the working models. This work shall be extended to explore and uncover the aspects that have an implicit and explicit role in demarcating the efficiency of the conventional and the hybrid working models; also, it gives the researchers the route to examine the impact of this working culture on the work–life balance of the professionals.

16.6 Conclusion and future work

This research work is a primary initiative of studying the efficacy of the hybrid work model on the work productivity of medical professionals using neutrosophic statistics. The existing neutrosophic paired t-test approach is used to draw results. This work gives more significance in applying neutrosophic statistics to study the consequential impacts of the hybrid working model rather than developing the concepts of neutrosophic statistical theory. This work shall be extended by applying other approaches of neutrosophic paired t-tests, and also, comparative analysis shall be made to test the efficiency of various neutrosophic approaches. The results obtained from the neutrosophic t-test agree with the classical data, but still, the neutrosophic data representation is more convenient and accommodating at times of dealing with an impact study.

References

[1] J. Oakman, N. Kinsman, R. Stuckey, M. Graham, V. Weale, A rapid review of mental and physical health effects of working at home: how do we optimise health? BMC Public Health 20 (1) (2020) 1–13.

[2] Y. Xiao, B. Becerik-Gerber, G. Lucas, S.C. Roll, Impacts of working from home during COVID-19 pandemic on physical and mental well-being of office workstation users, J. Occup. Environ. Med. 63 (3) (2021) 181.

[3] R.K. Delaney, A. Locke, M.L. Pershing, C. Geist, E. Clouse, M.P. Debbink, et al., Experiences of a health system's faculty, staff, and trainees' career development, work culture, and childcare needs during the COVID-19 pandemic, JAMA Netw. Open. 4 (4) (2021).

[4] M. Albassam, M. Aslam, Testing internal quality control of clinical laboratory data using paired-test under uncertainty, BioMed. Res. Int. 2021 (2021).

[5] M. Javad Koohsari, T. Nakaya, A. Shibata, K. Ishii, K. Oka, Working from home after the COVID-19 pandemic: do company employees sit more and move less? Sustainability 13 (2021) 939. 2021.

[6] C.M. Kalenkoski, S.W. Pabilonia, Initial impact of the COVID-19 pandemic on the employment and hours of self-employed coupled and single workers by gender and parental status, 2020.

[7] L.A. Zadeh, Fuzzy sets, in: Fuzzy Sets, Fuzzy Logic, and Fuzzy Systems: Selected Papers by Lotfi A Zadeh, 1996, pp. 394–432.

[8] F. Smarandache, Introduction to Neutrosophic Measure, Neutrosophic Integral, and Neutrosophic Probability, Infinite Study, 2013.

Chapter 17

Neutrosophic regression cum ratio estimators for the population mean: an application in medical science

Abhishek Singh[1], Muhammad Aslam[2], Gajendra K. Vishwakarma[1], Alok Dhital[3] and Ion Patrascu[4]

[1]Department of Mathematics and Computing, Indian Institute of Technology (ISM) Dhanbad, Dhanbad, Jharkhand, India, [2]Department of Statistics, Faculty of Science, King Abdulaziz University, Jeddah, Saudi Arabia, [3]Mathematics, Physical and Natural Science Division, University of New Mexico, Gallup, NM, United States, [4]Mathematics Department, Fratii Buzesti College, Craiova, Romania

17.1 Introduction

The researchers always intend to find the best estimation procedures for the estimation of population parameters in sample surveys concerning time, cost, and minimized sampling errors. This can be possible by using additional information which provides better efficacy of the estimators for the estimation of population parameters along with minimum sampling errors. The additional information is subsidiary (auxiliary) characteristics and is highly correlated with study characteristics or variables of the matter. Usually, the subsidiary information is available for each unit, if not then may be obtained from the past survey. If we consider the study character as Y and subsidiary character as X, then study character may be next-day global horizontal irradiance (ND-GHI) in solar energy data and temperature (T) may be subsidiary characteristics, next-day Global Horizontal Irradiance (ND-GHI) may be study characteristics in solar energy data and relative humidity (RH) may be subsidiary characteristics, study characteristics may be the number of COVID-19 cases in any area and subsidiary characteristics may be the number of vaccinated people in that area, study character may be temperature reading and subsidiary information may be the number of years and so on study character and subsidiary character can be defined. Cochran made a

Cognitive Intelligence with Neutrosophic Statistics in Bioinformatics.
DOI: https://doi.org/10.1016/B978-0-323-99456-9.00018-0

very important contribution to the modern sampling theory by suggesting the method of utilizing subsidiary information. Ratio estimator by [1] and product estimator by [2] are two well-known methods of estimation which characterize the subsidiary variables. Many authors have contributed to the sample survey using subsidiary information. A modified ratio estimator by utilizing coefficient of variation of subsidiary information is given by [3], and using a transformed subsidiary variable for estimating the population means is given by [4–8], having given an estimation of population parameter using subsidiary information.

Classical statistics (simple random sampling) has considered only single-valued or crisp data; that is, there is only certainty in data, but classical statistics fails if there is vagueness or uncertainty in data. The uncertainty in data means data contain some intervals or set type data. This type of uncertainty in data could be dealt with new method called fuzzy statistics based on fuzzy logic or sets and comes when the data are fuzzy, unclear, ambiguous, or hazy. [9,10] had given the first concept of fuzziness as a fuzzy set. The fuzzy logic is frequently developed by many authors and broadly employed in the decision-making environment. Our study is focused on neutrosophic logic and statistics, and the neutrosophic statistics based on neutrosophic logic or sets comes when fuzzy or intuitionistic statistics fails to calculate the indeterminacy of the fuzzy or vague data. If there is uncertainty or fuzziness in data, then fuzzy statistics is the solution for uncertain values of variables, but it has also a serious drawback as it does not count the measure of indeterminacy. The ignorance of indeterminacy by fuzzy statistics can be cured by only neutrosophic statistics which is a generalization of fuzzy statistics and facilitates to measure of indeterminacy along with the determinant part of hazy or uncertain data. The concept of neutrosophic was explained by Smarandache; for more literature on neutrosophic sets, logic, and statistics, see [11–18]. Neutrosophic statistics is a generalization of classical statistics as well as fuzzy and intuitionistic statistics [19,20]. There are lots of studies that have been done so far regarding neutrosophic statistics; in 2014, Smarandache had introduced neutrosophic statistics first as an introduction to neutrosophic statistics. The expressions of rock joint roughness coefficient using neutrosophic interval statistical numbers are given by [21,22], and the study on neutrosophic probability statistics is given by [23]. A study on robust single-valued neutrosophic soft aggregation operators in multi-criteria decision-making is given by [24]. Other authors like [25–46] and [47] have also contributed to the field of neutrosophic statistics through different estimation procedures. The study on classification of trapezoidal bipolar neutrosophic number, de-bipolarization technique, and its execution in cloud service-based MCGDM problem is given by [48]. The most recent study on neutrosophic statistics in sampling is neutrosophic ratio-type estimators for the population mean by [49] and generalized estimator for computation of population mean under neutrosophic ranked set technique: an application to solar energy data by [50]. The indeterminacy in data always indicates neutrosophic data and will be handled through neutrosophic statistics because of easy methods to deal with such types of indeterminant data. There are some articles where neutrosophic data have been applied; analyzing the neutrosophic number of rock joint roughness coefficient based on neutrosophic statistics by Chen et al. (2017) and wind speed data in [27–30,51] are some examples where neutrosophic data have been used; for more details of neutrosophic data, see the article by [49,50].

To fill the gap between crisp and neutrosophic data in sampling, [49] proposed neutrosophic ratio-type estimators for the population mean in sampling theory and studies based on this; in our study, we are proposing a neutrosophic generalized class of estimators for the population mean. For example, if we want to measure the monthly temperature of a city on an everyday basis, then we will not get a fixed and specific temperature on any particular day; that is, there will be interval type or indeterminacy in data. If we use classical statistics here, then it will cause loss of some information. So, dealing with this problem, we use neutrosophic statistics for getting interval-type results with the least MSEs. In our research, motivated by [49,50], we are giving a neutrosophic generalized class of estimators for estimating the population means. The estimators given by [49] are valid for either highly correlated or weak correlation coefficients, but our study is valid for all types of correlation coefficient whether it is a weak or high value of correlation coefficients. The neutrosophic observations can be shown in many ways, and the neutrosophic numbers may contain some unknown interval $[a, b]$. We present neutrosophic values as $Z_N = Z_L + Z_U I_N$ with $I_N \in [I_L, I_U]$, where N is for neutrosophic number. Thus, our neutrosophic observations will belong to an interval $Z_N \in [a, b]$, where "a" and "b" are lower and upper values of the neutrosophic data. For more novelty about neutrosophic statistics, one can visit the site http://fs.unm.edu/NS/NeutrosophicStatistics.htm.

The study in this manuscript followed by proposed neutrosophic estimators is given in Section 17.2 and proposed neutrosophic regression cum ratio estimators in Section 17.2.1. In Section 17.3, empirical study is given through data in medical science, while in 3.2 Monte Carlo simulation is given for raw data. Section 17.4 explains the results and discussion, and Section 17.5 gives the concluding remarks and the future study.

17.2 Proposed neutrosophic estimators

To estimate the population means through neutrosophic estimators, we draw total sample observations $n_N \in [n_L, n_U]$ from population N by the neutrosophic simple random sampling (NeSRS) method. Let $\bar{y}_{nN} \in [\bar{y}_{nL}, \bar{y}_{nU}]$ and $\bar{x}_{nN} \in [\bar{x}_{nL}, \bar{x}_{nU}]$ be the sample means of the neutrosophic study and ancillary characteristics, respectively, and also, let $\bar{Y}_N \in [\bar{Y}_L, \bar{Y}_U]$ and $\bar{X}_N \in [\bar{X}_L, \bar{X}_U]$ be the population means of the neutrosophic study and ancillary characteristics, respectively. The neutrosophic study and ancillary characteristics are $Y_N \in [Y_L, Y_U]$ and $X_N \in [X_L, X_U]$. The correlation coefficients between both neutrosophic study and subsidiary variables are $\rho_{yxN} \in \left[\rho_{yxL}, \rho_{yxU}\right]$, $C_{xN} \in [C_{xL}, C_{xU}]$, and $C_{yN} \in \left[C_{yL}, C_{yU}\right]$ where $Y_N$and X_N are the coefficient of variation of neutrosophic variables. The parameter $\beta_{2(x)N} \in \left[\beta_{2(x)L}, \beta_{2(x)U}\right]$ is the neutrosophic coefficient of kurtosis of a neutrosophic subsidiary variable X_N. Let the neutrosophic mean error terms be $e_{0N} \in [e_{0L}, e_{0U}]$ and $e_{1N} \in [e_{1L}, e_{1U}]$, where $e_{0N} = \left(\overline{y}_{nN} - \overline{Y}_N\right)/\overline{Y}_N$ and $e_{1N} = \left(\overline{x}_{nN} - \overline{X}_N\right)/\overline{X}_N$ are such that $E(e_{0N}) = E(e_{1N}) = 0$, $E\left(e_{0N}^2\right) = f_N C_{yN}^2$, $E\left(e_{1N}^2\right) = f_N C_{xN}^2$ and $E(e_{0N}e_{1N}) = f_N C_{yxN}$, where $f_N = \left(\frac{1}{n_N} - \frac{1}{N_N}\right)$, $S_{xN}^2 \in \left[S_{xL}^2, S_{xU}^2\right]$; $S_{yN}^2 \in \left[S_{yL}^2, S_{yU}^2\right]$; $S_{xyN} \in \left[S_{xyL}, S_{xyU}\right]$. $f_N \in [f_L, f_U]$; $e_{0N}^2 \in \left[e_{0L}^2, e_{0U}^2\right]$; $e_{1N}^2 \in \left[e_{1L}^2, e_{1U}^2\right]$; $e_{0N}e_{1N} \in [e_{0L}e_{1L}, e_{0U}e_{1U}]$; $C_{yN}^2 \in \left[C_{yL}^2, C_{yU}^2\right]$; $C_{xN}^2 \in \left[C_{xL}^2, C_{yU}^2\right]$; $C_{yxN} \in \left[C_{yxL}, C_{yxU}\right] = \rho_{yxN} C_{yN} C_{xN}$; $S_{xN} \in [S_{xL}, S_{xU}]$; $S_{yN} \in \left[S_{yL}, S_{yU}\right]$; $\rho_{yxN} \in \left[\rho_{yxL}, \rho_{yxU}\right]$; $C_{xN} \in [C_{xL}, C_{xU}]$; $C_{yN} \in \left[C_{yL}, C_{yU}\right]$.

To estimate the population means under indeterminacy, we have several existing regression cum ratio types of estimators which we have converted into neutrosophic regression cum ratio types estimators. Here, by utilizing the neutrosophic subsidiary information and motivated by [1], we have proffered neutrosophic ratio and regression estimators for population mean as, respectively

$$\overline{y}_{RN} = \overline{y}_{nN}\left(\frac{\overline{X}_N}{\overline{x}_{nN}}\right) = T_{RN} \tag{17.1}$$

$$\overline{y}_{\text{reg}N} = \overline{y}_{nN} + b_N\left(\overline{X}_N - \overline{x}_{nN}\right) = T_{\text{reg}N} \tag{17.2}$$

where $\overline{y}_{nN} = \frac{1}{n_N}\sum_{i=1}^{n_N} y_{iN}$ and $\overline{x}_{nN} = \frac{1}{n_N}\sum_{i=1}^{n_N} x_{iN}$

to obtain Bias and MSE, we expand the neutrosophic ratio estimator from Eq. (17.1) in neutrosophic mean error terms as

$$T_{RN} = \overline{Y}_N(1 + e_{0N})\left(\frac{\overline{X}_N}{\overline{X}_N(1 + e_{1N})}\right);$$

$$\approx \overline{Y}_N(1 + e_{0N})[1 + e_{1N}]^{-1}$$

$$T_{RN} \approx \overline{Y}_N\left(1 + e_{0N} - e_{1N} + e_{1N}^2 - e_{1N}e_{0N}\right)$$

$$\left(T_{RN} - \overline{Y}_N\right) \approx \overline{Y}_N\left(e_{0N} - e_{1N} + e_{1N}^2 - e_{1N}e_{0N}\right) \tag{17.3}$$

Taking expectations on both sides of Eq. (17.3), we have Bias of the ratio estimator under NeSRS as

$$\text{Bias}(T_{RN}) = \overline{Y}_N f_N \left(C_{xN}^2 - C_{yxN}\right) \tag{17.4}$$

Squaring and taking expectations on both sides of Eq. (17.3) and having a degree not more than 2, we have MSE of ratio estimator under NeSRS as

$$\text{MSE}(T_{RN}) = \overline{Y}_N^2 f_N \left(C_{yN}^2 + C_{xN}^2 - 2C_{yxN}\right) \tag{17.5}$$

to obtain Bias and MSE, we expand the neutrosophic product estimator from Eq. (17.2) in the neutrosophic mean error terms up to first-order approximation as

$$T_{\text{reg}N} = \overline{Y}_N(1 + e_{0N}) + b_N\left(\overline{X}_N - \overline{X}_N(1 + e_{1N})\right);$$

$$\approx \overline{Y}_N(1 + e_{0N}) - b_N\overline{X}_N e_{1N}$$

$$\left(T_{\text{reg}N} - \overline{Y}_N\right) \approx \overline{Y}_N e_{0N} - b_N\overline{X}_N e_{1N} \tag{17.6}$$

Taking expectations on both sides of Eq. (17.6), we have Bias of regression estimator under NeSRS as

$$\text{Bias}(T_{\text{REGN}}) = 0 \tag{17.7}$$

Squaring and taking expectations on both sides of Eq. (17.6) and having a degree not more than 2, we have MSE of regression estimator under NeSRS as

$$\text{MSE}(T_{\text{REGN}}) = f_N\left(\overline{Y}_N^2 C_{yN}^2 + b_N^2\overline{X}_N^2 C_{xN}^2 - 2b_N\overline{Y}_N\overline{X}_N C_{yxN}\right) \tag{17.8}$$

where b_N a constant is such that the MSE of the estimator and T_{REGN} is minimized. To minimize the MSE in Eq. (17.8), we are differentiating MSE in Eq. (17.8) with respect to b_N, and equating it to zero, we have

$$\frac{\partial(\text{MSE}(T_{\text{REGN}}))}{\partial b_N} = 0 \quad \Rightarrow \quad b_N = \frac{\overline{Y}_N}{\overline{X}_N}\left(\frac{C_{YXN}}{C_{XN}^2}\right) \tag{17.9}$$

Putting the value of b_N from Eq. (17.9) to Eq. (17.10), we have MSE of the estimator T_{REGN} as

$$\text{MSE}(T_{\text{REGN}}) = f_N\overline{Y}_N^2 C_{yN}^2\left(1 - \rho_{yxN}^2\right) \tag{17.10}$$

Proposed neutrosophic regression cum ratio estimators

By utilizing the neutrosophic coefficient of variation, neutrosophic coefficient of kurtosis of a neutrosophic subsidiary variable, a correlation coefficient of both neutrosophic study and auxiliary characteristics, and also motivated by [6–8,49,50], we have neutrosophic regression cum ratio-type estimators in presence of neutrosophic subsidiary information as follows

$$T_{0N} = \overline{y}_{nN} + b_N\left(\overline{X}_N - \overline{x}_{nN}\right)\left(\frac{\overline{X}_N}{\overline{x}_{nN}}\right) \tag{17.11}$$

$$T_{1N} = \overline{y}_{nN} + b_N\left(\overline{X}_N - \overline{x}_{nN}\right)\left(\frac{\overline{X}_N + C_{xN}}{\overline{x}_{nN} + C_{xN}}\right) \tag{17.12}$$

$$T_{2N} = \overline{y}_{nN} + b_N\left(\overline{X}_N - \overline{x}_{nN}\right)\left(\frac{\overline{X}_N + \beta_{2(x)N}}{\overline{x}_{nN} + \beta_{2(x)N}}\right) \tag{17.13}$$

$$T_{3N} = \overline{y}_{nN} + b_N\left(\overline{X}_N - \overline{x}_{nN}\right)\left(\frac{\beta_{2(x)N}\overline{X}_N + C_{xN}}{\beta_{2(x)N}\overline{x}_{nN} + C_{xN}}\right) \tag{17.14}$$

$$T_{4N} = \overline{y}_{nN} + b_N\left(\overline{X}_N - \overline{x}_{nN}\right)\left(\frac{C_{xN}\overline{X}_N + \beta_{2(x)N}}{C_{xN}\overline{x}_{nN} + \beta_{2(x)N}}\right) \tag{17.15}$$

$$T_{5N} = \overline{y}_{nN} + b_N\left(\overline{X}_N - \overline{x}_{nN}\right)\left(\frac{\overline{X}_N + \rho_{yxN}}{\overline{x}_{nN} + \rho_{yxN}}\right) \tag{17.16}$$

$$T_{6N} = \overline{y}_{nN} + b_N\left(\overline{X}_N - \overline{x}_{nN}\right)\left(\frac{\beta_{2(x)N}\overline{X}_N + \rho_{yxN}}{\beta_{2(x)N}\overline{x}_{nN} + \rho_{yxN}}\right) \tag{17.17}$$

$$T_{7N} = \overline{y}_{nN} + b_N\left(\overline{X}_N - \overline{x}_{nN}\right)\left(\frac{C_{xN}\overline{X}_N + \rho_{yxN}}{C_{xN}\overline{x}_{nN} + \rho_{yxN}}\right) \tag{17.18}$$

where $T_{0N} \in [T_{0L}, T_{0U}]$; $T_{1N} \in [T_{1L}, T_{1U}]$; $T_{2N} \in [T_{2L}, T_{2U}] T_{3N} \in [T_{3L}, T_{3U}]$; $T_{4N} \in [T_{4L}, T_{4U}]$; $T_{5N} \in [T_{5L}, T_{5U}]$; $T_{6N} \in [T_{6L}, T_{6U}] T_{7N} \in [T_{7L}, T_{7U}]$.

To obtain Bias and MSE, we expand the estimator T_{0N} from Eq. (17.11) in neutrosophic mean error terms up to a first-order approximation

$$T_{0N} = \overline{Y}_N(1 + e_{0N}) + b_N\left(\overline{X}_N - \overline{X}_N(1 + e_{1N})\right)\left(\frac{\overline{X}_N}{\overline{X}_N(1 + e_{1N})}\right)$$

$$\approx \left(\overline{Y}_N + \overline{Y}_N e_{0N} - b_N\overline{X}_N e_{1N}\right)(1 + e_{1N})^{-1}$$

$$\approx \left(\overline{Y}_N + \overline{Y}_N e_{0N} - b_N\overline{X}_N e_{1N}\right)\left(1 - e_{1N} + e_{1N}^2\right)$$

$$\approx \left(\overline{Y}_N + \overline{Y}_N e_{0N} - b_N\overline{X}_N e_{1N} - \overline{Y}_N e_{1N} + \overline{Y}_N e_{1N}^2 + b_N\overline{X}_N e_{1N}^2 - \overline{Y}_N e_{0N} e_{1N}\right)$$

$$\left(T_{0N} - \overline{Y}_N\right) \approx \left[\overline{Y}_N e_{0N} - \left(b_N\overline{X}_N + \overline{Y}_N\right)e_{1N} + \left(\overline{Y}_N + b_N\overline{X}_N\right)e_{1N}^2 - \overline{Y}_N e_{0N}e_{1N}\right] \tag{17.19}$$

Taking expectations on both sides of Eq. (17.19), we have Bias of the estimator T_{0N} under NeSRS as

$$\text{Bias}(T_{0N}) = f_{1N}\left[\left(\overline{Y}_N + b_N\overline{X}_N\right)C_{xN}^2 - \overline{Y}_N C_{yxN}\right] \tag{17.20}$$

Squaring and taking expectations on both sides of Eq. (17.19) and having a degree not more than 2, we have MSE of the estimator T_{0N} under NeSRS as

$$\text{MSE}(T_{0N}) = f_{1N}\left[\overline{Y}_N^2 C_{yN}^2 + \left(b_N\overline{X}_N + \overline{Y}_N\right)^2 C_{xN}^2 - 2\overline{Y}_N\left(b_N\overline{X}_N + \overline{Y}_N\right)C_{yxN}\right] \tag{17.21}$$

On simplification of Eq. (17.21), we have MSE of the estimator T_{0N} under NeSRS as

$$\text{MSE}(T_{0N}) = f_{1N}\overline{Y}_N^2\left[C_{xN}^2 + C_{yN}^2\left(1 - \rho_{yxN}^2\right)\right] \tag{17.22}$$

where $\text{Bias}(T_{0N}) \in [\text{Bias}(T_{0L}), \text{Bias}(T_{0U})]$;

$\text{MSE}(T_{0N}) \in [\text{MSE}(T_{0L}), \text{MSE}(T_{0U})]$.

To obtain Bias and MSE, we expand the estimator T_{1N} from Eq. (17.12) in neutrosophic mean error terms up to a first-order approximation

$$T_{1N} = \overline{Y}_N(1 + e_{0N}) + b_N\left(\overline{X}_N - \overline{X}_N(1 + e_{1N})\right)\left(\frac{\overline{X}_N + C_{xN}}{\overline{X}_N(1 + e_{1N}) + C_{xN}}\right)$$

$$\approx \left(\overline{Y}_N + \overline{Y}_N e_{0N} - b_N\overline{X}_N e_{1N}\right)(1 + c_{1N}e_{1N})^{-1}; C_{1N} = \frac{\overline{X}_N}{\overline{X}_N + C_{xN}}, \; C_{1N} \in [C_{1L}, C_{1U}]$$

$$\approx \left(\overline{Y}_N + \overline{Y}_N e_{0N} - b_N\overline{X}_N e_{1N}\right)\left(1 - c_{1N}e_{1N} + c_{1N}^2 e_{1N}^2\right)$$

$$\approx \left(\overline{Y}_N + \overline{Y}_N e_{0N} - b_N\overline{X}_N e_{1N} - \overline{Y}_N c_{1N}e_{1N} + \overline{Y}_N c_{1N}^2 e_{1N}^2 + \overline{X}_N b_N c_{1N} e_{1N}^2 - \overline{Y}_N c_{1N} e_{0N} e_{1N}\right)$$

$$\left(T_{1N} - \overline{Y}_N\right) \approx \left[\overline{Y}_N e_{0N} - \left(b_N\overline{X}_N + c_{1N}\overline{Y}_N\right)e_{1N} + \left(\overline{Y}_N c_{1N}^2 + \overline{X}_N b_N c_{1N}\right)e_{1N}^2 - \overline{Y}_N e_{0N}e_{1N}\right] \tag{17.23}$$

Taking expectations on both sides of Eq. (17.23), we have Bias of the estimator T_{1N} under NeSRS as

$$\text{Bias}(T_{1N}) = f_{1N}\left[\left(\overline{Y}_N c_{1N}^2 + \overline{X}_N b_N c_{1N}\right)C_{xN}^2 - \overline{Y}_N C_{yxN}\right] \tag{17.24}$$

Squaring and taking expectations on both sides of Eq. (17.23) and having a degree not more than 2, we have MSE of the estimator T_{1N} under NeSRS as

$$\text{MSE}(T_{1N}) = f_{1N}\left[\overline{Y}_N^2 C_{yN}^2 + \left(b_N\overline{X}_N + c_{1N}\overline{Y}_N\right)^2 C_{xN}^2 - 2\overline{Y}_N\left(b_N\overline{X}_N + c_{1N}\overline{Y}_N\right)C_{yxN}\right] \tag{17.25}$$

On simplification of Eq. (17.25), we have MSE of the estimator T_{1N} under NeSRS as

$$\text{MSE}(T_{1N}) = f_{1N}\overline{Y}_N^2\left[c_{1N}^2 C_{xN}^2 + C_{yN}^2\left(1 - \rho_{yxN}^2\right)\right] \tag{17.26}$$

where $C_{1N} = \frac{\overline{X}_N}{\overline{X}_N + C_{xN}}$ and $C_{1N} \in [C_{1L}, C_{1U}]$;

$\text{Bias}(T_{1N}) \in [\text{Bias}(T_{1L}), \text{Bias}(T_{1U})]$;

$$\text{MSE}(T_{1N}) \in [\text{MSE}(T_{1L}), \text{MSE}(T_{1U})].$$

To obtain Bias and MSE, we expand the estimator T_{2N} from Eq. (17.13) in neutrosophic mean error terms up to a first-order approximation

$$T_{2N} = \overline{Y}_N(1 + e_{0N}) + b_N\left(\overline{X}_N - \overline{X}_N(1 + e_{1N})\right)\left(\frac{\overline{X}_N + \beta_{2(x)N}}{\overline{X}_N(1 + e_{1N}) + \beta_{2(x)N}}\right)$$

$$\approx \left(\overline{Y}_N + \overline{Y}_N e_{0N} - b_N\overline{X}_N e_{1N}\right)(1 + c_{2N}e_{1N})^{-1};\ C_{2N} = \frac{\overline{X}_N}{\overline{X}_N + \beta_{2(x)N}},\ C_{2N} \in [C_{2L}, C_{2U}]$$

$$\approx \left(\overline{Y}_N + \overline{Y}_N e_{0N} - b_N\overline{X}_N e_{1N}\right)\left(1 - c_{2N}e_{1N} + c_{2N}^2 e_{1N}^2\right)$$

$$\approx \left(\overline{Y}_N + \overline{Y}_N e_{0N} - b_N\overline{X}_N e_{1N} - \overline{Y}_N c_{2N}e_{1N} + \overline{Y}_N c_{2N}^2 e_{1N}^2 + \overline{X}_N b_N c_{2N} e_{1N}^2 - \overline{Y}_N c_{2N} e_{0N} e_{1N}\right)$$

$$\left(T_{2N} - \overline{Y}_N\right) \approx \left[\overline{Y}_N e_{0N} - \left(b_N\overline{X}_N + c_{2N}\overline{Y}_N\right)e_{1N} + \left(\overline{Y}_N c_{2N}^2 + \overline{X}_N b_N c_{2N}\right)e_{1N}^2 - \overline{Y}_N e_{0N} e_{1N}\right] \tag{17.27}$$

Taking expectations on both sides of Eq. (17.27), we have Bias of the estimator T_{2N} under NeSRS as

$$\text{Bias}(T_{2N}) = f_{1N}\left[\left(\overline{Y}_N c_{2N}^2 + \overline{X}_N b_N c_{2N}\right)C_{xN}^2 - \overline{Y}_N C_{yxN}\right] \tag{17.28}$$

Squaring and taking expectations on both sides of Eq. (17.27) and having a degree not more than 2, we have MSE of the estimator T_{2N} under NeSRS as

$$\text{MSE}(T_{2N}) = f_{1N}\left[\overline{Y}_N^2 C_{yN}^2 + \left(b_N\overline{X}_N + c_{2N}\overline{Y}_N\right)^2 C_{xN}^2 - 2\overline{Y}_N\left(b_N\overline{X}_N + c_{2N}\overline{Y}_N\right)C_{yxN}\right] \tag{17.29}$$

On simplification of Eq. (17.29), we have MSE of the estimator T_{2N} under NeSRS as

$$\text{MSE}(T_{2N}) = f_{1N}\overline{Y}_N^2\left[c_{2N}^2 C_{xN}^2 + C_{yN}^2\left(1 - \rho_{yxN}^2\right)\right] \tag{17.30}$$

where $C_{2N} = \frac{\overline{X}_N}{\overline{X}_N + \beta_{2(x)N}}$ and $C_{2N} \in [C_{2L}, C_{2U}]$;

$\text{Bias}(T_{2N}) \in [\text{Bias}(T_{2L}), \text{Bias}(T_{2U})]$;

$$\text{MSE}(T_{2N}) \in [\text{MSE}(T_{2L}), \text{MSE}(T_{2U})].$$

To obtain Bias and MSE, we expand the estimator T_{3N} from Eq. (17.14) in neutrosophic mean error terms up to a first-order approximation

$$T_{3N} = \overline{Y}_N(1 + e_{0N}) + b_N\left(\overline{X}_N - \overline{X}_N(1 + e_{1N})\right)\left(\frac{\beta_{2(x)N}\overline{X}_N + C_{xN}}{\beta_{2(x)N}\overline{X}_N(1 + e_{1N}) + C_{xN}}\right)$$

$$\approx \left(\overline{Y}_N + \overline{Y}_N e_{0N} - b_N\overline{X}_N e_{1N}\right)(1 + c_{3N}e_{1N})^{-1};\, C_{3N} = \frac{\beta_{2(x)N}\overline{X}_N}{\beta_{2(x)N}\overline{X}_N + C_{xN}},\ C_{3N} \in [C_{3L}, C_{3U}]$$

$$\approx \left(\overline{Y}_N + \overline{Y}_N e_{0N} - b_N\overline{X}_N e_{1N}\right)\left(1 - c_{3N}e_{1N} + c_{3N}^2 e_{1N}^2\right)$$

$$\approx \left(\overline{Y}_N + \overline{Y}_N e_{0N} - b_N\overline{X}_N e_{1N} - \overline{Y}_N c_{3N}e_{1N} + \overline{Y}_N c_{3N}^2 e_{1N}^2 + \overline{X}_N b_N c_{3N} e_{1N}^2 - \overline{Y}_N c_{3N} e_{0N} e_{1N}\right)$$

$$\left(T_{3N} - \overline{Y}_N\right) \approx \left[\overline{Y}_N e_{0N} - \left(b_N\overline{X}_N + c_{3N}\overline{Y}_N\right)e_{1N} + \left(\overline{Y}_N c_{3N}^2 + \overline{X}_N b_N c_{3N}\right)e_{1N}^2 - \overline{Y}_N e_{0N} e_{1N}\right] \tag{17.31}$$

Taking expectations on both sides of Eq. (17.31), we have Bias of the estimator T_{3N} under NeSRS as

$$\text{Bias}(T_{3N}) = f_{1N}\left[\left(\overline{Y}_N c_{3N}^2 + \overline{X}_N b_N c_{3N}\right)C_{xN}^2 - \overline{Y}_N C_{yxN}\right] \tag{17.32}$$

Squaring and taking expectations on both sides of Eq. (17.31) and having a degree not more than 2, we have MSE of the estimator T_{3N} under NeSRS as

$$\text{MSE}(T_{3N}) = f_{1N}\left[\overline{Y}_N^2 C_{yN}^2 + \left(b_N\overline{X}_N + c_{3N}\overline{Y}_N\right)^2 C_{xN}^2 - 2\overline{Y}_N\left(b_N\overline{X}_N + c_{3N}\overline{Y}_N\right)C_{yxN}\right] \tag{17.33}$$

On simplification of Eq. (17.33), we have MSE of the estimator T_{3N} under NeSRS as

$$\text{MSE}(T_{3N}) = f_{1N}\overline{Y}_N^2\left[c_{3N}^2 C_{xN}^2 + C_{yN}^2\left(1 - \rho_{yxN}^2\right)\right] \tag{17.34}$$

where $C_{3N} = \frac{\beta_{2(x)N}\overline{X}_N}{\beta_{2(x)N}\overline{X}_N + C_{xN}}$ and $C_{3N} \in [C_{3L}, C_{3U}]$;

$\text{Bias}(T_{3N}) \in [\text{Bias}(T_{3L}), \text{Bias}(T_{3U})]$;

$$\text{MSE}(T_{3N}) \in [\text{MSE}(T_{3L}), \text{MSE}(T_{3U})].$$

To obtain Bias and MSE, we expand the estimator T_{4N} from Eq. (17.15) in neutrosophic mean error terms up to a first-order approximation

$$T_{4N} = \overline{Y}_N(1 + e_{0N}) + b_N\left(\overline{X}_N - \overline{X}_N(1 + e_{1N})\right)\left(\frac{C_{xN}\overline{X}_N + \beta_{2(x)N}}{C_{xN}\overline{X}_N(1 + e_{1N}) + \beta_{2(x)N}}\right)$$

$$\approx \left(\overline{Y}_N + \overline{Y}_N e_{0N} - b_N\overline{X}_N e_{1N}\right)(1 + c_{4N}e_{1N})^{-1}; C_{4N} = \frac{C_{xN}\overline{X}_N}{C_{xN}\overline{X}_N + \beta_{2(x)N}},\ C_{4N} \in [C_{4L},\ C_{4U}]$$

$$\approx \left(\overline{Y}_N + \overline{Y}_N e_{0N} - b_N\overline{X}_N e_{1N}\right)\left(1 - c_{4N}e_{1N} + c_{4N}^2 e_{1N}^2\right)$$

$$\approx \left(\overline{Y}_N + \overline{Y}_N e_{0N} - b_N\overline{X}_N e_{1N} - \overline{Y}_N c_{4N}e_{1N} + \overline{Y}_N c_{4N}^2 e_{1N}^2 + \overline{X}_N b_N c_{4N} e_{1N}^2 - \overline{Y}_N c_{4N} e_{0N} e_{1N}\right)$$

$$\left(T_{4N} - \overline{Y}_N\right) \approx \left[\overline{Y}_N e_{0N} - \left(b_N\overline{X}_N + c_{4N}\overline{Y}_N\right)e_{1N} + \left(\overline{Y}_N c_{4N}^2 + \overline{X}_N b_N c_{4N}\right)e_{1N}^2 - \overline{Y}_N e_{0N} e_{1N}\right] \tag{17.35}$$

Taking expectations on both sides of Eq. (17.35), we have Bias of the estimator T_{4N} under NeSRS as

$$\text{Bias}(T_{4N}) = f_{1N}\left[\left(\overline{Y}_N c_{4N}^2 + \overline{X}_N b_N c_{4N}\right)C_{xN}^2 - \overline{Y}_N C_{yxN}\right] \tag{17.36}$$

Squaring and taking expectations on both sides of Eq. (17.35) and having a degree not more than 2, we have MSE of the estimator T_{4N} under NeSRS as

$$\text{MSE}(T_{4N}) = f_{1N}\left[\overline{Y}_N^2 C_{yN}^2 + \left(b_N\overline{X}_N + c_{4N}\overline{Y}_N\right)^2 C_{xN}^2 - 2\overline{Y}_N\left(b_N\overline{X}_N + c_{4N}\overline{Y}_N\right)C_{yxN}\right] \tag{17.37}$$

On simplification of Eq. (17.37), we have MSE of the estimator T_{4N} under NeSRS as

$$\text{MSE}(T_{4N}) = f_{1N}\overline{Y}_N^2\left[c_{4N}^2 C_{xN}^2 + C_{yN}^2\left(1 - \rho_{yxN}^2\right)\right] \tag{17.38}$$

where $C_{4N} = \frac{C_{xN}\overline{X}_N}{C_{xN}\overline{X}_N + \beta_{2(x)N}}$ and $C_{4N} \in [C_{4L},\ C_{4U}]$;

$\text{Bias}(T_{4N}) \in [\text{Bias}(T_{4L}),\ \text{Bias}(T_{4U})]$;

$$\text{MSE}(T_{4N}) \in [\text{MSE}(T_{4L}),\ \text{MSE}(T_{4U})].$$

To obtain Bias and MSE, we expand the estimator T_{5N} from Eq. (17.16) in neutrosophic mean error terms up to a first-order approximation

$$T_{5N} = \overline{Y}_N(1 + e_{0N}) + b_N\left(\overline{X}_N - \overline{X}_N(1 + e_{1N})\right)\left(\frac{\overline{X}_N + \rho_{yxN}}{\overline{X}_N(1 + e_{1N}) + \rho_{yxN}}\right)$$

$$\approx \left(\overline{Y}_N + \overline{Y}_N e_{0N} - b_N\overline{X}_N e_{1N}\right)(1 + c_{5N}e_{1N})^{-1}; C_{5N} = \frac{\overline{X}_N}{\overline{X}_N + \rho_{yxN}},\ C_{5N} \in [C_{5L},\ C_{5U}]$$

$$\approx \left(\overline{Y}_N + \overline{Y}_N e_{0N} - b_N\overline{X}_N e_{1N}\right)\left(1 - c_{5N}e_{1N} + c_{5N}^2 e_{1N}^2\right)$$

$$\approx \left(\overline{Y}_N + \overline{Y}_N e_{0N} - b_N \overline{X}_N e_{1N} - \overline{Y}_N c_{5N} e_{1N} + \overline{Y}_N c_{5N}^2 e_{1N}^2 + \overline{X}_N b_N c_{5N} e_{1N}^2 - \overline{Y}_N c_{5N} e_{0N} e_{1N}\right)$$

$$\left(T_{5N} - \overline{Y}_N\right) \approx \left[\overline{Y}_N e_{0N} - \left(b_N \overline{X}_N + c_{5N} \overline{Y}_N\right) e_{1N} + \left(\overline{Y}_N c_{5N}^2 + \overline{X}_N b_N c_{5N}\right) e_{1N}^2 - \overline{Y}_N e_{0N} e_{1N}\right] \tag{17.39}$$

Taking expectations on both sides of Eq. (17.39), we have Bias of the estimator T_{5N} under NeSRS as

$$\text{Bias}(T_{5N}) = f_{1N}\left[\left(\overline{Y}_N c_{5N}^2 + \overline{X}_N b_N c_{5N}\right) C_{xN}^2 - \overline{Y}_N C_{yxN}\right] \tag{17.40}$$

Squaring and taking expectations on both sides of Eq. (17.39) and having a degree not more than 2, we have MSE of the estimator T_{5N} under NeSRS as

$$\text{MSE}(T_{5N}) = f_{1N}\left[\overline{Y}_N^2 C_{yN}^2 + \left(b_N \overline{X}_N + c_{5N} \overline{Y}_N\right)^2 C_{xN}^2 - 2\overline{Y}_N \left(b_N \overline{X}_N + c_{5N} \overline{Y}_N\right) C_{yxN}\right] \tag{17.41}$$

On simplification of Eq. (17.41), we have MSE of the estimator T_{5N} under NeSRS as

$$\text{MSE}(T_{5N}) = f_{1N} \overline{Y}_N^2 \left[c_{5N}^2 C_{xN}^2 + C_{yN}^2 \left(1 - \rho_{yxN}^2\right)\right] \tag{17.42}$$

where $C_{5N} = \frac{\overline{X}_N}{\overline{X}_N + \rho_{yxN}}$ and $C_{5N} \in [C_{5L}, C_{5U}]$;

$\text{Bias}(T_{5N}) \in [\text{Bias}(T_{5L}), \text{Bias}(T_{5U})]$;

$$\text{MSE}(T_{5N}) \in [\text{MSE}(T_{5L}), \text{MSE}(T_{5U})].$$

To obtain Bias and MSE, we expand the estimator T_{6N} from Eq. (17.17) in neutrosophic mean error terms up to a first-order approximation

$$T_{6N} = \overline{Y}_N(1 + e_{0N}) + b_N\left(\overline{X}_N - \overline{X}_N(1 + e_{1N})\right)\left(\frac{\beta_{2(x)N}\overline{X}_N + \rho_{yxN}}{\beta_{2(x)N}\overline{X}_N(1 + e_{1N}) + \rho_{yxN}}\right)$$

$$\approx \left(\overline{Y}_N + \overline{Y}_N e_{0N} - b_N \overline{X}_N e_{1N}\right)(1 + c_{6N} e_{1N})^{-1}; C_{6N} = \frac{\beta_{2(x)N}\overline{X}_N}{\beta_{2(x)N}\overline{X}_N + \rho_{yxN}}, \ C_{6N} \in [C_{6L}, C_{6U}]$$

$$\approx \left(\overline{Y}_N + \overline{Y}_N e_{0N} - b_N \overline{X}_N e_{1N}\right)\left(1 - c_{6N} e_{1N} + c_{6N}^2 e_{1N}^2\right)$$

$$\approx \left(\overline{Y}_N + \overline{Y}_N e_{0N} - b_N \overline{X}_N e_{1N} - \overline{Y}_N c_{6N} e_{1N} + \overline{Y}_N c_{6N}^2 e_{1N}^2 + \overline{X}_N b_N c_{6N} e_{1N}^2 - \overline{Y}_N c_{6N} e_{0N} e_{1N}\right)$$

$$\left(T_{6N} - \overline{Y}_N\right) \approx \left[\overline{Y}_N e_{0N} - \left(b_N \overline{X}_N + c_{6N} \overline{Y}_N\right) e_{1N} + \left(\overline{Y}_N c_{6N}^2 + \overline{X}_N b_N c_{6N}\right) e_{1N}^2 - \overline{Y}_N e_{0N} e_{1N}\right] \tag{17.43}$$

Taking expectations on both sides of Eq. (17.43), we have Bias of the estimator T_{6N} under NeSRS as

$$\mathrm{Bias}(T_{6N}) = f_{1N}\left[\left(\overline{Y}_N c_{6N}^2 + \overline{X}_N b_N c_{6N}\right) C_{xN}^2 - \overline{Y}_N C_{yxN}\right] \tag{17.44}$$

Squaring and taking expectations on both sides of Eq. (17.43) and having a degree not more than 2, we have MSE of the estimator T_{6N} under NeSRS as

$$\mathrm{MSE}(T_{6N}) = f_{1N}\left[\overline{Y}_N^2 C_{yN}^2 + \left(b_N\overline{X}_N + c_{6N}\overline{Y}_N\right)^2 C_{xN}^2 - 2\overline{Y}_N\left(b_N\overline{X}_N + c_{6N}\overline{Y}_N\right) C_{yxN}\right] \tag{17.45}$$

On simplification of Eq. (17.45), we have MSE of the estimator T_{6N} under NeSRS as

$$\mathrm{MSE}(T_{6N}) = f_{1N}\overline{Y}_N^2\left[c_{6N}^2 C_{xN}^2 + C_{yN}^2\left(1 - \rho_{yxN}^2\right)\right] \tag{17.46}$$

where $C_{6N} = \frac{\beta_{2(x)N}\overline{X}_N}{\beta_{2(x)N}\overline{X}_N + \rho_{yxN}}$ and $C_{6N} \in [C_{6L}, C_{6U}]$;

$\mathrm{Bias}(T_{6N}) \in [\mathrm{Bias}(T_{6L}), \mathrm{Bias}(T_{6U})]$;

$$\mathrm{MSE}(T_{6N}) \in [\mathrm{MSE}(T_{6L}), \mathrm{MSE}(T_{6U})].$$

To obtain Bias and MSE, we expand the estimator T_{7N} from Eq. (17.18) in neutrosophic mean error terms up to a first-order approximation

$$T_{7N} = \overline{Y}_N(1 + e_{0N}) + b_N\left(\overline{X}_N - \overline{X}_N(1 + e_{1N})\right)\left(\frac{C_{xN}\overline{X}_N + \rho_{yxN}}{C_{xN}\overline{X}_N(1 + e_{1N}) + \rho_{yxN}}\right)$$

$$\approx \left(\overline{Y}_N + \overline{Y}_N e_{0N} - b_N\overline{X}_N e_{1N}\right)(1 + c_{7N}e_{1N})^{-1};\ C_{7N} = \frac{C_{xN}\overline{X}_N}{C_{xN}\overline{X}_N + \rho_{yxN}},\ C_{7N} \in [C_{7L}, C_{7U}]$$

$$\approx \left(\overline{Y}_N + \overline{Y}_N e_{0N} - b_N\overline{X}_N e_{1N}\right)\left(1 - c_{7N}e_{1N} + c_{7N}^2 e_{1N}^2\right)$$

$$\approx \left(\overline{Y}_N + \overline{Y}_N e_{0N} - b_N\overline{X}_N e_{1N} - \overline{Y}_N c_{7N}e_{1N} + \overline{Y}_N c_{7N}^2 e_{1N}^2 + \overline{X}_N b_N c_{7N} e_{1N}^2 - \overline{Y}_N c_{7N} e_{0N} e_{1N}\right)$$

$$\left(T_{7N} - \overline{Y}_N\right) \approx \left[\overline{Y}_N e_{0N} - \left(b_N\overline{X}_N + c_{7N}\overline{Y}_N\right)e_{1N} + \left(\overline{Y}_N c_{7N}^2 + \overline{X}_N b_N c_{7N}\right)e_{1N}^2 - \overline{Y}_N e_{0N} e_{1N}\right] \tag{17.47}$$

Taking expectations on both sides of Eq. (17.47), we have Bias of the estimator T_{7N} under NeSRS as

$$\mathrm{Bias}(T_{7N}) = f_{1N}\left[\left(\overline{Y}_N c_{7N}^2 + \overline{X}_N b_N c_{7N}\right) C_{xN}^2 - \overline{Y}_N C_{yxN}\right] \tag{17.48}$$

Squaring and taking expectations on both sides of Eq. (17.47) and having a degree not more than 2, we have MSE of the estimator T_{7N} under NeSRS as

$$\mathrm{MSE}(T_{7N}) = f_{1N}\left[\overline{Y}_N^2 C_{yN}^2 + \left(b_N\overline{X}_N + c_{7N}\overline{Y}_N\right)^2 C_{xN}^2 - 2\overline{Y}_N\left(b_N\overline{X}_N + c_{7N}\overline{Y}_N\right) C_{yxN}\right] \tag{17.49}$$

On simplification of Eq. (17.49), we have MSE of the estimator T_{7N} under NeSRS as

$$\mathrm{MSE}(T_{7N}) = f_{1N}\overline{Y}_N^2\left[c_{7N}^2 C_{xN}^2 + C_{yN}^2\left(1-\rho_{yxN}^2\right)\right] \tag{17.50}$$

where $C_{7N} = \frac{C_{xN}\overline{X}_N}{C_{xN}\overline{X}_N + \rho_{yxN}}$ and $C_{7N} \in [C_{7L}, C_{7U}]$;

$\mathrm{Bias}(T_{7N}) \in [\mathrm{Bias}(T_{7L}), \mathrm{Bias}(T_{7U})]$;

$$\mathrm{MSE}(T_{7N}) \in [\mathrm{MSE}(T_{7L}), \mathrm{MSE}(T_{7U})]$$

17.3 Empirical study

To illustrate the properties of the neutrosophic estimators numerically, we have taken real-life indeterminate blood pressure data from the Punjab Province of Pakistan. The indeterminate data have four neutrosophic variables date-wise, but in our research, we are considering pulse versus lower blood pressure (BPL). Here, pulse is the neutrosophic subsidiary variable $X_N \in [X_L, X_U]$, and BPL is a neutrosophic study variable $Y_N \in [Y_L, Y_U]$. The parameters for our study are listed in Table 17.1.

Further, we have drawn total $n_N = 25$ samples from the population of size 70 by the method of NeSRS. With the help of this sample and from parameters given in Table 17.1 for both pulse versus BPL, we have calculated MSEs of the proposed neutrosophic regression cum ratio estimators. The results of MSEs by the proposed neutrosophic estimators are shown in Table 17.2.

TABLE 17.1 Description of the parameters for the estimation of means under neutrosophic simple random sampling.

Parameters	Neutrosophic values	Parameters	Neutrosophic values
N_N	[70, 70]	S_{yN}	[7.2678, 6.3637]
n_N	[25, 25]	C_{xN}	[0.03361405, 0.08981015]
$\overline{X}_N$	[72.37143, 79.31429]	C_{yN}	[0.1099991, 0.0798315]
$\overline{Y}_N$	[66.07143, 79.71429]	$\beta_{2(x)N}$	[4.094008, 7.748748]
S_{xN}	[2.432697, 7.123228]	ρ_{yxN}	[−0.03922896, −0.1930166]

TABLE 17.2 Mean square errors of the proposed estimators under neutrosophic simple random sampling.

Estimators	T versus ND-GHI	Estimators	T versus ND-GHI
	MSEs		*MSEs*
T_{RN}	[1.518, 2.812]	T_{3N}	[1.483, 2.320]
T_{reN}	*[1.356, 1.003]*	T_{4N}	*[1.374, 1.305]*
T_{0N}	[1.483, 2.320]	T_{5N}	[1.483, 2.327]
T_{1N}	[1.483, 2.318]	T_{6N}	[1.483, 2.321]
T_{2N}	[1.470, 2.096]	T_{7N}	[1.487, 2.395]

Monte Carlo simulation

Again, to illustrate the properties of the neutrosophic estimators numerically, we have carried a Monte Carlo simulation method by Singh and Vishwakarma (2019), [52], and the simulation study is done under a neutrosophic environment. As we know, neutrosophic random variables (NRVs) will also follow a neutrosophic normal distribution (NND), that is,

$$(X_N, Y_N) \sim NN\left[\left(\mu_{xN},\ \sigma^2_{xN}\right), \left(\mu_{yN},\ \sigma^2_{yN}\right)\right],\quad X_N \in [X_L,\ X_U];\quad Y_N \in [Y_L,\ Y_U];$$

$\mu_{xN} \in \left[\mu_{xL}, \mu_{xU}\right];\ \ \mu_{yN} \in \left[\mu_{yL}, \mu_{yU}\right];\ \ \sigma^2_{xN} \in \left[\sigma^2_{xL}, \sigma^2_{xU}\right];\ \ \sigma^2_{xN} \in \left[\sigma^2_{xL}, \sigma^2_{xU}\right]$. The neutrosophic data are drawn through a four-variate multivariate normal distribution with means $\left(\mu_{xL},\ \mu_{yL},\ \mu_{xU},\ \mu_{yU}\right)$ and covariance matrix

$$\begin{pmatrix} \sigma^2_{xL} & \rho_{yxL}\sigma_{xL}\sigma_{yL} & 0 & 0 \\ \rho_{yxL}\sigma_{xL}\sigma_{yL} & \sigma^2_{yL} & 0 & 0 \\ 0 & 0 & \sigma^2_{xU} & \rho_{yxU}\sigma_{xU}\sigma_{xU} \\ 0 & 0 & \rho_{yxU}\sigma_{xU}\sigma_{yU} & \sigma^2_{yU} \end{pmatrix}.$$

The required parameters for simulation of neutrosophic data are stated in Tables 17.3 and 17.4.

Further, we have drawn total $n_N = 35$ samples from the population of size 100 by the method of neutrosophic simple random sampling for both positive and negative correlation coefficients. With the help of these samples, we have calculated MSEs of the proposed neutrosophic regression cum ratio-type estimators. The whole process of getting MSEs of the neutrosophic estimators through the neutrosophic simple random sampling method is repeated 7000 times. The same process is done for the classical method of estimation too. The results of MSEs by the proposed neutrosophic estimators are shown in Tables 17.5 and 17.6.

TABLE 17.3 Simulated data description of the parameters for the estimation of means under neutrosophic simple random sampling.

Parameters	Neutrosophic values	Parameters	Neutrosophic values
N_N	[100, 100]	S_{yN}	[8.37, 10.15]
n_N	[35, 35]	C_{xN}	[0.1447, 0.1418]
$\overline{X}_N$	[58.73, 69.07]	C_{yN}	[0.1472, 0.1481]
$\overline{Y}_N$	[59.01, 68.54]	$\beta_{2(x)N}$	[3.584544, 2.686217]
S_{xN}	[8.50, 10.2]	ρ_{yxN}	[0.9, 0.9], [−0.9, −0.9], [0.7, 0.7], [−0.7, −0.7]

TABLE 17.4 Simulated data description of the parameters for the estimation of means under classical simple random sampling.

Parameters	Classical values	Parameters	Classical values
N	100	S_y	10.58873
n	35	C_x	0.174167
$\overline{X}$	65.96693	C_y	0.160516
$\overline{Y}$	64.75843	$\beta_{2(x)}$	2.982329
S_x	11.28	ρ_{yx}	0.70, −0.70

17.4 Results and discussion

The mathematical expression for the neutrosophic proposed regression cum ratio-type estimators is obtained up to the first order of approximation. Further, to explore the properties of the neutrosophic estimators numerically, we have conducted an empirical study on neutrosophic blood pressure data and a Monte Carlo simulation study on artificial neutrosophic data for both positive and negative values of the correlation coefficient. The MSEs of the proposed neutrosophic regression cum ratio estimators are calculated for C_{xN} (coefficient of variation), $\beta_{2(x)N}$ (kurtosis), and correlation coefficient (ρ_{yxN}) and are shown in Tables 17.2, 17.5, and 17.6.

In Table 17.2, the MSEs of the proposed neutrosophic regression cum ratio-type estimators are given through neutrosophic blood pressure data. The highlighted bold font in Table 17.2 shows the least MSE of the proposed neutrosophic regression cum ratio-type estimators. Similarly, in Table 17.5,

TABLE 17.5 Mean square errors of the proposed estimators under neutrosophic simple random sampling.

Estimators	Mean square errors (MSEs)		MSEs	
	$\rho_{yxN} = [0.7,\ 0.7]$	$\rho_{yxN} = [0.9,\ 0.9]$	$\rho_{yxN} = [-0.7,\ -0.7]$	$\rho_{yxN} = [-0.9,\ -0.9]$
T_{RN}	[1.303, 2.157]	[0.372, 0.468]	[7.005, 13.00]	[6.141, 10.66]
T_{regN}	*[1.109, 1.837]*	*[0.368, 0.460]*	*[1.037, 1.634]*	*[0.312, 0.567]*
T_{0N}	[2.629, 4.769]	[2.138, 3.784]	[3.042, 5.473]	[1.821, 3.351]
T_{1N}	[2.618, 4.750]	[2.124, 3.760]	[3.025, 5.442]	[1.810, 3.332]
T_{2N}	[2.439, 4.552]	[1.919, 3.437]	[2.823, 5.149]	[1.650, 3.106]
T_{3N}	[2.626, 4.761]	[2.134, 3.777]	[3.036, 5.461]	[1.817, 3.344]
T_{4N}	*[1.918, 3.894]*	*[1.352, 2.561]*	*[2.252, 4.345]*	*[1.147, 2.394]*
T_{5N}	[2.594, 4.707]	[2.075, 3.685]	[3.101, 5.571]	[1.877, 3.436]
T_{6N}	[2.619, 4.743]	[2.119, 3.755]	[3.062, 5.509]	[1.839, 3.381]
T_{7N}	[2.452, 4.481]	[1.852, 3.368]	[3.347, 5.912]	[2.177, 3.819]

TABLE 17.6 Mean square errors of the proposed estimators (neutrosophic vs classical).

Estimators	Mean square errors (MSEs)		MSEs	
	$\rho_{yxn} = [0.70, 0.70]$	$\rho_{yx} = 0.70$	$\rho_{yx} = [-0.70, -0.70]$	$\rho_{yx} = -0.70$
	Neutrosophic	Classical	Neutrosophic	Classical
T_{RN}	[1.303, 2.157]	1.841	[7.005, 13.00]	10.54
T_{regN}	*[1.109, 1.837]*	*1.705*	*[1.037, 1.634]*	*1.547*
T_{0N}	[2.629, 4.769]	4.892	[3.042, 5.473]	4.941
T_{1N}	[2.618, 4.750]	4.875	[3.025, 5.442]	4.923
T_{2N}	[2.439, 4.552]	4.653	[2.823, 5.149]	4.649
T_{3N}	[2.626, 4.761]	4.885	[3.036, 5.461]	4.935
T_{4N}	*[1.918, 3.894]*	*3.828*	*[2.252, 4.345]*	*3.670*
T_{5N}	[2.594, 4.707]	4.821	[3.101, 5.571]	5.014
T_{6N}	[2.619, 4.743]	4.864	[3.062, 5.509]	4.965
T_{7N}	[2.452, 4.481]	4.516	[3.347, 5.912]	5.391

the MSEs of the proposed neutrosophic regression cum ratio-type estimators are given through Monte Carlo simulation on artificial neutrosophic data for both positive and negative values of the correlation coefficient. Alike Table 17.2, also in Table 17.5, the highlighted bold font shows the least MSE of the proposed neutrosophic regression cum ratio-type estimators. We also see the MSEs of the proposed estimators decrease as the values of correlation coefficients increase for positive and negative.

In Table 17.6, a comparison is given for neutrosophic versus classical by the MSEs of the proposed estimators for the estimation of population parameters. We can see from Table 17.6 that neutrosophic results are more efficient and reliable than classical results. The results through classical estimators failed in giving the best estimation for vague or indeterminate data since it provides crisp or single-valued results only. The neutrosophic type results will provide the best estimation for uncertain or interval types data rather than single or crisp-type results. Hence, the neutrosophic-type results shown in Tables 17.2 and 17.5 are the best estimation for the neutrosophic-type data as it contains true values for uncertain data over the classical-type results.

17.5 Conclusion

Motivated by [49,50], and by utilizing neutrosophic subsidiary variables in our research, we have proposed neutrosophic regression cum ratio-type estimators. The mathematical expression of Biases and MSEs for the neutrosophic proposed regression cum ratio-type estimators is obtained up to the first order of approximation. Further, to explore the properties of the neutrosophic estimators numerically, we have carried out an empirical study on neutrosophic blood pressure data and a Monte Carlo simulation study on neutrosophic artificial data. The proposed neutrosophic estimators with the least MSEs have shown supremacy among the other proposed estimators. We have also shown the neutrosophic-type results obtained are more reliable and efficient than classical ones. Thus, we conclude, that for the neutrosophic data, the neutrosophic method of estimation is more reliable than the classical method of estimation. The initial neutrosophic work in sampling theory for the estimation of population means is given by [49,50], and our neutrosophic work is given for different estimators with the same concept. Further study can be in neutrosophic stratified sampling, cluster sampling, systematic sampling, successive sampling, finding missing values, estimation of sensitive information, or by applying different estimators better than this.

Acknowledgments

Authors heartily thank editors and anonymous learned reviewers for their valuable comments which have made substantial improvement to bring the original manuscript to its present form.

References

[1] G. Cochran, Some properties of estimators based on sampling scheme with varying probabilities, Aust. J. Stat. 17 (1940) 22–28.

[2] M.N. Murthy, Product method of estimation, Sankhya: Indian J. Stat. Ser. A (1964) 69–74.

[3] B.V.S. Sisodia, V.K. Dwivedi, A modified ratio estimator using co-efficient of variation of auxiliary variable, J. Indian Soc. Agric. Stat. 33 (1) (1981) 13–18.

[4] L.N. Upadhyaya, H.P. Singh, Use of transformed auxiliary variable in estimating the finite population mean, Biom. J. 41 (5) (1999) 627–636.

[5] H.P. Singh, R. Tailor, R. Tailor, M.S. Kakran, An improved estimator of population mean using Power transformation, J. Indian Soc. Agric. Stat. 58 (2) (2004) 223–230.

[6] C. Kadilar, H. Cingi, Ratio estimators in simple random sampling, Appl. Math. Comput. 151 (2004) 893–902.

[7] C. Kadilar, H. Cingi, An improvement in estimating the population mean by using the correlation co-efficient, Hacet. J.Math. Stat. Volume 35 (1) (2006) 103–109.

[8] Z. Yan, B. Tian, Ratio method to the mean estimation using co-efficient of skewness of auxiliary variable, ICICA 2010, Part II, CCIS 106 (2010) 103–110.

[9] L.A. Zadeh, Fuzzy sets, Inf. Control 8 (3) (1965) 338–353.

[10] Zadeh, L.A. Fuzzy sets, in: Fuzzy Sets, Fuzzy Logic, and Fuzzy Systems: Selected Papers by Lotfi A Zadeh (pp. 394–432), 1996.

[11] F. Smarandache, Neutrosophy: neutrosophic probability, set, and logic: analytic synthesis & synthetic analysis, ProQuest Information & Learning, Ann Arbor, MI, USA 105 (1998) 118–123.

[12] F. Smarandache, A unifying field in logics: neutrosophic logic, Philosophy (London, England), American Research Press, 1999, pp. 1–141.

[13] Smarandache, F., A unifying field in logics: neutrosophic logic, neutrosophic set, neutrosophic probability, and statistics. arXiv preprint math/0101228, 2001.

[14] F. Smarandache, Neutrosophic set a generalization of the intuitionistic fuzzy set, Int. J. Pure Appl. Math. 24 (3) (2005) 287.

[15] F. Smarandache, Neutrosophic logic-a generalization of the intuitionistic fuzzy logic. Multispace & multi structure, Neutrosoph. Transdiscipl. (100 Collected Pap. Sci.) 4 (2010) 396.

[16] Smarandache, F. Introduction to neutrosophic measure, neutrosophic integral, and neutrosophic probability. Infinite Study, 2013.

[17] Smarandache, F. Introduction to Neutrosophic Statistics: Infinite Study, 2014.

[18] F. Smarandache, Neutrosophic set is a generalization of intuitionistic fuzzy set, inconsistent intuitionistic fuzzy set (picture fuzzy set, ternary fuzzy set), pythagorean fuzzy set, spherical fuzzy set, and q-rung orthopair fuzzy set, while neutrosophication is a generalization of regret theory, grey system theory, and three-ways decision (revisited), J. New Theory 29 (2019) 1–31.

[19] K. Atanassov, Intuitionistic fuzzy sets, Fuzzy Sets Syst. 20 (1) (1986) 87–96.

[20] K. Atanassov, New operations defined over the intuitionistic fuzzy sets, Fuzzy Sets Syst. 61 (2) (1999) 137–142.

[21] J. Chen, J. Ye, S. Du, Scale effect and anisotropy analyzed for neutrosophic numbers of rock joint roughness coefficient based on neutrosophic statistics, Symmetry 9 (10) (2017) 208.

[22] J. Chen, J. Ye, S. Du, R. Yong, Expressions of rock joint roughness coefficient using neutrosophic interval statistical numbers, Symmetry 9 (7) (2017) 123.

[23] R. Alhabib, M.M. Ranna, H. Farah, A.A. Salama, Some neutrosophic probability distributions, Neutrosophic Sets Syst. 22 (2018) 30–38.

[24] C.T. Chen, Extensions of the TOPSIS for group decision-making under fuzzy environment, Fuzzy Sets Syst. 114 (1) (2000) 1–9.

[25] C. Jana, M. Pal, A robust single-valued neutrosophic soft aggregation operators in multi-criteria decision making, Symmetry 11 (1) (2019) 110.

[26] N.A. Nabeeh, F. Smarandache, M. Abdel-Basset, H.A. El-Ghareeb, A. Aboelfetouh, An integrated neutrosophic-topsis approach and its application to personnel selection: a new trend in brain processing and analysis, IEEE Access 7 (2019) 29734–29744.

[27] M. Aslam, O.H. Arif, R.A.K. Sherwani, New diagnosis test under the neutrosophic statistics: an application to diabetic patients, BioMed Res. Int. 00 (2020) 7. Available from: https://doi.org/10.1155/2020/2086185.

[28] M. Aslam, A. Algarni, Analyzing the solar energy data using a new anderson-darling test under indeterminacy, Int. J. Photoenergy 2020 (2020). Available from: https://doi.org/10.1155/2020/6662389.

[29] M. Aslam, Neutrosophic analysis of variance: application to university students, Complex Intell. Syst. 5 (2019) 403–407. Available from: https://doi.org/10.1007/s40747-019-0107-2.

[30] M. Aslam, Monitoring the road traffic crashes using NEWMA chart and repetitive sampling, Int.J. Inj. Control Saf. Promot. 28 (1) (2020) 39–45. Available from: https://doi.org/10.1080/17457300.2020.1835990.

[31] M. Aslam, A study on skewness and kurtosis estimators of wind speed distribution under indeterminacy, Theor. Appl. Climatol. 143 (3) (2021) 1227–1234.

[32] M. Aslam, Analyzing gray cast iron data using a new Shapiro-Wilks test for normality under indeterminacy, Int. J. Cast Met. Res. 34 (1) (2021) 1–5.

[33] M. Aslam, Testing average wind speed using sampling plan for Weibull distribution under indeterminacy, Sci. Rep. 11 (1) (2021) 1–9.

[34] M. Aslam, Clinical laboratory medicine measurements correlation analysis under uncertainty, Ann. Clin. Biochem. 58 (4) (2021) 377–383.

[35] M. Aslam, Radar data analysis in the presence of uncertainty, Eur. J. Remote Sens. 54 (1) (2021) 140–144.

[36] M. Aslam, Chi-square test under indeterminacy: an application using pulse count data, BMC Med. Res. Methodol. 21 (1) (2021) 1–5.

[37] M. Aslam, On testing autocorrelation in metrology data under indeterminacy, MAPAN 36 (3) (2021) 515–519.

[38] M. Aslam, Neutrosophic statistical test for counts in climatology, Sci. Rep. 11 (1) (2021) 1–5.

[39] M. Aslam, A. Shafqat, M. Albassam, J.C. Malela-Majika, S.C. Shongwe, A new CUSUM control chart under uncertainty with applications in petroleum and meteorology, PLoS One 16 (2) (2021) e0246185.

[40] M. Aslam, R.A.K. Sherwani, M. Saleem, Vague data analysis using neutrosophic Jarque–Bera test, PLoS One 16 (12) (2021) e0260689.

[41] M. Aslam, N. Khan, Normality test of temperature in Jeddah city using Cochran's test under indeterminacy, MAPAN 36 (3) (2021) 589–598.

[42] M. Aslam, M. Saleem, Radar circular data analysis using a new Watson's goodness of test under complexity, J. Sens. 2021 (2021).

[43] M. Aslam, M. Albassam, Monitoring road accidents and injuries using variance chart under resampling and having indeterminacy, Int. J. Env. Res. Public Health 18 (10) (2021) 5247.

[44] M. Albassam, M. Aslam, Testing internal quality control of clinical laboratory data using paired-test under uncertainty, BioMed Res. Int. (2021). Available from: https://doi.org/10.1155/2021/5527845.

[45] O.H. Arif, M. Aslam, A new sudden death chart for the Weibull distribution under complexity, Complex Intell. Syst. 7 (4) (2021) 2093–2101.

[46] A.M. Almarashi, M. Aslam, Correlated proportions test under indeterminacy, J. Math. 2021 (2021).

[47] G.S. Rao, M. Aslam, Inspection plan for COVID-19 patients for Weibull distribution using repetitive sampling under indeterminacy, BMC Med. Res. Methodol. 21 (1) (2021) 1–15.

[48] A. Chakraborty, S.P. Mondal, S. Alam, A. Dey, Classification of trapezoidal bipolar neutrosophic number, de-bipolarization technique and its execution in cloud service-based MCGDM problem, Complex Intell. Syst. 7 (1) (2021) 145–162.

[49] Z. Tahir, H. Khan, M. Aslam, J. Shabbir, Y. Mahmood, F. Smarandache, Neutrosophic ratio-type estimators for estimating the population mean, Complex Intell. Syst. (2021) 1–11.

[50] G.K. Vishwakarma, A. Singh, Generalized estimator for computation of population mean under neutrosophic ranked set technique: an application to solar energy data, Comput. Appl. Math. 41 (4) (2022) 1–29.

[51] M. Almaraashi, Short-term prediction of solar energy in Saudi Arabia using automated-design fuzzy logic systems, PLoS One 12 (8) (2017) e0182429.

[52] G.K. Vishwakarma, A. Singh, N. Singh, Calibration under measurement errors, J. King Saud Univ. Sci. 32 (7) (2020) 2950–2961.

Chapter 18

Effectiveness on impact of COVID vaccines on correlation coefficients of pentapartitioned neutrosophic pythagorean statistics

R. Radha[1], A. Stanis Arul Mary[1], Said Broumi[2,3], S. Jafari[4] and S.A. Edalatpanah[5]

[1]Department of Mathematics, Nirmala College for Women, Coimbatore, Tamil Nadu, India, [2]Laboratory of Information Processing, Faculty of Science, Ben M'Sik, University Hassan II, Casablanca, Morocco, [3]Regional Center for the Professions of Education and Training (C.R.M.E.F), Casablanca-Settat, Morocco, [4]Department of Mathematics, College of Vestsjaelland South, Slagelse, Denmark, [5]Department of Applied Mathematics, Ayandegan Institute of Higher Education, Tonekabon, Iran

18.1 Introduction and motivation

A coronavirus is a kind of common virus that causes an infection in your nose, sinuses, or upper throat. Coronaviruses are named for their appearance: "corona" means "crown." The virus's outer layers are covered with spike proteins that surround them like a crown. The main symptoms include fever, coughing, shortness of breath, trouble in breathing, fatigue, chills, sometimes shaking, body aches, headache, sore throat, congestion/runny nose, loss of smell or taste, nausea, and diarrhea.

The virus can lead to pneumonia, respiratory failure, heart problems, liver problems, septic shock, and death. Many COVID-19 complications may be caused by a condition known as cytokine release syndrome or a cytokine storm. This is when an infection triggers your immune system to flood your bloodstream with inflammatory proteins called cytokines. They can kill tissue and damage your organs. In some cases, lung transplants have been needed.

The global outbreak of the COVID-19 pandemic has spread worldwide, affecting almost all countries and territories. The outbreak was first identified

Cognitive Intelligence with Neutrosophic Statistics in Bioinformatics.
DOI: https://doi.org/10.1016/B978-0-323-99456-9.00001-5

in December 2019 in Wuhan, China. Countries around the world cautioned the public to take responsive care. The public care strategies have included handwashing, wearing face masks, physical distancing, and avoiding mass gatherings and assemblies. Lockdown and staying home strategies have been put in place as the needed action to flatten the curve and control the transmission of the disease.

Covaxin is India's first indigenous COVID-19 vaccine, manufactured by Hyderabad-based Bharat Biotech in collaboration with the Indian Council of Medical Research (ICMR), National Institute of Virology (NIIV). It uses an inactivated form of the coronavirus that destroys the ability of the virus to multiply in the human body and increases the immunity system to fight the virus. Covishield vaccine has been developed by AstraZeneca with Oxford University in the UK and is being manufactured by the Serum Institute of India (SII) in Pune, and the approval was granted by Drugs Controller General of India (DCGI).

The Russian COVID-19 vaccine Sputnik V (Gam-COVID-Vac) is an adenoviral-based, two-part vaccine against the SARS-CoV-2 coronavirus. Initially produced in Russia, Sputnik V uses a weakened virus to deliver small parts of a pathogen and stimulate an immune response.

There are so many vaccines now under development, and each of them is being given at slightly different dosing schedules. Most of the vaccines that are being developed need at least two doses, but there are some single-dose vaccine candidates as well. The interval between the doses depends on which vaccine you are getting, and the local authorities and the government would have made a guideline and would inform you about when the second dose is due. Most of those vaccines are currently being given between 3 and 4 weeks between the first and the second dose. But there is some data from some vaccines like the AstraZeneca vaccine, where delaying the second dose up to 12 weeks actually gives a better immune boost. Now, in terms of missing the second dose or being delayed, it is important to get the second dose if the vaccine is a two-dose schedule. It does not matter if it is early by a few days or late by a few days or even a couple of weeks. It is important to go back and get that second dose because the first dose actually presents this new antigen to the immune system to prime it. The second dose is the one that really gives a boost to the immune system so that the antibody response and T cell–mediated response are very strong and they also develop a memory response, which then lasts for a long time, so that when the body sees this antigen again, this virus protein again knows that it needs to react quickly. The vaccine is used for preventing COVID-19 infection. Moreover, even if you get infected, the vaccine may help to reduce the severity of that infection. Additionally, take all the necessary precautions to ensure your safety, such as wearing a face mask, avoiding crowded places, washing your hands or using a hand sanitizer at regular intervals, and maintaining social distancing whenever possible.

Zadeh [1] introduced the idea of fuzzy sets in 1965 that permits the membership to perform valued within the interval [0,1], and set theory is an extension of classical pure mathematics. Intuitionistic fuzzy set was first introduced by K. T. Atanassov [2] in 1983. After that, he introduced the concept of intuitionistic sets as a generalization of fuzzy sets.

Florentin Smarandache [3] introduced the idea of the neutrosophic set in 1995 that provides the information on neutral thought by introducing the new issue referred to as uncertainty within the set. Thus neutrosophic set was framed, and it includes the parts of truth membership function (T), indeterminacy membership function (I), and falsity membership function (F) severally. Neutrosophic sets deal with non-normal interval of $]-0\ 1+[$. Since the neutrosophic set deals the indeterminateness effectively, it plays a very important role in several applications areas embracing info technology, decision web, electronic database systems, diagnosis, multicriteria higher cognitive process issues, etc. Pentapartitioned neutrosophic set and its properties were introduced by Rama Malik and Surpati Pramanik [4]. In this case, indeterminacy is divided into three components: contradiction, ignorance, and an unknown membership function. Further, R. Radha and A. Stanis Arul Mary [5] outlined a brand new hybrid model of pentapartitioned neutrosophic Pythagorean sets (PNPSs) and quadripartitioned neutrosophic Pythagorean sets in 2021. The pentapartitioned neutrosophic Pythagorean topological spaces [6] were introduced, and their properties were investigated in 2021.

Correlation coefficients are beneficial tools used to determine the degree of similarity between objects. The importance of correlation coefficients in fuzzy environments lies in the fact that these types of tools can feasibly be applied to problems of pattern recognition, MADM, medical diagnosis, and clustering. In other researches, Ye [7] proposed the three vector similarity measures for single-valued neutrosophic sets and interval-valued neutrosophic set including Jaccard, [8] Dice, and cosine similarity measure and applied them to multicriteria decision-making problems with simplified neutrosophic information. Hanafy et al. introduced and studied the concepts of correlation and correlation coefficients of neutrosophic set.

In this paper, we have defined the concept of interval-valued pentapartitioned neutrosophic Pythagorean sets, and its properties were discussed. We have applied the concepts of correlation coefficients to pentapartitioned neutrosophic Pythagorean and interval-valued pentapartitioned neutrosophic Pythagorean sets, and some of its properties are obtained. The decision-making problem can be studied with a COVID injection.

18.2 Background, definitions, and notations

18.2.1 Definition [9]

Let X be a universe. A neutrosophic set A on X can be defined as follows:

$$A = \{ < x, T_A(x), I_A(x), F_A(x) > : x \in X\}$$

where

$$T_A, I_A, F_A: U \rightarrow [0,1] \quad \text{and} \quad 0 \leq T_A(x) + I_A(x) + F_A(x) \leq 3$$

Here, $T_A(x)$ is the degree of membership, $I_A(x)$ is the degree of indeterminacy, and $F_A(x)$ is the degree of nonmembership.

18.2.2 Definition [10]

Let X be a universe. A pentapartitioned neutrosophic Pythagorean [PNP] set A with T, F, C, and U as dependent neutrosophic components and I as an independent component for A on X is an object of the form

$$A = \{< x, T_A, C_A, I_A, U_A, F_A >: x \in X\}$$

where

$$T_A + F_A \leq 1, C_A + U_A \leq 1 \quad \text{and}$$

$$(T_A)^2 + (C_A)^2 + (I_A)^2 + (U_A)^2 + (F_A)^2 \leq 3$$

Here, $T_A(x)$ is the truth membership, $C_A(x)$ is contradiction membership, $U_A(x)$ is ignorance membership, $F_A(x)$ is the false membership, and $I_A(x)$ is an unknown membership.

18.2.3 Definition [11]

Let P be a nonempty set. A pentapartitioned neutrosophic set A over P characterizes each element p in P a truth membership function T_A, a contradiction membership function C_A, an ignorance membership function G_A, unknown membership function U_A, and a false membership function F_A, such that for each p in P

$$T_A + C_A + G_A + U_A + F_A \leq 5$$

18.2.4 Definition [10]

The complement of a pentapartitioned neutrosophic Pythagorean set A on R is denoted by A^C or A^* and is defined as

$$AC = \{< x, F_A(x), U_A(x), 1 - G_A(x), C_A(x), T_A(x) >: x \in X\}$$

18.2.5 Definition [10]

Let $A = < x, T_A(x), C_A(x), G_A(x), U_A(x), F_A(x) >$ and
$B = < x, T_B(x), C_B(x), G_B(x), U_B(x), F_B(x) >$ are pentapartitioned neutrosophic Pythagorean sets. Then

$$A \cup B = < x, \max(T_A(x), T_B(x)), \max(C_A(x), C_B(x)), \min(G_A(x), G_B(x)), \\ \min(U_A(x), U_B(x)), \min(F_A(x), F_B(x)), >$$

$$A \cap B = < x, \min(T_A(x), T_B(x)), \min(C_A(x), C_B(x)), \max(G_A(x), G_B(x)),$$
$$\max(U_A(x), U_B(x)), \max(F_A(x), F_B(x)) >$$

18.2.6 Definition [3]

A PNP topology on a nonempty set R is a family of a PNP sets in R satisfying the following axioms

1) $0, 1 \in \tau$
2) $R_1 \cap R_2 \in \tau$ for any $R_1, \quad R_2 \in \tau$
3) $\bigcup R_i \in \tau$ for any $R_i{:}i \in I \subseteq \tau$

The complement R^* of PNP open set (PNPOS, in short) in PNP topological space [PNPTS] (R,τ) is called a PNP closed set [PNPCS].

18.2.7 Definition [2]

Let (X, τ) be a PNPTS and $A = < x, T_A(x), C_A(x), G_A(x), U_A(x), F_A(x) >$ be a PNPS in X. Then, the interior and the closure of A are denoted by PNPInt(A) and PNPCl(A) and are defined as follows.

$\text{PNPCl}(A) = \cap \{K|K \text{ is a PNPCS and } A \subseteq K\}$ and

$\text{PNPInt}(A) = \cup \{G|G \text{ is a PNPOS and } G \subseteq A\}$

Also, it can be established that PNPCl(A) is a PNPCS and PNPInt(A) is a PNPOS, A is a PNPCS if and only if PNPCl(A) = A and A is a PNPOS if and only if PNPInt(A) = A. We say that A is PNP-dense if PNPCl(A) = X.

18.3 Correlation measure of pentapartitioned neutrosophic pythagorean sets

18.3.1 Definition

Let R be a nonempty set and I be the unit interval [0,1]. A pentapartitioned neutrosophic Pythagorean set with T, C, U, and F as dependent neutrosophic components (PNP) P and Q of the form

$$A = \{(x, TR_A(X), CR_A(X), GR_A(X), UR_A(X), FR_A(X)) : x \in X\} \text{ and}$$
$$B = \{(x, TR_B(X), CR_B(X), GR_B(X), UR_B(X), FR_B(X)) : x \in X\}.$$

Then, the correlation coefficient of A and B

$$\rho(A, B) = \frac{K(A, B)}{\sqrt{K(A, A).K(B, B)}}$$

$$K(A,B) = \sum_{i=1}^{n}(((TR_A(x_i))^2.((TR_B(x_i))^2 + ((CR_A(x_i))^2.((CR_B(x_i))^2 + ((GR_A(x_i))^2.((GR_B(x_i))^2 + ((UR_A(x_i))^2.((UR_B(x_i))^2 + ((FR_A(x_i))^2.((FR_B(x_i))^2)$$

$$K(A,A) = \sum_{i=1}^{n}(((TR_A(x_i))^2.((TR_A(x_i))^2 + ((CR_A(x_i))^2.((CR_A(x_i))^2 + ((GR_A(x_i))^2.((GR_A(x_i))^2 + ((UR_A(x_i))^2.((UR_A(x_i))^2 + ((FR_A(x_i))^2.((FR_A(x_i))^2)$$

$$K(A,B) = \sum_{i=1}^{n}(((TR_B(x_i))^2.((TR_B(x_i))^2 + ((CR_B(x_i))^2.((CR_B(x_i))^2 + ((GR_B(x_i))^2.((GR_B(x_i))^2 + ((UR_B(x_i))^2.((UR_B(x_i))^2 + ((FR_B(x_i))^2.((FR_B(x_i))^2)$$

18.3.2 Theorem

The defined correlation measure between PNP set A and B satisfies the following properties.

1) $0 \leq \rho(A,B) \leq 1$
2) $\rho(A,B) = 1$ if and only if $A = B$
3) $\rho(A,B) = \rho(B, A)$.

Proof:

(1) $0 \leq \rho(A,B) \leq 1$

As the truth membership, contradiction membership, ignorance membership, false membership, and unknown membership of the PNP set lie between 0 and 1, $\rho(A,B)$ also lies between 0 and 1.

We will prove

$$K(A,B) = \sum_{i=1}^{n}(((TR_A(x_i))^2.((TR_B(x_i))^2 + ((CR_A(x_i))^2.((CR_B(x_i))^2 + ((GR_A(x_i))^2.((GR_B(x_i))^2 + ((UR_A(x_i))^2.((UR_B(x_i))^2 + ((FR_A(x_i))^2.((FR_B(x_i))^2)$$
$$= ((TR_B(x_1))^2.((TR_B(x_1))^2 + ((CR_A(x_1))^2.((CR_B(x_1))^2 + ((GR_A(x_1))^2.((GR_B(x_1))^2 + ((UR_A(x_1))^2.((UR_B(x_1))^2 + ((FR_A(x_1))^2.((FR_B(x_1))^2) + ((TR_B(x_2))^2.((TR_B(x_2))^2 + ((CR_A(x_2))^2.((CR_B(x_2))^2 + ((GR_A(x_2))^2.((GR_B(x_2))^2 + ((UR_A(x_2))^2.((UR_B(x_2))^2 + ((FR_A(x_2))^2.((FR_B(x_2))^2) + \ldots + ((TR_B(x_n))^2.((TR_B(x_n))^2 + ((CR_A(x_n))^2.((CR_B(x_n))^2 + ((GR_A(x_n))^2.((GR_B(x_n))^2 + ((UR_A(x_n))^2.((UR_B(x_n))^2 + ((FR_A(x_n))^2.((FR_B(x_n))^2)$$

By Cauchy–Schwarz inequality, $(x_1y_1+x_2y_2+\ \ldots\ +x_ny_n)^2 \leq (x_1^2 + x_2^2 + \ \ldots + x_n^2).(y_1^2 + y_2^2 + \ \ldots\ + y_n^2)$, where $(x_1 + x_2 + \ \ldots\ + x_n) \in R^n$ and $(y_1 + y_2 + \ \ldots + y_n) \in R^n$, we get

$$\left(K\{A,B)\right)^2 = ((TR_A(x_1))^4 + (CR_A(x_1))^4 + (GR_A(x_1))^4 + (UR_A(x_1))^4 + (FR_A(x_1))^4 + \left((TR_A(x_2))^4 + (CR_A(x_2))^4 + (GR_A(x_2))^4 + (UR_A(x_2))^4 + (FR_A(x_2))^4 + \left((TR_A(x_3))^4 + CR_A(x_3)\right)^4 + (GR_A(x_3))^4 + (UR_A(x_3))^4 + (FR_A(x_3))^4 + \ldots + \left((TR_A(x_n))^4 + CR_A(x_n)\right)^4 + GR_A(x_n))^4$$

$$+ (UR_A(n))^4 + (FR_A(x_n))^4.\left(\left((TR_B(x_1))^4 + (CR_B(x_1))^4 + (GR_B(x_1))^4 + (UR_B(x_1))^4 + (FR_B(x_1))^4 + \left((TR_B(x_2))^4 + (CR_B(x_2))^4 + (GR_B(x_2))^4 + (UR_B(x_2))^4 + (FR_B(x_2))^4 + \left((TR_B(x_3))^4 + CR_B(x_3)\right)^4 + (GR_B(x_3))^4 + (UR_B(x_3))^4 + (FR_B(x_3))^4 + \ldots. + \left((TR_B(x_n))^4 + CR_B(x_n)\right)^4 + GR_B(x_n))^4 + (UR_B(x_n))^4 + (FR_B(x_n))^4$$

$$= (TR_A(x_1))^2.((TR_A(x_1))^2 + ((CR_A(x_1))^2.((CR_A(x_1))^2 + ((GR_A(x_1))^2.((GR_A(x_1))^2 + ((UR_A(x_1))^2.((UR_A(x_1))^2 + ((FR_A(x_1))^2.\left((FR_A(x_1))^2\right) + (TR_A(x_2))^2.((TR_A(x_2))^2 + \left((CR_A(x_2))^2.\left((CR_A(x_2))^2 + ((GR_A(x_2))^2.((GR_A(x_2))^2 + ((UR_A(x_2))^2.((UR_A(x_2))^2 + ((FR_A(x_2))^2.((FR_A(x_2))^2) + \ldots + (TR_A(x_n))^2.((TR_A(x_n))^2 + ((CR_A(x_n))^2.((CR_A(x_n))^2 + ((GR_A(x_n))^2.((GR_A(x_n))^2 + ((UR_A(x_n))^2.((UR_A(x_n))^2 + ((FR_A(x_n))^2.((FR_A(x_n))^2).((TR_B(x_1))^2.((TR_B(x_1))^2 + ((CR_B(x_1))^2.((CR_B(x_1))^2 + ((GR_B(x_1))^2.((GR_B(x_1))^2 + ((UR_B(x_1))^2.((UR_B(x_1))^2 + ((FR_A(x_B))^2.((FR_B(x_1))^2) + (TR_B(x_2))^2.((TR_B(x_2))^2 + ((CR_B(x_2))^2.((CR_B(x_2))^2 + ((GR_B(x_2))^2.((GR_B(x_2))^2 + (UR_B(x_2))^2.((UR_B(x_2))^2 + \left((FR_B(x_2))^2.\left((FR_B(x_2))^2\right) + \ldots. + (TR_B(x_n))^2.((TR_B(x_n))^2 + ((CR_B(x_n))^2.\left((CR_B(x_n))^2 + ((GR_B(x_n))^2.\left((GR_B(x_n))^2 + ((UR_B(x_n))^2.((UR_B(x_n))^2 + ((FR_B(x_n))^2.\left((FR_B(x_n))^2\right) = (A,A).K(B,B).$$

Therefore $(K(A, B))^2 \leq K(A, A).\ K(A, B)$ and thus $\rho(A,B) \leq 1$.

Hence, we obtain the following property $0 \leq \rho(A,B) \leq 1$

(2) $\rho(A,B) = 1$ if and only if $A = B$

Let the two PNP sets A and B be equal (i.e.,) $A = B$.

Hence for any $TR_A(x_i) = TR_B(x_i)$, $CR_A(x_i) = CR_B(x_i)$, $GR_A(x_i) = GR_B(x_i)$, $UR_A(x_i) = UR_B(x_i)$ and $FR_A(x_i) = FR_B(x_i)$

Then $K(A, A) = K(B, B) = \sum_{i=1}^{n} (((TR_A(x_i))^2.\ ((TR_A\ (x_i))^2 + ((CR_A(x_i))^2.\ ((CR_A(x_i))^2 + ((GR_A(x_i))^2.\ ((GR_A(x_i))^2 + ((UR_A(x_i))^2.((UR_A(x_i))^2 + ((FR_A(x_i))^2.((FR_A(x_i))^2)$ And

$$K(A,B) = \sum_{i=1}^{n}(((TR_A(x_i))^2.((TR_B(x_i))^2 + ((CR_A(x_i)\,)^2.((CR_B(x_i))^2 + ((GR_A(x_i))^2.((GR_B(x_i))^2 + ((UR_A(x_i))^2.((UR_B(x_i))^2 + ((FR_A(x_i))^2.((FR_B(x_i))^2)$$

$$K(A,B)=\sum_{i=1}^{n}(((TR_A(x_i))^2.((TR_B(x_i))^2 + ((CR_A(x_i))^2.((CR_B(x_i))^2 + ((GR_A(x_i))^2.((GR_B(x_i))^2 + ((UR_A(x_i))^2.((UR_B(x_i))^2 + ((FR_A(x_i))^2.((FR_B(x_i))^2)$$

Hence

$$\rho(A,B)=\frac{K(A,B)}{\sqrt{K(A,A).K(B,B)}} = \frac{K(A,A)}{\sqrt{K(A,A).K(A,A)}} = 1$$

Let $\rho(A,B)=1$. Then, the unit measure is possible only if

$$\frac{K(A,B)}{\sqrt{K(A,A).K(B,B)}} = 1$$

This refers that $TR_A(x_i)=TR_B(x_i)$, $CR_A(x_i)=CR_B(x_i)$, $GR_A(x_i)=GR_B(x_i)$, $UR_A(x_i)=UR_B(x_i)$ and $FR_A(x_i)=FR_B(x_i)$ for all i. Hence $A=B$.

(3) If $\rho(A,B)=\rho(B,A)$, it is obvious that

$$\rho(A,B)=\frac{K(A,B)}{\sqrt{K(A,A).K(B,B)}} = \frac{K(B,A)}{\sqrt{K(B,B).K(A,A)}} = \rho(B,A).$$

$$K(A,B)=\sum_{(i=1)}^{n}(((TR_A(x_i))^2.((TR_B(x_i))^2 + ((CR_A(x_i))^2.((CR_B(x_i))^2 + ((GR_A(x_i))^2.((GR_B(x_i))^2 + ((UR_A(x_i))^2.((UR_B(x_i))^2 + ((FR_A(x_i))^2.((FR_B(x_i))^2)$$

$$=\sum_{(i=1)}^{n}(((TR_B(x_i))^2.((TR_A(x_i))^2 + ((CR_B(x_i))^2.((CR_A(x_i))^2 + ((GR_B(x_i))^2.((GR_A(x_i))^2 + ((UR_B(x_i))^2.((UR_A(x_i))^2 + ((FR_B(x_i))^2.((FR_A(x_i))^2)$$

$$=K(B,A)$$

Hence the proof.

18.3.3 Definition

Let A and B be two PNP sets, then the correlation coefficient is defined as

$$\rho' = \frac{K(A,B)}{\max\{K(A,A),K(B,B)\}}$$

18.3.4 Theorem

The defined correlation measure between PNP sets A and B satisfies the following properties

1) $0 \leq \rho'(A,B) \leq 1$
2) $\rho'(A,B) = 1$ if and only if $A = B$
3) $\rho'(A,B) = \rho'(B,A)$.

Proof:

The property (1) and (2) is straightforward, so omit them here. Also $\rho'(A,B) \geq 0$ is evident. We now prove only

$$\rho'(A,B) \leq 1.$$

From the above theorem, we have $(K(A, B))^2 \leq K(A, A). K(B, B)$.

Therefore, $K(A, B) \leq \max \{K(A, A). K(B.B)\}$ and thus $\rho'(A,B) \leq 1$.

However, in many practical situations, the different sets may have taken different weights, and thus, weight w_i of the element x_i $x_i \in R(i = 1,2...n)$ should be taken into account. In the following, we develop a weighted correlation coefficient between PNP sets. Let $\omega = \{\omega_1, \omega_2, \ldots, \omega_n\}$ be the weight vector of the elements $x_i(i = 1, 2 \ldots, n)$ with $\omega_i \geq 0$ and $\sum_{i=1}^{n} \omega_i = 1$, then we have extended the above correlation coefficient $\rho(A,B)$ and $\rho'(A,B)$ to weighted correlation coefficients as follows.

$$\rho'' = \frac{K_\omega(A,B)}{\sqrt{K_\omega(A,A).K_\omega(A,A)}}$$

$$\begin{aligned}K_\omega(A,B) = \sum_{i=1}^{n} \omega_i(((TR_A(x_i))^2.((TR_B(x_i))^2 + ((CR_A(x_i))^2.((CR_B(x_i))^2 \\ + ((GR_A(x_i))^2.((GR_B(x_i))^2 \\ + ((UR_A(x_i))^2.((UR_B(x_i))^2 + ((FR_A(x_i))^2.((FR_B(x_i))^2)\end{aligned}$$

$$\begin{aligned}K_\omega(\mathrm{A},\mathrm{A}) = \sum_{i=1}^{n} \omega_i(((TR_A(x_i))^2.((TR_A(x_i))^2 + ((CR_A(x_i))^2.((CR_A(x_i))^2 \\ + ((GR_A(x_i))^2.((GR_A(x_i))^2 \\ + ((UR_A(x_i))^2.((UR_A(x_i))^2 + ((FR_A(x_i))^2.((FR_A(x_i))^2)\end{aligned}$$

$$K_\omega(A,B) = \sum_{i=1}^{n} \omega_i(((TR_B(x_i))^2.((TR_B(x_i))^2 + ((CR_B(x_i))^2.((CR_B(x_i))^2 + ((GR_B(x_i))^2.((GR_B(x_i))^2 + ((UR_B(x_i))^2.((UR_B(x_i))^2 + ((FR_B(x_i))^2.((FR_B(x_i))^2)$$

$$\rho''' = \frac{K_\omega(A,B)}{\max\{K_\omega(A,A), K_\omega(B,B)\}}$$

It can be easy to verify that if $\omega = \left(\frac{1}{n}, \frac{1}{n}, \ldots, \frac{1}{n}\right)^T$, then weighted correlation coefficient becomes correlation coefficient respectively.

18.3.5 Theorem

Let $\omega = (\omega_1, \omega_2, \ldots, \omega_n)^T$ be the weight vector of $x_i (i = 1,2,.n)$ with $\omega_i \geq 0$ and $\sum_{i=1}^{n} \omega_i = 1$, then the weighted correlation coefficient between PNP set A and B satisfies the following properties.

1) $0 \leq \rho'''(A,B) \leq 1$
2) $\rho'''(A,B) = 1$ if and only if $A = B$
3) $\rho'''(A,B) = \rho'''(B,A)$.

Proof:

$$\omega_1((TR_B(x_1))^2.((TR_B(x_1))^2 + ((CR_A(x_1))^2.((CR_B(x_1))^2 + ((GR_A(x_1))^2.((GR_B(x_1))^2 + ((UR_A(x_1))^2.((UR_B(x_1))^2 + ((FR_A(x_1))^2.((FR_B(x_1))^2) + \omega_2((TR_B(x_2))^2.((TR_B(x_2))^2 + ((CR_A(x_2))^2.((CR_B(x_2))^2 + ((GR_A(x_2))^2.((GR_B(x_2))^2 + ((UR_A(x_2))^2. ((UR_B(x_2))^2 + ((FR_A(x_2))^2.((FR_B(x_2))^2) + \ldots + \omega_n((TR_B(x_n))^2.((TR_B(x_n))^2 + ((CR_A(x_n))^2.((CR_B(x_n))^2 + ((GR_A(x_n))^2.((GR_B(x_n))^2 + ((UR_A(x_n))^2.((UR_B(x_n))^2 + ((FR_A(x_n))^2.((FR_B(x_n))^2)$$

$$\sum_{i=1}^{n} \omega_i\big(((TR_A(x_i))^2.\big((TR_B(x_i))^2 + \big((CR_A(x_i))^2.((CR_B(x_i))^2 + ((GR_A(x_i))^2.\big((GR_B(x_i))^2 + ((UR_A(x_i))^2.((UR_B(x_i))^2 + \big((FR_A(x_i))^2.\big((FR_B(x_i))^2\big)$$

$$= \omega_1((TR_B(x_1))^2.((TR_B(x_1))^2 + ((CR_A(x_1))^2.((CR_B(x_1))^2 + ((GR_A(x_1))^2.((GR_B(x_1))^2 + ((UR_A(x_1))^2.((UR_B(x_1))^2 + ((FR_A(x_1))^2.((FR_B(x_1))^2) + \omega_2$$

$$((TR_B(x_2))^2.((TR_B(x_2))^2 + ((CR_A(x_2))^2.((CR_B(x_2))^2$$
$$+ ((GR_A(x_2))^2.((GR_B(x_2))^2$$
$$+ ((UR_A(x_2))^2.((UR_B(x_2))^2 + ((FR_A(x_2))^2.\left((FR_B(x_2))^2\right) + \ldots$$
$$+ \omega_n((TR_B(x_n))^2.((TR_B(x_n))^2 + ((CR_A(x_n))^2.((CR_B(x_n))^2$$
$$+ ((GR_A(x_n))^2.((GR_B(x_n))^2$$
$$+ ((UR_A(x_n))^2.((UR_B(x_n))^2 + ((FR_A(x_n))^2.\left((FR_B(x_n))^2\right)$$
$$= (\sqrt{\omega_1}(TR_B(x_1))^2.(\sqrt{\omega_1}(TR_B(x_1))^2 + (\sqrt{\omega_1}(CR_A(x_1))^2.(\sqrt{\omega_1}(CR_B(x_1))^2$$
$$+ (\sqrt{\omega_1}(GR_A(x_1))^2.(\sqrt{\omega_1}(GR_B(x_1))^2 + (\sqrt{\omega_1}(UR_A(x_1))^2.(\sqrt{\omega_1}(UR_B(x_1))^2$$
$$+ (\sqrt{\omega_1}(FR_A(x_1))^2.(\sqrt{\omega_1}(FR_B(x_1))^2) + (\sqrt{\omega_2}(TR_B(x_2))^2.(\sqrt{\omega_2}(TR_B(x_2))^2$$
$$+ (\sqrt{\omega_2}(CR_A(x_2))^2.(\sqrt{\omega_2}(CR_B(x_2))^2 + (\sqrt{\omega_2}(GR_A(x_2))^2.(\sqrt{\omega_2}(GR_B(x_2))^2$$
$$+ (\sqrt{\omega_2}(UR_A(x_2))^2.(\sqrt{\omega_2}(UR_B(x_2))^2 + (\sqrt{\omega_2}(FR_A(x_2))^2.(\sqrt{\omega_2}(FR_B(x_2))^2)$$
$$+ \ldots + (\sqrt{\omega_n}(TR_B(x_n))^2.(\sqrt{\omega_n}(TR_B(x_n))^2 + (\sqrt{\omega_n}(CR_A(x_n))^2.(\sqrt{\omega_n}(CR_B(x_n))^2$$
$$+ (\sqrt{\omega_n}(GR_A(x_n))^2.(\sqrt{\omega_n}(GR_B(x_n))^2 + (\sqrt{\omega_n}(UR_A(x_n))^2.(\sqrt{\omega_n}(UR_B(x_n))^2$$
$$+ (\sqrt{\omega_n}(FR_A(x_n))^2.\left(\sqrt{\omega_n}(FR_B(x_n))^2\right)$$

By using Cauchy–Schwarz inequality, we get

$$(K_\omega(A,B))^2 \le (\omega_1((TR_B(x_1))^2.((TR_B(x_1))^2 + ((CR_A(x_1))^2.((CR_B(x_1))^2$$
$$+ ((GR_A(x_1))^2.((GR_B(x_1))^2$$
$$+ ((UR_A(x_1))^2.((UR_B(x_1))^2 + ((FR_A(x_1))^2.((FR_B(x_1))^2)$$
$$+ \omega_2((TR_B(x_2))^2.((TR_B(x_2))^2$$
$$+ ((CR_A(x_2))^2.((CR_B(x_2))^2 + ((GR_A(x_2))^2.((GR_B(x_2))^2$$
$$+ ((UR_A(x_2))^2.((UR_B(x_2))^2$$
$$+ ((FR_A(x_2))^2.\left((FR_B(x_2))^2\right) + \ldots + \omega_n((TR_B(x_n))^2.((TR_B(x_n))^2$$
$$+ ((CR_A(x_n))^2.((CR_B(x_n))^2$$
$$+ ((GR_A(x_n))^2.((GR_B(x_n))^2 + ((UR_A(x_n))^2.\left((UR_B(x_n))^2\right.$$

$$+\left((FR_A(x_n))^2.\left((FR_B(x_n))^2\right)\right)$$

$$=\sum_{i=1}^{N}\omega_i(((TR_A(x_i))^2.((TR_A(x_i))^2+((CR_A(x_i))^2.((CR_A(x_i))^2$$

$$+((GR_A(x_i))^2.((GR_A(x_i))^2$$

$$+((UR_A(x_i))^2.((UR_A(x_i))^2+((FR_A(x_i))^2.\left((FR_A(x_i))^2\right).$$

$$\sum_{i=1}^{n}\omega_i(((TR_B(x_i))^2.((TR_B(x_i))^2$$

$$+((CR_B(x_i))^2.((CR_B(x_i))^2+((GR_B(x_i))^2.\left((GR_B(x_i))^2+(UR_B(x_i))^2.(UR_B(x_i))^2\right.$$

$$+\left((FR_B(x_i))^2.\left((FR_B(x_i))^2\right)\right)$$

$$=K_\omega(A,A).\ K_\omega(B,B)$$

Therefore,
$(K_\omega(A,B))^2\le\sqrt{K_\omega(A,A)K_\omega(B,B)}$ and hence $0\le\rho''(A,B)\le 1$.
Theorem
The correlation coefficient of two PNP sets A and B as defined, that is, $\rho'''(A,B)$, satisfies the same properties as those in Theorem 3.7.

Proof:

The proof of this theorem is similar to that of the theorem.

18.4 Interval-valued pentapartitioned neutrosophic Pythagorean set

18.4.1 Definition

An interval-valued pentapartitioned neutrosophic Pythagorean set A in R is characterized by a truth membership function, a contradiction membership function, an unknown membership function, and an ignorance membership function. For each point k in $R, TR_A=\left[TR_A^L(k),TR_A^U(k)\right], CR_A=[CR_A^L(k),CR_A^U(k)],\ GR_A=[GR_A^L(k),GR_A^U(k)],\ \ UR_A=[UR_A^L(k),UR_A^U(k)]$ and $FR_A=[FR_A^L(k),FR_A^U(k)]$. Therefore an IVPNS can be denoted as

$$\begin{aligned}A&=\{k,T\,R_A^L(k),CR_A^L(k),GR_A^L(k),UR_A^L(k),FR_A^L(k)|k\in R\}\\&=\{(k,\left[TR_A^L(k),TR_A^U(k)\right],\left[CR_A^L(k),CR_A^U(k)\right],GR_A=\left[GR_A^L(k),GR_A^U(k)\right],\\&\quad\left[UR_A^L(k),UR_A^U(k)\right],FR_A^L(k),FR_A^U(k)|k\in R)\}\end{aligned}$$

Then, the sum of TR_A(k), $CR_A(k),GR_A(k),UR_A(k),FR_A(k)$ satisfies the condition

$$0\le(TR_A^L(k))^2+(CR_A^L(k))^2+(GR_A^L(k))^2+(UR_A^L(k))^2+(FR_A^L(k))^2\le 3.$$

The upper and lower ends of interval values of $TR_A^L(k)$, $CR_A^L(k), GR_A^L(k)$, $UR_A^L(k), FR_A^L(k)$ in an interval-valued pentapartitioned neutrosophic Pythagorean sets reduce to pentapartitioned neutrosophic Pythagorean sets.

18.4.2 Example

Let $R=\{a\}$ and $A=\{(a,\ [0.2,0.5],\ [0.3,0.7],\ [0.1,0.4],\ [0.1,0.7],\ [0.2,0.3]\}$ be IVPNPS on R.

Then $\tau=\{0,1,\ A\}$ is a topology on R, and A is an interval-valued pentapartitioned neutrosophic Pythagorean set.

18.4.3 Definition

The complement of an interval-valued pentapartitioned neutrosophic Pythagorean set

$$A=\{(k,\left[TR_A^L(k),TR_A^U(k)\right],\left[CR_A^L(k),CR_A^U(k)\right],\left[GR_A^L(k),GR_A^U(k)\right],$$

$$\left[UR_A^L(k),UR_A^U(k)\right],FR_A^L(k),FR_A^U(k)|k\in R)\}$$

is denoted by A^C and is defined as

$$A^C=\{(k,\left[FR_A^L(k),FR_A^U(k)\right],\left[UR_A^L(k),UR_A^U(k)\right],\left[GR_A^L(k),GR_A^U(k)\right],$$

$$\left[CR_A^L(k),UC(k)\right],TR_A^L(k),TR_A^U(k)|k\in R)\}$$

18.4.4 Example

Let $R=\{a\}$ and $A=\{(a,\ [0.2,0.5],\ [0.3,0.7],\ [0.1,0.4],\ [0.1,0.7],\ [0.2,0.3]\}$ be IVPNPS on R. Then the complement of A on R is

$$A^C=\left\{(a,[0.2,0.3],[0.1,0.7],[0.1,0.4],[0.3,0.7],[0.2,0.5]\right\}$$

18.4.5 Definition

An IVPNPS A is contained in the other IVPNPS B, $A\subseteq B$ if and only if

$TR_A^L(k)\le TR_B^U(k)$, $CR_A^L(k)\le CR_B^U(k)$, $GR_A^L(k)\le GR_B^U(k)$, $UR_A^L(k)\le UR_B^U(k)$ and $FR_A^L(k)\le FR_B^U(k)$ for any k in K.

18.4.6 Example

Let $R=\{a\}$ and $A=\{(a,\ [0.2,0.5],\ [0.3,0.7],\ [0.3,0.4],\ [0.3,0.7],\ [0.2,0.5]\}$ and

$B=\{(a,\ [0.3,0.6],\ [0.5,0.7],\ [0.2,0.4],\ [0.2,0.7],\ [0.1,0.5]\}$ be IVPNPS on R be IVPNPS on R.

18.4.7 Definition

Two IVPNPS A and B are equal, that is, $A = B$ if and only if $A \subseteq B$ and $B \subseteq A$.

18.4.8 Definition

Let $A = \{([TR_A^L, TR_A^U], [CR_A^L, CR_A^U], [GR_A^L, GR_A^U], [UR_A^L, UR_A^U], [FR_A^L, FR_A^U]\}$ and

$B = \{[TR_B^L, TR_B^U], [CR_B^L, CR_B^U], [GR_B^L, GR_B^U], [UR_B^L, UR_B^U], [FR_B^L, FR_B^U]\}$ are two interval-valued pentapartitioned neutrosophic Pythagorean sets. Then, the union and intersection of two interval-valued pentapartitioned neutrosophic Pythagorean sets A and B are denoted by

$$A \cup B = \{[\max(TR_A^L, TR_B^L), \max(TR_A^U, TR_B^U)], [\max(CR_A^L, CR_B^L),$$

$$\max(CR_A^U, CR_B^U)], [\min(GR_A^L, GR_B^L), \min(GR_A^U, GR_B^U)],$$

$$[\min(UR_A^L, UR_B^L), \min(UR_A^U, UR_B^U)], [\min(FR_A^L, FR_B^L), \min(FR_A^U, FR_B^U)]$$

$$A \cap B = \{[\min(TR_A^L, TR_B^L), \min(TR_A^U, TR_B^U)],$$

$$[\min(CR_A^L, CR_B^L), \min(CR_A^U, CR_B^U)], [\max(GR_A^L, GR_B^L), \max(GR_A^U, GR_B^U)],$$

$$[\max(UR_A^L, UR_B^L), \max(UR_A^U, UR_B^U)],$$

$$[\max(FR_A^L, FR_B^L), \max(FR_A^U, FR_B^U)]$$

18.4.9 Example

Let $R = \{a\}$ and $A = \{(a, [0.2,0.5], [0.3,0.7], [0.1,0.4], [0.1,0.7], [0.2,0.3]\}$ and

$B = \{(a, [0.1,0.6], [0.2,0.6], [0.2,0.4], [0.2,0.7], [0.1,0.5]\}$ be IVPNPS on R be IVPNS on R.

Then, the union and intersection of two IVPNPS A and B are denoted by

$$A \cup B = \{(a, [0.2, 0.6], [0.3, 0.7], [0.1, 0.4], [0.1, 0.7], [0.1, 0.3]\}$$

$$A \cap B = \{(a, [0.1, 0.5], [0.2, 0.6], [0.1, 0.4], [0.3, 0.7], [0.2, 0.5]\}$$

18.4.10 Definition

Let R be a nonempty set and I be the unit interval [0,1]. An interval-valued pentapartitioned neutrosophic Pythagorean set [IVPNPS] with T, C, U, and F

as dependent neutrosophic components (PNP) P and Q of the form

$A = \{(x, [TR_A^L(x), TR_A^U(x)], [CR_A^L(x), CR_A^U(x)], [GR_A^L(x), GR_A^U(x)],$

$[UR_A^L(x), UR_A^U(x)], [FR_A^L(x), FR_A^U(x)],):x \in X\}$ and

$B = \{(x, [TR_B^L(x), TR_B^U(x)], [CR_B^L(x), CR_B^U(x)], [GR_B^L(x), GR_B^U(x)],$

$[UR_B^L(x), UR_B^U(x)], [FR_B^L(x), FR_B^U(x)],):x \in X\}$

Then, the correlation coefficient of A and B

$$\gamma(A, B) = \frac{K(A, B)}{\sqrt{K(A, A).K(B, B)}}$$

$K(A,\ B) = \sum_{i=1}^{n}((TR_A^L((x_i))^2.((TR_B^L(x_i))^2 + ((CR_A^L(x_i))^2.((CR_B^L(x_i))^2$

$+ ((GR_A^L(x_i))^2.((GR_B^L(x_i))^2 + ((UR_A^L(x_i))^2.((UR_B^L(x_i))^2 + ((FR_A^L(x_i))^2.((FR_B^L(x_i))^2)$

$+ ((TR_A^U(x_i))^2.((TR_B^L(x_i))^2 + ((CR_A^L(x_i))^2.$

$((CR_B^L(x_i))^2 + ((GR_A^L(x_i))^2.((GR_B^L(x_i))^2$

$+ ((UR_A^L(x_i))^2.((UR_B^L(x_i))^2 + ((FR_A^L(x_i))^2.((FR_B^L(x_i))^2)$

$K(A, A) = \sum_{i=1}^{n}((TR_A^L((x_i))^2.((TR_A^L(x_i))^2 + ((CR_A^L(x_i))^2.((CR_A^L(x_i))^2$

$+ ((GR_A^L(x_i))^2.((GR_A^L(x_i))^2$

$+ ((UR_A^L(x_i))^2.((UR_A^L(x_i))^2 + ((FR_A^L(x_i))^2.((FR_A^L(x_i))^2) + ((TR_A^U(x_i))^2.$

$((TR_A^L(x_i))^2 + ((CR_A^L(x_i))^2.$

$((CR_A^L(x_i))^2 + ((GR_A^L(x_i))^2.((GR_A^L(x_i))^2$

$+ ((UR_A^L(x_i))^2.((UR_A^L(x_i))^2 + ((FR_A^L(x_i))^2.((FR_A^L(x_i))^2)$

$K(B,\ B) = \sum_{i=1}^{n}((TR_B^L((x_i))^2.((TR_B^L(x_i))^2 + ((CR_B^L(x_i))^2.((CR_B^L(x_i))^2$

$+ ((GR_B^L(x_i))^2.((GR_B^L(x_i))^2 +((UR_B^L(x_i))^2.((UR_B^L(x_i))^2 + ((FR_B^L(x_i))^2.((FR_B^L(x_i))^2)$

$+ ((TR_B^U(x_i))^2.((TR_B^L(x_i))^2 + ((CR_B^L(x_i))^2.\ ((CR_B^L(x_i))^2 + ((GR_B^L(x_i))^2.((GR_B^L(x_i))^2$

$+ ((UR_B^L(x_i))^2.((UR_B^L(x_i))^2 + ((FR_B^L(x_i))^2.((FR_B^L(x_i))^2)$

18.4.11 Theorem

The defined correlation measure between IVPNP set A and B satisfies the following properties.

1) $0 \leq \gamma(A, B) \leq 1$
2) $\gamma(A, B) = 1$ if and only if $A = B$
3) $\gamma(A, B) = \gamma(B, A)$.

18.4.12 Definition

Let A and B be two IVPNP sets, then the correlation coefficient is defined as

$$\gamma' = \frac{K(A, B)}{\max\{K(A, A), K(B, B)\}}$$

18.4.13 Theorem

The defined correlation measure between IVPNP sets A and B satisfies the following properties

1) $0 \leq \gamma'(A, B) \leq 1$
2) $\gamma'(A, B) = 1$ if and only if $A = B$
3) $\gamma'(A, B) = \gamma'(B, A)$.

However, in many practical situations, the different sets may have taken different weights, and thus, weight w_i of the element x_i $x_i \in R(i = 1, 2...n)$ should be taken into account. In the following, we develop a weighted correlation coefficient between IVPNP sets. Let $\omega = \{\omega_1, \omega_2, ..., \omega_n\}$ be the weight vector of the elements $x_i(i = 1, 2..., n)$ with $\omega_i \geq 0$ and $\sum_{i=1}^{n} \omega_i = 1$, then we have extended the above correlation coefficient $\gamma(A, B)$ and $\gamma'(A, B)$ to weighted correlation coefficients as follows.

$$\gamma'' = \frac{K_\omega(A, B)}{\sqrt{K_\omega(A, A).K_\omega(A, A)}}$$

$$K_\omega(A, B) = \sum_{i=1}^{n} \omega_i((TR_A^L((x_i))^2.((TR_B^L(x_i))^2 + ((CR_A^L(x_i))^2.\ ((CR_B^L(x_i))^2$$

$$+((GR_A^L(x_i))^2.\ ((GR_B^L(x_i))^2 + ((UR_A^L(x_i))^2.\ ((UR_B^L(x_i))^2 + ((FR_A^L(x_i))^2.((FR_B^L(x_i))^2)$$

$$+ ((TR_A^U(x_i))^2.((TR_B^L(x_i))^2 + ((CR_A^L(x_i))^2.\ ((CR_B^L(x_i))^2 + ((GR_A^L(x_i))^2.$$

$$((GR_B^L(x_i))^2 + ((UR_A^L(x_i))^2.\ ((UR_B^L(x_i))^2 + ((FR_A^L(x_i))^2.((FR_B^L(x_i))^2)$$

$$K_\omega(A, A) = \sum_{i=1}^{n} \omega_i((TR_A^L((x_i))^2.((TR_A^L(x_i))^2 + ((CR_A^L(x_i))^2$$

$$. ((CR_A^L(x_i))^2 + ((GR_A^L(x_i))^2. ((GR_A^L(x_i))^2$$

$$+ ((UR_A^L(x_i))^2. ((UR_A^L(x_i))^2 + ((FR_A^L(x_i))^2.((FR_A^L(x_i))^2) + ((TR_A^U(x_i))^2.((TR_A^L(x_i))^2$$

$$+ ((CR_A^L(x_i))^2. ((CR_A^L(x_i))^2 + ((GR_A^L(x_i))^2((GR_A^L(x_i))^2 + ((UR_A^L(x_i))^2. ((UR_A^L(x_i))^2$$

$$+ ((FR_A^L(x_i))^2.((FR_A^L(x_i))^2)$$

$$K_\omega(B,B) = \sum_{i=1}^n \omega_i((TR_B^L((x_i))^2.((TR_B^L(x_i))^2 + ((CR_B^L(x_i))^2$$

$$. ((CR_B^L(x_i))^2 + ((GR_B^L(x_i))^2.((GR_B^L(x_i))^2$$

$$+ ((UR_B^L(x_i))^2. ((UR_B^L(x_i))^2 + ((FR_B^L(x_i))^2. ((FR_B^L(x_i))^2)$$

$$+ ((TR_B^U(x_i))^2. ((TR_B^L(x_i))^2 + ((CR_B^L(x_i))^2. ((CR_B^L(x_i))^2 + ((GR_B^L(x_i))^2. ((GR_B^L(x_i))^2$$

$$+ ((UR_B^L(x_i))^2. ((UR_B^L(x_i))^2 + ((FR_B^L(x_i))^2.((FR_B^L(x_i))^2)$$

$$\gamma''' = \frac{K_\omega(A,B)}{\max\{K_\omega(A,A), K_\omega(B,B)\}}$$

It can be easy to verify that if $\omega = \left(\frac{1}{n}, \frac{1}{n}, \ldots, \frac{1}{n}\right)^T$, then weighted correlation coefficient becomes correlation coefficient respectively.

18.4.14 Theorem

Let $\omega = (\omega_1, \omega_2, \ldots, \omega_n)^T$ be the weight vector of x_i ($i = 1,2, \ldots, n$) with $\omega_i \geq 0$ and $\sum_{i=1}^n \omega_i = 1$, then the weighted correlation coefficient between IVPNP set A and B satisfies the following properties.

1) $0 \leq \gamma'''(A,B) \leq 1$
2) $\gamma'''(A,B) = 1$if and only if $A = B$
3) $\gamma'''(A,B) = \gamma'''(B,A)$.

18.4.15 Theorem

The correlation coefficient of two IVPNP sets A and B as defined, that is, $\rho'''(A, B)$, satisfies the same properties as those in the above theorem.

18.5 Application

In this section, we give some applications of PNPS in the COVID problem using correlation measure.

Vaccines are now authorized to prevent infection with SARS-CoV-2, the coronavirus that causes COVID-19. But until more is understood about how the vaccines affect a person's ability to transmit the virus, precautions such as mask-wearing, physical distancing, and hand hygiene should continue regardless of a person's vaccination status to help prevent the spread of COVID-19. In some practical problems, there is the possibility of each element having truth membership function, contradiction membership function, unknown membership function, ignorance membership function, and falsity membership functions. The proposed correlation measure among the patients versus symptoms and symptoms versus COVID injections gives the proper diagnosis. Now, an example of the above problem will be presented as follows.

18.5.1 Example for pentapartitioned neutrosophic Pythagorean sets

Let $R = \{R1, R2, R3\}$ be a set of COVID patients, $M = \{\text{Fever, Headache, Tiredness}\}$ be a set of symptoms, and $N = \{\text{Covishield, Covaxin, Sputnik}\}$ be a set of COVID injections. The relation between patients and symptoms is given in Table 18.1. The relation between symptoms is given in Table 18.2.

The correlation measures are calculated and shown in Tables 18.3 and 18.4. On the other hand, if we assign weights 0.33,0.34, and 0.33,

TABLE 18.1 A (the relation between patients and symptoms).

A	Fever	Headache	Tiredness
*R*1	(0.5,0.4,0.6,0.3,0.2)	(0.4,0.1,0.2,0.6,0.5)	(0.5,0.6,0.6,0.1,0.2)
*R*2	(0.6,0.3,0.7,0.5,0.4)	(0.3,0.2,0.5,0.4,0.2)	(0.2,0.1,0.7,0.6,0.3)
*R*3	(0.2,0.7,0.8,0.1,0.6)	(0.3,0.4,0.5,0.2,0.3)	(0.3,0.4,0.5,0.4,0.5)

TABLE 18.2 B (the relation between symptoms and disease).

B	Covishield	Sputnik	Covaxin
Fever	(0.7,0.4,0.6,0.3,0.1)	(0.1,0.3,0.5,0.7,0.9)	(0.3,0.4,0.5,0.6,0.7)
Headache	(0.2,0.3,0.5,0.4,0.6)	(0.2,0.4,0.6,0.2,0.1)	(0.2,0.3,0.4,0.5,0.6)
Tiredness	(0.1,0.4,0.7,0.5,0.6)	(0.4,0.5,0.7,0.3,0.2)	(0.1,0.4,0.7,0.5,0.2)

TABLE 18.3 A and B correlation measure.

ρ	Covishield	Sputnik	Covaxin
$R1$	**0.7699**	0.5457	0.7006
$R2$	**0.9075**	0.6617	0.7749
$R3$	0.67706	0.6585	**0.6801**

TABLE 18.4 A and B correlation measure.

ρ'	Covishield	Sputnik	Covaxin
$R1$	**0.6475**	0.3878	0.6079
$R2$	**0.9949**	0.5317	0.9448
$R3$	0.6718	0.5631	**0.6735**

TABLE 18.5 A and B correlation measure.

ρ''	Covishield	Sputnik	Covaxin
$R1$	**0.7535**	0.5791	0.6115
$R2$	**0.8559**	0.6717	0.8050
$R3$	0.6937	0.6669	**0.6952**

TABLE 18.6 A and B correlation measure.

ρ'''	Covishield	Sputnik	Covaxin
$R1$	**0.6337**	0.4439	0.4924
$R2$	**0.7952**	0.5538	0.7828
$R3$	0.6626	0.5654	**0.6951**

respectively, then we can calculate weighted correlation coefficients. By using weights, the weighted correlation are shown in Tables 18.5 and 18.6.

The highest correlation measure from Tables 18.5–18.8 gives the diagnosis of COVID patients. Therefore $R1$ and $R2$ suffer from Covishield and $R3$

TABLE 18.7 A (the relation between patients and symptoms).

A	Fever	Headache	Tiredness
*R*1	([0.1,0.3], [0.2,0.4], [0.3,0.6], [0.2,0.5], [0,0.3]	([0.1,0.4], [0.2,0.5], [0.1,0.5], [0.2,0.4], [0.1,0.3]))	([0.2,0.5], [0.1,0.4], [0.2,0.7], [0.2,0.4], [0.1,0.4])
*R*2	([0.2,0.4], [0,0.3], [0.1,0.5], [0.1,0.3], [0.1,0.4])	([0.2,0.4], [0.3,0.5], [0.1,0.6], [0.1,0.2], [0.2,0.5])	([0.1,0.4], [0.2,0.5], [0.1,0.6], [0.2,0.4], [0.1,0.3])
*R*3	([0.1,0.3], [0.2,0.4], [0.2,0.6], [0.1,0.5], [0.2,0.4])	([0.1,0.3], [0.2,0.3], [0.3,0.5], [0.2,0.5], [0.1,0.4])	([0.2,0.3], [0.1,0.3], [0.1,0.6], [0.2,0.4], [0.1,0.5])

TABLE 18.8 B (the relation between symptoms and COVID injections).

B	Covaxin	Covishield	Sputnik
Fever	([0.2,0.4], [0.1,0.2], [0.2,0.8], [0.1,0.3], [0.2,0.4])	([0.1,0.3], [0.2,0.3], [0.3,0.5], [0.2,0.5], [0.1,0.4])	([0.4,0.5], [0.1,0.3], [0.5,0.6], [0.2,0.4], [0.1,0.2])
Headache	([0.4,0.5], [0.1,0.3], [0.3,0.6], [0.2,0.5], [0.1,0.3])	([0.2,0.3], [0.2,0.4], [0.4,0.7], [0.1,0.3], [0.1,0.4])	([0.1,0.2], [0.3,0.5], [0.4,0.6], [0.1,0.2], [0.2,0.4])
Tiredness	([0.1,0.3], [0.2,0.5], [0.5,0.6], [0.3,0.4], [0.1,0.5])	([0.3,0.4], [0,0.3], [0,0.5], [0,0.3], [0.1,0.4])	([0.3,0.4], [0,0.2], [0.1,0.5], [0.3,0.4], [0.1,0.3])

suffers from Covaxin. Hence, we can see from the above three kinds of correlation coefficient indices that the results are the same.

18.5.2 Example for interval-valued pentapartitioned neutrosophic Pythagorean set

Let $R = \{R1, R2, R3\}$ be a set of COVID patients, $M = \{$Fever, Headache, Tiredness$\}$ be a set of symptoms, and $N = \{$Covishield, Covaxin, Sputnik$\}$ be a set of COVID injections. The relation between patients and symptoms is given in Table 18.6. The relation between symptoms and disease is given in Table 18.7. The correlation measures calculated are given in Tables 18.9 and 18.10.

TABLE 18.9 A and B correlation measure.

γ	Covaxin	Covishield	Sputnik
R1	0.8580	**0.8668**	0.8249
R2	0.7825	**0.8746**	0.8459
R3	**0.8981**	0.8732	0.8852

TABLE 18.10 A and B correlation measure.

γ'	Covaxin	Covishield	Sputnik
$R1$	0.6830	**0.8534**	0.7772
$R2$	0.6092	**0.8703**	0.8132
$R3$	**0.8809**	0.7782	0.8246

TABLE 18.11 A and B correlation measure.

γ''	Covaxin	Covishield	Sputnik
$R1$	0.8581	**0.8670**	0.8245
$R2$	0.7825	**0.8748**	0.8458
$R3$	**0.8985**	0.8735	0.8856

TABLE 18.12 A and B correlation measure.

γ'''	Covaxin	Covishield	Sputnik
$R1$	0.6832	**0.8543**	0.7769
$R2$	0.6092	**0.8704**	0.8131
$R3$	**0.8810**	0.7784	0.8249

On the other hand, if we assign weights 0.33, 0.34, and 0.33, respectively, then we can calculate weighted correlation coefficients. By using weights, the weighted correlation measures are shown in Tables 18.11 and 18.12.

The highest correlation measure from Tables 18.9–18.12 gives the diagnosis of COVID patients. Therefore $R1$ and $R2$ effects more from Covishield and $R3$ effects more from Covaxin. Hence, we can see from the above three kinds of correlation coefficient indices that the results are the same.

18.6 Conclusion

In this paper, we found the correlation measure of PNP sets and some of its properties are proved. An interval-valued PNP set has been introduced, and its characteristics of the respective set have been discussed. Also, we have extended the concept of correlation measure from neutrosophic Pythagorean set to PNP set and IVPNP set. Illustrative examples have been handled in the situation where the elements in a set are correlative, and weighted correlation coefficients have been defined. We studied an application of PNP statistics in COVID diagnosis using correlation measures. Vaccines are now authorized to prevent infection with SARS-CoV-2, the coronavirus that causes COVID-19. But until more is understood about how the vaccines affect a person's ability to transmit the virus, precautions such as wearing mask, physical distancing, and hand hygiene should continue regardless of a person's vaccination status to help prevent the spread of COVID-19. In the future, we can extend the concept to bipolar PNP sets.

References

[1] L. Zadeh, Fuzzy sets, Inf. Control. 8 (1965) 87–96.

[2] K. Atanassov, Intuitionistic fuzzy sets, Fuzzy Sets Syst. 20 (1986) 87–96.

[3] F. Smarandache, A unifying field in logics, Neutrosophy: Neutrosophic Probability, Set and Logic, American Research Press, Rehoboth, 1998.

[4] R. Malik, S. Pramanik, Pentapartitioned neutrosophic set and its properties, Neutrosophic Sets Syst. 36 (2020) 184–192.

[5] R. Radha, A. Stanis Arul Mary, Pentapartitioned neutrosophic Pythagorean set, IRJASH 3 (2021) 62–82.

[6] R. Radha, A. Stanis Arul Mary, Pentapartitioned neutrosophic pythagorean Topological spaces, JXAT XIII (2021) 1–6.

[7] J. Ye, Vector similarity measures of simplified neutrosophic sets and their application in multicriteria decision making, Int. J. Fuzzy Syst. 16 (2014) 204–215.

[8] S. Ye, J. Ye, Dice similarity measure between single valued neutrosophic multisets and its application in medical diagnosis, Neutrosophic Sets Syst. 6 (2014) 48–53.

[9] G.W. Wei, H.J. Wang, R. Lin, Application of correlation coefficient to interval-valued intuitionistic fuzzy multiple attribute decision-making with incomplete weight information, Knowl. Inf. Syst. 26 (2011) 337–349.

[10] R. Chatterjee, P. Majumdar, S.K. Samanta, On some similarity measures and entropy on quadripartitioned single valued neutrosophic sets, J. Intell. Fuzzy Syst. 302 (2016) 475–2485.

[11] H. Wang, F. Smarandache, Y.Q. Zhang, R. Sunderraman, Single valued neutrosophic sets, Multispace Multistruct 4 (2010) 410–413.

Chapter 19

Neutrosophic statistics and the medical data: a systematic review

Amna Riaz[1], Rehan Ahmad Khan Sherwani[2], Tahir Abbas[3] and Muhammad Aslam[4]

[1]Department of Statistics, University of Gujrat, Gujrat, Pakistan, [2]College of Statistical and Actuarial Sciences, University of the Punjab, Lahore, Pakistan, [3]Department of Statistics, Government College University, Lahore, Pakistan, [4]Department of Statistics, Faculty of Science, King Abdulaziz University, Jeddah, Saudi Arabia

19.1 Neutrosophic statistics

The knowledge of statistical applications is necessary to apply in the health sciences for health promotion, protective activities, and clinical studies. It helps how to face the contents and courses which are more complex and formulate the improved scientific principles to analyze and develop healthcare and research activities. Due to the significant developments in technology, accessibility of information and research from the last decades, it is need of time that the health professional need to be maturely modernized and adopt the procedure and techniques which may help them to face the recent professional challenges. The requirement of statistical methods is acknowledged now in practice in the departments which are dealing to establish health. The statistical approach is as well important as the social and scientific approach to analyze the health situation and identify the problems related to individuals, their families, and the general public with possible solutions. So it is very important for the professionals of public health to understand the knowledge of statistical methods and data interpretation for positive and smooth health exercise in the society [1].

Though the purpose to track the challenging problems of health for the public and to apply the possible treatments on patients have been achieved, still in the analyzed data there is indeterminacy of some levels which create arbitrary thoughts and are not within the scope of classical statistics and neutrosophic statistics which may handle it in some better way. Neutrosophic

Cognitive Intelligence with Neutrosophic Statistics in Bioinformatics.
DOI: https://doi.org/10.1016/B978-0-323-99456-9.00004-0

statistics is a generalization of classical statistics. Neutrosophic data is a set of data that consist of some level of indeterminacy (data that may be unknown, not sure, imprecise, and incomplete), and neutrosophic descriptive statistics describes and summarizes its characteristics [1]. The concept of the neutrosophic set was introduced as an extension of the classical set and fuzzy set with some degree of indeterminacy. As a part of neutrosophy, it studied the nature, origin, and scope of neutralities and solved many problems having uncertainty, vagueness, or indeterminacy [2–4].

Classical statistical methods can only be applied when the data and observations are determinate and certain. Practically sometimes, it is not always necessary to have determined observations, and classical statistics cannot be used to analyze them. Neutrosophic statistics as an extension of classical statistics can be applied to analyze indeterminate, fuzzy, imprecise, or uncertain observations [5]. Neutrosophic statistical methods are used for the presentation and interpretation of neutrosophic data and to analyze the neutrosophic numbers and events, neutrosophic regression, neutrosophic probability distribution, and neutrosophic estimation. Neutrosophic numbers are written as $N = a_o + I^* b_o$; here, a_o, b_o may be real/complex numbers, and I is its indeterminate part. Neutrosophic statistics is a generalization of classical statistics and the same as the neutrosophic probabilities are of the classical imprecise probabilities. Neutrosophic probability (NP) of any event "*A*" is defined as the probability that event "*A*" (1) will occur, (2) will not occur, and (3) not sure whether it will occur or not, that is, chance of indeterminacy. Neutrosophic distribution (ND) is a function which models the NP of a random variable "*x*": $\text{NP}(x) = \{R(x), I(x), S(x)\}$, and here (1) $R(x)$ = probability that x occurs, (2) $S(x)$ = probability that x not occurs, and (3) $I(x)$ = indeterminate probability of x [1].

Aslam proposed a sampling plan under the exponential distribution using neutrosophic statistics. This proposed neutrosophy sampling plan was an alternative sampling plan of classical statistics in the presence of uncertain, fuzzy, indeterminate values of parameters or observations. Important measures of neutrosophy were defined in the proposed sampling plans, and it was proposed that the parameters of the neutrosophic sampling plan will be found by the nonlinear optimization of neutrosophy [5].

Cisneros et al. used neutrosophic statistics to analyze a variable among many variables which interact in the health field to deal the discretionary and arbitrary opinions and achieve consensus between them, as indeterminate (unknown) elements for the criteria evaluated in the scientific framework. The main focus of their study was to find the level of evidence about recommendations to track the problem of health in the medical community based on the suggested analysis of the study and the indeterminacy level in the neutrosophic statistical data. Further, the focus was to find out the factors and levels of recommendations presented by the expert's clinical pictures for safety perspectives, measure, and model the neutrosophic variable and suggest suitable alternatives while assessing the existing uncertainty or indeterminacies of the examined variable [1].

Statistics helps in analyzing the situations where variability occurs in the obtained data due to the random element of the sample. In the health community, difficulties in evaluating the health determinants and situation of disease are random elements among other reasons, and variability is the response of patients under the same treatment. To develop the neutrosophic statistical study, experts suggested analyzing the levels of evidence of recommendations to track the problem of health on the basis of statistics in the medical community. The recommendation depends upon the advancements and updates of clinical research and statistical evidence at the international level [1].

To do neutrosophic statistical modeling on the clinical decisions taken by health experts, five clinical decisions on the basis of the typical situation of society in the health arena were taken as factors. And neutrosophic frequencies of these factors were analyzed for consensus on clinical decisions on the basis of the patient's conclusive diagnosis. Analysis of clinical decision was performed before the patient's conclusive diagnosis, and a set of clinical decisions were made up for treatment in a group of patients for every factor. The level of evidence about recommendations to track the problem of health in the medical community was analyzed for each factor from 100 patients in public, and the acceptance level of the experts of health for the clinical decision of treatment was measured. It was observed from the neutrosophic frequencies that in evaluating the complex clinical pictures there was level of indeterminacy for every clinical picture on days with acceptance level of clinical decision about treatment. Classical statistics was not appropriate with indeterminate levels, so the use of neutrosophic statistics was necessary [1].

In the health community, neutrosophic statistics is more appropriate in the presence of indeterminacy among the variables which influence health. The neutrosophic coefficient of variation determined the factor which required evidence level about recommendations to track the problems of health in the medical community. This study evaluated the evidence level about recommendations to track the problems of health in the medical community and found the screening factor for pancreatic cancer in asymptomatic adults using abdominal palpation, ultrasound, or serological markers with coefficient of variation of 69.9% indeterminacy. It influences the diagnosis moments and the right level of recommendation. Alternatives were proposed on the basis of the results about initial clinical studies of patients along with their presented clinical pictures. Every alternative responded to every neutrosophic state of the variable in the neutrosophic set. Neutrosophic statistics exists in every dimension of the health field. It is used to determine the variables in which there is some level of indeterminacy [1].

Culture is a process of production, formation, and interchanging the human effort with its several manifestations, and its characteristics have its effects in the field of health; so direct relation may exist between health and interculturality. So the health personnel must have the awareness about treatments for diseases. The concept of intercultural was not incorporated adequately in the matters of health systems in the region of Latin America, and in the systems of national health, traditional and inherited ways have been adopted by patients for the treatments of the diseases in the local communities until recently. Chico et al. addressed the characteristics of intercultural medical care in their study by taking into view the procedural limitations affecting its execution in Ecuador. A diagnosis was carried out to identify the relevance level of personnel of medical care concerning interculturality in health. In this research, theoretical and empirical methods were used, and the sample was calculated by using neutrosophic statistics. Neutrosophic sample resulted as $n = [27.46, 33.44]$ which indicated that the size of sample may be between 27 and 33. Neutrosophic Likert scale including the indeterminate part was used to assess the relevance. Pearson correlation was used to test, and significant correlation was observed between knowledge of assessing the relevance of intercultural medical care and efficient medical treatments of different diseases [6].

Neutrosophic genetics is introduced as a study about genetics based on neutrosophy concepts of sets, logic, statistics, measure, probability, and neutrosophic methods and techniques. Smarandache described the applicability of neutrosophic genetic on the basis of Neutrosophic Theory of Evolution in their study. Neutrosophic Theory of Evolution is the extension of Darwin's evolution theory, and it consists of three degrees: (1) evolution, (2) involution, and (3) neutrality or indeterminacy (uncertain whether the change is toward the involution or evolution). The concepts of classical mutation, speciation, and coevolution were extended to neutrosophic mutation, speciation, and coevolution, respectively, in the structure of neutrosophic genetics. Neutral mutation, degree of continuation, and degree of neutrality were the extended concepts in classical mutation, speciation, and coevolution, respectively, which consist of indeterminacy [7].

Data Envelopment Analysis (DEA) as a linear programming operates the input and output data to measure the effectiveness of homogeneous decision-making units comparatively without the information of production functions. DEA methods are now effectively connected in the services of health care. Crisp values of input and output data are required for classical models of DEA which are not usually available in the applications of real world [8,9]. The observed input and output values are fuzzy sometime in genuine cases. Experts used their knowledge to handle fuzzy data, and the concept of fuzzy set was applied in many researches to solve different problems. Several methods were considered to deal the indeterminacy in the inputs and outputs of DEA, but these methods could not be applied to solve DEA with imprecise, indeterminate, and incomplete information due to the constraints involved in them. To handle the indeterminate DEA and fuzzy model with DEA, Edalatpanah (2018) presented a novel model of DEA on the basis of neutrosophic inputs and outputs [10]. Analysts can improve the procedure of decision-making by using the neutrosophic statistics in DEA. Yang et al. designed a simple and innovative DEA model, and all values of input and output of it are triangular single-valued numbers of neutrosophy. The new DEA model assessed the performance of 13 hospitals (from Tehran University of Medical Sciences) of Iran and concluded that DEA model under neutrosophy is more useful and effective for decision-makers when input and output values are not exact or have some indeterminacy or uncertainty. To improve the effective working of health sector, assessing its performance is very important for policymakers [11].

Statistical procedures play a significant role in data analyzing, estimation, and forecasting in medical sciences. It is important to screen out the data and detect the outliers for effective forecasting by the experts in this field. The existence of outliers may affect the estimation and forecasting, so Grubbs's test was introduced and used widely to detect and remove outliers. Generally, the observations in the data are imprecise, fuzzy, or in the form of intervals. Aslam designed Grubbs's test under the concepts of neutrosophic statistics as the generalization of Grubbs's test of classical statistics to detect the outliers in data with the presence of indeterminacy. The operational method of the novel test was presented under the interval method of neutrosophic statistics, and its practical implementation was also presented in the medical field. A medical specialist examined the heart conditions of 11 patients, such as blood systolic pressure, diastolic pressure, and pulse rate, to identify the risk of stroke or heart disease. It is important for a medical specialist to identify and remove the outlier from the data in the prediction of the patient's risk for stroke or heart disease. The existence of outlier may mislead a medical expert to predict the chance of heart disease and its diagnosis. As the data were in the form of interval with some indeterminacy so Grubbs's test under classical statistics was not an appropriate approach and medical specialist is attracted to Grubbs's test under neutrosophic statistics to analyze the data. Novel Grubbs's test was compared to the existing Grubbs's test and found as more flexible, informative, and effective as compared to the existing test under the indeterminate atmosphere and was recommended to apply it in biostatistics and medical science for analysis purposes [12].

Further, the authors presented a study in which they introduced the tests of kurtosis and skewness under neutrosophic statistics to test the normality of neutrosophic data related to heart disease. The proposed tests provided information about indeterminate measures and were found as more informative and flexible as compared to tests based on classical statistics in the presence of vagueness and uncertainty [13].

Statistical techniques help to highlight the useful information of the data and analyze its different properties by using the hypothesis testing. Most of the statistical techniques of hypothesis testing require the assumption of normality and homogeneity of variances. But in real life, usually these assumptions of parametric tests do not hold, and their violations lead toward nonparametric tests [14]. Classical Kruskal–Wallis test is nonparametric and used as an alternative test for one-way analysis of variance. It is used to assess the equivalence of numerous means when the form of observations is exact, but it fails to apply when the observations are in the form of interval and fuzzy or have some level of indeterminacy. Sherwani et al. proposed a neutrosophic Kruskal–Wallis test for fuzzy or indeterminate data, and a complete methodology of the novel form of the classical Kruskal–Wallis test and its implementations were described in their research. The novel test will be appropriate when the data have fuzzy or neutrosophic observations, and the samples under the neutrosophy must be random and mutually independent [15].

To apply the novel proposed test, data of daily occupancy of ICU by the patients of COVID-positive in Pakistan on the basis of three age groups during December 2020 in Pakistan were observed. The three age groups of patients were taken as 35 and below, 35–55, and 55 and above. Neutrosophy was introduced in the data, and both parts (indeterminate and determinate) of the data were recorded. The null hypothesis of no difference in the average daily occupancy of ICU by the patients with COVID-positive among the three groups of age was tested statistically against the alternative hypothesis of difference in the average daily occupancy of ICU by the patients of COVID-positive for at least two groups of age. Neutrosophic Kruskal–Wallis H was applied, and significant difference among daily occupancy of ICU by the patients of COVID-positive for different groups of age was found at 1% level of significance. Results of the novel test on the data of COVID-19 were more adequate and recommended that the neutrosophic Kruskal–Wallis H test was more applicable and suitable for the fuzzy data. The efficiency of the novel form and classical form of Kruskal–Wallis H test was compared, and the classical Kruskal–Wallis H test became helpless with the data based on the interval form and may have given misleading results. Neutrosophic Kruskal–Wallis H test was found as flexible, efficient, relevant, and appropriate than its classical form in the existence of neutrosophy [15].

Sign test is also nonparametric, and it is used as an alternative test of one-sample or paired *t*-test under the classical statistics. Sign test is only applicable to the data in exact form, and it fails when the data is interval-valued or fuzzy or has some level of indeterminacy and vagueness. A modified sign test named as neutrosophic sign test was proposed by Sherwani et al. to deal with the fuzzy or uncertain state of data, and its methodology was designed both for one and two samples in their study. Data comprise the 79 observations of the reproduction rate of COVID-19 in Pakistan during the period October 2020–December 2020, and neutrosophy was introduced in it to evaluate the performance of the proposed neutrosophic sign test for the one sample. Also, the gender-wise neutrosophic data of daily occupancy of ICU by the patients of COVID-positive in Pakistan were used. Further, the applicability of two samples neutrosophic sign test was given on the neutrosophic data of COVID-19 recorded in Pakistan. Significant difference between males and females' daily occupancy of ICU by the patients with COVID-positive was tested, and the performance of the proposed sign test on two samples is also evaluated. It is suggested on the basis of their findings that proposed novel methodologies are more appropriate in making the decisions of nonparametric problems having interval form, indeterminate, or uncertain data, and under these conditions, classical nonparametric tests may provide misleading results [14].

Neutrosophic sign and Kruskal–Wallis H tests are strongly recommended to apply in other fields of biomedical sciences in the presence of indeterminacy [14,15]. In the real life, neutrosophy exists in many scenarios, and classical tests of statistics fail and cannot be applied with the fuzzy observations [16]. Neutrosophic methods and tests were suggested in dealing with fuzzy and indeterminate observations by recent studies [17,18].

Globally, all the countries have been facing the challenges of loss such as human lives, increasing rate of unemployment, and lower growth of GDP since the outbreak of the pandemic period of COVID-19. The only key factor to be away from these difficulties is the vaccine invention and its prime distribution. A lot of companies all over the world are in competition to get a vaccine for COVID-19, but it is not possible to fulfill the needs of vaccine for about 8 billion population in this period after being approved. Some lack may occur in fair distribution of vaccine doses between countries at global level or within country among the society at local level due to the limited doses of available vaccine. And to be satisfied with the COVID vaccine for all the population in a country made this competition more intense. So governments need to identify and classify priority groups in order to allocate the doses of the COVID vaccine first, and the experiences of the experts must be utilized by decision-makers for it. The approach of multiple criteria decision-making (MCDM) is very important to determine the optimum alternatives. Analytic hierarchy process (AHP) and Technique for Order Preference by Similarity to Ideal Solution (TOPSIS) were the most frequently used methods of MCDM in many studies of real applications. Neutrosophy concept is now extensively used to deal with ambiguity and uncertainty of data and contributed to analyze and understand COVID-19 too [19].

Hezam et al. established a study and classified the priority groups for the allocation of the COVID-19 vaccine by using the approach of neutrosophic MCDM. In this approach, 15 sub-criteria and 4 principal criteria both under the neutrosophy scale were identified by using three experts on the basis of age, nature of job, woman's status, and health status so that the most deserving groups of the society need to be vaccinated preliminarily. As the input values were fuzzy or indeterminate and in the presence of uncertain and fuzzy inputs, the use and importance of neutrosophy are inescapable, so AHP under neutrosophy was proposed to evaluate weights of these criteria and their ranking. Further, neutrosophic TOPSIS method was applied to rank and determine the alternatives to the COVID-19 vaccine. Results indicated that the most prioritized people of the society for the preliminary vaccination are personnel of health care, people of older age with health of high risk, lactating mothers, and pregnant women. Further, it was indicated that patients and workers of health care have priority for the most effective vaccine over any other vaccine. The proposed technique is more effective and flexible for the neutrosophic inputs and is recommended for the applications of treatment priority [19].

Further, Khalifa et al. introduced the concept of neutrosophy on the model of deep transfer learning where the original and neutrosophic images were considered in order to control the pandemic of COVID-19 [20]. M-polar neutrosophic numbers (MPNNs) were used in the operators of generalized Einstein weighted aggression and generalized weighted aggression to diagnose and examine COVID-19 [21]. And a neutrosophic classifier-based HealthFog framework was proposed for the diagnosis of COVID-19 and its prevention and treatment [22]. Chou et al. also conducted a research on the methodology of multi-period health diagnostics under the neutrosophy set based on the single-valued [23].

In the modern era, cancer is spreading very rapidly in developing and developed countries. Globally, prostate cancer is observed as the second most common type of cancer in men and resulting in high rate of death. In biological and social sciences, there have been many researches on the prostate cancer under the classical statistics, but these studies have no impact in the presence of indeterminacy. Aslam and Albassam presented an epidemiological research to study the relation between dietetic fat and prostate cancer. The level of dietetic fat in daily food is not a determined or exact value but is in the form of an interval. Under the indeterminate atmosphere, classical statistics cannot be applied to study the relation between dietetic fat and prostate cancer. Fuzzy logic was useful to apply in the presence of indeterminacy among the parameter or observations, and it is commonly used in epidemiological studies. Many researchers used the fuzzy logic to determine the prostate cancer [24]. Fuzzy logic was generalized in the sense of neutrosophy by considering the data of interval-valued and measure of vagueness or indeterminacy by Smarandache [25], and neutrosophic statistics was introduced as a generalization of classical statistics for the data based on neutrosophic number. And the efficiency of neutrosophy logic was verified [26]. Neutrosophic logic has been implemented by many authors for fuzzy, uncertain, and vague data. Smarandache introduced neutrosophic statistics as an extension of classical statistics and used to do analysis under the indeterminate and uncertain atmosphere. Neutrosophic regression is defined to apply when there is some level of indeterminacy in the observed data [27]. Aslam and Albassam took the data of death rate of prostate cancer and the level of dietetic fat in food from the thirty countries to study the relation between dietetic fat and prostate cancer under the uncertain atmosphere. Neutrosophic regression line was fitted on the observed data by taking the level of dietetic fat as explanatory and death rate as explained variables. It was observed from the analysis of neutrosophy that as the level of dietetic fat among people increases, death rate of prostate cancer also increases, and neutrosophic regression confirmed it. Further, neutrosophic correlation found a significant correlation between dietetic fat and risk of prostate cancer. It was concluded that under the uncertain and indeterminate atmosphere neutrosophic regression model was more effective and appropriate than classical regression model for estimation and forecasting which will be helpful to control the rate of death. Further, it is recommended that under the uncertain atmosphere health manager must use neutrosophy regression modeling for effective and better analyses [24].

There are numerous active studies concerning the developments in medical field. People of developed cultures are very much concerned about fitness and health [28]. Advancements in the Internet of Things technology are widely used in the health care [28,29]. Medical statistics indicated that heart disease is causing a high rate of illness and death globally despite all developments and advancements in the field of medicine. Experts are facing a lot of difficulties in diagnosing the heart disease due to analysis of numerous factors and having immersed and incomplete information [28]. Also, cancer is another worldwide challenge for any healthcare system and country. It is a silent killer as mostly patients with cancer are unaware of the changings occurring in the body [29]. Traditional MCDM techniques are not efficient in the presence of uncertainty and vagueness of disease symptoms, so novel computer-based frameworks were proposed and neutrosophic MCDM method was suggested to help the medical experts and patients for accurately diagnosing and monitor diseased patients with heart failure and cancer by handling vagueness and uncertainty of disease symptoms. A method of simple additive weighting and neutrosophy theory was integrated to handle the uncertain, incomplete, and vague information in diagnosing the diseases by using the membership degrees of truth (T), indeterminacy (I), and falseness (F). Mobile application and wireless body area network were used to capture symptoms of disease and social interactions in the body of user. The neutrosophic MCDM technique based on neutrosophy numbers handled the vagueness, incompleteness, and uncertainty in symptoms and examined the differences of diseases on the basis of reported symptoms for a person who was not a patient but having the similar symptoms of the disease. Experimental results of neutrosophic MCDM indicated that the proposed method was valid and provided a worthy solution in the presence of uncertain and vague information [28,29]. Further, the triangular neutrosophy scale forecasts the disease that a patient may have a chance to be infected by creating a matrix of decision criteria upon the suggestions of experts [29].

Many researchers examined the techniques of decision analysis in previous studies and found that the most problems in making decision are linked to imprecise, multi-polar, and uncertain information and fuzzy sets failed to tackle it properly. In another study, authors combined the concepts of neutrosophic set and m-polar fuzzy set and introduced the novel m-polar neutrosophic set (MPNS), its various characterizations, and structure of topology on MPNS. Techniques of decision analysis are very important in medical diagnosis. Methods of MCDM were handled by the use of neutrosophic set with levels of falseness, truth, and indeterminacy for corresponding alternatives. Further, algorithms for MCDM based on m-polar neutrosophic topology and MPNS under certainty were presented in medical diagnosis. The better flexibility and validity of the proposed algorithm were observed as compared to existing methods [30].

As we know that different diagnostic tests in medical sciences were proposed on the theory of classical statistics and used widely under the assumption that data comprise the determined observations, these diagnostic tests are unable to analyze the data having vague or indeterminate observations. Neutrosophic statistics defined on the neutrosophic numbers and logic can be applied to analyze the data based on indeterminate, fuzzy, or vague observations. Aslam et al. proposed the neutrosophic diagnosis test under the neutrosophic statistics which can be applied to the data having indeterminacy to diagnose a disease in medical science. For example, the practitioner considers the diabetes test to diagnose the sugar level status of patients, and its positive result indicates them as diabetic patients. But in some cases, the diagnosis test comes up with indeterminacy about disease or medical expert is uncertain about the correct diagnosis. In this indeterminate situation, the neutrosophic diagnosis test is more suitable, adequate, and effective to diagnose the disease status rather than the diagnosis test under the classical statistics [31].

Soft computing techniques are widely used to encode uncertainty and indeterminacy in the applications of the real world. The incorporation of these techniques with medical applications may help to solve many insoluble problems in the medical system. The required data about patients are mostly inaccessible to make decision in the environment of medical diagnosis. Fuzzy methodologies provided solution extensively for uncertain and incomplete information, and assimilation of neutrosophic methods is an evidence of its superiority in the medical field. Previous studies proved significance role of neutrosophic sets to de-noise, cluster, and segmentation of medical image and suggested the usefulness of integrated methods of neutrosophic sets in the medical domain and diagnosis based on neutrosophic models for future scope. Neutrosophic set is widely used now for making decisions in medical diagnosis. Nguyen et al. reported the importance of neutrosophic sets and medical applications of diagnosis grounded on them. Their research highlighted the processing of medical image (de-noising, classification, clustering, segmentation, etc.) under neutrosophy. To process the medical image under neutrosophy, a framework consisting of three components (including the indeterminate part) was introduced and medical image was presented in the domain of neutrosophic set. The value of uncertain or indeterminate subset must be minimized to achieve the superior performance for the processing of medical image. The use of neutrosophy is highly recommended for accurate diagnosis [32].

Natural products are usually economical, with less side effects and increasing demand specially the nutritious supplements, health products, cosmetics, and medical products based on herbs. According to the World Health Organization, almost 80% population of the world are still using the plant-based traditional medicines, and leaves are their important ingredients. Leaf's quality describes its excellence level, being flawless, and extensive variations. Further, leaf disease causes threats to the status of production and economy in the global industry of agriculture. Detection of leaf disease with the use of digital image processing can reduce the dependence on agriculturalists for the safety of products. The authors of the present study presented a novel approach based on computer vision to recognize the status of herb leaves as diseased or healthy. The segmentation process traces the suspicious parts in diagnosing a disease. A segmentation method based on the neutrosophic logic was designed, and a novel feature set was extracted for the detection and classification of leaf disease. The three segmented elements of neutrosophic image were as indeterminate (I), false (F), and true (T). True part (T) and false part (F) were defined as diseased and healthy areas of leaf, and indeterminate part (I) was the uncertain part, that is, neither diseased nor healthy. Neutrosophic logic provided indeterminate part to handle the presence of uncertainty efficiently. In order to identify the status of leaf as healthy or diseased, new feature sets using histogram, color, texture, and region of disease sequence were evaluated on the basis of segmented regions. Four hundred cases (200 diseased/infected and 200 healthy) were validated for the proposed method. Discriminant power for the effectiveness of combined feature was demonstrated and monitored by using nine classifiers, and the accuracy of novel features was measured. Classification accuracy of novel features set was 98.4%, and it can be used effectively in order to identify leaf disease [33]. Guo and Cheng also proposed the segmentation technique under neutrosophy logic and used the neutrosophy approach in the segmentation of image and pattern recognition [34].

Another computer vision-based novel approach was proposed for the feature selection algorithm under neutrosophy logic to classify the disease of plant leaf. The presence of disease in plant leaf is alarming for the production of food and its quality. Diagnosing the leaf diseases at early stage plays an important role in the growth of agricultural production. Feature selection is defined as method of choosing the best feature among feature set from input data. The role of feature selection is very important in pattern recognition and image processing and to improve the performance and reduce the computational cost. Techniques of feature selection look forward for highly correlated features [35].

Neutrosophic Cognitive Map (NCM) as a relationship graph was formed among features and used to handle the situation of uncertainty that occurs due to unknown relation between features of numerous images and to choose the best feature subsets. Neutrosophic methodology was applied to the databases of diseased and healthy leaf images of apple taken online from repository of plant village and efficiently identified a diseased leaf by using indeterminacy or vagueness between features of image. In neutrosophy, a logical statement is transmuted in a space of truth (T), false (F), and indeterminacy (I), and NCM method handles the neutralities between false and true. Indeterminate part created a smooth and accurate identification of leaf disease. Eight advanced existing techniques of feature selection were compared with the proposed method, and this novel method proved its capability on the databases of this study and outperformed the existing techniques. Neural network yielded the highest classification accuracy (99.8%) of the feature selection for the leaf images based on the novel NCM approach by using eleven features only [35].

Medical experts are interested in correct and accurate decisions for the human health, and handling the uncertainty and indeterminacy is a serious challenge in this field. Gynecologists use cardiotocography (CTG) for early monitoring of the distress of the fetus graphically. To diagnose the condition of fetus heart rate, algorithms of machine learning are very important. Indeterminacy in the CTG data is crucial thing for the classification, but neutrosophy theory precisely estimates the indeterminate boundaries of data. The authors of the present study proposed the framework of Interval Neutrosophic Rough Neural Network (IN-RNN) as a diagnostic framework under the concept of neutrosophy to classify the indeterminate medical data of CTG. Three functions defined in neutrosophy classifiers are false, true, and indeterminacy degrees to deal uncertainty. The dataset of CTG was downloaded from the repository of the University of California. It consists of 2126 instances, 21 attributes for input to find the condition of fetus heart rate and contraction of uterine together, and 3 classes of output as normal, pathologic, and suspicious. The IN-RNN combined the concepts of rough neural network (RNN) and interval neutrosophy to handle the indeterminacy in the data of CTG [36].

This technique was based on two backpropagation RNNs, and the use of neutrosophy set theory improved the performance of RNN. First, RNN predicted the membership values as true (T), but the other RNN predicted membership values as false (F); these results produced the uncertain boundary zone, and the values of indeterminacy (I) can be estimated by the difference between the values of true and false memberships using interval neutrosophy set. If the difference is low between them, the uncertainty will be high. The decision class under interval neutrosophy was formed due to indeterminate membership in the evaluation of false and true memberships. The proposed framework of IN-RNN was simulated by using python language. To measure the efficiency of the proposed IN-RNN, cross-validation was used, performance measures such as precision, accuracy rate, and sensitivity were derived, and ROC was employed to the hidden cases in CGT data.

Algorithms of IN-RNN model were compared with the algorithms of other machine learning models by using the WEKA application, and the results showed that its performance was more effective and feasible to classify the CGT data. Neutrosophic techniques are very useful in handling the indeterminate data and help in making the most accurate decision, so these techniques are widely used in the biomedical field [36].

Diagnosing the category of a tumor at early stage may save a human life. As medical images are very important in the disease analysis, so many systems were established in this domain which mainly depends on the feature extraction for accurate understanding and classification of medical images. Smarandache introduced the neutrosophy theory in which degrees of falsity (F), truth (T), and indeterminacy (I) are associated to every event of a set, and the degree of indeterminacy handles the uncertainty. In this study, a new smart system of Slantlet transform (SLT) under composite neutrosophy was proposed for the feature extraction of statistical texture from MR images in order to diagnose the malignity of brain tumor accurately and precisely. MR images of the patients were acquired, three sets of memberships such as false (F), true (T), and indeterminate (I) were defined under neutrosophy, and SLT system was applied on each set of memberships for feature extraction. Statistical techniques such that one-way ANOVA was applied to reduce the quantity of extracted features, and four neural network techniques were applied on the extracted features to predict the type of tumor in brain. The performance of the novel model was assessed on a dataset of images taken from the MICCAI Brain Tumor Segmentation Challenge 2017 (BraTS 2017). Experimental results indicated high efficiency and accuracy of the proposed method to diagnose the brain tumors by deriving features of the Gray-Level Run-Length Matrix (GLRLM) from composite neutrosophic SLT technique [37]. In other studies, authors proposed hybrid technique based on convolutional neural network and neutrosophy for the detection of brain tumor [38] and combined the approaches expert maximum fuzzy-sure entropy (EMFSE) and neutrosophic set (NS) and proposed NS-EMFSE for the segmentation of brain tumor and detected the enhancing portion of the tumor. The performance of the proposed approach was evaluated using 100 images of MRI and found as an effective approach [39].

Irena and Sethukarasi proposed a method NCMAW-KELM based neutrosophic C-mean-based attribute weighting (NCMAW) and Kernel extreme learning machine (KELM) for the classification of medical data by handling the presence of indeterminacy [40].

To recognize the patterns of disease, segmentation accuracy is very important. Segmentation including fuzzy clustering is useful in medical imaging to solve the problems related to common boundaries among clusters. In a study, a novel clustering algorithm on the basis of neutrosophic orthogonal matrices was proposed to enhance the accuracy of segmentation from the images of dental X-ray. Image data were transformed to a neutrosophic set by using this algorithm, and the orthogonal principal segmented the pixels to form clusters. The validation of the novel approach was tested on the real dental datasets taken from the Hospital of Hanoi Medical University and found more accurate and superior approach as compared to other relevant approaches for the clustering quality [41].

The techniques of deep learning played an important role to diagnose and analyze a medical image and extracted the data features successfully. The performance of deep learning was affected due to the indeterminate information in the data of medical images. Previous studies reported that the hybridization of deep learning with neutrosophy approach improved its performance by handling the ambiguity and indeterminacy of medical data. The authors of the present study also investigated that the use of neutrosophy systems enhanced the deep learning techniques in the processes of classification, segmentation, and clustering of medical images. The transformation of simple images into neutrosophic images was under the same equation of memberships (T, I, F) in the analysis of medical images using the neutrosophic sets. Further, it was found that the integration of neutrosophic systems with support vector machine and convolution neural network precisely detected the disease, and its integration with LASTM achieved highest accuracy to classify the cardio visions [42].

In the light of these studies, the role of neutrosophy statistics is very essential to analyze the neutrosophic data in the medical field. The use of neutrosophy in medical data analysis needs to be acknowledged, and more comprehensive studies need to be developed to enforce the neutrosophic statistics in the presence of indeterminacy or uncertainty of medical data.

References

[1] J.E.L. Cisneros, C.L. Barrionuevo, E.L. González, A.G.L. Jácome, Handling of indeterminacy in statistics. Application in community medicine, Neutrosophic Sets Syst. 44 (2021) 235–244.

[2] F. Smarandache, A unifying field in logics: neutrosophic logic, Philos.: Am. Res. Press. (1999) 1–141.

[3] F. Smarandache, Neutrosophic set-a generalization of the intuitionistic fuzzy set, Int. J. Pure Appl. Math 24 (3) (2005) 287.

[4] F. Smarandache (Ed.), Neutrosophy and neutrosophic logic, in: First International Conference on Neutrosophy, Neutrosophic Logic, Set, Probability, and Statistics University of New Mexico, Gallup, NM, 2002.

[5] M. Aslam, Design of sampling plan for exponential distribution under neutrosophic statistical interval method, IEEE Access. 6 (2018) 64153–64158.

[6] M.G.G. Chico, N.H. Bandera, S.H. Lazo, N.L. Sailema, Assessment of the relevance of intercultural medical care. Neutrosophic sampling, Neutrosophic Sets Syst. 44 (2021) 420–426.

[7] F. Smarandache, Introduction to neutrosophic genetics, Int. J. Neutrosophic Sci. 13 (1) (2020) 23.

[8] R.D. Banker, A. Charnes, W.W. Cooper, Some models for estimating technical and scale inefficiencies in data envelopment analysis, Manag. Sci. 30 (9) (1984) 1078–1092.
[9] A. Charnes, W.W. Cooper, E. Rhodes, Measuring the efficiency of decision making units, Eur. J. Operat. Res. 2 (6) (1978) 429–444.
[10] S.A. Edalatpanah, Neutrosophic perspective on DEA, J. Appl. Res. Ind. Eng. 5 (4) (2018) 339–345.
[11] W. Yang, L. Cai, S.A. Edalatpanah, F. Smarandache, Triangular single valued neutrosophic data envelopment analysis: application to hospital performance measurement, Symmetry. 12 (4) (2020) 588.
[12] M. Aslam, Introducing grubbs's test for detecting outliers under neutrosophic statistics–an application to medical data, J. King Saud. Univ.-Sci 32 (6) (2020) 2696–2700.
[13] M. Aslam, M. Albassam, Testing the normality of heart associated variables having neutrosophic numbers, J. Intell. Fuzzy Syst. (2021) 1–7 (Preprint).
[14] R.A.K. Sherwani, H. Shakeel, M. Saleem, W.B. Awan, M. Aslam, M. Farooq, A new neutrosophic sign test: an application to COVID-19 data, PLoS One 16 (8) (2021) e0255671.
[15] R.A.K. Sherwani, H. Shakeel, W.B. Awan, M. Faheem, M. Aslam, Analysis of COVID-19 data using neutrosophic Kruskal Wallis H test, BMC Med. Res. Methodol. 21 (1) (2021) 1–7.
[16] J.J. Buckley, Fuzzy statistics: hypothesis testing, Soft Comput. 9 (7) (2005) 512–518.
[17] J. Chen, J. Ye, S. Du, R. Yong, Expressions of rock joint roughness coefficient using neutrosophic interval statistical numbers, Symmetry. 9 (7) (2017) 123.
[18] J.P. Meyer, M.A. Seaman, A comparison of the exact Kruskal-Wallis distribution to asymptotic approximations for all sample sizes up to 105, J. Exp. Educ. 81 (2) (2013) 139–156.
[19] I.M. Hezam, M.K. Nayeem, A. Foul, A.F. Alrasheedi, COVID-19 vaccine: a neutrosophic MCDM approach for determining the priority groups, Results Phys. 20 (2021) 103654.
[20] N.E.M. Khalifa, F. Smarandache, G. Manogaran, M. Loey, A study of the neutrosophic set significance on deep transfer learning models: an experimental case on a limited covid-19 chest x-ray dataset, Cogn. Comput. (2021) 1–10.
[21] M. Raza Hashmi, M. Riaz, F. Smarandache, m-polar neutrosophic generalized weighted and m-polar neutrosophic generalized Einstein weighted aggregation operators to diagnose coronavirus (COVID-19), J. Intell. Fuzzy Syst. (2020) 1–21 (Preprint).
[22] I. Yasser, A. Twakol, A. El-Khalek, A. Samrah, A. Salama, COVID-X: novel health-fog framework based on neutrosophic classifier for confrontation covid-19, Neutrosophic Sets Syst. 35 (1) (2020) 1.
[23] J.C.-s Chou, Y.-F. Lin, S.S.-C. Lin, A further study on multiperiod health diagnostics methodology under a single-valued neutrosophic set, Comput. Math. Methods Med. (2020) 2020.
[24] M. Aslam, M. Albassam, Application of neutrosophic logic to evaluate correlation between prostate cancer mortality and dietary fat assumption, Symmetry. 11 (3) (2019) 330.
[25] F. Smarandache, Neutrosophic logic-a generalization of the intuitionistic fuzzy logic, Multispace Multistruct. Neutrosophic Transdisciplinarity (100 collected Pap. Sci.) 4 (2010) 396.
[26] F. Smarandache, H.E. Khalid, A.K. Essa, Neutrosophic Logic: The Revolutionary Logic in Science and Philosophy, Infinite Study, 2018.
[27] F. Smarandache, Introduction to Neutrosophic Statistics, Infinite Study, 2014.
[28] M. Abdel-Basset, A. Gamal, G. Manogaran, L.H. Son, H.V. Long, A novel group decision making model based on neutrosophic sets for heart disease diagnosis, Multimed. Tools Appl. 79 (15) (2020) 9977–10002.

[29] M. Abdel-Basset, M. Mohamed, A novel and powerful framework based on neutrosophic sets to aid patients with cancer, Fut. Gener. Comput. Syst. 98 (2019) 144–153.

[30] M.R. Hashmi, M. Riaz, F. Smarandache, m-Polar neutrosophic topology with applications to multi-criteria decision-making in medical diagnosis and clustering analysis, Int. J. Fuzzy Syst. 22 (1) (2020) 273–292.

[31] M. Aslam, O.H. Arif, R.A.K. Sherwani, New diagnosis test under the neutrosophic statistics: an application to diabetic patients, BioMed. Res. Int. 2020 (2020).

[32] G.N. Nguyen, L.H. Son, A.S. Ashour, N. Dey, A survey of the state-of-the-arts on neutrosophic sets in biomedical diagnoses, Int. J. Mach. Learn. Cybern. 10 (1) (2019) 1–13.

[33] G. Dhingra, V. Kumar, H.D. Joshi, A novel computer vision based neutrosophic approach for leaf disease identification and classification, Measurement 135 (2019) 782–794.

[34] Y. Guo, H.-D. Cheng, New neutrosophic approach to image segmentation, Pattern Recognit. 42 (5) (2009) 587–595.

[35] F.D. Shadrach, G. Kandasamy, Neutrosophic cognitive maps (NCM) based feature selection approach for early leaf disease diagnosis, J. Ambient. Intell. Humaniz. Comput. 12 (5) (2021) 5627–5638.

[36] B. Amin, A. Salama, I.M. El-Henawy, K. Mahfouz, M.G. Gafar, Intelligent neutrosophic diagnostic system for cardiotocography data, Comput. Intell. Neurosci. (2021) 2021.

[37] S.H. Wady, R.Z. Yousif, H.R. Hasan, A novel intelligent system for brain tumor diagnosis based on a composite neutrosophic-slantlet transform domain for statistical texture feature extraction, BioMed. Res. Int. 2020 (2020).

[38] F. Özyurt, E. Sert, E. Avci, E. Dogantekin, Brain tumor detection based on convolutional neural network with neutrosophic expert maximum fuzzy sure entropy, Measurement 147 (2019) 106830.

[39] E. Sert, D. Avci, Brain tumor segmentation using neutrosophic expert maximum fuzzy-sure entropy and other approaches, Biomed. Signal. Process. Control. 47 (2019) 276–287.

[40] D.S. Irene, T. Sethukarasi, Efficient kernel extreme learning machine and neutrosophic C-means-based attribute weighting method for medical data classification, J. Circuits Syst. Comput. 29 (16) (2020) 2050260.

[41] M. Ali, M. Khan, N.T. Tung, Segmentation of dental X-ray images in medical imaging using neutrosophic orthogonal matrices, Expert. Syst. Appl. 91 (2018) 434–441.

[42] N. Mostafa, K. Ahmed, I. El-Henawy, Hybridization between deep learning algorithms and neutrosophic theory in medical image processing: a survey, Neutrosophic Sets Syst. 45 (2021) 378–401.

Chapter 20

Comparative behavior of systolic blood pressure under indeterminacy

Azhar Ali Janjua[1], Muhammad Aslam[2] and Zahid Ali[3]
[1]Government Graduate College, Hafizabad, Pakistan, [2]Department of Statistics, Faculty of Science King Abdulaziz University, Jeddah, Saudi Arabia, [3]Services Hospital, Lahore, Pakistan

20.1 Introduction

Cardiovascular is the global top cause of death which is responsible for about 16 percent world's total deaths, arising from two million deaths to 8.9 million deaths[1]. Blood pressure and pulse rate are major signals of threat for stroke and cardiovascular [1,2]. Globally, stroke and cardiovascular are the prominent origins of morbidity and mortality [3,4]. Traditionally, monitoring blood pressure and heart rate is the two important quantification dossiers that a physician may practice to visualize the health status during primary medical examination in a cost-effective manner [5]. Blood pressure is the measure of pressure in the arteries. Pulse rate which is also known as heart rate denotes the number of times in a minute the heart beats. Usually, pulse readings remain between 60 and 100 beats. Both measures evolve together, but their combination conveys different symptoms which arise within the human body [6,7]. Blood pressure and pulse imbalance may result in straining the heart with a variety of symptoms, like dizziness, fatigue, difficulty in exercising, fainting, weakness, confusion, shortness of breath, and severe conditions like low heart rate and high blood pressure may lead to cardiac arrest [8].

The heart pumps out blood into arteries during heartbeat. Blood flow remains maximum during heartbeat which is called systolic blood pressure (SBP). Table A1 shows the systolic blood pressure classification suggested by American Heart Association (AHA). It is the highest measure of blood pressure. The heart refills with blood between the beats, and the heartbeat

1. Retrieved on 30 November 2021: https://www.who.int/news-room/fact-sheets/detail/the-top-10-causes-of-death.

Cognitive Intelligence with Neutrosophic Statistics in Bioinformatics.
DOI: https://doi.org/10.1016/B978-0-323-99456-9.00012-X

remains lowest during that time [9]. Blood flow during the pause before the next beat is called diastolic blood pressure (DBP). Table A2 depicts the classification of DBP suggested by AHA. Normally, DBP in relaxation is 80 mm Hg or a bit lower [10]. Blood pressure measurement consists of these readings which vary due to many factors, such as the pulse of bloodstream and the force it applies adjusted continuously. Sakhuja et al. [11] called systolic and diastolic blood pressure greater than 140 and 90 mm Hg, respectively. An investigation from the Australian National University established that an optimal blood pressure level (110/70 mm Hg) seems to retain the brain 6 months younger compared with (120/80 mm Hg) chronological age [12]. Owing to the association between SBP and DBP, with prevailing symptoms of high SBP, the DBP seems greater even at rest. Low DBP is predicted with dehydration and severe bleeding, or it may occur with widening of arteries [10].

High SBP is generally initiated with contraction of the arteries by making it difficult for the heart to push blood into vessels which is called hypertension, and SBP lower than normal is called hypotension [13]. Blood pressure readings are not free from errors or inaccuracies [14]. During exercising, heart muscle pushes the blood with greater power or at times when the heart rate is amplified which increased SBP but is normal. However, high SBP while resting is considered problematic in high blood pressure to diagnose hypertension.

Sometimes, low blood pressure takes place by changing positions abruptly, with standing up due to gravity pulling down the blood, commonly called orthostatic hypotension [8]. Too low SBP measurements (called hypotension) are serious enough, it possibly may cause fainting, dizziness, or light-headedness, and ignoring it may lead to organs such as kidneys to pursue shutting down [13,15]. Hypotension may occur due to various reasons, for example low amount of blood flowing into the body, serious dehydration, major bleeding may leave small amount of blood to push, heart muscle becomes too weak to push blood as it happens in cardiomyopathy or arteries widen as in vasovagal syncope. Many factors may lead to combination of high blood pressure with low heart rate, like the thickened heart tissue may remodel to beat harder electrical impulses and the pulse may slow down to transmit the impulses. Blood pressure medication, for example, beta-blockers, may weaken the pulse to reduce the workload on the heart. Traumatic injuries and bleeding enhance pressure on the brain called the Cushing reflex.

Pulse pressure is the difference between systolic and diastolic blood pressure to clue for heart situation [9]. A wider pulse pressure shows aortic valve regurgitation while the narrow pulse pressure is the sign of aortic stenosis. High pulse pressure may sign stiffness in the arterial walls and risk of coronary artery and myocardial infarction [16,17]. It may be associated with condensed coronary perfusion and therefore, an incidence cardiovascular is forecasted which rolled out the reducing arterial stiffness.

Ultimately, the objective should be at systolic and diastolic rather than reducing pulse pressure.

In Indian Ayurveda and Traditional Chinese Medicine (TCM) estimated wrist pulse is used to examine the health status in a cost-effective way [5]. The association between SBP and DBP imitates arterial properties. Single-cuff enumeration of blood pressure is not free from errors, and the dual-cuff method proves to be more reliable [14]. The prevailing techniques for measuring blood pressure (sphygmomanometry and oscillometry) and heart rate (electrocardiography) necessitate an expandable cuff, which are prominent and troublesome for intensive care [18,19]. Heart rate [20,21] and blood pressure [22,23] are evaluated using signal handling learning algorithms. These methods [21,23] have concentrated on nursing a single physiological parameter, that is, blood pressure or heart rate. Panwar et al. [24] suggested sensor technology to evaluate the physiological parameter, suggested electro-optical technology for monitoring, and developed framework for simultaneous enumeration of heart rate and blood pressure with single channel. Major risk factor for heart disease is hypertension and stroke [6,7,25]. Blood pressure and pulse are two important readings for healthy body [6,7]. Non-diabetes patients constraining SBP less than 120 mm Hg rather than 140 mm Hg show lesser cardiovascular events [26]. Low DBP is unfavorable with identified coronary artery [27]. Yang et al. [28] asserted conventional measure of DBP in association with all-cause of mortality. Hayase [29] developed a dynamic model with input of pulse rate for daily continuous blood pressure which shows positive impact of a healthy society. Myocardial infarction risk is a particular subject of DBP as it is the major motive for coronary artery filling [4].

This study is conducted to evaluate the comparative association between SBP, DBP, heart rate, and body temperature. The literature is evident in utilizing the classical statistical analysis by using crisp numbers while the biological variables such as used in this study vary in boundaries and remained uncertain within these boundaries. Classical approach is incapable to capture the indeterminacy and, ultimately, ignores it, which renders the capacity vulnerable to deal with the biological behavior of these variables. A new approach which incorporates indeterminacy and considers the random variables in boundaries is called the neutrosophic approach by considering the neutrosophic random variable which is extended from classical statistics. In this regard, classical statistics is comparatively analyzed with neutrosophic statistics by displaying the formation of the data, enumeration of summary statistics, correlation coefficients, and regression analysis with calculation of indeterminacy. The rest of the study is organized as follows: the next section shows the background and motivation and methodology; then interpretation and discussion are presented, and afterward, conclusion is drawn. In the last, references and appendix are provided.

20.2 Background and motivation

The validation of a logical response to some erraticism may be authenticated on the basis of some statistic(s). Sample represents population and, however, may not exactly determine the characteristics under study, and the approximation is used. Classical statistics provides the determined estimates and bypasses the indeterminacy which alienates neutrosophic statistics from its counterpart. Further, estimates based on some sample indeterminacy are referred to neutrosophic statistics. Classical statistics deals with randomness only and is ignorant of indeterminacy, while neutrosophic statistics is the extension of classical statistics [30] and considers the indeterminacy part as well along with randomness. In classical statistics, the analysis is based on determined numbers, while in neutrosophic statistics, data remain in intervals and estimated values lie within the interval. Indeterminacy of expected values between lower and upper limits is ignored in classical statistics. In the case of zero indeterminacy, neutrosophic statistics approaches to classical statistics. Smarandache [31] explained the distinction between randomness and indeterminacy. Table 20.1 explains the differences in related approaches.

TABLE 20.1 Clarification of statistical approaches.

Classical approach	Fuzzy approach	Intuitionistic fuzzy approach	Neutrosophic approach
Classical approach is applied for the analysis of the data when all observations/ parameters in the sample or the population are precise and determined	The fuzzy approach is applied for the analysis of the data having imprecise, uncertain, and fuzzy observations/ parameters. The statistics is based on fuzzy approach and does not consider the measure of indeterminacy	Intuitionistic fuzzy is the extension of the classical fuzzy logic. The IF is considered membership and nonmembership, which belong to the real unit interval. Therefore, the statistics is based on IF and will be the extension of fuzzy approach	The neutrosophic approach is based on the idea of neutrosophic logic. The neutrosophic logic is the extension of the fuzzy logic and is considered the measure of indeterminacy. Therefore, the neutrosophic approach is the extension of classical approach which deals to analyze under uncertainty

(*Continued*)

TABLE 20.1 (Continued)

Classical approach	Fuzzy approach	Intuitionistic fuzzy approach	Neutrosophic approach
Limitations			
Classical approach can only be applied when all data or parameters are determined and precise	The fuzzy approach will be applied only when some observations/ parameters are fuzzy	The IF will be applied only when membership and nonmembership, which belong to the real unit interval	The neutrosophic approach is applied under the uncertainty environment. It reduces to classical approach when all observations/ parameters are determined

Source: From Aslam, M., Neutrosophic analysis of variance: application to university students. Complex. Intell. Syst. 5 (4) (2019) 403–407.

Classical statistics practices average statistic assessment to assess the control of a variable on other variables, while the impact of biological observations on human health varies from negative to positive or vice versa beyond some boundaries. The application of classical methodology in investigations concerning biological observations seems misrepresentative as it hindered the amplitude where the observations exceeding limits may cause unfavorable consequences. The way the classical approach mishandled the biological control of human health ultimately hinders for the adaptation of more suitable policy description.

Deneutrosophication is the process of transforming the neutrosophic data into classical form. Smarandache [32] explained the process either by considering the average value or the midpoint of interval data. This transformation may specifically be the requirement of the researcher. Here, deneutrosophication is used for the comparison of analysis between classical statistics and neutrosophic statistics.

20.3 Methodology

Smarandache [33] explained neutrosophic logic as the generalization of fuzzy logic that considers indeterminacy in intervals. As the neutrosophic statistics is the extension of classical statistics worthwhile under uncertainty, neutrosophic regression is the extended form of classical regression appropriate under an indeterminate environment. For further details about the solicitation of neutrosophic regression, it may be comprehended from Smarandache [31–33],

Aslam and Albassam [34], Aslam [35,36], and Janjua et al. [37,38]. Application of classical approach is traditionally applied to investigate the association of biological observations. However, recent literature witnessed the suitability of neutrosophic approach for investigation in biological sciences. Smarandache [39] explained the study of genetics using neutrosophic, Al Shumrani et al. [40] developed neutrosophic triplet topology which suggested multi-criteria decision-making of enzymes on selected DNA, Saeed et al. [41] probed infectious syndromes and suggested that the neutrosophic approach is the most effective for dealing with multi-criteria decision-making complications which is flexible, robust, and simple, Khalifa et al. [42] investigated COVID-19 X-ray based on neutrosophic theory to transfigure the images to grayscale spatial medical domain to the neutrosophic domain which attains highest possible accuracy and suggested to achieve better testing using same methodology, Basha et al. [43] investigated for cost-effective RT-PCR test based on chest X-ray using neutrosophic logic and suggested the model a powerful, highly sensitive, and simple for COVID-19 acknowledgment, Aslam et al. [36] presented the diagnostic tests using this approach and proposed for effective use in medical science and biostatistics, and Mukhamediyeva and Egamberdiev [44] used neutrosophic numbers' approach and developed a decision-making algorithm for ordering of medical emblems.

Assume that $S_N \epsilon [S_L, S_U]$ is a neutrosophic random variable, where S_L and S_U are the lower value and upper value of the indeterminacy interval. The neutrosophic form of $S_N \epsilon [S_L, S_U]$ can be written as $S_N = S_L + S_U I_N; I_N \epsilon [I_L, I_U]$, where S_L is the determined part and $S_U I_N$ is indeterminate part of the neutrosophic variable, and $I_N \epsilon [I_L, I_U]$ presents the indeterminacy interval. Note here that the neutrosophic random variable $S_N = S_L + S_U I_N; I_N \epsilon [I_L, I_U]$ reduces to variable S_L under classical statistic if $I_L = 0$ [32,34].

Neutrosophic random variable is derived from a neutrosophic normal distribution with neutrosophic mean, say $\mu_N \in [\mu_L, \ \mu_U]$, and neutrosophic standard deviation, say $\sigma_N \in [\sigma_L, \sigma_U]$ [30]. Classical analysis of variance (ANOVA) test is incapable to deal with fuzzy, uncertain, and imprecise observations. The neutrosophic approach is suggested to investigate the data having uncertainty [30]. By following Smarandache [32] and Aslam and Albassam [34], the neutrosophic correlation $r_N \epsilon [r_L, r_U]$ between two random variables $S_N = S_L + S_U I_N; I_N \epsilon [I_L, I_U]$ and $P_N = P_L + P_U I_N; I_N \epsilon [I_L, I_U]$ is defined as follows

$$r_N = \frac{n_N \sum sp - \sum s \sum p}{\sqrt{\left\{n_N \sum s^2 - \left(\sum s\right)^2\right\}\left\{n_N \sum p^2 - \left(\sum p\right)^2\right\}}}; \ n_N \epsilon [n_L, n_U] \tag{20.1}$$

The neutrosophic form $r_N \epsilon [r_L, r_U]$ may be written as

$$r_N = r_L + r_U I_N; I_N \epsilon [I_L, I_U] \tag{20.2}$$

Note that in case of zero indeterminacy, $r_N \epsilon [r_L, r_U]$ will approach to correlation under classical statistics. Under neutrosophic statistics, simple linear regression may be defined as

$$P_N = a_N + b_N S_N; a_N \epsilon [a_L, a_U], b_N \epsilon [b_L, b_U] \tag{20.3}$$

where $a_N \epsilon [a_L, a_U]$ and $b_N \epsilon [b_L, b_U]$ are intercept and slope, respectively, under neutrosophic approach.

The neutrosophic form of $P_N \epsilon [P_L, P_U]$ can be written as follows

$$P_N = P_L + P_U I_N; I_N \epsilon [I_L, I_U] \tag{20.4}$$

Eq. (20.4) approaches to classical statistic if no indeterminate value is recorded.

20.4 Sources of data

Primary and secondary healthcare department is responsible for the delivery of essential and effective health services in Punjab Province, Pakistan. Punjab Health Facilities Management Company (PHFMC) on behalf of the health department engages in providing the required services. Basic Health Unit (BHU) is the first-level healthcare unit under the supervision of qualified doctors which usually covers around 10,000 to 25,000 population. Three months (June 2021 to August 2021) of patient's daily data who visited BHU with reporting gastritis (authenticated and reported by a qualified medical doctor) are collected and arranged in neutrosophic formation.

20.5 Data description

Table 20.2 presents the observations of sample data after deneutrosophication. It shows the crisp observations which are used for further analysis under classical statistics.

TABLE 20.2 Sample data under classical analysis.

Obs.	SYS	DST	HBT	TMP	Obs.	SYS	DST	HBT	TMP
1	115	75	78	99	36	100	65	78	100
2	115	70	80	99	37	105	70	73	99
3	115	65	84	99	38	115	80	72	99
4	115	70	78	98	39	100	75	78	99
5	120	65	80	99	40	110	70	71	99

(Continued)

TABLE 20.2 (Continued)

Obs.	SYS	DST	HBT	TMP	Obs.	SYS	DST	HBT	TMP
6	125	80	80	98	41	115	75	71	98
7	115	80	72	98	42	100	75	78	100
8	110	80	74	99	43	105	70	73	99
9	105	70	77	99	44	110	80	72	98
10	110	65	78	99	45	105	80	71	99
11	115	70	73	100	46	105	70	78	99
12	115	70	79	100	47	95	65	80	99
13	90	75	84	100	48	100	70	78	99
14	105	70	76	100	49	115	70	71	98
15	110	75	75	99	50	120	80	72	98
16	115	75	75	100	51	100	70	78	100
17	115	70	72	99	52	105	80	77	99
18	110	70	75	99	53	110	75	71	98
19	110	75	73	99	54	95	70	78	100
20	110	60	76	100	55	110	80	73	99
21	113	65	75	100	56	110	75	73	99
22	100	73	76	100	57	105	70	77	100
23	120	75	78	99	58	110	65	79	99
24	130	80	76	99	59	105	80	76	99
25	105	80	80	101	60	95	70	81	98
26	115	75	59	99	61	95	70	81	99
27	95	75	82	100	62	110	70	78	99
28	100	70	78	99	63	110	85	72	98
29	105	80	74	99	64	105	65	77	99
30	120	75	72	99	65	100	70	79	99
31	110	75	75	99	66	105	40	78	100
32	105	75	73	99	67	105	75	76	99
33	100	75	81	100	68	105	70	74	99
34	110	75	73	99	69	100	70	80	99
35	120	80	70	99	70	100	70	81	98

TABLE 20.3 Sample data under neutrosophic analysis.

Obs.	SYS_N	DST_N	HBT_N	TMP_N	Obs.	SYS_N	DST_N	HBT_N	TMP_N
1	[140, 90]	[80, 70]	[70, 85]	[98, 99]	36	[110, 90]	[80, 50]	[72, 84]	[98, 102]
2	[140, 90]	[80, 60]	[70, 90]	[98, 100]	37	[110, 100]	[80, 60]	[72, 74]	[99, 98]
3	[150, 80]	[70, 60]	[78, 90]	[98, 100]	38	[130, 100]	[90, 70]	[72, 72]	[99, 98]
4	[130, 100]	[70, 70]	[70, 85]	[98, 98]	39	[110, 90]	[80, 70]	[72, 84]	[98, 100]
5	[150, 90]	[70, 60]	[78, 82]	[100, 98]	40	[120, 100]	[80, 60]	[72, 70]	[98, 99]
6	[160, 90]	[90, 70]	[78, 82]	[98, 98]	41	[130, 100]	[80, 70]	[70, 72]	[98, 98]
7	[120, 110]	[80, 80]	[74, 70]	[98, 98]	42	[110, 90]	[80, 70]	[72, 84]	[98, 101]
8	[120, 100]	[90, 70]	[73, 75]	[99, 99]	43	[110, 100]	[70, 70]	[72, 74]	[98, 99]
9	[120, 90]	[80, 60]	[72, 82]	[98, 99]	44	[120, 100]	[70, 90]	[72, 72]	[98, 98]
10	[130, 90]	[70, 60]	[70, 85]	[98, 100]	45	[110, 100]	[90, 70]	[72, 70]	[99, 98]
11	[130, 100]	[80, 60]	[70, 75]	[99, 101]	46	[120, 90]	[80, 60]	[72, 84]	[98, 100]
12	[140, 90]	[80, 60]	[72, 85]	[98, 101]	47	[100, 90]	[70, 60]	[74, 85]	[98, 100]
13	[100, 80]	[80, 70]	[78, 90]	[99, 100]	48	[110, 90]	[80, 60]	[72, 84]	[98, 100]
14	[120, 90]	[80, 60]	[72, 80]	[99, 101]	49	[130, 100]	[70, 70]	[70, 72]	[98, 98]
15	[130, 90]	[80, 70]	[69, 80]	[99, 99]	50	[140, 100]	[90, 70]	[70, 74]	[98, 98]
16	[140, 90]	[80, 70]	[68, 81]	[98, 101]	51	[110, 90]	[80, 60]	[72, 84]	[98, 101]
17	[130, 100]	[70, 70]	[70, 74]	[98, 99]	52	[110, 100]	[90, 70]	[80, 74]	[99, 98]

(*Continued*)

TABLE 20.3 (Continued)

Obs.	SYS_N	DST_N	HBT_N	TMP_N	Obs.	SYS_N	DST_N	HBT_N	TMP_N
18	[130, 90]	[80, 60]	[70, 80]	[99, 98]	53	[120, 100]	[80, 70]	[72, 70]	[98, 98]
19	[120, 100]	[80, 70]	[72, 74]	[98, 99]	54	[100, 90]	[80, 60]	[72, 84]	[99, 100]
20	[130, 90]	[60, 60]	[70, 82]	[99, 100]	55	[120, 100]	[90, 70]	[72, 74]	[99, 99]
21	[135, 90]	[70, 60]	[70, 79]	[98, 101]	56	[120, 100]	[80, 70]	[72, 74]	[99, 98]
22	[110, 90]	[80, 65]	[70, 82]	[99, 101]	57	[120, 90]	[80, 60]	[72, 82]	[98, 101]
23	[150, 90]	[80, 70]	[70, 85]	[98, 100]	58	[130, 90]	[70, 60]	[72, 85]	[98, 99]
24	[160, 100]	[90, 70]	[78, 74]	[98, 100]	59	[110, 100]	[80, 80]	[74, 78]	[99, 98]
25	[120, 90]	[90, 70]	[74, 85]	[100, 101]	60	[100, 90]	[80, 60]	[76, 85]	[98, 98]
26	[130, 100]	[80, 70]	[72, 46]	[98, 99]	61	[100, 90]	[80, 60]	[76, 86]	[98, 100]
27	[100, 90]	[80, 70]	[74, 90]	[99, 100]	62	[130, 90]	[80, 60]	[72, 84]	[98, 100]
28	[110, 90]	[80, 60]	[72, 84]	[98, 100]	63	[120, 100]	[80, 90]	[70, 74]	[98, 98]
29	[110, 100]	[80, 80]	[72, 76]	[98, 99]	64	[120, 90]	[70, 60]	[72, 82]	[98, 100]
30	[140, 100]	[80, 70]	[70, 74]	[98, 100]	65	[110, 90]	[80, 60]	[74, 84]	[98, 100]
31	[130, 90]	[90, 60]	[70, 80]	[98, 100]	66	[120, 90]	[60, 100]	[72, 84]	[98, 101]
32	[110, 100]	[80, 70]	[72, 74]	[99, 98]	67	[120, 90]	[90, 60]	[72, 80]	[98, 99]
33	[110, 90]	[80, 70]	[74, 88]	[98, 101]	68	[110, 100]	[80, 60]	[74, 74]	[100, 98]
34	[120, 100]	[80, 70]	[74, 72]	[98, 100]	69	[110, 90]	[80, 60]	[74, 85]	[98, 100]
35	[140, 100]	[90, 70]	[70, 70]	[98, 100]	70	[110, 90]	[80, 60]	[74, 88]	[98, 98]

TABLE 20.4 Data description under classical statistics.

Variables	Obs.	Mean	Std. Dev.	Min.	Max	Skew.	Kurt.
SYS	70	108.1071	7.8301	90	130	0.1410	2.9047
DST	70	72.8929	5.1365	60	85	0.0293	2.4941
HBT	70	75.8429	3.9467	59	84	-0.8977	6.1396
TMP	70	98.8786	0.5863	98	100.5	0.3180	2.7322

where
SYS = systolic blood pressure, DST = diastolic blood pressure
HBT = heartbeat rate, TMP = body temperature
Neutrosophic variables are represented with "N" subscription.

Table 20.3 depicts the data observations in neutrosophic format which are used for further analysis under neutrosophic approach.

Table 20.4 presents the data description under classical approach. The deviation of skewness from zero and kurtosis from three signals the departure from symmetry and normality, respectively. The usage of single estimated (average) value in classical statistics to determine the relationship among variables may be inaccurate. The amplitude/range of the variables varies; thus ignoring the potential variability seems a significant potential for indeterminacy which conventional approach is incapable to capture. Similarly, other variables are conventional and are self-explanatory with said reservations.

Table 20.5 explains the summary statistics of variables under neutrosophic approach. The description of variables comprises of two parts; first, a determinate part, and second, the indeterminate part. As the indeterminate part approaches zero, the neutrosophic statistics approaches classical. The inclusion of indeterminacy empowers neutrosophic approach to provide the estimates in boundaries rather than the crisp values. In neutrosophic approach, the skewness and kurtosis point out the departure from symmetry and normality along with the direction to this departure. The classical approach is incompetent to engulf this dynamism to indeterminacy.

Table 20.6 shows the correlation matrix under classical analysis where a single statistics is estimated to analyze the association of variables of the study. Ignoring the indeterminacy is limiting the scope which may lead to misleading conclusion(s).

Table 20.7 explains the correlation coefficient matrix using neutrosophic statistics. The coefficients are estimated within intervals by keeping

TABLE 20.5 Data description under neutrosophic statistics.

Var.	Obs.	Mean	Std. Dev.	Min.	Max	Skew.	Kurt.
SYS_N	[70]	$[94 + 122.2\ I_N]$	$[5.8 + 14.4\ I_N]$	$[80 + 100\ I_N]$	$[110 + 160\ I_N]$	$[0.2 + 0.6\ I_N]$	$[2.5 + 2.9\ I_N]$
DST_N	[70]	$[66.1 + 79.7\ I_N]$	$[7.3 + 6.4\ I_N]$	$[50 + 60\ I_N]$	$[90 + 90\ I_N]$	$[1.0 - 0.3\ I_N]$	$[4.7 + 3.5\ I_N]$
HBT_N	[70]	$[79.3 + 72.4\ I_N]$	$[7.1 + 2.4\ I_N]$	$[46 + 68\ I_N]$	$[90 + 80\ I_N]$	$[-1.4 + 1.1\ I_N]$	$[7.9 + 4.2\ I_N]$
TMP_N	[70]	$[99.4 + 98.3\ I_N]$	$[1.1 + 0.6\ I_N]$	$[98 + 98\ I_N]$	$[102 + 100\ I_N]$	$[0.1 + 1.4\ I_N]$	$[1.8 + 3.9\ I_N]$

TABLE 20.6 Correlation matrix under classical analysis.

Variables	Systolic	Diastolic	Heartbeat	Temp.
SYS	1	..	..	..
DST	0.2327	1	..	..
HBT	−0.4266	−0.326	1	..
TMP	−0.2836	−0.2907	0.3486	1

TABLE 20.7 Correlation matrix under neutrosophic statistics.

Variables	SYS_N	DST_N	HBT_N	TMP_N
SYS_N	$[1 + 1\ I_N]$	..	..	..
DST_N	$[0.433 + 0.015\ I_N]$	$[1 + 1\ I_N]$	..	..
HBT_N	$[-0.792-0.140\ I_N]$	$[-0.338 + 0.166\ I_N]$	$[1 + 1\ I_N]$	..
TMP_N	$[-0.546-0.149\ I_N]$	$[-0.406 + 0.149\ I_N]$	$[0.433 + 0.202\ I_N]$	$[1 + 1\ I_N]$

indeterminacy in the calculation which enables greater flexibility and reliable association between variables under study. Classical approach restricts this uncertainty and distorts the analysis capacity.

Table 20.8 expresses the comparative analysis of systolic blood pressure from diastolic heart rate and body temperature. The comparison elucidates the dynamics of neutrosophic approach by incorporating indeterminacy in the estimation process. Classical approach enumerates on average and estimates a single crisp value and ignores the prevailing uncertainty "I_N." Neutrosophic approach provides estimates in intervals which offer flexibility and elaborate the relationship of variables from a broader perspective. The explanation of the two approaches carries differences as the classical approach provides single coefficient while the neutrosophic approach estimates coefficients in intervals and is different from the first one because of the inclusion of indeterminacy. The magnitude of the coefficients and the explanatory power of the estimated model vary as well. The newly developed approach seems more appropriate and reliable; however, it is in the initial stage which requires further post-estimation tests which fall outside the scope of this study.

TABLE 20.8 Comparison of regression analysis.

Variables	Classical	Neutrosophic
DST	0.10	$[0.17 + 0.12 I_N]$; $I_N \in [0, 0.417]$
HBT	−0.71	$[-0.50 - 0.72\, I_N]$; $I_N \in [0, 0.306]$
TMP	−1.88	$[-0.93 - 3.40\, I_N]$; $I_N \in [0, 0.726]$
Constant	339.79	$[215.06 + 498.58\, I_N]$; $I_N \in [0, 0.569]$
R-squared	0.21	$[0.71 + 0.04\, I_N]$; $I_N \in [0, 16.750]$

20.6 Interpretation and discussion

Neutrosophic statistics is the extension of classical statistics, and the consideration of indeterminacy led to this extension. The provision of estimates within an interval is conducive to creating flexibility. Table 20.2 presents the classical sample data, and Table 20.3 explains the neutrosophic sample data. Neutrosophic data are in intervals which clarify the movement of these observations within the range. However, the indeterminacy exists about the occurrence of true value within the range. The values beyond the interval, which a variable may not opt for, may be called false value.

Table 20.4 shows the classical sample data under consideration consisting of 70 observations and is balanced. The average value (mean) of "SYS" is 108.107 with 7.83 standard deviations, and the variable ranges from 90 to 130 (minimum to maximum). Table 20.5 explains that "SYS_N" consists of 70 observations with indeterminate mean value which varies from 94 to 122.21 with indeterminacy of "I_N" (Ref. Table A3). The standard deviation ranges from 5.8 to 14.4 which shows that the variation in lower boundaries of systolic blood pressure is smaller compared with fluctuations in its upper limit. The plausible scenario may be that the prevailing health-related activities are stimulating the systolic blood pressure, and mostly, the patients with higher systolic blood pressure may be observed, and the preparations to deal with higher blood pressure patients are preferable. Minimum and maximum values of "SYS_N" interval are 80 to 100, while the maximum values interval are 110 to 160. Skewness shows departure from symmetry, and the direction of this major departure is inclined toward the upper limits of the variable. Similarly, the kurtosis shows a smaller departure from normality while inclined to the upper boundaries of the variable. "DST_N," "HBT_N", and "TMP_N" shows greater variation in lower limits relative to upper limits.

Table 20.7 shows classical correlation coefficient matrix which conveys the association among the variables. "HBT" and "TMP" show negative association among "SYS" and "DST." Table 20.7 explains neutrosophic correlation coefficient matrix where in addition to the negative association among

"SYS_N," "HBT_N", and "TMP_N," the variation in magnitudes between lower and upper limits is also clearly varied. Greater negative association among lower limits and smaller negative association (compared with classical and lower limits) are estimated. In the case of "DST_N", lower limits show negative association while upper limits explain the positive association with "HBT_N" and "TMP_N." In addition to greater explanation, indeterminacy is also measurable (Table A4) in neutrosophic approach.

Table 20.8 illuminates the comparative regression analysis of the two approaches. In classical approach, systolic blood pressure is influenced by diastolic blood pressure which brings about 0.1 units change in the same direction while heartbeat and body temperature bring about 0.71 and 1.88 units change in inverse direction. This model explains about 21% variation in systolic blood pressure. Neutrosophic approach estimates an interval for which the expected value may fall with estimation of an indeterminacy in spite of a conventional crisp value, which may wider the confidence for policy insinuation. One unit change in diastolic may bring a change in systolic blood pressure within an interval of 0.17 to 0.12 units. For lower values of systolic blood pressure, diastolic blood pressure may bring a change of about 0.17 units, while for upper limit, this impact reduces about 0.12 units, with indeterminacy of "I_N" (Ref. Table 20.8). Both the approaches enumerate similar directions whereas the elaborations of estimates bring wider changes in information, interpretation, and in determining the policy. Neutrosophic approach explains that this model explains about 71% variation in lower values while just 4% variation for larger values of systolic blood pressure. It may be perceived that the model is unsuitable for determining the upper values of systolic blood pressure and further investigation is required.

20.7 Conclusion

The study is conducted to comparatively analyze the association of systolic blood pressure, diastolic blood pressure, heart rate, and body temperature using classical and neutrosophic approaches. Classical analysis shows the crispy rigid statistics which proves inelastic and may lead to inaccurate policy prescription, while the new approach provides the wider insides of the analysis by estimating the lower and upper boundaries of the statistics. These statistics showed new dimensions for policy prescription while the behaviors of the biological variables alter from lower to upper boundaries. Neutrosophic approach shows a wider scope and is reliable when compared with classical approach. The dynamic behavior of the variables leads classical approach being incapable to draw policy prescription while neutrosophic approach is empowered with its indeterminacy enumeration and occurrence of expected values in boundaries. Neutrosophic approach is found more sensitive than outliers, and further development of post-estimation tests is suggested which are out of the scope of this study.

Appendix

TABLE A1 Classification of systolic blood pressure.

Systolic blood pressure range	Classification
Up to 120 mmHg	Normal
120–129 mmHg	Elevated
129–139 mmHg	High blood pressure stage 1
139–180 mmHg	High blood pressure stage 2
180–200 mmHg	Hypertensive crisis

Source: Based on American Heart Association (AHA), Heart disease and stroke statistics 2018 at-a-glance. On-line at: http://www.heart.org/idc/groups/ahamahpublic/@wcm/@sop/@smd/documents/downloadable/ucm_491265.pdf, 2018.

TABLE A2 Classification of diastolic blood pressure.

Systolic blood pressure range	Classification
Up to 80 mmHg	Normal
80–89 mmHg	High blood pressure stage 1
89–120 mmHg	High blood pressure stage 2
120–140 mmHg	Hypertensive crisis

Source: Based on American Heart Association (AHA), Heart disease and stroke statistics 2018 at-a-glance. On-line at: http://www.heart.org/idc/groups/ahamahpublic/@wcm/@sop/@smd/documents/downloadable/ucm_491265.pdf, 2018.

TABLE A3 Indeterminacy of data description.

Var.	Mean	Std. Dev.	Min.	Max	Skew.	Kurt.
SYS_N	$I_N \in [0,0.231]$	$I_N \in [0,0.597]$	$I_N \in [0,0.200]$	$I_N \in [0,0.313]$	$I_N \in [0,0.667]$	$I_N \in [0,0.138]$
DST_N	$I_N \in [0,0.171]$	$I_N \in [0,0.141]$	$I_N \in [0,0.167]$	$I_N \in [0,0.000]$	$I_N \in [0,2.333]$	$I_N \in [0,0.343]$
HBT_N	$I_N \in [0,0.095]$	$I_N \in [0,1.958]$	$I_N \in [0,0.324]$	$I_N \in [0,0.125]$	$I_N \in [0,0.273]$	$I_N \in [0,0.881]$
TMP_N	$I_N \in [0,0.011]$	$I_N \in [0,0.833]$	$I_N \in [0,0.000]$	$I_N \in [0,0.020]$	$I_N \in [0,0.929]$	$I_N \in [0,0.538]$

TABLE A4 I_N **determ** I_N **acy of correlation matrix.**

Variables	SYS_N	DST_N	HBT_N	TMP_N
SYS_N	$I_N \in [0,0]$	..	..	..
DST_N	$I_N \in [0,27.867]$	$I_N \in [0,0]$	..	..
HBT_N	$I_N \in [0,4.657]$	$I_N \in [0,1.0361]$	$I_N \in [0,0]$	..
TMP_N	$I_N \in [0,2.664]$	$I_N \in [0,1.725]$	$I_N \in [0,1.144]$	$I_N \in [0,0]$

References

[1] S. Datta, A.D. Choudhury, A. Chowdhury, R. Banerjee, T. Banerjee, A. Pal, et al., U.S. Patent Application No. 15/900,774, 2019.

[2] R. Mukkamala, J.O. Hahn, Toward ubiquitous blood pressure monitoring via pulse transit time: Predictions on maximum calibration period and acceptable error limits, IEEE Trans. Biomed. Eng. 65 (6) (2017) 1410–1420. Available from: https://doi.org/10.1109/TBME0.2017.2756018.

[3] American Heart Association (AHA), Heart disease and stroke statistics 2018 at-a-glance. on-line at: http://www.heart.org/idc/groups/ahamahpublic/@wcm/@sop/@smd/documents/downloadable/ucm_491265.pdf, 2018.

[4] M. Arvanitis, G. Qi, D.L. Bhatt, W.S. Post, N. Chatterjee, A. Battle, et al., Linear and nonlinear mendelian randomization analyses of the association between diastolic blood pressure and cardiovascular events: the J-curve revisited, Circulation 143 (9) (2021) 895–906. Available from: https://doi.org/10.1161/CIRCULATIONAHA.120.049819.

[5] G.C. Suguna, S.T. Veerabhadrappa, A review of wrist pulse analysis, Biomed. Res. 30 (4) (2019) 538–545.

[6] H.K. Wall, M.D. Ritchey, C. Gillespie, J.D. Omura, A. Jamal, M.G. George, Vital signs: prevalence of key cardiovascular disease risk factors for Million Hearts 2022—United States, 2011–2016, MMWR Morb. Mortal. Wkly. Rep. 67 (2018) 983–991. Available from: https://doi.org/10.15585/mmwr.mm6735a4.

[7] M.D. Ritchey, H.K. Wall, M.G. George, J.S. Wright, US trends in premature heart disease mortality over the past 50 years: where do we go from here? Trends Cardiovasc. Med. (2019). Available from: https://doi.org/10.1016/j.tcm.2019.09.005.

[8] U.S. Department of Health & Human Services. National Institutes of Health (NIH), Orthostatic hypotension. Genetics Home Reference. Updated August 18, 2020, 2020.

[9] K.S. Tang, E.D. Medeiros, A.D. Shah, Wide pulse pressure: a clinical review, J. Clin. Hypert 22 (11) (2020) 1960–1967. Available from: https://doi.org/10.1111/jch.14051.

[10] Centers for Disease Control and Prevention (CDC & P), Measure your blood pressure. Updated November 30, 2020, 2020.

[11] S. Sakhuja, C.L. Colvin, O.P. Akinyelure, B.C. Jaeger, K. Foti, S. Oparil, et al., Reasons for uncontrolled blood pressure among US adults: Data from the US national health and nutrition examination survey, Hypertension 78 (5) (2021) 1567–1576. Available from: https://doi.org/10.1161/HYPERTENSIONAHA.121.17590.

[12] W.S. Cherbuin, et al., Optimal blood pressure keeps our brains younger, Front. Aging Neurosci, 2021. Available from: https://doi.org/10.3389/fnagi.2021.694982.

[13] M. Böhm, H. Schumacher, K.K. Teo, E. Lonn, F. Mahfoud, J.F. Mann, et al., Achieved diastolic blood pressure and pulse pressure at target systolic blood pressure (120–140 mmHg) and cardiovascular outcomes in high-risk patients: results from on target and transcend trials, Eur. Heart J. 39 (33) (2018) 3105–3114. Available from: https://doi.org/10.1093/eurheartj/ehy287.

[14] J. Jilek, M. Stork, Determination of systolic, mean and diastolic blood pressures with Dual cuff system is based on physiology, in: 2016 International Conference on Applied Electronics (AE) (pp. 117–120). IEEE, 2016. https://doi.org/10.1109/AE.2016.7577254.

[15] H. Kanegae, T. Oikawa, Y. Okawara, S. Hoshide, K. Kario, Which blood pressure measurement, systolic or diastolic, better predicts future hypertension in normotensive young adults? J. Clin. Hypertens. 19 (6) (2017) 603–610.

[16] S.S. Franklin, M.G. Larson, S.A. Khan, N.D. Wong, E.P. Leip, W.B. Kannel, et al., Does the relation of blood pressure to coronary heart disease risk change with aging? The Framingham Heart Study, Circulation 103 (2001) 1245–1249. Available from: https://doi.org/10.1161/01.CIR.103.9.1245.

[17] W.B. White, The systolic blood pressure versus pulse pressure controversy, Am. J. Cardiol. 87 (2001) 1278–1281. Available from: https://doi.org/10.1016/S0002-9149(01)01519-3.

[18] D. Biswas, N. Simões-Capela, C. Van Hoof, N. Van Helleputte, Heart rate estimation from wrist-worn photoplethysmography: A review, IEEE Sens. J. 19 (16) (2019) 6560–6570. Available from: https://doi.org/10.1109/JSEN.2019.2914166.

[19] A. Chandrasekhar, C.S. Kim, M. Naji, K. Natarajan, J.O. Hahn, R. Mukkamala, Smartphone-based blood pressure monitoring via the oscillometric finger-pressing method, Sci. Transl. Med. 10 (431) (2018). Available from: https://doi.org/10.1126/scitranslmed.aap8674.

[20] A. Reiss, I. Indlekofer, P. Schmidt, K. Van Laerhoven, Deep PPG: large-scale heart rate estimation with convolutional neural networks, Sensors 19 (14) (2019) 3079.

[21] Z. Zhang, P. Zhouyue, L. Benyuan, Troika: A general framework for heart rate monitoring using wrist-type photoplethysmographic signals during intensive physical exercise, IEEE Trans. Biomed. Eng. 62 (2) (2015) 522–531.

[22] C. Fischer, T. Penzel, Continuous non-invasive determination of the nocturnal blood pressure variation using photoplethysmographic pulse wave signals-comparison of pulse propagation time, pulse transit time and RR-interval. Doctoral Thesis; 2019.

[23] S.S. Mousavi, M. Firouzmand, M. Charmi, M. Hemmati, M. Moghadam, Y. Ghorbani, Blood pressure estimation from appropriate and inappropriate PPG signals using A whole-based method, Biomed. Signal. Process. Control. 47 (2019) 196–206.

[24] M. Panwar, A. Gautam, D. Biswas, A. Acharyya, PP-Net: A deep learning framework for PPG-based blood pressure and heart rate estimation, IEEE Sens. J. 20 (17) (2020) 10000–10011.

[25] A.S. Go, M.A. Bauman, S.M. Coleman King, et al., An effective approach to high blood pressure control: a science advisory from the American Heart Association, the American College of cardiology, and the centers for disease control and prevention, J. Am. CollCardiol 63 (2014) 1230–1238. Available from: https://doi.org/10.1016/j.jacc.2013.11.007.

[26] SPRINT Research Group, A randomized trial of intensive versus standard blood-pressure control, N. Engl. J. Med. 373 (22) (2015) 2103–2116. Available from: https://www.nejm.org/doi/full/10.1056/nejmoa1511939.

[27] A.M. Dart, Should pulse pressure influence prescribing? Aust. Prescr. 40 (1) (2017) 26. Available from: https://www.ncbi.nlm.nih.gov/pmc/articles/PMC5313243/.

[28] L.T. Yang, P.A. Pellikka, M. Enriquez-Sarano, C.G. Scott, R. Padang, S.V. Mankad, et al., Diastolic blood pressure and heart rate are independently associated with mortality in chronic aortic regurgitation, J. Am. Coll. Cardiol. 75 (1) (2020) 29–39.

[29] T. Hayase, Blood pressure estimation based on pulse rate variation in a certain period, Sci. Rep. 10 (1) (2020) 1–14.

[30] M. Aslam, Neutrosophic analysis of variance: application to university students, Complex. Intell. Syst. 5 (4) (2019) 403–407. Available from: https://doi.org/10.1007/s40747-019-0107-2.

[31] F. Smarandache, Introduction to neutrosophic measure, neutrosophic integral, and neutrosophic probability, Infinite Study (2013).

[32] F. Smarandache, Introduction to Neutrosophic Statistics, *Sitech & Education Publishing, Craiova*, 2014.

[33] F. Smarandache, Neutrosophic logic-a generalization of the intuitionistic fuzzy logic. Multispace & multistructure, Neutrosophic Transdisciplinarity 4 (2010) 396.

[34] M. Aslam, M. Albassam, Application of neutrosophic logic to evaluate correlation between prostate cancer mortality and dietary fat assumption, Symmetry 11 (3) (2019) 330. Available from: https://doi.org/10.3390/sym11030330.

[35] M. Aslam, A new sampling plan using neutrosophic process loss consideration, Symmetry 10 (5) (2018) 132. Available from: https://doi.org/10.3390/sym10050132.

[36] M. Aslam, O.H. Arif, R.A.K. Sherwani, New diagnosis test under the neutrosophic statistics: an application to diabetic patients, Bio Med. Res. Int. (2020) 2020. Available from: https://doi.org/10.1155/2020/2086185.

[37] A.A. Janjua, M. Aslam, N. Sultana, Evaluating the relationship between climate variability and agricultural crops under indeterminacy, Theor. Appl. Climatol. 142 (3) (2020) 1641–1648.

[38] A.A. Janjua, M. Aslam, N. Sultana, Z. Batool, Identification of climate induced optimal rice yield and vulnerable districts rankings of the Punjab, Pakistan, Sci. Rep. 11 (1) (2021) 1–15.

[39] F. Smarandache, Introduction to Neutrosophic Genetics, Int. J. Neutrosophic Sci. 13 (1) (2020) 23.

[40] M.A. Al Shumrani, M. Gulistan, F. Smarandache, Further theory of neutrosophic triplet topology and applications, Symmetry 12 (8) (2020) 1207. Available from: https://doi.org/10.3390/sym12081207.

[41] M. Saeed, M. Ahsan, A. Mehmood, M.H. Saeed, J. Asad, Infectious diseases diagnosis and treatment suggestions using complex neutrosophic hypersoft mapping, IEEE Access. (2021).

[42] N.E.M. Khalifa, F. Smarandache, G. Manogaran, M. Loey, A study of the neutrosophic set significance on deep transfer learning models: An experimental case on a limited covid-19 chest x-ray dataset, Cognit. Comput. (2021) 1–10.

[43] S.H. Basha, A.M. Anter, A.E. Hassanien, A. Abdalla, Hybrid intelligent model for classifying chest X-ray images of COVID-19 patients using genetic algorithm and neutrosophic logic, Soft Comput. (2021) 1–16. Available from: https://doi.org/10.1007/s00500-021-06103-7.

[44] D. Mukhamediyeva, N. Egamberdiev, Algorithm of classification of medical objects on the basis of neutrosophic numbers, InterConf (2021).

Chapter 21

Effects of wild medicinal edible plant to prevent COVID-19: a survey by neutrosophic statistics in tribal people of Kokrajhar, Assam, India

Bhimraj Basumatary[1], Jeevan Krishna Khaklary[2], Nijwm Wary[1] and Giorgio Nordo[3]

[1]Department of Mathematical Sciences, Bodoland University, Kokrajhar, Assam, India, [2]Department of Mathematics, CIT, Kokrajhar, Assam, India, [3]MIFT-Department of Mathematical and Computer Science, Physical Sciences and Earth Sciences, Messina University, Italy

21.1 Introduction and motivation

The long history of humans' ability to adapt to natural environments and to interact with nature and social circumstances is profoundly devoted to edible wild plants and animals. From the early hunter-gatherers and across different adaptation stages, plants have assumed great importance in human societies, and many people all over the world have depended on many wild species, particularly for food and medicines.

Wild edible plants are those plants with edible parts that grow naturally on farmland and fallow or uncultivated land [1]. Different wild edible plants have played a significant role in different geographical regions of the world throughout human history. Wild vegetables contribute to people's food security and health in many rural areas of the world. They may have remarkable nutrient values and can be an important source of vitamins, fibers, minerals, and fatty acids; they may also show important medicinal properties. Wild edible plants have always been an essential and widespread food source for food-insecure families living in poverty in developing countries. Wild edible plants (Figs. 21.1–21.8) are nutritionally rich and can especially supplement vitamins and micronutrients to the human body.

Cognitive Intelligence with Neutrosophic Statistics in Bioinformatics.
DOI: https://doi.org/10.1016/B978-0-323-99456-9.00011-8

FIGURE 21.1 Kaila (*Gymnopetalum cochinchinense*). (freshly plucked *Kaila* fruits).

FIGURE 21.2 Elangsi (*Alternanthera philoxeroides*). (freshly plucked alligator weeds).

FIGURE 21.3 Karokandai (*Oroxylum indicum*). (freshly plucked Indian trumpet flowers).

FIGURE 21.4 Burithokon (*Cheilocostus speciosus*). (stems of tender crepe ginger).

FIGURE 21.5 Kuntai-nara (*Solanum torvum*). (freshly plucked turkey berry fruits).

FIGURE 21.6 Maisundri (*Houttuynia cordata*). (freshly plucked fish mint leaves).

FIGURE 21.7 Sibung (*Sphenoclea zeylanica*). (freshly plucked wedgewort sprouts).

FIGURE 21.8 Sibru (*Lasia spinosa*). (freshly plucked Spiny Arum stalks).

The food and nutritional contribution and the medicinal value of wild edible plants have not been investigated fully in Kokrajhar. Therefore, the objective of this review is to explore available information about wild edible plants' nutritional contribution, supplementary role, marketability, and medicinal value, as well as to determine consumer perception toward wild edible plants in Kokrajhar.

The recent and ongoing outbreak of coronavirus disease (COVID-19) is a huge global challenge. The outbreak, which first occurred in Wuhan City, Hubei Province, China, and then rapidly spread to other provinces and more than 200 countries abroad, has been declared a global pandemic by the WHO. Those with compromised immune systems and existing respiratory, metabolic, or cardiac problems are more susceptible to the infection and are at higher risk of serious illness or even death. The present review was designed to report important functional food plants with immunomodulatory and antiviral properties. The functional food plants herein documented might not only enhance the immune system and cure respiratory tract infections but can also greatly impact the overall health of the general public. As many people in the world are now confined to their homes, the inclusion of these easily accessible plants in the daily diet may help to strengthen the immune system and guard against infection by SARS-CoV-2. This might reduce the risk of COVID-19 and initiate a rapid recovery in cases of SARS-CoV-2 infection.

21.2 Novelty of the work

The study has been taken up to analyze the effect of the current (COVID-19) situation in the tribal areas and non-tribal areas by observing the use of wild medicinal plants in different areas of the Kokrajhar district, Assam, India. It is observed by a survey that tribal people (especially those who use wild medicinal plants) in the Kokrajhar district are affected less than the non-tribal people. It has also been seen that the tribal people who frequently use wild edible medicinal plants though they have been affected by COVID-19 their recovery rate is quite well in comparison to non-tribal people. In the event that such a distinguishing situation prevails in the different areas of the district, a comparative study would be helpful in adopting effective policy and essential measures separately for such people who use wild edible medicinal plants and who do not use them, by observing different areas of Kokrajhar district. By observing this, we forwarded our study and used neutrosophic statistics and the VICKOR method to study the COVID-19 situation of the Kokrajhar district.

Das et al. [2] surveyed medicinal plants used by different tribes of Cachar district, Assam. Bhattacharya et al. [3] studied rare medicinal plants of Assam. Li et al. [4] searched Edible and Herbal Plants for the Prevention and Management of COVID-19. Duguma [5] searched Wild Edible Plant Nutritional Contribution and Consumer Perception in Ethiopia. Patel et al. [6] studied therapeutic opportunities of edible antiviral plants for COVID-19. Khadka et al. [7] studied the use of medicinal plants to prevent COVID-19 in Nepal. Yang Fan [8] worked on Food as medicine: A possible preventive measure against coronavirus disease (COVID-19). Lim et al. [9] studied Medicinal Plants in COVID-19: Potential and Limitations. We examine and compare the current growth pattern of COVID-19 between tribal and non-tribal people in different regions of Kokrajhar, Assam, India, where during the COVID-19 pandemic, tribes have increased their use of wild medicinal edible herbs and the recovery rate by COVID-19 is high in comparison to non-tribal people of the same area to describe the current state of wild medicinal edible plant use and provide critical recommendations to the relevant authorities. In addition, this study presents a multi-criteria decision-making (MCDM) procedure based on the neutrosophic VIKOR method to survey the region where most wild medicinal edible plants are used and found. Also, with the help of the neutrosophic VIKOR method, we can identify the regions where most people of the Kokrajhar district are affected and recovered.

21.3 A classification of the wild edible medicinal plant of Kokrajhar

Kokrajhar is one of the 33 districts of Assam. Kokrajhar is the gateway to Assam and other NE states of India by road and railways. The Kokrajhar district is situated on the north bank of the Brahmaputra River, sharing an international border with Bhutan to the north and an interstate border with West Bengal to the west. The district is bordered on the south by the Dhubri district and on the east by Chirang and Bongaigaon districts. The district is roughly located between the longitudes of 89°46' E and 90°38' E and the latitudes of 26°19' N and 26°54' N. According to the Census 2011, the district has a total area of 3169.22 sq. km and a population of 8,86,999 people. The district's primary ethnic groups include Bodo, Rava, Garo, Nath/Yogi, Bengali, Nepali, Santhal, Rajbongshi, and others.

The lower Brahmaputra Valley in Assam has a humid subtropical climate, which is the characteristic of the Kokrajhar district. The humidity is high, and there is a lot of rain. The district of Kokrajhar is known for its forests. The district is home to the state's most densely forested area. The currently estimated area under reserved forests is 1636.26 sq. km, indicating that reserved forests cover 51.63% of the district's entire geographical area. The district's soil is rich in nutrients and ideal for paddy farming. Forest resources abound throughout the district. A great range of wild edible medicinal plants grow well in their native habitats in the forests of Kokrajhar district and are largely utilized by Bodo people and other tribal populations of this region in their daily diet since the time unknown.

The Bodo tribes used to harvest several varieties of wild edible vegetables from the forest during the Bwisagu festival (a Bodo tribe's New Year ceremony) and then eat them cooked with meat or as mixed veggies. Many of the indigenous tree and shrub species in this area are still unknown and underutilized, and as a result, there is a pressing need to investigate them as a source of food for humans. The current research aims to identify, document, and preserve traditional knowledge of plant species, mostly wild edible vegetables utilized by Bodo tribes in the Kokrajhar district.

The present study in the district of Kokrajhar revealed that a total of 50 wild plant species mainly used for human consumption belonging to tribal families have been reported. The majority of these plants are eaten cooked as a vegetable, and some of them are eaten fried and raw or chutney by the Bodo tribe. Out of the documented plant species, the edible parts used are leaves or young shoot, tubers, petiole, stem, flower, fruit, rhizome, and root. Among these, the most commonly used edible part is leaves or young shoot. The wild edible vegetables used by the tribe of the study area are arranged with their scientific names, local name, time of availability, partly used, and medicinal values/use as shown in Table 21.1, and a total of 8 photographs of some wild medicinal edible plants are shown in this paper.

Kokrajhar has a rich reserve of varieties of medicinal plants and herbs. Most of those plants and their medicinal application are known only to the natives and the tribes residing in various parts of Kokrajhar for ages. Most of those medicinal plants have not yet been explored experimentally. The active ingredients present in these plants of the state of Kokrajhar may be used for designing some new drugs and pharmaceutical agents which can pave some new alleys in the world of pharmaceutical sciences and be a blessing for mankind. Plant-derived pharmaceutical formulations used to treat diseases are termed alternative medicine. Alternative medicine is better than our conventional allopathic medication and can enhance the impact of conventional drugs if used properly along with them. Nature-derived phytochemicals that constitute alternative drugs do not have any side effects reported to date if used in a specific dose. Though there have been some studies on these medicinal plants of the state of Kokrajhar, and some databases of the plants have been created, fingerprinting and isolation of the active principles and the medicinally potent phytochemicals from these plants need to be performed. Some of the medicinal plants work miraculously in certain diseased conditions according to the tribal people of Kokrajhar. Maybe while hunting for drugs in laboratories for certain deadly diseases day and night, researchers and scientists are missing some miraculous and potent phytochemical constituents which could be modified for formulating the drug, which is present in the plants grown in the wild and ignorance on the road side, backyards, and valleys of Kokrajhar. Extensive experimental indulgence is essential for exploring the indigenous medicinal plants of the state of Kokrajhar and for fingerprinting their phytoconstituents for steps ahead in modern pharmaceutical research and sciences.

21.4 Data set preparation

The research for treatments and vaccines for novel coronavirus disease (COVID-19) is still going on. Desperation in the community, particularly among middle- and low-income groups who have been hit hard by the

TABLE 21.1 List of some wild medicinal edible plants.

Sl. no.	Scientific name	Common name	Local name	Part of the plant eaten	Medicinal values
1	*Houttuynia cordata*	Fish mint	Maisundri	Whole plant	Antiviral, antibacterial, anti-inflammatory, etc.
2	*Apium graveolens var. secalinum*	Chinese celery or leaf celery	Dao penda	Stem and leaves	Antiviral, anti-inflammatory, antibacterial, antioxidant
3	*Paederia foetida*	Skunk vine or stink vine	Kipibendwng	Leaves	Antiviral, antidiarrheal, anti-inflammatory, etc.
4	*Corchorus olitorius*	Jute leaves	Patw bilai	Leaves	Antiviral, antioxidant, anti-inflammatory, etc.
5	*Hibiscus cannabinus*	Kenaf or Java jute	Mwita bangal	Leaves	Antiviral, anti-inflammatory, anticancer, antioxidants, analgesic, etc.
6	*Mentha spicata*	Mint	Pudina	Stem and leaves	Antiviral, antioxidant, antibacterial, antiseptic, anti-inflammatory, etc.
7	*Moringa oleifera*	Drumstick	Swrjina	Leaves	Antiviral, antifungal, antidepressant, anti-inflammatory, antioxidant, etc.
8	*Alternanthera philoxeroides*	Alligator weed	Elangsi	Stems and leaves	Antiviral, antibacterial, etc.
9	*Artocarpus heterophyllus*	Jackfruit	Kantal Begor	Shell removed seeds	Antiviral, anti-inflammatory, antimicrobial, antioxidant, etc.

(*Continued*)

TABLE 21.1 (Continued)

Sl. no.	Scientific name	Common name	Local name	Part of the plant eaten	Medicinal values
10	*Nyctanthes arbor-tristis*	Night jasmine	Sepali Bibar	Flowers	Antiviral, antioxidant, antibacterial, antifungal, etc.
11	*Cayratia trifolia*	Three-leaf cayratia	Dousrem	Leaves	Antiviral, antibacterial, diuretic, anticancer, etc.
12	*Spondias pinnata*	Forest mango, wild mango, amra	Taisuri	Outer flesh of the fruit	Antiviral, antioxidant, antibacterial, antimicrobial, etc.
13	*Solanum torvum*	Turkey berry	Kuntai-nara	Fruits	Antiviral, anti-inflammatory, analgesic, antimicrobial, antioxidant, etc.
14	*Basella alba*	Malabar spinach	Mwikrai/ Mwiprai	Leaves and stem	Antiviral, antioxidant, anti-inflammatory, anticancer, etc.
15	*Ipomoea aquatica*	Water Spinach	Mandey	Stems and leaves	Antioxidant, anti-inflammatory, antimicrobial, etc.
16	*Bambusa balcooa*	Spiny bamboo	Meoai	Shoots	Antioxidant, anti-thyroidal, antitumor, etc.
17	*Diplazium esculentum*	Vegetable ferns	Dingkia	Fiddle heads	Anti-inflammatory, antioxidant, etc. (16)
18	*Gmelina abrorea*	Gambhari	Gambari bibar	Flowers	Astringent, absorbent, coolant, etc.
19	*Stellaria media*	Chickweed	Tuntini/ Nabiki	Leaves and petioles	Antifungal, antibacterial, antioxidant, anti-inflammatory, analgesic, etc.

20	*Lasia spinosa*	Spiny Arum or Spiny Lasia	Sibru	Stalks	Antioxidant, antibacterial, anti-inflammatory, etc.
21	*Blumea lanceolaria*	Chapa	Jwglaori	Leaves	Antioxidant, antipyretic, antibacterial
22	*Dillenia indica*	Elephant apple	Taigir	Outer petals of the fruit	Antioxidant, anti-inflammatory, antimicrobial, antileukemic, etc.
23	*Garcinia atroviridis*	Asam gelugur	Taika	Whole fruit sliced and dried	Antioxidant, antimicrobial, antifungal, antitumor, etc.
24	*Garcinia xanthochymus*	Sour mangosteen, gamboge	Tengfwr	Outer flesh of the fruit	Antioxidant, antimicrobial, antidiabetic, anti-inflammatory, etc.
25	*Antidesma acidum*	Sour Currant Shrub, Bignay	Lapasaikho	Leaves	Antioxidant, antidiabetic, antibacterial, etc.
26	*Lippia alba*	Bushy matgrass	Ontaibajab	Leaves	Antioxidant, analgesic, anti-inflammatory, antipyretic, antihypertensive, etc.
27	*Persicaria perfoliata*	Asiatic tearthumb	Mwita-sikla	Leaves	Antioxidant, anti-inflammatory, diuretic, etc.
28	*Monochoria vaginalis*	Oval-leaf pondweed	Ajwnai	Buds of the flower	Antioxidant, anti-inflammatory, etc.
29	*Spilanthes paniculata*	Toothache plant	Usumwi	Leaves	Antioxidant, anti-inflammatory, diuretic, antimicrobial, antifungal, etc.
30	*Malva neglecta*	Common mallow, dwarf mallow	Lapa	Leaves and stem	Antimicrobial, antioxidant, anti-inflammatory, etc.

(*Continued*)

TABLE 21.1 (Continued)

Sl. no.	Scientific name	Common name	Local name	Part of the plant eaten	Medicinal values
31	*Hibiscus sabdariffa*	Roselle	Mwita gwja	Younger leaves	Antioxidant, anti-inflammatory, antimicrobial, antibacterial, etc.
32	*Carica papaya*	Papaya	Mudupul bibar	Flowers	Antioxidant, antibacterial, antiviral, anti-inflammatory, antidiabetic, etc.
33	*Centella asiatica*	Indian pennywort, Asiatic pennywort	Mani-muni	Leaves and petioles	Antimicrobial, antioxidant, anticancer, anti-inflammatory, etc.
34	*Alpinia nigra/ Allughas*	Black Galangal	Tharai	Soft inner stalk/aerial stem	Antioxidant, antifungal, anti-inflammatory, analgesic, antibacterial, etc.
35	*Oxalis corniculata*	Changeri	Singri	Aerial part	Anti-inflammatory, antifungal, anticancer, antidiabetic, antimicrobial, etc.
36	*Eryngium foetidum*	Culantro	Gongar Dundia	Leaves	Antioxidant, anti-inflammatory, antibacterial, anticonvulsant, etc.
37	*Colocasia esculenta*	Taro stolon	Tasoaiting	Stolon	Antihypertensive, anticancer, etc.
38	*Chenopodium album*	Goosefoot, Bathua	Butua	Leaves	Antiviral, antioxidant, antimicrobial, antifungal, etc.
39	*Leucas aspera*	Thumbai	Kansingsia	Leaves	Antifungal, antioxidant, antimicrobial, antipyretic, etc.

40	*Solanum nigrum*	Black nightshade	Mwisung	Young sprouts	Antioxidant, anti-inflammatory, antimicrobial, etc.
41	*Clerodendrum paniculatum*	Glory bower	Lwkwnapul	Young leaves	Antioxidant, anti-inflammatory, etc.
42	*Amorphophallus*	Voodoo lily, Elephant yam	Olodor	Inner stem	Antioxidant, anti-inflammatory, antibacterial, antifungal, etc.
43	*C. esculenta*	Taro plant	Taso bipang	Young aerial part	Analgesic, anti-inflammatory, anticancer, antidiarrheal, etc.
44	*Amaranthus palmeri*	Palmer amaranth, pigweed	Kuduna	Whole young plant	Anti-inflammatory,laxative, diuretic, antidiabetic, antipyretic, etc.
45	*Sphenoclea zeylanica*	Gooseweed or wedgewort	Sibung	Aerial part	Antimicrobial, antifungal, antibacterial, etc.
46	*Premna herbacea*	Stemless Premna	Keradapini	Leaves	Antipyretic, antinociceptive, anti-inflammatory, etc.
47	*Chrysophyllum cainito*	Star apple, cainito	Penel	Fruits	Antiviral, antifungal, antioxidant, antibacterial, etc.
48	*Oroxylum indicum*	Midnight horror or broken bone plant or Indian trumpet flower	Karokandai	Flowers, leaves	No specific studies found
49	*Cheilocostus speciosus*	Crepe ginger	Burithokon	Young stalks	No specific studies found
50	*Gymnopetalum cochinchinense*	Unknown	Kaila	Whole fruit	No specific studies found

economic effects of forced lockdowns, has sparked a surge in interest in alternative medical plant-based therapies. Our team of researchers will provide evidence summaries examining the potential of complementary therapies in COVID-19 management in response to inquiries received. We show and discuss the findings of some wild medicinal edible plants that have been reported to have antiviral, anti-inflammatory, and immunomodulatory activities and may be worth investigating further. In the tribal people of Kokrajhar, India, wild medicinal plants represent the essential unit of the traditional medicine system. Traditional medicine edible plants are abundant among tribal people, notably in folk medicine. During the COVID-19 pandemic, tribes in Kokrajhar have increased their use of wild medicinal edible herbs. Our finding shows that less than 25% (from Table 21.2) of tribal people of the area are affected by COVID-19, and the recovery rate is high in comparison to non-tribal people of the area. In this situation, a study was needed to describe the wild medicinal edible plants used, their social importance, their cultivation status, and the sources of information people used to use them. The goal of this research was to describe the current state of wild medicinal edible plant use and provide critical recommendations to the relevant authorities. This study presents a MCDM procedure based on the neutrosophic VIKOR method to survey the region where most wild medicinal edible plants are used and found. Also, with the help of the neutrosophic VIKOR method, we can identify the regions where most people of the Kokrajhar district are affected and recovered.

21.5 About VIKOR strategy

VIKOR strategy was created for multistandards optimization of complex frameworks. It decides the trade-off positioning list, the trade-off arrangement, and the weight dependability spans for inclination soundness of the trade-off arrangement acquired with the underlying (given) loads. VIKOR centers around positioning and choosing from a lot of options within the sight of clashing measures. Opricovic et al. [10] considered two MCDM strategies, VIKOR method (VM) and TOPSIS method (TM) which are looked at, zeroing in on demonstrating the accumulating capacity and normalization, to uncover and analyze the procedural premise of these two MCDM techniques. VM strategy presents the positioning list depending on the specific proportion of "closeness" to the ideal arrangement by utilizing direct standardization. Liou et al. [11] used VM to analyze the management level of Taiwan's domestic carriers and to identify the gaps between what aircraft delivers and what consumers seek, while Sanayei et al. [12] used VM to position providers in a flexible chain framework. Recently in [13–15], the neutrosophic VICKOR [16] method is discussed.

TABLE 21.2 COVID-19 cases reported in the district daily broken up in ST and non-ST.

Sl. no.								Deaths reported	
	Date	Total cases	ST	Non-ST	Below 18 years	18–44 years	Above 45 years	ST	Non-ST
1.	09th–03th April 2021	24	2	22	1	7	16		1
2.	14th–18th April 2021	47	3	44	1	36	10		
3.	19th–23rd April 2021	103	17	86	10	66	27	1	
4.	24th–28th April 2021	157	24	133	16	97	44	2	
5.	29th April–03rd May 2021	193	48	145	12	108	73	1	2
6.	04th–08th May 2021	199	47	152	13	119	67	2	1
7.	09th–13th May 2021	312	93	219	20	200	92	1	4
8.	14th–18th May 2021	327	93	237	21	189	117	2	4
9.	19th–23rd May 2021	406	109	297	36	253	117	2	5
10.	24th–28th May 2021	408	108	299	31	263	114		
11.	29th May–02nd June 2021	273	61	212	31	184	55		3
12.	03rd–07th June 2021	189	46	143	19	112	58	1	2
13.	08th–12th June 2021	198	49	149	14	120	64	1	3
14.	13th–17th June 2021	201	40	161	21	121	59		1
15.	18th–22nd June 2021	234	54	180	30	146	58	3	

(*Continued*)

TABLE 21.2 (Continued)

Sl. no.								Deaths reported	
	Date	Total cases	ST	Non-ST	Below 18 years	18–44 years	Above 45 years	ST	Non-ST
16.	23rd–27th June 2021	193	41	152	22	117	54		2
17.	28th June–02nd July 2021	190	57	133	21	115	54	1	2
18.	03rd–07th July 2021	143	35	108	17	80	46	1	1
19.	08th–12th July 2021	159	55	104	16	106	37	1	2
20.	13th–17th July 2021	124	28	96	12	60	52		2
21.	18th–22ndJuly 2021	88	40	48	9	53	26		2
22.	23rd–26th July 2021	108	29	79	10	57	41	1	1
	Total	4276	1079	3199	383	2609	1281	20	38

21.6 Case study

21.6.1 Step 1

By observing the spread of COVID-19, and the use of the wild edible medicinal plant in various regions of the Kokrajhar district, we have observed three alternatives, namely, environmental factors, social factors, and medical factors. To locate the weakest options, a specialist board of trustees of three experts, E_1, E_2, and E_3, has been shaped. These specialists are from various departments, one is doctors, one is professor, and one is a research scholar. In the light of the writing (survey), in regard to the assessment of the affected area of coronavirus and uses of wild medicinal edible plants on the models and sub-measures things were examined with the specialists. By observing the most common hydrologic vulnerability in the COVID-19 approach, the COVID-19 basin of the state is divided into three sub-basin regions as follows:

1. A_1-Village area (where most wild medicinal edible plants are available and used and maximum tribal people)
2. A_2-Town area (less tribal people)
3. A_3-Suburban area (mixed people)

21.6.2 Step 2

Orchestrating the dynamic gathering and describing a lot of pertinent ascribe. Idea plan determination requires recognizable proof of choice models, and afterward assessment scales are set up to rank the ideas. These rules must be characterized by corporate techniques (Table 21.3).

TABLE 21.3 Linguistic variable table for each criterion (in triangular neutrosophic number).

Very low (VL)	(0.10, 0.90, 0.90)
Between VL and L	(0.20, 0.75, 0.80)
Low (L)	(0.30, 0.70, 0.70)
Between L and M	(0.40, 0.65, 0.60)
Medium (M)	(0.50, 0.50, 0.50)
Between M and H	(0.60, 0.35, 0.40)
High (H)	(0.70, 0.30, 0.30)
Between H and VH	(0.80, 0.25, 0.20)
Very high (VH)	(0.90, 0.10, 0.10)

A committee of three experts E_1, E_2, and E_3 have been formed to select the most assessment of coronavirus vulnerability. The following criteria have been defined

1. C_1 = Immune system
2. C_2 = Tension/hypertension
3. C_3 = Physical exercise/work (labor)
4. C_4 = Non/non-vegetable
5. C_5 = Wild medicinal edible plant
6. C_6 = Social distance

21.6.3 Step 3

We define the appropriate etymological factors for model significance weights and neutrosophic ratings for choices for each measure, and then these semantic factors can be presented as neutrosophic numbers. Three experts utilized the phonetic weighting factors to survey the significance of the models. Experts have controlled the significant loads of the measurements, which are calculated.

21.6.4 Step 4

Assume that the nth expert's neutrosophic rating and weight are $\chi_{pqr} = (\chi_{pqr_1}, \chi_{pqr_2}, \chi_{pqr_3})$, and $\omega_{qr} = (\omega_{qr_1}, \omega_{qr_2}, \omega_{qr_3})$. As a result, the aggregated neutrosophic rating χ_{pq} of alternatives for each criterion can be determined as in Table 21.4.

TABLE 21.4 Aggregated neutrosophic weight.

	Weight	A_1	A_2	A_3
C_1	(0.10, 0.46, 0.90)	(0.70, 0.25, 0.30)	(0.10, 0.70, 0.90)	(0.50, 0.43, 0.50)
C_2	(0.50, 0.40, 0.50)	(0.50, 0.43, 0.50)	(0.50, 0.36, 0.50)	(0.50, 0.43, 0.50)
C_3	(0.30, 0.36, 0.70)	(0.50, 0.23, 0.50)	(0.30, 0.56, 0.70)	(0.50, 0.30, 0.50)
C_4	(0.30, 0.36, 0.70)	(0.50, 0.30, 0.50)	(0.30, 0.50, 0.70)	(0.50, 0.30, 0.50)
C_5	(0.10, 0.54, 0.90)	(0.10, 0.83, 0.90)	(0.50, 0.23, 0.50)	(0.30, 0.56, 0.50)
C_6	(0.10, 0.45, 0.90)	(0.10, 0.76, 0.90)	(0.50, 0.23, 0.50)	(0.30, 0.36, 0.50)

21.6.5 Step 5

Weight of each criterion is shown in Table 21.5.

21.6.6 Step 6

All criterions' best G_q^* and worstG_q^- values are given in Table 21.6.

21.6.7 Step 7

The values of **S, R**, and **Q** are shown in Table 21.7.

TABLE 21.5 Weight of each criterion.

	C_1	C_2	C_3	C_4	C_5	C_6
Weight	0.486	0.466	0.453	0.453	0.513	0.483
A_1	0.416	0.476	0.410	0.433	0.610	0.586
A_2	0.566	0.453	0.520	0.500	0.410	0.410
A_3	0.476	0.476	0.433	0.433	0.453	0.386

TABLE 21.6 Best and worst values.

	C_1	C_2	C_3	C_4	C_5	C_6
G_q^*	0.566	0.476	0.520	0.500	0.610	0.586
G_q^-	0.416	0.453	0.410	0.433	0.410	0.386

TABLE 21.7 S, R, and Q values for all alternatives.

	A_1	A_2	A_3
S	1.392	1.40404	1.988587
R	0.486	0.513	0.483
Q	0.05	0.510091	0.5

21.6.8 Step 8

Table 21.8 shows the ranking of the alternatives.

21.6.9 Step 9

Table 21.9 shows the ranking of the alternatives.

21.7 Result and discussion

Table 21.9 shows the criteria-wise ranking of the vulnerability of four alternatives discussed in this study. The result shows that the alternative A_1(rural Area) is the highest rank in the criteria C_5, and C_6, alternative $A_2$is the highest rank in the criteriaC_1,C_3, and C_4, alternative $A_3$is the highest rank in the criteriaC_2, respectively, and alternativeA_1 (rural Area) is the lowest rank in the criteria C_1,C_2, and C_3, alternative A_2(urban area) is the lowest rank in the criteria C_5, and alternative$A_3$is the lowest rank in the criteria C_4 and C_6.

Table 21.10 shows ranking criteria weights, criteria C_5 (wild medicinal edible plant) is the highest weights as per the experts followed by C_6 (social distance), C_1(immune system), C_3 (physical exercise/work (labor), and

TABLE 21.8 Alternatives are ranked in ascending order by S, R, and Q.

Rank	1	2	3
S	A_1	A_2	A_3
R	A_3	A_1	A_2
Q	A_1	A_3	A_2

TABLE 21.9 Criteria-wise ranking of four alternatives.

Ordering of alternatives from high to low				
1.	C_5	A_1	A_3	A_2
2.	C_6	A_1	A_2	A_3
3.	C_1	A_2	A_3	A_1
4.	C_3	A_2	A_3	A_1
5.	C_4	A_2	A_1	A_3
6.	C_2	A_3	A_2	A_1

TABLE 21.10 Ranking of criteria weights.

1.	C_5
2.	C_6
3.	C_1
4.	C_3
5.	C_4
6.	C_2

C_4(non-wild medicinal edible plant), respectively, while criteria C_2(tension/hypertension) has the least criteria weights.

21.8 Conclusion

In this investigation, we evaluated the COVID-19 weakness area in Kokrajhar with respect to the consumption of wild edible medicinal plant of Kokrajhar by using neutrosophic VIKOR. Neutrosophic VIKOR technique is a useful apparatus in multi-standards dynamic bargained arrangement which gives the greatest gathering utility of the larger part, and at least the individual lament of the adversary. In this paper, we proposed an altered neutrosophic VIKOR that was upheld by the OWA administrator and decided loads of rules. From the study, it is clear that the use of wild edible medicinal plants and their physical labor is helping the tribal people of Kokrajhar in increasing their immune systems.

Tribal people of Kokrajhar prepared a soup consisting of Chapa, toothache plant, Indian pennywort, a black pepper, and Bird's eye chili (Banlubwrdwn) when they have a cough, from which they have relief from cough. Since last year, this soup has been in more frequent use as medicine than before. According to them, this may be the best medicine for relieving cough. In diarrhea or dysentery or stomach disorder, extract of the juice of young buds of goat weed, glory bower, lemon, Guava is taken to relieve the disorders. They also have an alternate method when diarrhea and vomiting problem are not treated in the first trial. They take an extract of wild edible onion and wild edible garlic (not the same as common onion or garlic) which is very effective. They also use steam of boiling water which is inhaled in case of tight/congested nose from which breathing problem is relieved. Garlic cloves and mustard oil are heated and applied to the nostrils, the throat, and the chest during fever and cold. They have different treatment methods for normal, medium, and extreme headaches. The paste of the fish mint and tulsi (basil) is applied on the forehead while suffering from a savior headache. A paste made from lentil, vermilion powder, wild fern, Datura, net mushroom, red tree fungus, and touch-me-not is applied to the head of the person who is suffering from a normal headache. Also, a paste made of common cocklebur (scientific name: Xanthium strumarium), red tree fungus (growing on Sal tree), net mushroom, and Bird's eye chili is applied to the head when suffering from an extreme headache. This may be the one reason that tribal people are affected less than 25% as compared to non-tribal people of Kokrajhar.

Acknowledgments

We would like to thank Gouri Rani Baglary, a research scholar from the Department of Food Engineering and Technology, CIT, Kokrajhar, for her valuable inputs in finding the scientific names of most of the uncommon wild vegetables in Table 21.1.

References

[1] E.C. Omenna, A.I. Ojo, Comparative phytochemical screening of kenaf and jute leaves, J. Nutr. Health Food Eng. 8 (5) (2018) 366–369.

[2] A.K. Das, B.K. Dutta, G.D. Sharma, Medicinal plants used by different tribes of Cachar district, Assam, Indian. J. Tradit. Knowl. 7 (3) (2018) 446–454.

[3] P.C. Bhattacharya, R. Muzumder, G.C. Devi Sarmah, Rare medicinal plants of Assam, Anc. Sci. Life 10 (4) (1991) 234–238.

[4] S. Li, C.S. Cheng, C. Zhang, G.Y. Tang, H.Y. Tan, H.Y. Chen, et al., Edible and herbal plants for the prevention and management of COVID-19, Front. Pharmacol. 12 (2021) 900. Apr 28.

[5] H.T. Duguma, Wild edible plant nutritional contribution and consumer perception in Ethiopia, Int. J. Food Sci. 2020 (2020). Sep 4.

[6] B. Patel, S. Sharma, N. Nair, J. Majeed, R.K. Goyal, M. Dhobi, Therapeutic opportunities of edible antiviral plants for COVID-19, Mol. Cell. Biochem. 476 (6) (2021) 2345–2364.

[7] D. Khadka, M.K. Dhamala, F. Li, P.C. Aryal, P.R. Magar, S. Bhatta, et al., The use of medicinal plants to prevent COVID-19 in Nepal, J. Ethnobiol. Ethnomed. 17 (1) (2021) 1–7.

[8] F. Yang, Y. Zhang, A. Tariq, X. Jiang, Z. Ahmed, Z. Zhihao, et al., Food as medicine: a possible preventive measure against coronavirus disease (COVID-19), Phytother. Res. 34 (12) (2020) 3124–3136.

[9] X.Y. Lim, B.P. Teh, T.Y. Tan, Medicinal plants in COVID-19: potential and limitations, Front. Pharmacol. 12 (2021) 355. Mar 24.

[10] Y. Ali, A.U. Khan, H.B. Hameed, Selection of sustainable mode of transportation based on environmental, economic, and social factors: Pakistan a case in point, Transp. Dev. Econ. 8 (1) (2022) 1–4.

[11] J.J. Liou, C.Y. Tsai, R.H. Lin, G.H. Tzeng, A modified VIKOR multiple-criteria decision method for improving domestic airlines service quality, J. Air Transp. Manag. 17 (2) (2011) 57–61.

[12] A. Sanayei, S.F. Mousavi, A. Yazdankhah, Group decision making process for supplier selection with VIKOR under fuzzy environment, Expert. Syst. Appl. 37 (1) (2010) 24–30.

[13] H. Eroğlu, R. Şahin, A neutrosophic VIKOR method-based decision-making with an improved distance measure and score function: case study of selection for renewable energy alternatives, Cognit. Comput. 12 (6) (2020) 1338–1355.

[14] J. Wang, G. Wei, M. Lu, An extended VIKOR method for multiple criteria group decision making with triangular fuzzy neutrosophic numbers, Symmetry 10 (10) (2018) 497.

[15] M. Abdel-Baset, V. Chang, A. Gamal, F. Smarandache, An integrated neutrosophic ANP and VIKOR method for achieving sustainable supplier selection: a case study in importing field, Comput. Ind. 106 (2019) 94–110.

[16] M. Abdel-Basset, Y. Zhou, M. Mohamed, V. Chang, A group decision making framework based on neutrosophic VIKOR approach for e-government website evaluation, J. Intell. Fuzzy Syst. 34 (6) (2018) 4213–4224.

Chapter 22

Generalized robust-type neutrosophic ratio estimators of pharmaceutical daily stock prices

Rajesh Singh[1], Florentin Smarandache[2] and Rohan Mishra[1]
[1]Department of Statistics, Institute of Science, Banaras Hindu University, Varanasi, Uttar Pradesh, India, [2]Mathematics, Physical and Natural Sciences Division, University of New Mexico, Gallup, NM, United States

22.1 Introduction

Classical statistics and its methods deal with randomness, but there are cases where the data at hand are indeterminate or vague or ambiguous rather than random. In such situations, estimation using classical statistical methods does not yield promising results. Fuzzy logic [1,2] is one solution to tackle such a problem, but still, it ignores indeterminacy. In such cases, neutrosophic methods are much more reliable. They deal with both randomness and more importantly indeterminacy.

Neutrosophic statistics refers to a set of data such that the data (or a part of it) are indeterminate and the methods to analyze such data [3]. Neutrosophic statistics is an extension of classical statistics [4], and when the indeterminacy is zero, neutrosophic statistics coincides with classical statistics [3]. Estimation through neutrosophic methods is not a new field yet it is still unexplored. Fig. 22.1 explains when to use neutrosophic probability sampling designs.

The data collected frequently contain one or more typical observations known as outliers. These outliers are well separated from the majority or bulk of the data or in some way deviate from the general pattern of the data [5]. The problem with the outliers is that they adversely influence the classical estimates like the sample mean, sample standard deviation, and correlation coefficient, to name a few. In such situations, researchers have to use robust measures to reach a reliable estimate.

Cognitive Intelligence with Neutrosophic Statistics in Bioinformatics.
DOI: https://doi.org/10.1016/B978-0-323-99456-9.00019-2

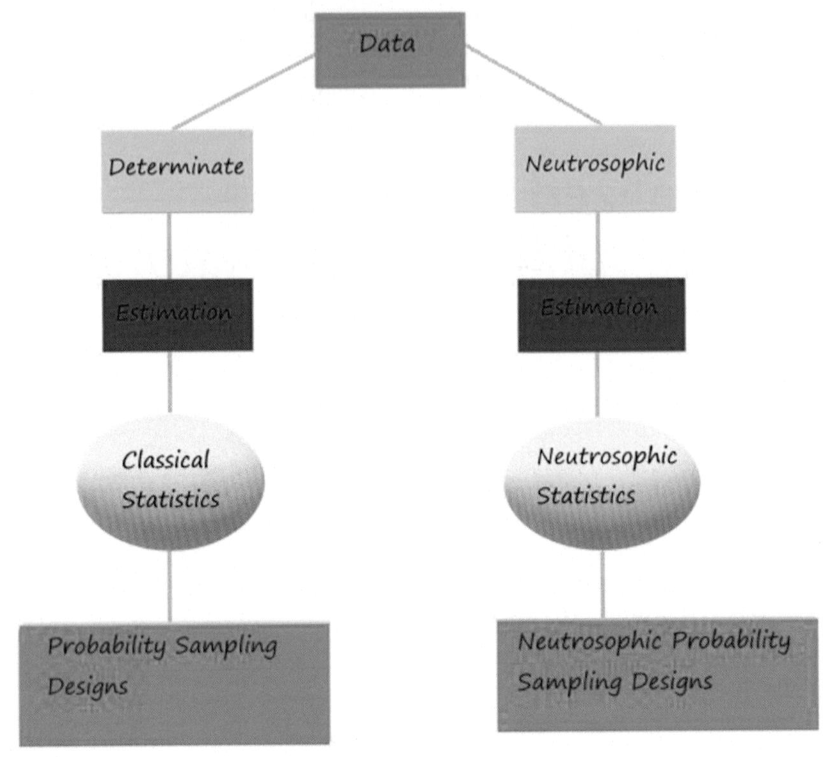

FIGURE 22.1 Flow diagram describing the application of classical statistics and neutrosophic statistics.

It is well established that the use of auxiliary information in sample surveys results in substantial improvement in the efficiency of the estimators of the population mean. This motivated researchers to use auxiliary information at the estimation stage. Ratio, product, and regression methods are common examples in this context. Khoshnevisan et al. [6], Singh and Kumar [7], Singh and Mishra [8], and Singh et al. [9] used auxiliary information to propose new improved estimators.

As it was proposed in [10], neutrosophic statistics has been used in a wide variety of fields [1,3]. Recently, neutrosophic analysis of variance technique, neutrosophic applied statistics, and neutrosophic statistical quality control have been developed [11–14]. But the application of neutrosophic simple random sampling is relatively new and still unexplored, and as a result, using auxiliary information only in some neutrosophic ratio-type estimators is in the literature [4,15,16].

The role of pharmaceutical companies is to enable people to fight illness, to make them feel better in sickness, and to keep them from getting sick in the first place through research and development of drugs. Moderna was one of the first pharmaceutical companies to produce vaccines for the SARS-CoV-2 virus. Years of its research in designing both mRNA therapeutics and mRNA vaccines particularly in manufacturing and formulating mRNA that can produce targeted proteins in the body made it ready to combat the SARS-CoV-2 virus [17], and as a result, they were one of the first to produce the much-needed vaccine.

There are many reasons why outliers occur, and it is highly likely that in a real survey, the sample might have some outliers which are non-noticeable. In such situations, using robust estimators becomes the only viable option. Since neutrosophic statistics deals with vague or ambiguous data, it is very likely that a neutrosophic sample might contain outliers. Still, there are absolutely no neutrosophic robust ratio estimators. This chapter is the first to address this void. In this chapter, we have proposed the first generalized robust-type neutrosophic ratio estimator for neutrosophic simple random sampling without replacement, using some robust measures, viz., tri-mean (TM), mid-range (Mr), and Hodges–Lehman (HL) along with some conventional non-robust measures, viz., coefficient of skewness, coefficient of kurtosis, and coefficient of variation of auxiliary variable. Since probability sampling in neutrosophy is unexplored, its terminologies are presented in Section 22.2. Some related neutrosophic ratio-type estimators are presented in Section 22.3. The proposed generalized robust-type neutrosophic ratio estimators along with their derivations of bias and MSE up to the first order of approximation are presented in Section 22.4. Their performance is assessed through a simulation study on a real neutrosophic population of the daily stock price of Moderna Inc. [18] in Section 22.5. The empirical MSE obtained shows that the proposed generalized estimators result in better estimation.

22.2 Terminology

A simple random neutrosophic sample of size n from a classical or neutrosophic population is a sample of n individuals such that at least one of them has some indeterminacy [3].

As presented in [4], a neutrosophic observation is of the form

$$Z_N = Z_L + Z_U I_N, \text{where } I_N \in [I_{N_L}, I_{N_U}] \text{ and } Z_N \in [Z_{N_L}, Z_{N_U}].$$

Now consider a simple random neutrosophic sample of size $n_N \in [n_{N_L}, n_{N_U}]$ drawn from a finite population of size N and $y_N(i) \in [y_{N_L}, y_{N_U}]$ and $x_N(i) \in [x_{N_L}, x_{N_U}]$ are i^{th} neutrosophic sample observation. Here, the population mean of neutrosophic survey and auxiliary variable are $\overline{Y}_N \in [\overline{Y}_{NL}, \overline{Y}_{NU}]$ and $\overline{X}_N \in [\overline{X}_{N_L}, \overline{X}_{N_U}]$, respectively.

$C_{yN} \in [C_{yNL}, C_{yNU}]$ and $C_{xN} \in [C_{xNL}, C_{xNU}]$ are population coefficients of variation of neutrosophic survey and auxiliary variables, respectively. In addition, $\rho xyN \in [\rho xyNL, \rho xyNU]$, $\beta 1(xN) \in [\beta 1(xNL), \beta 1(xNU)]$, and

$\beta 2(xN) \in [\beta 2(xNL), \beta 2(xNU)]$ are the correlation coefficient between the neutrosophic survey and auxiliary variables, coefficient of skewness, and coefficient of kurtosis of the neutrosophic auxiliary variable, respectively.

The MSE of a neutrosophic estimator is of the form $\mathrm{MSE}(\overline{y}_N) \in [\mathrm{MSE}_L, \mathrm{MSE}_U]$.

The error terms in neutrosophic statistics are: $\overline{e}_{yN} = \overline{y}_N - \overline{Y}_N$, $\overline{e}_{xN} = \overline{x}_N - \overline{X}_N$,

$$E(\overline{e}_{yN}) = E(\overline{e}_{xN}) = 0,$$

$$\overline{e}^2_{yN} = \frac{N-n}{Nn}\overline{Y}^2_N C^2_{yN},$$

$$\overline{e}^2_{xN} = \frac{N-n}{Nn}\overline{X}^2_N C^2_{xN},$$

$$\overline{e}_{xN}\overline{e}_{yN} = \frac{N-n}{Nn}\overline{X}_N\overline{Y}_N\rho_{xyN}C_{xN}C_{yN},$$

where

$$\overline{e}_{yN} \in \left[\overline{e}_{yN_L}, \overline{e}_{yN_U}\right],$$

$$\overline{e}_{xN} \in \left[\overline{e}_{xN_L}, \overline{e}_{xN_U}\right],$$

$$\overline{e}_{2yN} \in \left[\overline{e}_{2yN_L}, \overline{e}_{2yN_U}\right],$$

$$\overline{e}_{2xN} \in \left[\overline{e}_{2xN_L}, \overline{e}_{2xN_U}\right].$$

22.3 Some related neutrosophic estimators

Tahir et al. [4] proposed ratio-type estimators t_R and t_{US_r} given as:

$$t_R = \frac{\overline{y}N}{\overline{x}N}\overline{X}_N, \tag{22.1}$$

and

$$t_{USr} = \overline{y}N\frac{\overline{X}_N\beta_2(xN) + C_{xN}}{\overline{x}_N\beta_2(xN) + C_{xN}}, \tag{22.2}$$

where $\overline{y}_N \in [\overline{y}_{N_L}, \overline{y}_{N_U}]$ and $t_R \in [tR_L, tR_U]$, $t_{USr} \in [tUSr_L, tUSr_U]$.

The MSE expressions of the estimators t_R and t_{USr} are, respectively, given by:

$$\mathrm{MSE}(t_R) = \frac{N-n}{Nn}\overline{Y}^2_N\left[C^2_{yN} + C^2_{xN} - 2C_{xN}C_{yN}\rho_{xyN}\right], \tag{22.3}$$

and

$$\mathrm{MSE}(t_{USr}) = \frac{N-n}{Nn}\overline{Y}_N^2\left[C_{yN}^2 + \left(\frac{\overline{X}_N\beta_2(xN)}{\overline{X}_N\beta_2(xN)+C_{xN}}\right)\right.$$

$$\left. C_{xN}^2 - 2\left(\frac{\overline{X}_N\beta_2(xN)}{\overline{X}_N\beta_2(xN)+C_{xN}}\right)C_{xN}C_{yN}p_{xyN}\right], \tag{22.4}$$

where $CyN2 \in [CyN2\ L, CyN2\ U]$, $CxN2 \in [CxN2\ L, CxN2\ U]$ and $\rho xyN \in [\rho xyNL, \rho xyNU]$.

22.4 Generalized robust-type neutrosophic ratio estimator

The main objective of this chapter is to develop a robust-type neutrosophic ratio estimator for estimating the population mean of finite neutrosophic data. In this section, we propose a generalized robust-type neutrosophic ratio estimator following the concept of [15,19] for estimating the population mean of finite neutrosophic data as:

$$t = \overline{y}N\left(\frac{\alpha\overline{X}_N+\beta}{\alpha\overline{x}_N+\beta}\right) \tag{22.5}$$

where α and β are known generalized scalar constants. The proposed robust-type estimators along with their expressions of bias and MSE are presented in Tables 22.1 and 22.2.

Estimators $t_R$and t_{US_r} proposed by Tahir et al. [4] also belong to (5) for $\alpha = 1$, $\beta = 0$ and $\alpha = \beta_2(xN)$, $\beta = C_{xN}$, respectively.

For obtaining the expression of bias and MSE, we rewrite (5) in error terms as:

$$t = (\overline{y}N + \overline{e}_{yN})\left(\frac{\alpha\overline{X}_N+\beta}{\alpha\overline{X}_N+\alpha\overline{e}_{xN}+\beta}\right) \tag{22.6}$$

Simplifying (6), we get

$$t = (\overline{y}N + \overline{e}_{yN})\left(1+\frac{\beta}{\alpha\overline{X}_N}\right)\left(\frac{\overline{e}_{xN}}{\overline{X}_N}+1+\frac{\beta}{\alpha\overline{X}_N}\right)^{-1} \tag{22.7}$$

Taking

$$\psi = 1 + \frac{\beta}{\alpha\overline{X}_N}$$

TABLE 22.1 Proposed robust-type neutrosophic ratio estimators with their corresponding values of α, β, and bias.

Estimators	Form	α	β	Bias
t_1	$\bar{y}_N\left(\frac{MR\bar{X}_N+\beta_1(x_N)}{MR\bar{x}_N+\beta_1(x_N)}\right)$	MR	$\beta_1(x_N)$	$\frac{N-n}{Nn}\bar{Y}_N\left(\left(\frac{MR\bar{X}_N}{MR\bar{X}_N+\beta_1(x_N)}\right)^2 C_{xN}^2-\left(\frac{MR\bar{X}_N}{MR\bar{X}_N+\beta_1(x_N)}\right)\rho_{x_N y_N}C_{xN}C_{yN}\right)$
t_2	$\bar{y}_N\left(\frac{MR\bar{X}_N+\beta_2(x_N)}{MR\bar{x}_N+\beta_2(x_N)}\right)$	MR	$\beta_2(x_N)$	$\frac{N-n}{Nn}\bar{Y}_N\left(\left(\frac{MR\bar{X}_N}{MR\bar{X}_N+\beta_2(x_N)}\right)^2 C_{xN}^2-\left(\frac{MR\bar{X}_N}{MR\bar{X}_N+\beta_2(x_N)}\right)\rho_{x_N y_N}C_{xN}C_{yN}\right)$
t_3	$\bar{y}_N\left(\frac{TM\bar{X}_N+\beta_1(x_N)}{TM\bar{x}_N+\beta_1(x_N)}\right)$	TM	$\beta_1(x_N)$	$\frac{N-n}{Nn}\bar{Y}_N\left(\left(\frac{TM\bar{X}_N}{TM\bar{X}_N+\beta_1(x_N)}\right)^2 C_{xN}^2-\left(\frac{TM\bar{X}_N}{TM\bar{X}_N+\beta_1(x_N)}\right)\rho_{x_N y_N}C_{xN}C_{yN}\right)$
t_4	$\bar{y}_N\left(\frac{TM\bar{X}_N+\beta_2(x_N)}{TM\bar{x}_N+\beta_2(x_N)}\right)$	TM	$\beta_2(x_N)$	$\frac{N-n}{Nn}\bar{Y}_N\left(\left(\frac{TM\bar{X}_N}{TM\bar{X}_N+\beta_2(x_N)}\right)^2 C_{xN}^2-\left(\frac{TM\bar{X}_N}{TM\bar{X}_N+\beta_2(x_N)}\right)\rho_{x_N y_N}C_{xN}C_{yN}\right)$
t_5	$\bar{y}_N\left(\frac{HL\bar{X}_N+\beta_1(x_N)}{HL\bar{x}_N+\beta_1(x_N)}\right)$	HL	$\beta_1(x_N)$	$\frac{N-n}{Nn}\bar{Y}_N\left(\left(\frac{HL\bar{X}_N}{HL\bar{X}_N+\beta_1(x_N)}\right)^2 C_{xN}^2-\left(\frac{HL\bar{X}_N}{HL\bar{X}_N+\beta_1(x_N)}\right)\rho_{x_N y_N}C_{xN}C_{yN}\right)$
t_6	$\bar{y}_N\left(\frac{HL\bar{X}_N+\beta_2(x_N)}{HL\bar{x}_N+\beta_2(x_N)}\right)$	HL	$\beta_2(x_N)$	$\frac{N-n}{Nn}\bar{Y}_N\left(\left(\frac{HL\bar{X}_N}{HL\bar{X}_N+\beta_2(x_N)}\right)^2 C_{xN}^2-\left(\frac{HL\bar{X}_N}{HL\bar{X}_N+\beta_2(x_N)}\right)\rho_{x_N y_N}C_{xN}C_{yN}\right)$
t_7	$\bar{y}_N\left(\frac{MR\bar{X}_N+TM}{MR\bar{x}_N+TM}\right)$	MR	TM	$\frac{N-n}{Nn}\bar{Y}_N\left(\left(\frac{MR\bar{X}_N}{MR\bar{X}_N+TM}\right)^2 C_{xN}^2-\left(\frac{MR\bar{X}_N}{MR\bar{X}_N+TM}\right)\rho_{x_N y_N}C_{xN}C_{yN}\right)$
t_8	$\bar{y}_N\left(\frac{TM\bar{X}_N+MR}{TM\bar{x}_N+MR}\right)$	TM	MR	$\frac{N-n}{Nn}\bar{Y}_N\left(\left(\frac{TM\bar{X}_N}{TM\bar{X}_N+MR}\right)^2 C_{xN}^2-\left(\frac{TM\bar{X}_N}{TM\bar{X}_N+MR}\right)\rho_{x_N y_N}C_{xN}C_{yN}\right)$
t_9	$\bar{y}_N\left(\frac{HL\bar{X}_N+TM}{HL\bar{x}_N+TM}\right)$	HL	TM	$\frac{N-n}{Nn}\bar{Y}_N\left(\left(\frac{HL\bar{X}_N}{HL\bar{X}_N+TM}\right)^2 C_{xN}^2-\left(\frac{HL\bar{X}_N}{HL\bar{X}_N+TM}\right)\rho_{x_N y_N}C_{xN}C_{yN}\right)$
t_{10}	$\bar{y}_N\left(\frac{TM\bar{X}_N+HL}{TM\bar{x}_N+HL}\right)$	TM	HL	$\frac{N-n}{Nn}\bar{Y}_N\left(\left(\frac{TM\bar{X}_N}{TM\bar{X}_N+HL}\right)^2 C_{xN}^2-\left(\frac{TM\bar{X}_N}{TM\bar{X}_N+HL}\right)\rho_{x_N y_N}C_{xN}C_{yN}\right)$

TABLE 22.2 Proposed robust-type neutrosophic ratio estimators with their corresponding values of MSE.

Estimators	MSE
t_1	$\frac{N-n}{Nn}\overline{Y}_N^2\left(C_{yN}^2+\left(\frac{MR\overline{X}_N}{MR\overline{X}_N+\beta_1(x_N)}\right)^2C_{xN}^2-2\left(\frac{MR\overline{X}_N}{MR\overline{X}_N+\beta_1(x_N)}\right)\rho_{x_Ny_N}C_{xN}C_{yN}\right)$
t_2	$\frac{N-n}{Nn}\overline{Y}_N^2\left(C_{yN}^2+\left(\frac{MR\overline{X}_N}{MR\overline{X}_N+\beta_2(x_N)}\right)^2C_{xN}^2-2\left(\frac{MR\overline{X}_N}{MR\overline{X}_N+\beta_2(x_N)}\right)\rho_{x_Ny_N}C_{xN}C_{yN}\right)$
t_3	$\frac{N-n}{Nn}\overline{Y}_N^2\left(C_{yN}^2+\left(\frac{TM\overline{X}_N}{TM\overline{X}_N+\beta_1(x_N)}\right)^2C_{xN}^2-2\left(\frac{TM\overline{X}_N}{TM\overline{X}_N+\beta_1(x_N)}\right)\rho_{x_Ny_N}C_{xN}C_{yN}\right)$
t_4	$\frac{N-n}{Nn}\overline{Y}_N^2\left(C_{yN}^2+\left(\frac{TM\overline{X}_N}{TM\overline{X}_N+\beta_2(x_N)}\right)^2C_{xN}^2-2\left(\frac{TM\overline{X}_N}{TM\overline{X}_N+\beta_2(x_N)}\right)\rho_{x_Ny_N}C_{xN}C_{yN}\right)$
t_5	$\frac{N-n}{Nn}\overline{Y}_N^2\left(C_{yN}^2+\left(\frac{HL\overline{X}_N}{HL\overline{X}_N+\beta_1(x_N)}\right)^2C_{xN}^2-2\left(\frac{HL\overline{X}_N}{HL\overline{X}_N+\beta_1(x_N)}\right)\rho_{x_Ny_N}C_{xN}C_{yN}\right)$
t_6	$\frac{N-n}{Nn}\overline{Y}_N^2\left(C_{yN}^2+\left(\frac{HL\overline{X}_N}{HL\overline{X}_N+\beta_2(x_N)}\right)^2C_{xN}^2-2\left(\frac{HL\overline{X}_N}{HL\overline{X}_N+\beta_2(x_N)}\right)\rho_{x_Ny_N}C_{xN}C_{yN}\right)$
t_7	$\frac{N-n}{Nn}\overline{Y}_N^2\left(C_{yN}^2+\left(\frac{MR\overline{X}_N}{MR\overline{X}_N+TM}\right)^2C_{xN}^2-2\left(\frac{MR\overline{X}_N}{MR\overline{X}_N+TM}\right)\rho_{x_Ny_N}C_{xN}C_{yN}\right)$
t_8	$\frac{N-n}{Nn}\overline{Y}_N^2\left(C_{yN}^2+\left(\frac{TM\overline{X}_N}{TM\overline{X}_N+MR}\right)^2C_{xN}^2-2\left(\frac{TM\overline{X}_N}{TM\overline{X}_N+MR}\right)\rho_{x_Ny_N}C_{xN}C_{yN}\right)$
t_9	$\frac{N-n}{Nn}\overline{Y}_N^2\left(C_{yN}^2+\left(\frac{HL\overline{X}_N}{HL\overline{X}_N+TM}\right)^2C_{xN}^2-2\left(\frac{HL\overline{X}_N}{HL\overline{X}_N+TM}\right)\rho_{x_Ny_N}C_{xN}C_{yN}\right)$
t_{10}	$\frac{N-n}{Nn}\overline{Y}_N^2\left(C_{yN}^2+\left(\frac{TM\overline{X}_N}{TM\overline{X}_N+HL}\right)^2C_{xN}^2-2\left(\frac{TM\overline{X}_N}{TM\overline{X}_N+HL}\right)\rho_{x_Ny_N}C_{xN}C_{yN}\right)$

On simplifying (7), we get

$$t=(\overline{y}N+\overline{e}_{yN})\psi\left(\frac{1}{\psi}-\frac{\overline{e}_xN}{\overline{X}_N}\frac{1}{\psi^2}+\frac{\overline{e}_{xN}^2}{\overline{X}_N^2}\frac{1}{\psi^3}\right) \tag{22.8}$$

$$t=\overline{Y}_N+\overline{e}_{yN}-\frac{1}{\psi}\frac{\overline{Y}_N\overline{e}_{xN}}{\overline{X}_N}-\frac{1}{\psi}\frac{\overline{e}_{xN}\overline{e}_{yN}}{\overline{X}_N}+\frac{1}{\psi^2}\frac{\overline{Y}_N}{\overline{X}_N^2}\overline{e}_{xN}^2 \tag{22.9}$$

Subtracting $\overline{Y}_N$from both sides in (9) and taking expectation, we get

$$Bias(t)=\frac{N-n_N}{Nn_N}\overline{Y}_N\left(\left(\frac{\alpha\overline{X}_N}{\alpha\overline{X}_N+\beta}\right)^2C_{xN}^2-\left(\frac{\alpha\overline{X}_N}{\alpha\overline{X}_N+\beta}\right)\rho_{xyN}C_{xN}C_{yN}\right) \tag{22.10}$$

Similarly, subtracting $\overline{Y}_N$from both sides in (9), squaring and taking expectation we get

$$\text{MSE}(t) = \frac{N - n_N}{Nn_N}\overline{Y}_N^2\left(C_{yN}^2 + \left(\frac{\alpha\overline{X}_N}{\alpha\overline{X}_N + \beta}\right)^2 C_{xN}^2 - 2\left(\frac{\alpha\overline{X}_N}{\alpha\overline{X}_N + \beta}\right)^2 \rho_{xyN}C_{xN}C_{yN}\right) \tag{22.11}$$

22.5 Simulation study

Moderna is one of the biggest US-based pharmaceutical companies which became a public company in December 2018 via the largest biotech initial public offering in history [20]. It finds its place in the list of the top 10 pharmaceutical companies in the world by market capitalization [21]. Due to its intensive and innovative research in mRNA technology and rapid response during the pandemic of 2019–20, its impact is on an increase and so is its stock price.

In this section, we have conducted a simulation study on the daily stock prices of Moderna Inc. from 1 September 2020 to 1 September 2021 [18].

The rationale behind taking daily stock prices as a neutrosophic variable is that the stock price of all the stocks at every trading day starts from an opening price and at the end of the trading day reaches a closing price at which the trading stops [22]. However, it always lies between a high price (the highest price of the day) and a low price (the lowest price of the day) which may or may not be the same as the opening or closing price.

We are estimating this high- and low-price interval within which the price of the stock lies using the daily opening price as an auxiliary variable which is not a neutrosophic variable since its value for each trading day is fixed and known in advance.

The simulation study is conducted with 10,000 replications for each sample. We have calculated the empirical indeterminacy intervals of MSE for each neutrosophic estimator presented in this article for sample sizes 120, 125, 130, and 135.

22.6 Discussion

To examine the performance of the proposed estimators, we conducted a simulation study to estimate the daily stock price of the pharmaceutical company, Moderna. From this simulation study, we obtained empirical indeterminacy intervals of MSE of all the neutrosophic estimators for four different sample sizes which are presented in Tables 22.3 and 22.4. From these tables, we can see that the MSE of the classical estimator is falling in the indeterminacy interval of the neutrosophic estimator. Further, we can see that all the proposed robust-type estimators perform better than the neutrosophic ratio

TABLE 22.3 MSE of all the neutrosophic estimators.

$n_N \in [20, 20]$				$n_N \in [25, 25]$			
	MSE[t_{*L},	t^*U]	Classical		MSE[t_{*L},	t^*U]	Classical
tR	1.8785	2.2873	1.89317	*tR*	1.5303	1.7879	1.6384
t1	1.8784	2.2875	1.8931	*t1*	1.5302	1.7881	1.6384
t2	1.86464	2.2875	1.89313	*t2*	1.5302	1.788	1.6384
tUSr	1.8677	2.3169	1.8927	*tUSr*	1.5222	1.8149	1.6423
t3	1.87833	2.2877	1.893	*t3*	1.5301	1.7883	1.6384
t4	1.87837	2.2876	1.8931	*t4*	1.5301	1.7882	1.6384
t5	1.87834	2.2877	1.893	*t5*	1.5301	1.7882	1.63845
t6	1.87837	2.2876	1.8931	*t6*	1.53017	1.78823	1.63844
t7	1.86727	2.3192	1.8929	*t7*	1.52197	1.81705	1.64282
t8	1.8736	2.4295	1.9265	*t8*	1.52807	1.90975	1.67908
t9	1.86461	2.3502	1.8992	*t9*	1.52024	1.8444	1.65114
t10	1.8646	2.3534	1.9	*t10*	1.5202	1.8472	1.65212

TABLE 22.4 MSE of all the neutrosophic estimators.

$n_N \in [30, 30]$				$n_N \in [35, 35]$			
	MSE[t_{*L},	t^*U]	Classical		MSE[t_{*L},	t^*U]	Classical
tR	1.1411	1.42543	1.1372	*tR*	1.104452	1.3086	1.3014
t1	1.13716	1.42562	1.13713	*t1*	1.04444	1.3087	1.2951
t2	1.13714	1.42558	1.40712	*t2*	1.04445	1.30868	1.2951
tUSr	1.13155	1.44841	1.41253	*tUSr*	1.03751	1.32192	1.29413
t3	1.13707	1.42576	1.40715	*t3*	1.04437	1.30878	1.29508
t4	1.1371	1.4257	1.40714	*t4*	1.0444	1.30875	1.29509
t5	1.13708	1.42576	1.40715	*t5*	1.04438	1.30878	1.29508
t6	1.1371	1.4257	1.40714	*t6*	1.0444	1.30875	1.29509
t7	1.13135	1.45017	1.41309	*t7*	1.03717	1.32297	1.29421
t8	1.13781	1.52721	1.4473	*t8*	1.03572	1.37177	1.30817
t9	1.13055	1.47315	1.42159	*t9*	1.03437	1.33709	1.29652
t10	1.13063	1.47545	1.42254	*t10*	1.03423	1.33853	1.29686

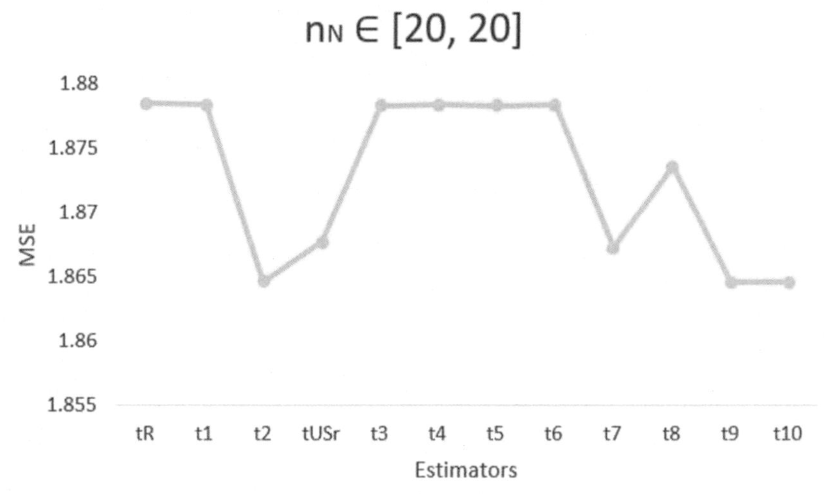

FIGURE 22.2 MSE of all neutrosophic estimators at lower values for sample size [20, 20].

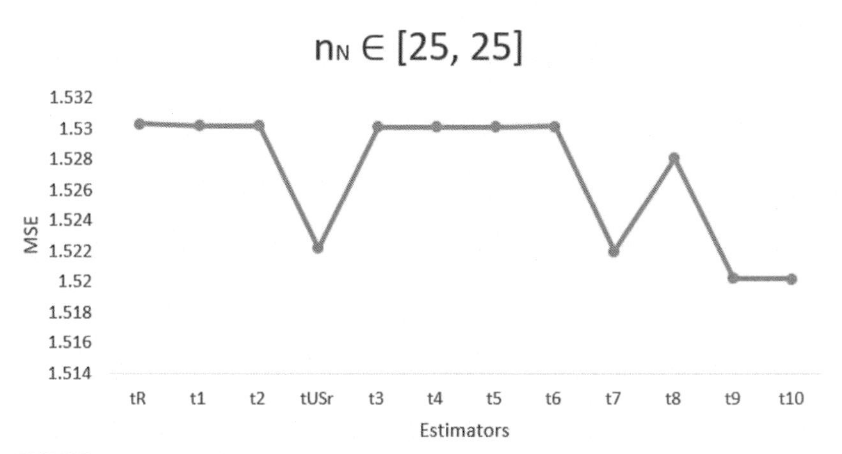

FIGURE 22.3 MSE of all neutrosophic estimators at lower values for sample size [25, 25].

estimator for all sample sizes (see Figs. 22.2–22.5). From the results obtained, we can say that the daily stock price of Moderna lies closer to its daily lower stock price as compared to its daily higher stock price.

22.7 Conclusion

In this chapter, we proposed the first neutrosophic robust ratio-type estimator and showed a real-life application of neutrosophic simple random sampling without replacement in the field of bioinformatics. We proposed a generalized

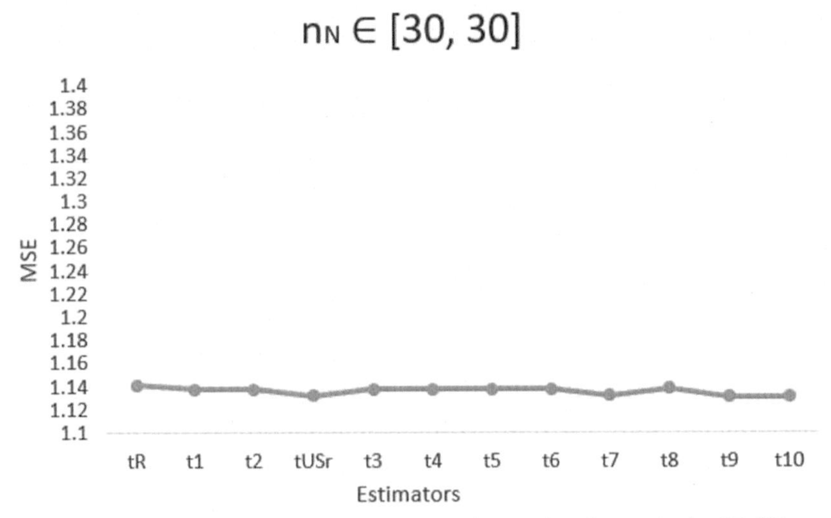

FIGURE 22.4 MSE of all neutrosophic estimators at lower values for sample size [30, 30].

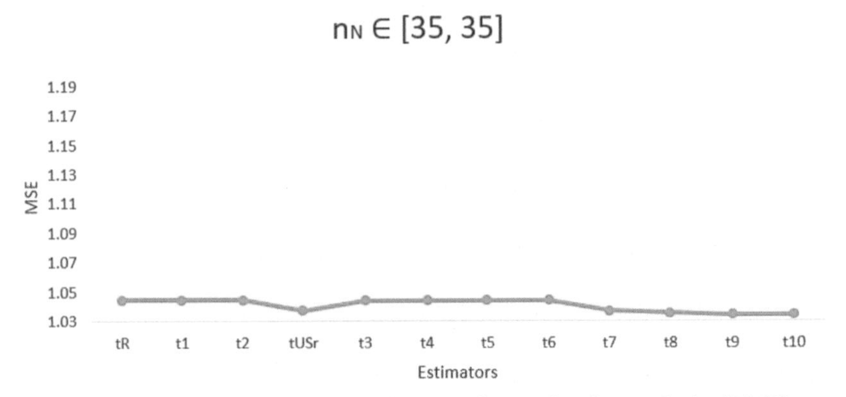

FIGURE 22.5 MSE of all neutrosophic estimators at lower values for sample size [35, 35].

robust-type neutrosophic ratio estimator, and from this generalized estimator, we proposed and studied the properties of ten robust-type neutrosophic ratio estimators. To examine their efficiencies on a real population and to compare them with existing ratio-type estimators, we applied the estimators to estimate the daily stock price of Moderna through a simulation study. Based on the results of the study presented in Tables 22.3 and 22.4, we can say that the proposed robust neutrosophic ratio-type estimators perform better than non-robust neutrosophic ratio estimators. Among the proposed robust neutrosophic ratio-type estimator, estimator t_{10} performs better as it gave the lowest MSE. We propose that robust-type neutrosophic ratio estimator t_{10} should be used for estimating the unknown neutrosophic population mean.

References

[1] N. Jan, L. Zedam, T. Mahmood, et al., Multiple attribute decision making method under linguistic cubic information, J. Intell. Fuzzy Syst. 36 (1) (2019) 253–269.

[2] D.F. Li, T. Mahmood, Z. Ali, et al., Decision making based on interval-valued complex single-valued neutrosophic hesitant fuzzy generalized hybrid weighted averaging operators, J. Intell. Fuzzy Syst. 38 (4) (2020) 4359–4401.

[3] F. Smarandache, Introduction to Neutrosophic Statistics, Sitech & Education Publishing, 2014. Available from. Available from: https://books.google.co.in/books?id = 3CLdBAAAQBAJ.

[4] Z. Tahir, H. Khan, M. Aslam, et al., Neutrosophic ratio-type estimators for estimating the population mean, Complex. Intell. Syst. (2021) 1–11.

[5] A. Ricardo, V.R. Maronna, M. Douglas, Robust statistics: Theory and Methods (with r), second ed., Wiley, 2018.

[6] M. Khoshnevisan, R. Singh, P. Chauhan, et al., A general family of estimators for estimating population mean using known value of some population parameter (s), Infinite Study (2007).

[7] P. Mishra, N.K. Adichwal, R. Singh, A new log-product-type estimator using auxiliary information, J. Sci. Res. 61 (1) (2017) 179–183.

[8] R. Singh, P. Mishra, C.N. Bouza-Herrera, Estimation of population mean using information on auxiliary attribute: a review, Ranked Set. Sampl. (2019) 239–249.

[9] R. Singh, M. Kumar, H.P. Singh, On estimation of poulation mean using information on auxiliary attribute, Pak. J. Stat. Oper. Res. (2013) 363–371.

[10] F. Smarandache, Neutrosophy: Neutrosophic Probability, Set and Logic: Analytic Synthesis & Synthetic Analysis, American Research Press, 1998.

[11] M. Aslam, Neutrosophic analysis of variance: application to university students, Complex. Intell. Syst. 5 (4) (2019) 3–407.

[12] M. Aslam, Monitoring the road traffic crashes using NEWMA chart and repetitive sampling, Int. J. Injury Control. Saf. Promot 28 (1) (2021) 39–45.

[13] M. Aslam, A new goodness of fit test in the presence of uncertain parameters, Complex. Intell. Syst. 7 (1) (2021) 359–365.

[14] M. Aslam, A new sampling plan using neutrosophic process loss consideration, Symmetry 10 (5) (2018) 132.

[15] R. Singh, R. Mishra, Neutrosophic transformed ratio estimators for estimating finite neutrosophic population mean, Recent Advancement in Sampling Theory & its Applications, MKES Publication, 2021, pp. 39–47.

[16] G.K. Vishwakarma, A. Singh, Generalized estimator for computation of population mean under neutrosophic ranked set technique: an application to solar energy data, Comput. Appl. Math 41 (4) (2022) 1–29.

[17] Moderna. Moderna Inc., Available from: https://www.modernatx.com/modernaswork-potential-vaccine-against-covid-19, 2021.

[18] YAHOO FINANCE. Moderna Inc. Available from: https://finance.yahoo.com/quote/MRNA/history/, 2021.

[19] M.N. Qureshi, C. Kadilar, M. Noor Ul Amin, et al., Rare and clustered population estimation using the adaptive cluster sampling with some robust measures, J. Stat. Comput. Simul. 88 (14) (2018) 2761–2774.

[20] Wikipedia. Moderna Inc, Available from: https://en.wikipedia.org/wiki/Moderna, 1999.

[21] Companies Market Cap, Moderna Inc. Available from: https://companiesmarketcap.com/pharmaceuticals/largest-pharmaceutical-companiesby-market-cap/, 2021.

[22] R. Mishra, B. Ram, Portfolio selection using R, Yugosl. J. Oper. Res. 30 (2) (2020) 137–146.

Chapter 23

Multi-attribute decision-making problem in medical diagnosis using neutrosophic probabilistic distance measures

M. Arockia Dasan[1], V.F. Little Flower[1], E. Bementa[2] and X. Tubax[3]
[1]Department of Mathematics, St. Jude's College, Manonmaniam Sundaranar University, Kanyakumari, Tamil Nadu, India, [2]PG and Research Department of Physics, Arulanandar College, Madurai, Tamil Nadu, India, [3]School of Information Technology, Cyryx College, Male, Maldives

23.1 Introduction

The Russian mathematician George Cantor is considered the father of set theory, who has defined the most basic concept in mathematics; that is, a set means a collection of "well-defined" objects. But in a real-life situation, we cannot collect the "beautiful flowers," "strong boys," etc., because the words "beautiful," "smell," "strong," etc., are not well-defined and are imprecise, uncertain words. Zadeh [1] introduced the concept of fuzzy set theory, which is the generalization of the crisp set to analyze imprecise mathematical information because the classic mathematical Cantor's set cannot handle the uncertainty concept in real-life situations. The original fuzzy set is characterized by the membership function whose value lies in the standard unit interval. The fuzzy set theory is applied in an extensive range over the world to many areas such as medical diagnosis, control systems, pattern recognition problems. The fuzzy similarity measure is a similarity measure defined on fuzzy sets which helps to study the relations between them. Pappis and Karacapilidis [2] introduced some fuzzy similarity measures with the comparative assessment of the similarity measures. The fuzzy distance number plays a vital role in fuzzy set theory, which depicts the distances of any two fuzzy sets. Using the fuzzy distance numbers, Voxman [3] insisted on some distance measures for modeling real-life problems in fuzzy sets. In 2015, Pramanik and Mondal [4] initiated the weighted fuzzy tangent similarity

Cognitive Intelligence with Neutrosophic Statistics in Bioinformatics.
DOI: https://doi.org/10.1016/B978-0-323-99456-9.00003-9

measures, and using these fuzzy tangent similarity measures, a medical application is formulated. Adlassing [5] applied fuzzy set theory to create medical relationships and fuzzy logic to computerize the diagnosis systems. Zadeh [6] further defined the fuzzy event and its probability, which is a generalization of the classical probability of an event.

In this unstable world, every person can have a chance to like or dislike an object, which leads to a new era in mathematics called intuitionistic fuzzy set theory by giving percentage for like and dislike at the same time. An intuitionistic fuzzy set is a generalized form of Zadeh's fuzzy set, developed by Atanassov [7] in the year 1986, which considers the degrees of both the membership and the nonmembership, such that the degree values are real numbers in $[0, 1]$, and its summation should be less than 1. The intuitionistic fuzzy set theory is applied to many research fields such as coding theory, control systems, data analysis, medical diagnosis, pattern recognition. Fan et al. [8] defined some new operators and intuitionistic fuzzy distance measures for aggregating intuitionistic fuzzy information with the solutions to decision-making problems. The intuitionistic fuzzy distance measure was introduced by Szmidt and Kacprzyk [9], and they applied these measures also to medical diagnosis problems. Dutta and Goala [10] introduced intuitionistic fuzzy distance measures and gave their application in medical diagnosis. Further, De et al. [11] initiated the intuitionistic fuzzy set applications in medical diagnosis. Moreover, Szmidt and Kacprzyk [12] introduced the concept of intuitionistic fuzzy events with their probability, which is a generalization of Zadeh's probability fuzzy events.

Most of the countries in this world choose their leader by conducting an election. The election gives some options for all the people such as some people who wish to vote for the contestant "X", some people who wish to vote for the opponent "Y", and some people who wish not to vote in the election. The percentage of people who wish not to vote in the election can be deciding the winner of the election if they vote for X or Y, which is called "neutral or indeterminist." These kinds of situations lead to a new era in mathematics called neutrosophic fuzzy set theory by giving percentages for membership, nonmembership, and indeterminacy. Smarandache [13] introduced the concept of a neutrosophic fuzzy set, which is the generalization of both fuzzy sets and intuitionistic fuzzy sets. The term "neutrosophy" means knowledge of neutral thought, and the concept "neutral" is the main distinction between fuzzy set theory or intuitionistic fuzzy set theory and neutrosophic set theory. The neutrosophic fuzzy set is an imprecise set to deal with the concept of uncertainty, vagueness, and irregularity, which consists of three independent components, such as truth membership, indeterminacy membership, and falsity membership, whose values are a subset of a real or nonstandard unit interval $]^{-}0, 1^{+}[$. The scientific and engineering fields face many difficulties in real life to use the neutrosophic set whose values are in the real or nonstandard unit interval $]^{-}0, 1^{+}[$. In this situation, we need some specified neutrosophic sets and operators. Wang et al. [14] structured the multi-spaces and multi-structure by defining a single-valued neutrosophic set to overcome this hurdle. A single-valued neutrosophic set is also a neutrosophic set in which the independent components are truth membership, indeterminacy membership, and falsity membership, whose values are from the unit interval [0, 1] such that their summation is less than 3. The distance measure on neutrosophic sets is an important tool in the applications of data mining, pattern recognition, and decision-making situations. Majumdar and Samantha [15] defined Hamming and Euclidean distance measures on single-valued neutrosophic sets. The tangent similarity measure along with the properties was introduced by Mondal et al. [16]. Hanafy et al. [17] investigated the properties of the neutrosophic correlation coefficient of the similarity measure. Biswas et al. [18] defined cosine similarity measures with the help of trapezoidal fuzzy neutrosophic numbers and solved a multi-attribute decision-making problem (MADM).

In this world, medical diagnosis is full of uncertainty which is made up of three types of sources that are technical, personal, and conceptual uncertainties. The sources of uncertainties come from probability, ambiguity, and complexity. The process of medical diagnosis involves investigating the symptoms and causes of certain diseases, either by oral examination or laboratory test. By understanding the diagnostic process, the physician can make the best decisions related to the patient's health. The probability of a common disease presenting itself is higher than that of a rare disorder. Decision-making is one of the most challenging areas of choosing two or more possible alternatives. The decision-makers have to decide by using complete or incomplete information. Shahzadi et al. [19] developed an application of single-valued neutrosophic sets in medical diagnosis using the Euclidean distance measure. In 2015, Mondal [20] presented the application of a multi-attribute neutrosophic decision-making model of school choice based on GRA with interval weight information. Jayaparthasarathy et al. [21] introduced neutrosophic supra topology with the application of MADM problems in the data mining process. Karaaslan and Khizar [22] developed a multi-criteria group decision-making method with the help of some neutrosophic operators and neutrosophic matrices. Karaaslan [23] also developed a medical diagnosis decision-making algorithm by defining Gaussian numbers and α-cut in single-valued neutrosophic numbers. Broumi and Smarandache [24] established the properties of distance and similarity measures on single-valued neutrosophic sets. Apart from this, Deli et al. [25] introduced neutrosophic parameterized soft relations along with the real-life application. Ye [26] defined the single-valued neutrosophic correlation coefficient, cross-entropy [27] in multi-criteria decision-making problems and similarity distance measures [28] in clustering methods. Moreover, Ye and Zhang [29] applied the single-valued neutrosophic similarity measures in MADM problems. Zavadskas et al. [30] developed a hybrid multiple-criteria decision-making method in the engineering field. In 1998, Smarandache [31] defined the neutrosophic probability which is a generalization of the intuitionistic fuzzy event and its probability. Shao et al. [32] developed a MADM problem by introducing the probabilistic single-valued neutrosophic hesitant fuzzy sets. Shao and Zhang [33] reduced the unnecessary evaluation process with the help of probabilistic neutrosophic measures. Peng et al. [34] introduced the probability multivalued neutrosophic sets and applied them in multi-criteria group decision-making problems. Arockia Dasan et al. [35] developed a method to solve multi-criterion decision-making problems in plant hybridization by using score functions on single-valued neutrosophic sets. Abhishek et al. [36] discussed the applications of classification problems in picture fuzzy sets by using a probabilistic distance measure.

23.1.1 Motivation and novelty of the work

On neutrosophic fuzzy sets and single-valued neutrosophic fuzzy sets, there are many distance measures and similarity measures [16,18–20,28,29,32–35] defined and applied in data analysis, medical diagnosis, and pattern recognition problems to deal with the multi-decision-making problems. These distance measures help to check whether the results are more convenient for our day-to-day life. Among these measures, only a few neutrosophic probabilistic distance measures [32,33] are applied in MADM problems. With these in mind, one of the novelties of this paper is motivated to define a general formula for probabilistic distance measures on single-valued neutrosophic sets using the probability of occurrences, nonoccurrences, and indeterminacy occurrences in a probabilistic nature. In the general formula, all the neutrosophic probabilistic distance measure values are gradually increasing if α values increase, which converge to 1. Next, we formulate a new mathematical model by the probabilistic distance measure on MADM problems and apply the proposed methodology in MADM problems for the medical diagnosis problems to find proper disease really suffered by the patient in a probabilistic neutrosophic environment.

23.2 Preliminaries

In this section, we review some basic definitions of the neutrosophic sets, single-valued neutrosophic sets, and neutrosophic distance measures.

Definition 23.2.1: [1] Let X be a non-empty set, and a fuzzy set A on X is of the form $A = \{(x, \mu_A(x)) : x \in X\}$, where $0 \leq \mu_A(x) \leq 1$ represents the degree of membership function of each $x \in X$ to the set A. For X, I^X denotes the collection of all fuzzy sets of X.

Definition 23.2.2: [7] Let X be a non-empty set. An intuitionistic set A is of the form $A = \{(x, \mu_A(x), \gamma_A(x)) : x \in X\}$, where $\mu_A(x)$ and $\gamma_A(x)$ represent the degree of membership and nonmembership function, respectively, of each $x \in X$ to the set A and $0 \leq \mu_A(x) + \gamma_A(x) \leq 1$ for all $x \in X$. The set of all intuitionistic fuzzy sets of X is denoted by $I(X)$.

Definition 23.2.3: [13] Let X be a non-empty set. A neutrosophic set A having the form $A = \{(x, \mu_A(x), \sigma_A(x), \gamma_A(x)) : x \in X\}$, where $\mu_A(x), \sigma_A(x)$, and $\gamma_A(x) \in]^-0, 1^+[$ represent the degree of membership (namely, $\mu_A(x)$), the degree of indeterminacy (namely, $\sigma_A(x)$) and the degree of nonmembership (namely, $\gamma_A(x)$), respectively, of each $x \in X$ to the set A such that $\mu_A(x) + \sigma_A(x) + \gamma_A(x) \in]^-0, 3^+[$for all $x \in X$. For $X, N(X)$ denotes the collection of all neutrosophic sets of X.

Definition 23.2.4: [21] A single-valued neutrosophic set (SVNS) A in X is a neutrosophic set which is of the form $A = \{(x, \mu_A(x), \sigma_A(x), \gamma_A(x)) : x \in X\}$ that is characterized by the degree of membership (namely, $\mu_A(x)$), the degree of indeterminacy (namely, $\sigma_A(x)$) and the degree of nonmembership (namely, $\gamma_A(x)$), where $\mu_A(x), \sigma_A(x), \gamma_A(x) \in [0, 1]$ such that $0 \leq \mu_A(x) + \sigma_A(x) + \gamma_A(x) \leq 3$, for all $x \in X$, respectively. For $X, SVNS(X)$ denotes the collection of all single-valued neutrosophic sets of X.

Two important concepts for SVNSs are the distance measure and similarity measure, which are applied to compare the neutrosophic fuzzy information.

Definition 23.2.5: [15] Let $X = \{x_1, x_2, \ldots, x_n\}$ be a discrete confined set. A mapping d:SVNS(X) × SVNS(X) → [0, 1] is said to be a distance measure between two single-valued neutrosophic sets if it satisfies the following axioms:

1. $d(A, B) \geq 0$ for all $A, B \in \text{SVNS}(X)$.
2. $d(A, B) = 0$ if and only if $A = B$ for all $A, B \in \text{SVNS}(X)$.
3. $d(A, B) = d(B, A)$ for all $A, B \in \text{SVNS}(X)$.
4. If$A \subseteq B \subseteq C$ for all $A, B, C \in \text{SVNS}(X)$, then $d(A, C) \geq d(A, B)$ and $d(A, C) \geq d(B, C)$.

If the mapping is defined as $d(A, B) = \max\{|\mu_A(x_i) - \mu_B(x_i)|, |\sigma_A(x_i) - \sigma_B(x_i)|, |\gamma_A(x_i) - \gamma_B(x_i)|\}, \forall x_i \in X$, then $d(A, B)$ satisfies axioms of distance measure and is called the extended Hausdorff distance measure between two single-valued neutrosophic sets A and B.

Definition 23.2.6: [15] The normalized Hamming distance between two single-valued neutrosophic sets A and B is defined by

$$d(A, B) = \frac{1}{3n} \sum_{j=1}^{n} \left(|\mu_A(x_j) - \mu_B(x_j)| + |\sigma_A(x_j) - \sigma_B(x_j)| + |\gamma_A(x_j) - \gamma_B(x_j)| \right).$$

Definition 23.2.7: [15] The normalized Euclidean distance between two single-valued neutrosophic sets A and B is defined by

$$d(A,B)=\left\{\frac{1}{3n}\sum_{j=1}^{n}\left(\mu_A(x_j)-\mu_B(x_j)\right)^2+\left(\sigma_A(x_j)-\sigma_B(x_j)\right)^2+\left(\gamma_A(x_j)-\gamma_B(x_j)\right)^2\right\}^{\frac{1}{2}}.$$

Definition 23.2.8: [24] A mapping S:SVNS(X) × SVNS(X) → [0, 1] is said to be a similarity measure between two single-valued neutrosophic sets if it satisfies the properties of axioms:

1. $S(A,B)\geq 0$ for all $A,B\in$ SVNS(X).
2. $S(A,B)=1$ if and only if $A=B$ for all $A,B\in$ SVNS(X).
3. $S(A,B)=S(B,A)$ for all $A,B\in$ SVNS(X).
4. If $A\subseteq B\subseteq C$ for all $A,B,C\in$ SVNS(X), then $S(A,C)\leq S(A,B)$ and $S(A,C)\leq S(B,C)$.

Definition 23.2.9: [35] Let $X=\{x_1,x_2,\ldots,x_n\}$ be a universal set. Let $A=\{(x_i,\mu_A(x_i),\sigma_A(x_i),$

$\gamma_A(x_i)):x_i\in X\}$ and $B=\left\{\left(x_i,\mu_B(x_i),\sigma_B(x_i),\gamma_B(x_i)\right):x_i\in X\right\}$ are two single-valued neutrosophic sets on X. Then, define a mapping d:SVNS(X) × SVNS(X) → [0, 1] as:

$$d(A,B)=\frac{5}{3n}\sum_{i=1}^{n}\frac{\sin\left\{\frac{\pi}{6}\left|\mu_A(x_i)-\mu_B(x_i)\right|\right\}+\sin\left\{\frac{\pi}{6}\left|\sigma_A(x_i)-\sigma_B(x_i)\right|\right\}+\sin\left\{\frac{\pi}{6}\left|\gamma_A(x_i)-\gamma_B(x_i)\right|\right\}}{1+\sin\left\{\frac{\pi}{6}\left|\mu_A(x_i)-\mu_B(x_i)\right|\right\}+\sin\left\{\frac{\pi}{6}\left|\sigma_A(x_i)-\sigma_B(x_i)\right|\right\}+\sin\left\{\frac{\pi}{6}\left|\gamma_A(x_i)-\gamma_B(x_i)\right|\right\}}.$$

23.3 Probabilistic distance measure on single-valued neutrosophic sets

This section develops the probability of occurrences and nonoccurrences of a neutrosophic event A using neutrosophic indeterminacy. We define a general formula for probabilistic distance measure on single-valued neutrosophic sets, and its different properties are studied.

Definition 23.3.1: Let X be the universe of discourse or sample space with a neutrosophic event $A=\{x_1,x_2,\ldots,x_n\}$. Consider the probability of occurrence of the neutrosophic event as $p(x_1),p(x_2),\ldots,p(x_n)$, respectively. Then, the minimal and maximal probabilities of occurrence of a neutrosophic event A, denoted by $p_{\min}(A)$, and $p_{\max}(A)$, are defined as $p_{\min}(A)=\sum_{i=1}^{n}p(x_i)\mu_A(x_i)$ and $p_{\max}(A)=p_{\min}(A)+\sum_{i=1}^{n}p(x_i)\sigma_A(x_i)$. The probability of occurrence of the neutrosophic event A, denoted by $p(A)$, is a single-valued neutrosophic number $p(A)=\frac{p_{\min}(A)+p_{\max}(A)}{2}$ which lies in the interval $[p_{\min}(A),p_{\max}(A)]$.

Definition 23.3.2: Let X be the universe of discourse or sample space with a neutrosophic event $A = \{x_1, x_2, ..., x_n\}$. Consider the probability of nonoccurrence of the neutrosophic event as $q(x_1), q(x_2), ..., q(x_n)$, respectively. Then, the minimal and maximal probabilities of nonoccurrence of a neutrosophic event A are defined as $q_{\min}(A) = \sum_{i=1}^{n} p(x_i)\gamma_A(x_i)$ and $q_{\max}(A) = q_{\min}(A) + \sum_{i=1}^{n} q(x_i)\sigma_A(x_i)$. Hence, the probability of nonoccurrence of the neutrosophic event A, denoted by $q(A)$, is a single-valued neutrosophic number $q(A) = \frac{q_{\min}(A) + q_{\max}(A)}{2}$ which lies in the interval $[q_{\min}(A), q_{\max}(A)]$.

Example 23.3.3: Let $X = \{x_1, x_2\}$ be the universal set or sample space and $A = \{(x_1, 0.3, 0.4, 0.6), (x_2, 0, 0.1, 0)\}$ as the single-valued neutrosophic event. Suppose the probability of the neutrosophic event is $p(x_1) = 0.7, p(x_2) = 0.6$, and the probability of nonoccurrence of the neutrosophic event is $q(x_1) = 0.3$, $q(x_2) = 0.4$. Then, $p_{\min}(A) = 0.21$, $p_{\max}(A) = 0.55$, $q_{\min}(A) = 0.18$, and $q_{\max}(A) = 0.34$. Therefore, $p(A) = 0.38$, $q(A) = 0.26$. Hence, $p(A) \in [0.21, 0.55]$, and $q(A) \in [0.18, 0.34]$.

Definition 23.3.4: Let $X = \{x_1, x_2, ..., x_n\}$ be a universal set. Let $A = \{(x_i, \mu_A(x_i), \sigma_A(x_i), \gamma_A(x_i)) : x_i \in X\}$ and $B = \{(x_i, \mu_B(x_i), \sigma_B(x_i), \gamma_B(x_i)) : x_i \in X\}$ be two single-valued neutrosophic sets on X. Then, define a mapping $d: \text{SVNS}(X) \times \text{SVNS}(X) \to [0, 1]$ as:

$$d(A,B) = \left\{\frac{1}{3n}\sum_{i=1}^{n} p(A,B)\left|\mu_A(x_i) - \mu_B(x_i)\right|^{\alpha} + s(A,B)\left|\sigma_A(x_i) - \sigma_B(x_i)\right|^{\alpha} + q(A,B)\left|\gamma_A(x_i) - \gamma_B(x_i)\right|^{\alpha}\right\}^{\frac{1}{\alpha}},$$

for $\alpha \geq 1$, where, $p(A,B) = \min\{p(A), p(B)\}$, $q(A,B) = \min\{q(A), q(B)\}$ and $s(A,B) = \max\{p(A,B), q(A,B)\}$.

Theorem 23.3.5: The following properties are true for the single-valued neutrosophic sets A, B, C.

1. $d(A,B) \geq 0$ for all $A, B \in \text{SVNS}(X)$.
2. $d(A,B) = 0$ if and only if $A = B$ for all $A, B \in \text{SVNS}(X)$.
3. $d(A,B) = d(B,A)$ for all $A, B \in \text{SVNS}(X)$.
4. If $A \subseteq B \subseteq C$ for all $A, B, C \in \text{SVNS}(X)$ then $d(A,C) \geq d(A,B)$ and $d(A,C) \geq d(B,C)$.

Proof: Part(i): Trivially, $d(A,B) \geq 0$ is true for single-valued neutrosophic sets A and B by Definition 3.4.

Part(ii): $d(A,B) = 0 \Leftrightarrow \left\{\frac{1}{3n}\sum_{i=1}^{n} p(A,B)\left|\mu_A(x_i) - \mu_B(x_i)\right|^{\alpha} + s(A,B)\left|\sigma_A(x_i) - \sigma_B(x_i)\right|^{\alpha} + q(A,B)\left|\gamma_A(x_i) - \gamma_B(x_i)\right|^{\alpha}\right\}^{\frac{1}{\alpha}} = 0 \Leftrightarrow \left|\mu_A(x_i) - \mu_B(x_i)\right|^{\alpha} = 0, \left|\sigma_A(x_i) - \sigma_B(x_i)\right|^{\alpha} = 0, \left|\gamma_A(x_i) - \gamma_B(x_i)\right|^{\alpha \geq 1} = 0 \ \forall i \Leftrightarrow \mu_A(x_i) = \mu_B(x_i), \sigma_A(x_i) = \sigma_B(x_i), \gamma_A(x_i) = \gamma_B(x_i) \forall i \Leftrightarrow A = B$.

Part(iii):

$$d(A,B)=\left\{\frac{1}{3n}\sum_{i=1}^{n}p(A,B)\left|\mu_A(x_i)-\mu_B(x_i)\right|^{\alpha}+s(A,B)\left|\sigma_A(x_i)\right.\right.$$
$$\left.\left.-\sigma_B(x_i)\right|^{\alpha}+q(A,B)\left|\gamma_A(x_i)-\gamma_B(x_i)\right|^{\alpha}\right\}^{\frac{1}{\alpha}}$$
$$=\left\{\frac{1}{3n}\sum_{i=1}^{n}p(B,A)\left|\mu_B(x_i)-\mu_A(x_i)\right|^{\alpha}+s(B,A)\left|\sigma_B(x_i)-\sigma_A(x_i)\right|^{\alpha}\right.$$
$$\left.+q(B,A)\left|\gamma_B(x_i)-\gamma_A(x_i)\right|^{\alpha}\right\}^{\frac{1}{\alpha}}=d(B,A).$$

Part(iv): If $A\subseteq B\subseteq C$ then $\mu_A(x_i)\leq\mu_B(x_i)\leq\mu_C(x_i),\sigma_A(x_i)\leq\sigma_B(x_i)\leq\sigma$ $C(x_i)$ and $\gamma_A(x_i)\geq\gamma_B(x_i)\geq\gamma_C(x_i)$. Now we have the following inequalities $\mu_A(x_i)-\mu_C(x_i)\geq\mu_A(x_i)-\mu_B(x_i),\mu_A(x_i)-\mu_C(x_i)\geq\mu_B(x_i)-\mu_C(x_i),\sigma_A(x_i)-\sigma_C(x_i)\geq\sigma_A(x_i)-\sigma_B(x_i),\sigma_A(x_i)-\sigma_C(x_i)\geq\sigma_B(x_i)-\sigma_C(x_i),\gamma_A(x_i)-\gamma_C(x_i)\geq\gamma_A(x_i)-\gamma_B(x_i),\gamma_A(x_i)-\gamma_C(x_i)\geq\gamma_B(x_i)-\gamma_C(x_i)$, and by the monotonicity property on probability theory, $p(A,C)\geq p(A,B)$, $s(A,C)\geq s(A,B)$ and $q(A,C)\geq q(A,B)$. Then we have

$$\left\{\frac{1}{3n}\sum_{i=1}^{n}p(A,C)\left|\mu_A(x_i)-\mu_c(x_i)\right|^{\alpha}+s(A,C)\left|\sigma_A(x_i)\right.\right.$$
$$\left.\left.-\sigma_C(x_i)\right|^{\alpha}+q(A,C)\left|\gamma_A(x_i)-\gamma_C(x_i)\right|^{\alpha}\right\}^{\frac{1}{\alpha}}$$
$$\geq\left\{\frac{1}{3n}\sum_{i=1}^{n}p(A,B)\left|\mu_A(x_i)-\mu_B(x_i)\right|^{\alpha}+s(A,B)\left|\sigma_A(x_i)\right.\right.$$
$$\left.\left.-\sigma_C(x_i)\right|^{\alpha}+q(A,B)\left|\gamma_A(x_i)-\gamma_B(x_i)\right|^{\alpha}\right\}^{\frac{1}{\alpha}}$$

and

$$\left\{\frac{1}{3n}\sum_{i=1}^{n}p(A,C)\left|\mu_A(x_i)-\mu_c(x_i)\right|^{\alpha}+s(A,C)(\sigma_A(x_i)\right.$$
$$\left.-\sigma_C(x_i))^{\alpha}+q(A,C)\left|\gamma_A(x_i)-\gamma_C(x_i)\right|^{\alpha}\right\}^{\frac{1}{\alpha}}$$
$$\geq\left\{\frac{1}{3n}\sum_{i=1}^{n}p(B,C)\left|\mu_B(x_i)-\mu_C(x_i)\right|^{\alpha}+s(B,C)\left|\sigma_B(x_i)\right.\right.$$
$$\left.\left.-\sigma_C(x_i)\right|^{\alpha}+q(B,C)\left|\gamma_B(x_i)-\gamma_C(x_i)\right|^{\alpha}\right\}^{\frac{1}{\alpha}}.$$

Hence $d(A,C)\geq d(A,B)$ and $d(A,C)\geq d(B,C)$.

Remark 23.3.6: From the above theorem, we can observe that the proposed mapping $d(A, B)$ satisfies the distance measure axioms [15], and we call it as single-valued neutrosophic probabilistic distance measure. If $\alpha = 2$, then $d(A,B) = \left\{ \frac{1}{3n} \sum_{i=1}^{n} p(A,B) \left| \mu_A(x_i) - \mu_B(x_i) \right|^2 + s(A,B) |\sigma_A(x_i) - \sigma_B(x_i)|^2 + q(A,B) \left| \gamma_A(x_i) - \gamma_B(x_i) \right|^2 \right\}^{\frac{1}{2}}$ is called the probabilistic normalized Euclidean distance measure on single-valued neutrosophic sets.

Theorem 23.3.7: $S(A, B) = 1 - d(A, B)$ is a similarity measure for every $A, B \in \mathrm{SVNS}(X)$, and call it as single-valued neutrosophic probabilistic similarity measure.

Proof: We can easily verify Definition 2.8 of similarity measure's axioms.

23.4 Methodologies in neutrosophic multi-attribute decision-making problems

In this section, we propose a mathematical model as a method for the MADM problem by the single-valued neutrosophic probabilistic distance measure to find the suitable decision attribute for the alternative from the information tables (see Tables 23.1 and 23.2). The following steps are the necessary steps for the proposed methodological approach to select the proper decision attributes for the alternative.

23.4.1 Problem field selection

Here, all the attributes a_{ij} and $d_{jk} (i = 1, 2, \ldots, m, j = 1, 2, \ldots, n$ and $k = 1, 2, \ldots, p)$ are all single-valued neutrosophic sets.

TABLE 23.1 Alternatives versus conditional attributes.

	S_1	S_2	.	.	.	S_n
A_1	(a_{11})	(a_{12})	.	.	.	(a_{1n})
A_2	(a_{21})	(a_{22})	.	.	.	(a_{2n})
.	.	.	.	.	.	.
.	.	.	.	.	.	.
.	.	.	.	.	.	.
A_m	(a_{m1})	(a_{m2})	.	.	.	(a_{mn})

TABLE 23.2 Conditional attributes versus decision attributes.

	D_1	D_2	.	.	.	D_p
S_1	(d_{11})	(d_{12})	.	.	.	(d_{1p})
S_2	(d_{21})	(d_{22})	.	.	.	(d_{2p})
.	.	.	.	.	.	.
.	.	.	.	.	.	.
.	.	.	.	.	.	.
S_n	(d_{n1})	(d_{n2})	.	.	.	(d_{np})

23.4.2 Algorithm for finding the probability $p(A_i, D_k)$

Step 1: Consider the multi-attribute decision-making problem with m alternatives $A_1, A_2, \ldots, A_m$, n conditional attributes $S_1, S_2, \ldots, S_n$, and p decision attributes are $D_1, D_2, \ldots, D_P$.

Step 2: Determine the minimum membership value for each conditional attribute $S_j$with respect to the alternatives A_i, that is, $\min_j \mu_{A_i}(a_{ij}) = \mu'(\alpha_j)$, and for each $j = 1, 2, \ldots, n$, collect $E_1 = \{\mu'(\alpha_1), \mu'(\alpha_2), \ldots, \mu'(\alpha_n)\}$. Similarly, determine the minimum membership value for each conditional attribute $S_j$with respect to the decision attributes D_k, that is, $\min_j \mu_{D_k}(d_{jk}) = \mu'(\beta_j)$, and for each $j = 1, 2, \ldots, n$, collect $E_2 = \{\mu'(\beta_1), \mu'(\beta_2), \ldots, \mu'(\beta_n)\}$.

Step 3: Calculate $R_1 = \sum_{j=1}^{n} \mu'(\alpha_j)$, and $R_2 = \sum_{j=1}^{n} \mu'(\beta_j)$.

Step 4: If $R_1 \neq 0$ and $R_2 \neq 0$, find the probability of occurrence of each conditional attribute S_j with respect to the alternatives A_i, $p(S_j, A_i) = \frac{\mu'(\alpha_j)}{R_1}$ and also find the probability of occurrence of each conditional attribute S_j with respect to the decision attributes D_k, $p(S_j, D_k) = \frac{\mu'(\beta_j)}{R_2}$. If $R_1, R_2 = 0$, then let us take $p(S_j, A_i) = 1$ and $p(S_j, D_k) = 1$.

Step 5: Find the minimum probability of occurrence for each alternative A_i, $p_{\min}(A_i) = \sum_{j=1}^{n} p(S_j, A_i)\mu_{A_i}(a_{ij})$ and the minimum probability of occurrence for each decision attribute D_k, $p_{\min}(D_k) = \sum_{j=1}^{n} p(S_j, D_k)\mu_{D_k}(d_{jk})$. Find the maximum probability of occurrence for each alternative A_i, $p_{\max}(A_i) = p_{\min}(A_i) + \sum_{j=1}^{n} p(S_j, A_i)\sigma_{A_i}(a_{ij})$ and the maximum probability of occurrence for each decision attribute D_k, $p_{\max}(D_k) = p_{\min}(D_k) + \sum_{j=1}^{n} p(S_j, D_k)\sigma_{D_k}(d_{kj})$.

Step 6: Calculate the probability of occurrence for each alternative A_i, $p(A_i) = \frac{p_{\min}(A_i) + p_{\max}(A_i)}{2}$ and the probability of occurrence for each decision attribute D_k, $p(D_k) = \frac{p_{\min}(D_k) + p_{\max}(D_k)}{2}$.

Step 7: Find the probability of occurrence of each alternative A_i corresponding to each decision attribute D_k using $p(A_i, D_k) = \min\{p(A_i), p(D_k)\}$, for each $i = 1, 2, \ldots, m$, and $k = 1, 2, \ldots, p$.

23.4.3 Algorithm for finding the probability $q(A_i, D_k)$

Step 8: Determine the minimum nonmembership value for each conditional attribute S_j with respect to the alternatives A_i, that is, $\min_j \gamma_{A_i}(a_{ij}) = \gamma'(\alpha_j)$, and for each $j = 1, 2, \ldots, n$, collect $E'_1 = \{\gamma'(\alpha_1), \gamma'(\alpha_2), \ldots \gamma'(\alpha_n)\}$. Similarly, determine the minimum nonmembership value for each conditional attribute S_j with respect to the decision attributes D_k, that is, $\min_j \gamma_{D_k}(d_{jk}) = \gamma'(\beta_j)$, and for each $j = 1, 2, \ldots, n$, collect $E'_2 = \{\gamma'(\beta_k), \gamma'(\beta_k), .., \gamma'(\gamma_n)\}$.

Step 9: Calculate $R_1{}' = \sum_{j=1}^{n} \gamma'(\alpha_j)$ and $R'_2 = \sum_{j=1}^{n} \gamma'(\beta_j)$.

Step 10: If $R'_1 \neq 0$ and $R'_2 \neq 0$, find the probability of nonoccurrence of each conditional attribute S_j with respect to the alternatives $A_i, q(S_j, A_i) = \frac{\gamma'(\alpha_j)}{R'_1}$, and find the probability of nonoccurrence of each conditional attribute S_j with respect to the decision attributes D_k, $q(S_j, D_k) = \frac{\gamma'(\beta_j)}{R_2{}'}$. If R'_1, $R'_2 = 0$, then let us take $q(S_j, A_i) = 1$ and $q(S_j, D_k) = 1$.

Step 11: Find the minimum probability of nonoccurrence for each alternative A_i, $q_{\min}(A_i) = \sum_{j=1}^{n} q(S_j, A_i)\gamma_{A_i}(a_{ij})$ and the minimum probability of nonoccurrence for each decision attribute D_k, $q_{\min}(D_k) = \sum_{j=1}^{n} q(S_j, D_k)\gamma_{D_k}(d_{jk})$. Find the maximum probability of nonoccurrence for each alternative A_i, $q_{\max}(A_i) = q_{\min}(A_i) + \sum_{j=1}^{n} q(S_j, A_i)\sigma_{A_i}(a_{ij})$ and the maximum probability of nonoccurrence for each decision attribute D_k, $q_{\max}(D_k) = q_{\min}(D_k) + \sum_{j=1}^{n} q(S_j, D_k)\sigma_{D_k}(d_{jk})$.

Step 12: Calculate the probability of nonoccurrence for each alternative A_i, $q(A_i) = \frac{q_{\min}(A_i) + q_{\max}(A_i)}{2}$ and the probability of nonoccurrence for each decision attribute D_k, $q(D_k) = \frac{q_{\min}(D_k) + q_{\max}(D_k)}{2}$.

Step 13: Find the probability of nonoccurrence of each alternative A_i corresponding to each decision attribute D_k using $q(A_i, D_k) = \min\{q(A_i), q(D_k)\}$, for each $i = 1, 2, \ldots, m$ and $k = 1, 2, \ldots, p$.

23.4.4 Algorithm for finding the probability $s(A_i, D_k)$

Step 14: Find the probability of indeterminacy occurrence of each alternative A_i corresponding to each decision attribute D_k using $s(A_i, D_k) = \max\{p(A_i, D_k), q(A_i, D_k)\}$, for each $i = 1, 2, \ldots, m$, and $k = 1, 2, \ldots, p$.

23.4.5 The distance measures of alternatives and attributes

Step 15: For $\alpha = 1, 2, 3, \ldots$, find the single-valued neutrosophic probabilistic distance measure of the alternative A_i and the decision attribute D_k using $d(A_i, D_k) = \{\frac{1}{3n}\sum_{j=1}^{n} p(A_i, D_k)|\mu_{A_i}(a_{ij}) - \mu_{d_k}(d_{jk})|^{\alpha} + s(A_i, D_k)$ $|\sigma_{A_i}(a_{ij}) - \sigma_{D_k}(d_{jk})|^{\alpha} + q(A_i, D_k)|\gamma_{A_i}(a_{ij}) - \gamma_{D_k}(d_{jk})|^{\alpha}\}^{\frac{1}{\alpha}}$, where $p(A_i, D_k) = \min\{p(A_i), p(D_k)\}$, $q(A_i, D_k) = \min\{q(A_i), q(D_k)\}$, and $s(A_i, D_k) = \max\{p(A_i, D_k), q(A_i, D_k)$, $i = 1, 2, \ldots, m$, and $k = 1, 2, \ldots, p$.

23.4.6 Tabulation

All the single-valued neutrosophic probabilistic distance measure values of the alternatives and the decision attributes for $\alpha = 1, 2, 3, \ldots$, are listed as follows.

23.4.7 Final decision

From the probabilistic distance measure table (see Table 23.3), identify each alternative A_i, $i = 1, 2, \ldots, m$ with the corresponding decision attribute D_k, $k = 1, 2, \ldots, p$ for each $\alpha = 1, 2, 3, \ldots$, by choosing the lowest probabilistic distance measure value, which is more similar between the alternatives A_i and the decision attributes D_k, and then conclude that the most repeated decision attribute D_k is the suitable decision attribute for the alternative A_i.

The following chart explains the summary of the above methodology in Fig. 23.1.

TABLE 23.3 Neutrosophic probabilistic distance measure table.

$\alpha = 1, 2, 3, \ldots$	D_1	D_2	.	.	.	D_p
A_1	$d(A_1, D_1)$	$d(A_1, D_2)$	.	.	.	$d(A_1, D_p)$
A_2	$d(A_2, D_1)$	$d(A_2, D_2)$	.	.	.	$d(A_2, D_p)$
.	.	.	.	.	.	.
.	.	.	.	.	.	.
.	.	.	.	.	.	.
A_m	$d(A_m, D_1)$	$d(A_m, D_2)$	.	.	.	$d(A_m, D_p)$

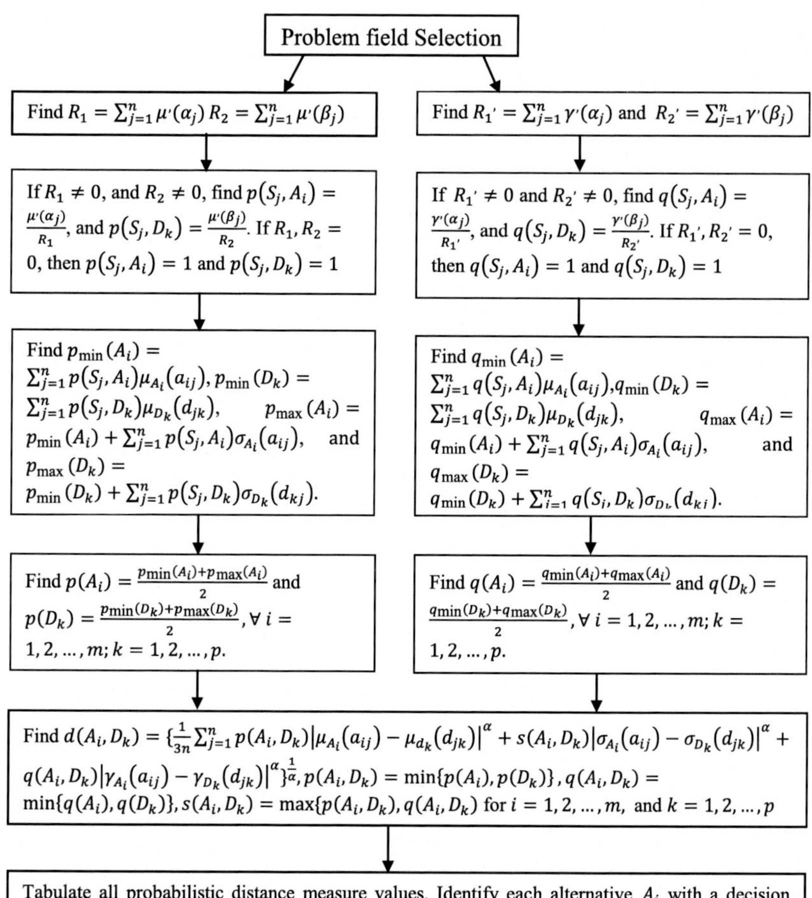

FIGURE 23.1 The summary of the proposed methodology.

23.5 Numerical example: application of single-valued neutrosophic probabilistic distance measures in medical diagnosis

Medical diagnosis is a process of determining the disease or condition that explains a person's symptoms. Diagnosis is often challenging because many symptoms are wandering off the point. The diagnosis will require synthesis of signs and symptoms in the case to report identify the core problems. The correct diagnosis sometimes follows a short and simple journey. In the diagnosis process, there is more chance of errors to occur. To deal with uncertainty, the doctors often overemphasize the importance of diagnostic tests, at the expense of history and physical examination, believing laboratory tests to be more accurate.

In this section, we demonstrate a numerical example in medical diagnosis to determine the diseases of the patient as a real-life application of the single-valued neutrosophic probabilistic distance measure for the above-proposed methodology is ineffective.

23.5.1 Problem field selection

Step 1: Let $A = \{A_1, A_2, A_3, A_4\}$ be the set of patients as the alternatives, $S = \{S_1, S_2, S_3, S_4, S_5\}$be the set of symptoms as the conditional attributes, and the decision attributes $D = \{D_1, D_2, D_3, D_4\}$ be the set of diseases. Our main aim is to find the patient who has disease such as D_1, D_2, D_3, and D_4. The following tables show that the membership, nonmembership, and indeterminacy values of the neutrosophic set (see Tables 23.4 and 23.5). For example, in Table 23.4, we can observe that the membership value for the patient A_1 who has symptom S_4 is 0.5, the nonmembership value is 0.6, and the indeterminacy value is 0.1, so

TABLE 23.4 Patients versus symptoms.

	S_1	S_2	S_3	S_4	S_5
A_1	(0.7, 0.3, 0.2)	(1, 0.3, 0.1)	(0.4, 0.3, 0.1)	(0.5, 0.1, 0.6)	(0.3, 0.2, 0.1)
A_2	(0.5, 0.1, 0)	(0.6, 0, 0.2)	(0.1, 0, 0)	(0.01, 0, 1)	(0.2, 0, 0.3)
A_3	(0, 0.3, 0.1)	(0.3, 0.1, 0.5)	(0.5, 0.6, 0.1)	(0.3, 0.1, 0)	(0.5, 0, 0.6)
A_4	(0.7, 0.3, 0.3)	(0.3, 0.3, 0.3)	(0.3, 0, 1)	(0.6, 0.3, 0.1)	(0.3, 0.2, 0.1)

TABLE 23.5 Symptoms versus diseases.

	D_1	D_2	D_3	D_4
S_1	(0.3, 0.5, 0.1)	(0.6, 0.5, 0.1)	(0.2, 0, 1)	(0.5, 0.3, 1)
S_2	(0.3, 0.3, 0)	(0, 0, 1)	(0, 0.6, 0.3)	(0.2, 0.1, 0.1)
S_3	(0.3, 0.2, 0.1)	(0.5, 0.3, 0.1)	(0.7, 0.8, 0.1)	(0.2, 0.2, 0.2)
S_4	(0.3, 0.4, 0.1)	(0.6, 0, 0.1)	(0.2, 0.5, 0.3)	(0.4, 0.3, 0.1)
S_5	(0.7, 0.8, 1)	(0.5, 0.5, 0.5)	(0, 0, 1)	(1, 0, 0)

neutrosophically we denoted as $(0.5, 0.1, 0.6)$. The membership value for the symptom S_5 who has disease D_2 in Table 23.5 is 0.5, the nonmembership value is 0.5, and the indeterminacy value is 0.5; it is also neutrosophically denoted as $(0.5, 0.5, 0.5)$.

23.5.2 Algorithm for finding the probability $p(A_i, D_k)$

Step 2: The minimum membership values for each conditional attribute S_j with respect to the alternatives A_i are $\min\mu_{A_i}(a_{i1}) = 0$, $\min\mu_{A_i}(a_{i2}) = 0.3$, $\min\mu_{A_i}(a_{i3}) = 0.1$, $\min\mu_{A_i}(a_{i4}) = 0.01$, $\min\mu_{A_i}(a_{i5}) = 0.2$. Therefore, $E_1 = \{0, 0.3, 0.1, 0.01, 0.2\}$. Similarly, the minimum membership values for each conditional attribute S_j with respect to the decision attributes D_k are $\min\mu_{D_k}(d_{1k}) = 0.2$, $\min\mu_{D_k}(d_{2k}) = 0$, $\min\mu_{D_k}(d_{3k}) = 0.2$, $\min\mu_{D_k}(d_{4k}) = 0.2$, $\min\mu_{D_k}(d_{5k}) = 0$. Therefore, $E_2 = \{0.2, 0, 0.2, 0.2, 0\}$.
Step 3: The sum $R_1 = \sum_{j=1}^{n} \mu'(\alpha_j) = 0.61$ and $R_2 = \sum_{j=1}^{n} \mu'(\beta_j) = 0.6$.
Step 4: Since $R_1 \neq R_2$, then the probabilities of occurrences of each conditional attribute S_j with respect to the alternatives A_i and with respect to the decision attributes D_k using $p(S_j, A_i) = \frac{\mu'(\alpha_j)}{R_1}$, and $p(S_j, D_k) = \frac{\mu'(\beta_j)}{R_2}$, respectively, are as follows: $p(S_1, A_i) = 0, p(S_2, A_i) = 0.4918, p(S_3, A_i) = 0.16393, p(S_4, A_i) = 0.01639, p(S_5, A_i) = 0.32787$, and $p(S_1, D_k) = 0.33$, $p(S_2, D_k) = 0, p(S_3, D_k) = 0.33$, $p(S_4, D_k) = 0.33$, $p(S_5, D_k) = 0$.
Step 5: The minimum probability of occurrence $p_{\min}(A_i) = \sum_{j=1}^{n} p(S_j, A_i)\mu_{A_i}(a_{ij})$, $p_{\min}(D_k) = \sum_{j=1}^{n} p(S_j, D_k)\mu_{D_k}(d_{jk})$ and the maximum probability of occurrence $p_{\max}(A_i) = p_{\min}(A_i) + \sum_{j=1}^{n} p(S_j, A_i)\sigma_{A_i}(a_{ij})$, $p_{\max}(D_k) = p_{\min}(D_k) + \sum_{k=1}^{p} p(S_j, D_k)\sigma_{D_k}(d_{jk})$ are, respectively, as follows: $p_{\min}(A_1) = 0.66393, p_{\min}(A_2) = 0.37769, p_{\min}(A_3) = 0.39836$, $p_{\min}(A_4) = 0.30491$, $p_{\min}(D_1) = 0.297, p_{\min}(D_2) = 0.561, p_{\min}(D_3) = 0.363$, $p_{\min}(D_4) = 0.363$, and $p_{\max}(A_1) = 0.92786, p_{\max}(A_2) = 0.37769, p_{\max}(A_3) = 0.54753, p_{\max}(A_4) = 0.52295$ and $p_{\max}(D_1) = 0.66, p_{\max}(D_2) = 0.825$, $p_{\max}(D_3) = 0.792, p_{\max}(D_4) = 0.627$.
Step 6: The probabilities of occurrence for each alternative A_i and for each decision attribute D_k using $p(A_i) = \frac{p_{\min}(A_i) + p_{\max}(A_i)}{2}$ and $p(D_k) = \frac{p_{\min}(D_k) + p_{\max}(D_k)}{2}$, respectively, are as follows: $P(A_1) = 0.79589$, $P(A_2) = 0.37769, P(A_3) = 0.47295, P(A_4) = 0.41393$ and $P(D_1) = 0.4785$, $P(D_2) = 0.693, P(D_3) = 0.5775, P(D_4) = 0.495$.
Step 7: The probabilities of occurrence of each alternative A_i corresponding to each decision attribute D_k using $p(A_i, D_k) = \min\{p(A_i), p(D_k)\}$ are as follows: $p(A_1, D_1) = 0.4785, p(A_1, D_2) = 0.693, p(A_1, D_3) = 0.5775$, $p(A_1, D_4) = 0.495, p(A_2, D_1) = 0.37769, p(A_2, D_2) = 0.37769, p(A_2, D_3) = 0.37769, p(A_2, D_4) = 0.37769, p(A_3, D_1) = 0.47295, p(A_3, D_2) = 0.47295$, $p(A_3, D_3) = 0.47295, p(A_3, D_4) = 0.47295, p(A_4, D_1) = 0.41393, p(A_4, D_2) = 0.41393, p(A_4, D_3) = 0.41393, p(A_4, D_4) = 0.41393$.

23.5.3 Algorithm for finding the probability $q(A_i, D_k)$

Step 8: The minimum nonmembership values for each conditional attribute S_j with respect to the alternatives A_i are $\min\gamma_{A_i}(a_{i1}) = 0$, $\min\gamma_{A_i}(a_{i2}) = 0.1$, $\min\gamma_{A_i}(a_{i3}) = 0$, $\min\gamma_{A_i}(a_{i4}) = 0, \min\gamma_{A_i}(a_{i5}) = 0.1$. Therefore, $E_1^{'} = \{0, 0.1, 0, 0, 0.1\}$. Similarly, the minimum nonmembership values for each conditional attribute S_j with respect to the decision attributes D_k are $\min\gamma_{D_k}(d_{1k}) = 0.1$, $\min\gamma_{D_k}(d_{2k}) = 0$, $\min\gamma_{D_k}(d_{3k}) = 0.1$, $\min\gamma_{D_k}(d_{4k}) = 0.1$, $\min\gamma_{D_k}(d_{5k}) = 0$. Therefore, $E_2^{'} = \{0.1, 0, 0.1, 0.1, 0\}$.

Step 9: The sum $R_1^{'} = \sum_{j=1}^{n} \gamma^{'}(\alpha_j) = 0.2$ and $R_2^{'} = \sum_{j=1}^{n} \gamma^{'}(\beta_j) = 0.3$.

Step 10: Since $R_1^{'} \neq R_2^{'}$, then the probabilities of nonoccurrences of each conditional attribute S_j with respect to the alternatives A_i and with respect to the decision attributes D_k using $q(S_j, A_i) = \frac{\gamma^{'}(\alpha_j)}{R_1^{'}}$, and $q(S_j, D_k) = \frac{\gamma^{'}(\beta_j)}{R_2^{'}}$, respectively, are as follows: $q(S_1, A_i) = 0, q(S_2, A_i) = 0.5$, $p(S_3, A_i) = 0, p(S_4, A_i) = 0, p(S_5, A_i) = 0.5$, and $q(S_1, D_k) = 0.33, p(S_2, D_k) = 0, q(S_3, D_k) = 0.33, q(S_4, D_k) = 0.33, q(S_5, D_k) = 0$.

Step 11: The minimum probability of nonoccurrence $q_{\min}(A_i) = \sum_{j=1}^{n} q(S_j, A_i)\gamma_{A_i}(a_{ij})$, $q_{\min}(D_k) = \sum_{j=1}^{n} q(S_j, D_k)\gamma_{D_k}(d_{jk})$ and the maximum probability of nonoccurrence $q_{\max}(A_i) = q_{\min}(A_i) + \sum_{j=1}^{n} q(S_j, A_i)\sigma_{A_i}(a_{ij})$, $q_{\max}(D_k) = q_{\min}(D_k) + \sum_{j=1}^{n} q(S_j, D_k)\sigma_{D_k}(d_{jk})$ are, respectively, as follows: $q_{\min}(A_1) = 0.1, q_{\min}(A_2) = 0.25, q_{\min}(A_3) = 0.55$, $q_{\min}(A_4) = 0.2$; $q_{\min}(D_1) = 0.099, q_{\min}(D_2) = 0.099, q_{\min}(D_3) = 0.462, q_{\min}(D_4) = 0.429$, and $q_{\max}(A_1) = 0.35$, $q_{\max}(A_2) = 0.25$, $q_{\max}(A_3) = 0.6$, $q_{\max}(A_4) = 0.45; q_{\max}(D_1) = 0.462$, $q_{\max}(D_2) = 0.3639$, $q_{\max}(D_3) = 0.891$, $q_{\max}(D_4) = 0.693$.

Step 12: The probabilities of nonoccurrence for each alternative A_i and for each decision attribute D_k using $q(A_i) = \frac{q_{\min}(A_i) + q_{\max}(A_i)}{2}$ and $q(D_k) = \frac{q_{\min}(D_k) + q_{\max}(D_k)}{2}$, respectively, are as follows: $q(A_1) = 0.225$, $q(A_2) = 0.25, q(A_3) = 0.575, q(A_4) = 0.325$, and $q(D_1) = 0.4785, q(D_2) = 0.2319, q(D_3) = 0.6765, q(D_4) = 0.561$.

Step 13: The probabilities of nonoccurrence of each alternative A_i corresponding to each decision attribute D_k using $q(A_i, D_k) = \min\{q(A_i), q(D_k)\}$are as follows: $q(A_1, D_1) = 0.225, q(A_1, D_2) = 0.2319$, $q(A_1, D_3) = 0.225, q(A_1, D_4) = 0.25, q(A_2, D_1) = 0.25, q(A_2, D_2) = 0.2319$, $q(A_2, D_3) = 0.25, q(A_2, D_4) = 0.25, q(A_3, D_1) = 0.4785, q(A_3, D_2) = 0.2319$, $q(A_3, D_3) = 0.575, q(A_3, D_4) = 0.561, q(A_4, D_1) = 0.325, q(A_4, D_2) = 0.2319$, $q(A_4, D_3) = 0.325, q(A_4, D_4) = 0.325$.

23.5.4 Algorithm for finding the probability $s(A_i, D_k)$

Step 14: The probabilities of indeterminacy of each alternative A_i corresponding to each decision attribute D_k using $s(A_i, D_k) = \max\{p(A_i, D_k), q(A_i, D_k)\}$ are as follows: $s(A_1, D_1) = 0.4785, s(A_1, D_2) = 0.693, s(A_1, D_3)$

TABLE 23.6 Neutrosophic probabilistic distance measure table.

$\alpha = 1$	D_1	D_2	D_3	D_4
A_1	0.12289	0.1789	0.19085	**0.1116**
A_2	0.12121	0.11279	0.13624	**0.10436**
A_3	0.09753	**0.08649**	0.13142	0.14101
A_4	0.10648	0.09472	0.15193	**0.09143**
$\alpha = 2$	D_1	D_2	D_3	D_4
A_1	0.24475	0.30079	0.32631	**0.23921**
A_2	0.24808	**0.24558**	0.29558	0.24750
A_3	0.20701	**0.18388**	0.21391	0.27505
A_4	0.23819	**0.19798**	0.28888	0.21251
$\alpha = 3$	D_1	D_2	D_3	D_4
A_1	**0.33092**	0.37435	0.42216	0.3396
A_2	0.34026	**0.33533**	0.39486	0.3597
A_3	0.28926	**0.24978**	0.26085	0.36553
A_4	0.3381	**0.28063**	0.38458	0.3844
$\alpha = 4$	D_1	D_2	D_3	D_4
A_1	**0.39595**	0.42816	0.49909	0.41193
A_2	0.41546	**0.40191**	0.46469	0.44608
A_3	0.35586	0.29765	**0.29365**	0.43378
A_4	0.41627	**0.35301**	0.45816	0.38052
$\alpha = 5$	D_1	D_2	D_3	D_4
A_1	**0.44737**	0.45427	0.561056	0.46524
A_2	0.46756	**0.45449**	0.51789	0.51253
A_3	0.410204	0.33407	**0.31848**	0.488285
A_4	0.47792	**0.4146**	0.51549	0.43419
$\alpha = 6$	D_1	D_2	D_3	D_4
A_1	**0.48938**	0.50795	0.61078	0.50563
A_2	0.5122	**0.49739**	0.55728	0.56448
A_3	0.4545	0.36269	**0.3380**	0.5329
A_4	0.5270	**0.46573**	0.5605	0.47499

(Continued)

TABLE 23.6 (Continued)

$\alpha=7$	D_1	D_2	D_3	D_4
A_1	**0.52453**	0.53946	0.65095	0.53715
A_2	0.54839	**0.53324**	0.59694	0.60598
A_3	0.49075	0.38574	**0.3540**	0.5699
A_4	0.5667	0.5079	0.59633	**0.5069**
$\alpha=8$	D_1	D_2	D_3	D_4
A_1	**0.55447**	0.56695	0.68382	0.56244
A_2	0.5782	**0.5635**	0.62796	0.63986
A_3	0.52065	0.40469	**0.3672**	0.5982
A_4	0.58913	0.54307	0.6255	**0.53260**
$\alpha=9$	D_1	D_2	D_3	D_4
A_1	**0.54218**	0.59108	0.71109	0.58319
A_2	0.60320	**0.5895**	0.6550	0.66809
A_3	0.54558	0.4205	**0.37832**	0.627148
A_4	0.62609	0.572631	0.6498	**0.5537**
$\alpha=10$	D_1	D_2	D_3	D_4
A_1	0.60294	0.61233	0.73405	**0.60053**
A_2	0.62445	**0.61924**	0.67886	0.69202
A_3	0.56659	0.43400	**0.38779**	0.64938
A_4	0.64879	0.5978	0.6695	**0.5152**
$\alpha=11$	D_1	D_2	D_3	D_4
A_1	0.62275	0.63114	0.7536	**0.61526**
A_2	0.64277	**0.63143**	0.7	0.71260
A_3	0.58449	0.44559	**0.39597**	0.66845
A_4	0.668	0.61950	0.68659	**0.58670**
$\alpha=12$	D_1	D_2	D_3	D_4
A_1	0.64034	0.64782	0.7705	**0.62729**
A_2	0.65857	**0.64853**	0.71882	0.73054
A_3	0.59989	0.45568	**0.40309**	0.6849
A_4	0.68474	0.68377	0.70126	**0.59986**

$= 0.5775$, $s(A_1, D_4) = 0.495$, $s(A_2, D_1) = 0.37769$, $s(A_2, D_2) = 0.37769$, $s(A_2, D_3) = 0.37769$, $s(A_2, D_4) = 0.37769$, $s(A_3, D_1) = 0.47295$, $s(A_3, D_2) = 0.47925$, $s(A_3, D_3) = 0.575$, $s(A_3, D_4) = 47925$, $s(A_4, D_1) = 0.41393$, $s(A_4, D_2) = 0.41393$, $s(A_4, D_3) = 0.41393$, $s(A_4, D_4) = 0.41393$.

23.5.5 The probabilistic distance measures of alternatives and attributes

Step 15: The probabilistic distance measure values for each alternative A_i and decision attribute $D_k, d(A_i, D_k) = \{\frac{1}{3n}\sum_{j=1}^{n} p(A_i, D_k)\left|\mu_{A_i}(a_{ij}) - \mu_{D_k}(d_{jk})\right|^{\alpha} + s(A_i, D_k)\left|\sigma_{A_i}(a_{ij}) - \sigma_{D_k}(d_{jk})\right|^{\alpha} + q(A_i, D_k)\left|\gamma_{A_i}(a_{ij}) - \gamma_{D_k}(d_{jk})\right|^{\alpha}\}^{\frac{1}{\alpha}}$, *for* $\alpha = 1, 2, \ldots, 10, 11, 12$, are shown in Table 23.6.

23.5.6 Tabulation

23.5.6.1 *Final decision*

From the above neutrosophic probabilistic distance measure table (see Table 23.6), for $\alpha = 1, 2, \ldots, 12$, the most repeated lowest distance measure for the patient A_1 is to the disease D_1, so the patient A_1 suffers from the disease D_1. The most repeated lowest distance measure for the patient A_2 is to the disease D_2, so the patient A_2 suffers from the disease D_2. For the patient A_3, the most repeated lowest distance measure is to the disease D_3, so the patient A_3 suffers from the disease D_3. The disease for the patient A_4 is D_4, because the most repeated lowest distance measure for the patient A_4 is to the disease D_4.

23.6 Results and discussion

In this section, we analyze the results of MADM problem in detail.

1. We can observe in Table 23.6 that all the neutrosophic probabilistic distance measure values are gradually increasing if α values increase, which tends to 1.
2. In the numerical example, we computed the neutrosophic probabilistic distance measures for $\alpha = 1, 2, \ldots, 12$.
3. For $\alpha = 1, 2, 10, 11, 12$, the lowest distance measure for the patient A_1 is to the disease D_4, and for $\alpha = 3, 4, 5, 6, 7, 8, 9$, the lowest distance measure for the patient A_1 is to the disease D_1. Therefore the most repeated disease for the patient A_1 is D_1, so the patient A_1 suffers from the disease D_1. Suppose the lowest distance measure for the patient A_1 is to the disease D_4 for $\alpha = 13, 14$, then the most repeated disease for the patient A_1 will be D_1 and D_4; so the patient A_1 will suffer from the disease D_1 and D_4.

4. For $\alpha = 1$, the lowest distance measure for the patient A_2 is to the disease D_4, and for $\alpha = 2, 3, \ldots, 12$, the lowest distance measure for the patient A_2 is to the disease D_2. Thus the disease D_2 is the most repeated disease for the patient A_2, so the patient A_2 suffers from the disease D_2. If for the remaining α's, the lowest distance measure for the patient A_2 is to other than D_2 and D_4, then there should be a change in the final decision.
5. For $\alpha = 1, 2, 3$, the lowest distance measure for the patient A_3 is to the disease D_2, and for $\alpha = 4, 5, 6, 7, 8, 9, 10, 11, 12$, the lowest distance measure for the patient A_3 is to the disease D_3 so the most repeated disease for the patient A_3 is D_3, and the patient A_3 suffers from the disease D_3.
6. For $\alpha = 1, 7, 8, 9, 10, 11, 12$, the lowest distance measure for the patient A_4 is to the disease D_4, and for $\alpha = 2, 3, 4, 5, 6$, the lowest distance measure for the patient A_4 is to the disease D_2. Therefore the most repeated disease for the patient A_4 is D_4, so the disease for the patient A_4 is D_4. Suppose the lowest distance measure for the patient A_4 is to the disease D_2 for $\alpha = 13, 14$, then the most repeated disease for the patient A_1 will be D_2, and D_4; so the patient A_2 will suffer from the disease D_2 and D_4.

23.7 Advantages and limitations

This section states some advantages and limitations of the proposed method.

1. The first novelty of the present chapter is to define the probabilistic distance measure on single-valued neutrosophic sets using the probability of occurrences, nonoccurrences, and indeterminacy occurrences.
2. The second novelty of this chapter is that the different types of proposed neutrosophic probabilistic distance measures for various $\alpha \geq 1$ defined.
3. Another novelty is the usage of all our proposed neutrosophic probabilistic distance measures into the medical diagnosis problem as a real-life application.
4. The first advantage is that for each $\alpha \geq 1$, the lowest probabilistic distance measure value of alternative A_i, $i = 1, 2, \ldots, m$ and decision attribute D_k, $k = 1, 2, \ldots, p$ gives more similarity between them.
5. The decision varies for each $\alpha \geq 1$, and then, the most repeated decision is the suitable decision. So this method can be considered as one of the common methods for solving multi-attribute decision-making problems in a neutrosophic environment.
6. One more advantage of this chapter is that the suitable decision attribute will get if α increases only.
7. Our proposed method gives the same decision when we compare with Ref. [36] for financial risk analysis of picture fuzzy set in classification problems.
8. Another advantage of this work is that this method can use the neutrosophic probabilistic similarity measures instead of the neutrosophic

probabilistic distance measures in multi-attribute decision-making problems (MADM).

9. This method can be applied to multi-criterion decision-making problems (MCDM).
10. One of the limitations of the present work is the number of decisions should not exceed the number of alternatives.
11. This new method in the MADM problem does not use any correlation coefficient [17,26], score functions [21], cosine similarity measures [18], tangent similarity measures [28], cotangent similarity measures [24], distance functions such as Euclidean distances [16], Hamming distance [15], rank matrices [22], cross-entropy [27].
12. Due to the monotonous process and loyalty, our method can be applied for any number of data in data analysis, medical diagnosis, recruitments, and many other real-life problems.

23.8 Conclusions and future work

The neutrosophic set is a mathematical tool to handle imprecise, incomplete, and inconsistent information in multi-decision-making problems. In multi-decision-making problems, probabilistic distance measures on neutrosophic sets play an important role in taking decisions regarding alternatives. This chapter defined a general formula of probabilistic distance measures on single-valued neutrosophic sets using the concepts of probability of occurrences, nonoccurrences, and indeterminacy occurrences. We also defined different types of probabilistic distance measures where $\alpha \geq 1$ and showed that the difference of this distance measure from unity is a similarity measure. In this general formula, the values of all neutrosophic probabilistic distance measures are gradually increasing if α values increase, which converge to 1.

A mathematical model is developed to solve MADM problems by choosing the lowest probabilistic distance measure values between the decision attributes and alternatives. A numerical example in medical diagnosis is also discussed as a real-life application. The lowest probabilistic distance measure gives more similarity between the patients and the disease. The proposed method gave the same decision when we compare with [36] for financial risk analysis of picture fuzzy set in classification problems. This newly developed probabilistic distance measure can be applied to solve multi-criterion decision-making problems (MCDM), and this method will be an eye-opener for the neutrosophic researchers to implement in other research areas of data analysis, medical diagnosis, recruitments, and many other real-life problems.

References

[1] L.A. Zadeh, Fuzzy sets, Inform. Control 8 (1965) 338–353.

[2] C.P. Pappis, N.I. Karacapilidis, A comparative assessment of measures of similarity of fuzzy values, Fuzzy Sets Syst. 56 (2) (1993) 171–174.

[3] W. Voxman, Some remarks on distance between fuzzy numbers, Fuzzy Sets Syst. 100 (1–3) (1998) 353–365.
[4] S. Pramanik, K. Mondal, Weighted fuzzy similarity measure based on tangent function and its application to medical diagnosis, Int. J. Innov. Res. Sci. Eng. Technol. 4 (2) (2015) 158–164.
[5] K.P. Adlassnig, Fuzzy set theory in medical diagnosis, IEEE Trans. Syst. Man Cybernet. 16 (2) (1986) 260–265.
[6] L.A. Zadeh, Probability measures of fuzzy events, J. Math. Anal. Appl. 23 (1968) 421–427.
[7] K. Atanassov, Intuitionistic fuzzy sets, Fuzzy Sets Syst 20 (1986) 87–96.
[8] C. Fan, Y. Song, Q. Fu, L. Lei, X. Wang, New operators for aggregating intuitionistic fuzzy information with their application in decision making, IEEE Access 6 (2018) 27214–27238.
[9] E. Szmidt, J. Kacprzyk, Intuitionistic fuzzy set in some medical applications, International Conference on Computational Intelligence, Springer, Berlin, Heidelberg, 2001, pp. 148–151.
[10] P. Dutta, S. Goala, Fuzzy decision making in medical diagnosis using an advanced distance measure on intuitionistic fuzzy sets, Open Cybernet. Syst. J. 12 (2018) 136–149.
[11] S.K. De, A. Biswas, R. Roy, An application of intuitionistic fuzzy sets in medical diagnosis, Fuzzy Sets Syst. 117 (2) (2001) 209–213.
[12] E. Szmidt, J. Kacprzyk, Remarks on some applications of intuitionistic fuzzy sets in decision making, NIFS 2 (3) (1996) 22–31.
[13] F. Smarandache, Neutrosophic set: a generalization of the intuitionistic fuzzy sets, Int. J. Pure. Appl. Math. 24 (2005) 287–297.
[14] H. Wang, F. Smarandache, Y. Zhang, R. Sunderraman, Single valued neutrosophic sets multi-space and multi-structure, Tech. Sci. Appl. Math. 4 (2010) 410–413.
[15] P. Majumdar, S.K. Samanta, On similarity and entropy of neutrosophic sets, J. Intell. Fuzzy Syst. 26 (3) (2014) 1245–1252.
[16] K. Mondal, S. Pramanik, Neutrosophic similarity measure and its application to multi attribute decision making, Neutrosophic Sets Syst. 9 (2015) 80–86.
[17] I.M. Hanafy, A.A. Salama, K. Mahfouz, Correlation of neutrosophic data, Int. Ref. J. Eng. Sci. 1 (2) (2012) 39–43.
[18] P. Biswas, S. Pramanik, B.C. Giri, Cosine similarity measure based multi-attribute decision-making with trapezoidal fuzzy neutrosophic numbers, Neutrosophic Sets Syst. 8 (2015) 47–57.
[19] G. Shahzadi, M. Akram, A.B. Saeid, An application of single-valued neutrosophic sets in medical diagnosis, Neutrosophic Sets Syst. 18 (2017) 80–87.
[20] K. Mondal, S. Pramanik, Neutrosophic decision making model of school choice, Neutrosophic Sets Syst. 9 (2015) 62–68.
[21] G. Jayaparthasarathy, V.F. Little Flower, M. Arockia Dasan, Neutrosophic supra topological applications in data mining process, Neutrosophic Sets Syst. 27 (2019) 80–97.
[22] F. Karaaslan, K. Hayat, Some new operations on single-valued neutrosophic matricesand their applications in multi-criteria group decision making, Appl. Intell. 48 (12) (2018) 4594–4614.
[23] F. Karaaslan, Gaussian single-valued neutrosophic numbers and its application in multi-attribute decision making, Neutrosophic Sets Syst. 22 (2018) 101–117.
[24] S. Broumi, F. Smarandache, Several similarity measures of neutrosophic sets, Neutrosophic Sets Syst. 1 (10) (2013) 54–62.
[25] I. Deli, Y. Toktas, S. Broumi, Neutrosophic parameterized soft relations and their application, Neutrosophic Sets Syst. 4 (2014) 25–34.

[26] J. Ye, Multi-criteria decision-making method using the correlation coefficient under single-valued neutrosophic environment, Int. J. General Syst. 42 (4) (2013) 386–394.

[27] J. Ye, Single-valued neutrosophic cross-entropy for multi-criteria decision-making problems, Appl. Math. Model. 38 (2014) 1170–1175.

[28] J. Ye, Clustering methods using distance-based similarity measures of single-valued neutrosophic sets, J. Intell. Syst. 23 (2014) 379–389.

[29] J. Ye, Q. Zhang, Single valued neutrosophic similarity measures for multiple attribute decision-making, Neutrosophic Sets Syst. 2 (2014) 48–54.

[30] E.K. Zavadskas, K. Govindan, J. Antucheviciene, Z. Turskis, Hybrid multiple-criteria decision-making methods: a review of applications in engineering, Sci. Iran 23 (1) (2016) 1–20.

[31] F. Smarandache, A Unifying Field of Logics. Neutrosophy. Neutrosophic Probability, Set and Logic, American Research Press, Rehoboth, 1998.

[32] S. Shao, X. Zhang, Y. Li, C. Bo, Probabilistic single-valued (interval) neutrosophic hesitant fuzzy sets and its application in multi-attribute decision-making, Symmetry 10 (2018) 419.

[33] S. Shao, X. Zhang, Measures of probabilistic neutrosophic hesitant fuzzy sets and the application in reducing unnecessary evaluation process, Mathematics 7 (2019) 649.

[34] J.J. Peng, J.Q. Wang, X.H. Wu, J. Wang, X.H. Chen, Multi-valued Neutrosophic Sets and Power Aggregation Operators with Their Applications in Multi-criteria Group, Decision-making problems, Int. J. Comp. Intel. Syst. 8 (2) (2015) 345–363.

[35] M. Arockia Dasan, E. Bementa, F. Smarandache, X. Tubax, Neutrosophical plant hybridization in decision-making problems, in: F. Smarandache, M. Abdel-Basset (Eds.), Neutrosophic Operational Research, Springer, Cham, 2021, pp. 1–17. Available from: https://doi.org/10.1007/978-3-030-57197-9_1.

[36] A. Guleria, R.K. Bajaj, A novel probabilistic distance measure for picture fuzzy sets with its application in classification problems, Hacettepe J. Math. Stat. 49 (6) (2020) 2134–2153.

Index

Note: Page numbers followed by "*f*" and "*t*" refer to figures and tables, respectively.

C

D

E

F

G

H

Q

Printed in the United States
by Baker & Taylor Publisher Services